SK360 ELECTRONICS I T-TH 6:30 - 9:30 P.M. WALTER HOFFMAN
8481718

ELECTRONICS: PRINCIPLES AND APPLICATIONS

Kamiran S. Badrkhan

N. David Larky

IE53

Published by

SOUTH-WESTERN PUBLISHING CO.

CINCINNATI   WEST CHICAGO, ILL.   DALLAS   PELHAM MANOR, N.Y.   PALO ALTO, CALIF.

ISBN: 0-538-33530-0

Library of Congress Catalog Card Number: 83-60656

1 2 3 4 5 6 7 8 9 D 9 8 7 6 5 4 3

Printed in the United States of America

CONTENTS

PREFACE

PERSPECTIVE

ELECTRONICS: Principles and Applications presents the theory and principles of electricity and electronics, with an emphasis on the problem-solving skills electronics technicians use daily in their jobs. Students in technical electronics programs will find its easy-to-follow format and practical approach both beneficial and interesting as they progress from basic work in simple circuits to more challenging assignments in complex electronic devices. Persons studying for entry into related fields, such as computer science, broadcasting, or audio/video recording, can also benefit from participation in a course built around the content of this book.

No previous courses or experience in the field of electricity or electronics are necessary. The content of this book is such that beginning students can understand and apply all of the information they need as they work through the material presented in normal sequence, from beginning to end. Further, since a series of reviews is incorporated within the continuity of the book, the content is both self-reinforcing and confidence-building. Students have a strong sense of their own progress—and testing components within the text support their feelings of achievement.

To gain full benefit from the content of this book, students should have enough background in mathematics to be able to read and execute simple algebraic equations. If students are not comfortable in applying the basic principles of algebra, a short review in the formatting and solving of equations is recommended.

CONTENT ORGANIZATION

The content of this book was assembled in specific recognition of the realities of the field of electronics. That is, electronics today is dominated by the design and use of solid-state, microcircuit devices. *ELECTRONICS: Principles and Applications* emphasizes the use of electronic theory to solve practical problems—the kinds of situations students will encounter as entry-level electronics technicians.

The content of this book is organized for both student motivation and for a modular approach to learning (including built-in reinforcement). To illustrate, the first chapter of the book presents an overview of career opportunities associated with the knowledge to be acquired through work in the book. Right from the outset, this builds an awareness of the goals toward which students may choose to direct their efforts.

As another example, each topic within every chapter is followed by a series of challenging, thought-provoking questions that provide an opportunity for reinforcing or remediating learning experiences through immediate feedback. Each significant segment of information in the book is covered by one or more questions in these in-chapter reviews. Students gain a mechanism for checkpointing their knowledge. They can tell, on their own, if further work is needed in any content area. Remediation can then be self-directed as appropriate—and at the time when progress is checked for individual units of acquired knowledge. By using the in-chapter checkpoints as a basis for class discussions, the instructor can also spot gaps in comprehension. These, in turn, can be dealt with through on-the-spot explanations, special lectures, additional reading assignments, or combinations of these methods.

In addition to the in-chapter questions, formal reviews are provided following key groups of chapters. These in-depth reviews summarize key information and provide a basis for both instructors and students to measure progress and to identify points at which special explanations or remediation may be necessary.

TEACHING AIDS

Several teaching aids are provided with *ELECTRONICS: Principles and Applications*.

Demonstration transparencies. These demonstration transparencies will assist the instructor in the presentation of basic technical electricity/electronics concepts.

Instructor's Manual and key. The convenient Instructor's Manual contains an overview of the text, teaching suggestions, lesson-plan outlines, answer keys for all questions, and chapter tests and answer keys.

AUTHORSHIP

ELECTRONICS: Principles and Applications has been written by professionals from both academic and industrial backgrounds to give students a balanced perspective of the theoretical and practical.

Kamiran S. Badrkhan is Assistant Dean, Electronics/Computer Science, Los Angeles Trade-Technical College, with teaching experience at both the post-secondary and secondary levels. He also acts as an industrial consultant.

N. David Larky has a practical electricity/electronics background and is currently Director of Strategic Development, TRW Electronics and Defense, Redondo Beach, California. He holds a number of patents, including basic patents on color television.

ACKNOWLEDGMENTS

The appropriateness of content and technical accuracy of this book owe much to the extensive efforts of reviewers who have read and commented upon the manuscript during development. In particular, in-depth reviews were provided by Edward F. Marsoobian, Ph.D., Professor, Industrial Studies Department, California State University, Los Angeles, and Everett Hedrick, TRW Customer Service Division, Hawthorne, CA.

OPPORTUNITIES IN ELECTRONICS

YOU AND ELECTRONICS

This book is designed for use as an introduction to the principles and applications of electronics. It will provide the information and learning experiences you need for virtually any options you choose:

- If you decide to take only one course in electronics during your career, this book will help you build a basic understanding of this important branch of technology. Further, the book is carefully organized and indexed so that you can use it as a continuing reference in later studies or work situations.

- If you are majoring in some branch of technology or engineering requiring a background in electronics, this book will give you a foundation upon which to build.

- If you are in a career-oriented program, possibly leading to a job as an electronics technician, this book will be especially valuable because it provides both theoretical and practical content.

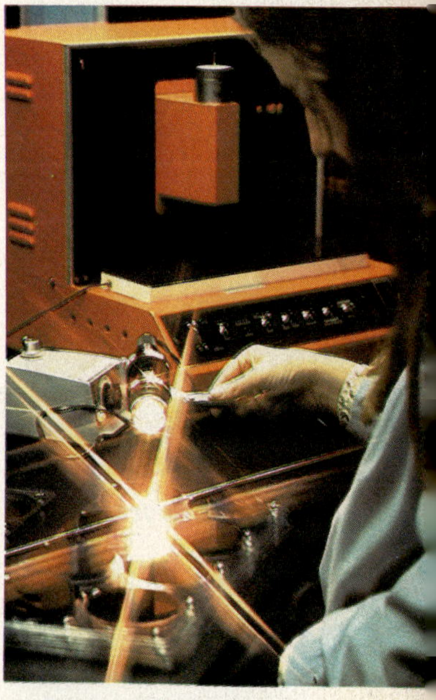

Miniature electronic devices have been made possible with the delopment of the semiconductor chip. Above, a silicon wafer used in making chips is being tested. (Photo by Liane Enkelis)

Miniaturization made possible by microcircuits is dramatized in the photo (opposite page) of a complete circuit passing through the eye of a needle.

SOME DEFINITIONS

To understand your opportunities, and to go on in the study of electronics, you will need to understand and be able to use certain basic terms. The explanations that follow are *not* a total set of definitions. They are enough to get you started.

Electricity is a form of energy. It is difficult to provide a more specific definition because the term is used so broadly. Electricity takes in many branches of science and technology.

The one common denominator in definitions of electricity is that this form of energy is based upon the movement of electrons.

Electrons are among the basic parts of an atom. An *atom* is the smallest unit of any substance (matter) that still retains its identity. All matter that composes the earth and the life it supports is built from 93 naturally occurring elements, each having a specific form of atom.

An atom is so small that it can't be seen, even by a powerful microscope. Each atom consists of a series of even smaller particles that are bound together by electrical force. The electrical forces that bind the atoms are called *charges*. Electrical charges are identified as *positive* and *negative*. It is a basic rule that opposite charges attract each other; like charges repel each other, or exert force to drive each other apart.

An electron is a negatively charged subatomic particle. Within an atom, an electron is in constant motion around a positively charged nucleus. Electrons are actually groups of particles kept together through a balance of electrical charges.

The structure of an atom has been compared with the major units of our solar system. The sun at the center of the solar system corresponds with the nucleus of the atom. The planets revolving in orbits around the sun are like electrons.

Electronic grocery sales. Electronically controlled laser beams beneath the surface of the checkstand read coded labels on packages. This data is computer-processed and sales are rung up automatically. Customers see amounts of purchases on the digital display of the electronic cash register.

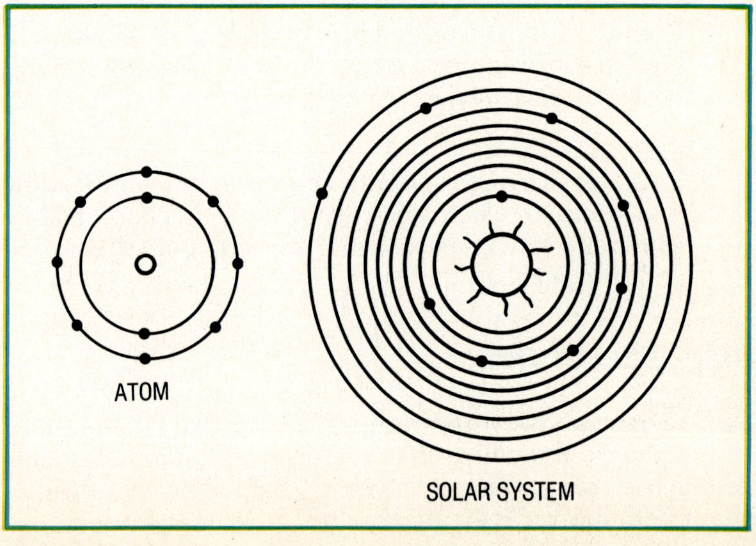

Same principle, different sizes. The structure of a microscopic-sized atom resembles that of the entire solar system.

Some electrons leave their atoms and are attracted toward positive forces in surrounding matter. Electrons that are not orbiting within atoms are called free electrons. The atoms of some kinds of materials have free electrons that move quite easily from one location to another among a group of atoms. The continuous movement of electrons is the energy we know as electricity.

Electric current, also referred to as just *current,* is electrons moving along a conductor. A conductor is any material through which electrons can flow easily. Electrons always move *from* a negative source (or terminal) *toward* a positive sink (or terminal) that receives the electrons.

Jobs in electronics are everywhere. In the photo above, keyboards are being assembled for computer terminals. (Photo by Liane Enkelis)

Electronics is the technology involving the application of electricity through use of special devices and materials.

OPPORTUNITIES IN ELECTRONICS

Electronics is a rapidly growing technology. For evidence, just look around yourself. Read almost any magazine or watch television. Hundreds of millions of radios are in use. More than 70 million homes in the United States and Canada have at least one TV set. Stereo and video recording or playback systems are found in millions of other homes. Millions of computers are in use by businesses. Hundreds of thousands of computers are finding their way into homes for personal use. Computers housed on electronic chips are integral parts of automobiles. Electronics, in short, is everywhere.

Video games have become the basis for a multibillion-dollar industry. Devices, such as the one in the photo at the left, that receive heavy use at the hands of young people will need skilled service personnel to keep them going.

Assembly jobs in electronics can be skilled positions. The assembler in this manufacturing facility must be able to read manufacturing prints and to perform a number of diverse operations, including soldering. (Photo by Liane Enkelis)

To bring this vast field into some sort of focus, these major activities are reviewed:

- Electronics manufacturing
- Electrical and electronics servicing
- Electrical construction
- Technical occupations involving electronics.

Electronics Manufacturing

This category encompasses all organizations that manufacture electrical and electronics equipment. Included are appliances, radios, television sets, complete telephone systems, test equipment, and entire defense and military electronic systems. Some occupational opportunities in electronics manufacturing are described below.

Assemblers. These are people with entry-level skills. Some knowledge of electricity and electronics is helpful. However, most assemblers are trained on the job. Their work consists

Quality inspection after final assembly of electronic devices requires special care and skills. (Photo by Liane Enkelis)

of specific tasks of connecting and placing parts into machines or appliances as they pass down assembly lines.

Processors. A process is an operation that goes on continuously. Most processes are at least partly automatic. In electronics, processing skills are applied in making the basic printed circuit boards used in electronic devices. These boards hold and connect the working parts of electronics equipment. Processes within manufacturing organizations can also include such continuous operations as wave soldering the connections in assembled boards.

Testers. Reliability is critical in electronic devices. When you plug in your television set, you want to push the button and see a picture. When you fly in an airplane, you want to feel sure that the navigational equipment keeps it on course. Equipment must be designed from the very beginning so it will be reliable. Also, it is important to be sure equipment is built according to its design. So, virtually every step of assembly and other manufacturing operations in an electronics plant is followed by a test. Specialized test equipment is

Electronics on wheels. Ignition and other control systems on most new cars have some electronic components. Use of microprocessors in motor vehicles will increase, opening up many new and profitable servicing positions. (Courtesy TRW, Inc.)

Routine inspections and servicing on heavy electrical equipment provide jobs for skilled technicians. (Courtesy TRW, Inc.)

used to make sure that all of the connections and parts are functioning properly, and that they are working together as they should.

Repair and Rework. Testing is designed to find problems. Once the problems are identified, technically qualified people must make the adjustments necessary to bring products up to quality standards. Experienced people are usually given special training to qualify them for these positions. Sometimes the design must be changed. Sometimes variations are needed in the way products are manufactured.

Electrical and Electronics Servicing

Anything mechanical or electronic will, at some time, malfunction or break down. Because there are so many electrical and electronic devices in use, keeping them repaired is an important part of the electronics industry. Some of the special requirements and occupations associated with electrical and electronics servicing are described below.

Appliance and consumer electronics equipment repair. All companies that manufacture electrical and electronic equipment for use by consumers give warranties. When products under warranty develop troubles, their manufacturers must provide service. Some manufacturers operate their own service centers and employ staff technicians to operate them. Others contract for warranty service. These occupations include both shop jobs and "outside" assignments. Shop personnel work on equipment that is brought in to them. Outside service personnel visit the homes or offices of customers to repair equipment.

Heavy electrical equipment repair. Separate organizations and technical specialties are usually set up to maintain and repair heavy electrical equipment used in offices, factories, or on industrial sites. Much of this work involves motors, air conditioners, and materials handling equipment. Note the difference between maintenance and repair. On heavy equipment, routine inspections, lubrication, and new parts are often required. This maintenance is in addition to repair following breakdowns.

Communications equipment. The maintenance and repair of communications equipment has become a series of separate specialties. Each type of communications industry employs and trains people who specialize in keeping the equipment operating and in fixing it when it breaks down.

Electric circuits and switching equipment are necessities for communication and control. Specialists will continue to be needed by facilities that manufacture this type of equipment. (Courtesy TRW, Inc.)

Types of equipment involved include telephone, broadcasting, microwave, satellite ground station, and laser signal transmission systems.

Electrical utilities. Electric power must be available on demand, around the clock. Therefore, electric utilities that sell power have 24-hour staffs of specialized technicians who keep their equipment running. These people monitor the networks that deliver power through transformers and power lines. They are responsible for service to businesses and homes. The demand is so great that utility maintenance and repair has become a specialized occupation in itself.

There are many facilities, most of them smaller shops, that specialize in fixing small electric motors used in commercial, industrial, and manufacturing operations. Included are the motors that run heavy appliances, such as washing machines or refrigerators. Also included are the industrial motors used for air compressors, drill presses, sewing machines, or other devices. Automotive electric service is a specialty that is usually included in this same category. Automotive electrical shops repair generators, alternators,

Computer maintenance and repair. Computers are also called *data processing systems*. The word *system* implies that there are a number of parts or devices functioning together to produce a single end result. Modern computer systems are complex. This means that they require really comprehensive maintenance activity. Computers rarely break down. For the entire industry, customer availability of equipment usually

Reliability of computers is vital to the continued operation of the organizations that use these electronic devices. Guarantees about reliability are usually part of sales contracts covering computers. Electronic maintenance specialists are essential to meeting these reliability standards. (Photo by Liane Enkelis)

averages more than 95 percent of the contracted operating time. However, to achieve this record, maintenance is necessary. In large computer installations, the manufacturer often has a resident technician who works full time at maintaining units of equipment. Smaller computer systems are usually maintained by having technicians visit installation sites. This is done either on a regular basis or when there is a problem.

Emerging new specialties are opening up in electronics manufacturing. The quality control technician in this photo is checking substrates for integrated circuits. (Courtesy TRW, Inc.)

Office equipment. Modern office equipment uses the same technology as computers. Therefore, maintenance is necessary for the same reasons, except that needs are scaled down to the size of the individual equipment units. The equipment involved includes the many millions of office copiers, electric typewriters, electronic calculators, postage meters, dictation units, and other equipment necessary for modern business.

Electrical Construction

Millions of housing units and hundreds of thousands of industrial and business facilities are constructed each year. The needs of the construction industry are met by electricians who specialize in installing the wires and outlets required to support the living or working functions of these structures. There are excellent employment opportunities for persons with backgrounds in electricity and electronics who have learned the specialized skills of this trade.

Technical Occupations Involving Electronics

Occupations in this category cut across all of the other categories described above. These are technical specialties found in manufacturing, maintenance, service, repair, and at some construction sites.

Engineering. Engineers are the top professionals of the electrical and electronics industries. To qualify for work as an engineer, a person needs at least a four-year college degree. Many engineers have advanced, graduate degrees. Many also have certificates as professional engineers. Within the electrical and electronics industries, engineers fill a number of key functions, such as designing, preparing manufacturing specifications, and supervising manufacturing. They also bring added knowledge and insight to a number of management jobs.

Technicians. Technicians are found throughout electrical and electronics industries. They operate in close support of

engineers. For example, the end result of an engineer's design of a new product might be a series of drawings and specifications. The technicians would then take over and build prototypes of the device. Technicians would also monitor the project to make sure that the units work properly.

Education. The teaching of electricity and electronics has become a major part of vocational, technical, and higher education. Electricity and electronics are taught at junior high schools, high schools, community colleges, four-year colleges, graduate schools, and at many private technical schools. For secondary school and community college positions, basic teaching credentials are needed, as well as technical knowledge and background. At four-year colleges and universities, advanced degrees are needed. At private technical schools, extensive experience as well as certain academic training is required.

Broadcasting. Technicians who work in the broadcasting industry form a specialized group. Part of the specialization lies in the fact that these persons need a qualifying license,

Careful testing is an integral part of electronic manufacturing operations. Each circuit board in this plant is computer-checked to verify that all circuits are complete. Testing determines whether boards are used or returned for rework. (Photo by Liane Enkelis)

Electronic devices in space must be extremely compact, highly reliable, and rugged. Navigation and information systems placed in space by missiles provide vital information and functioning could become a life-or-death matter. (Courtesy TRW, Inc.)

awarded on the basis of tests administered by the Federal Communications Commission (FCC). In their jobs, these people frequently monitor the schedules and the quality of broadcast transmissions. In addition, many broadcast technicians work closely with performers and directors who are putting on news or entertainment shows. The technician's work can thus go beyond routine material learned in a classroom.

Quality of sound recording depends upon the technicians who work directly with performers and directors. (Photo by Liane Enkelis)

HISTORICAL BACKGROUND

The first recorded evidence on the discovery and study of electricity dates to approximately 600 B.C. A Greek, Thales of Miletus, recorded the thought that there might be a relationship between electricity and magnetism.

Thales of Miletus also provided the basis for giving electricity and electronics their names. One of his experiments dealt with amber. Amber is a crystal-like mineral that looks something like yellowish glass. Thales rubbed a piece of amber material with a silk cloth. He noticed a change in the cloth which we would now recognize as static electricity. The Greek word for amber is *electrum*. Later scientists adapted the writings of Thales of Miletus to call the electrical particles of atoms electrons. The words *electricity* and *electronic* come from the same root.

The first evidence of the use of batteries dates to about 220 B.C. Equipment discovered consisted of crude batteries used for electroplating gold and silver.

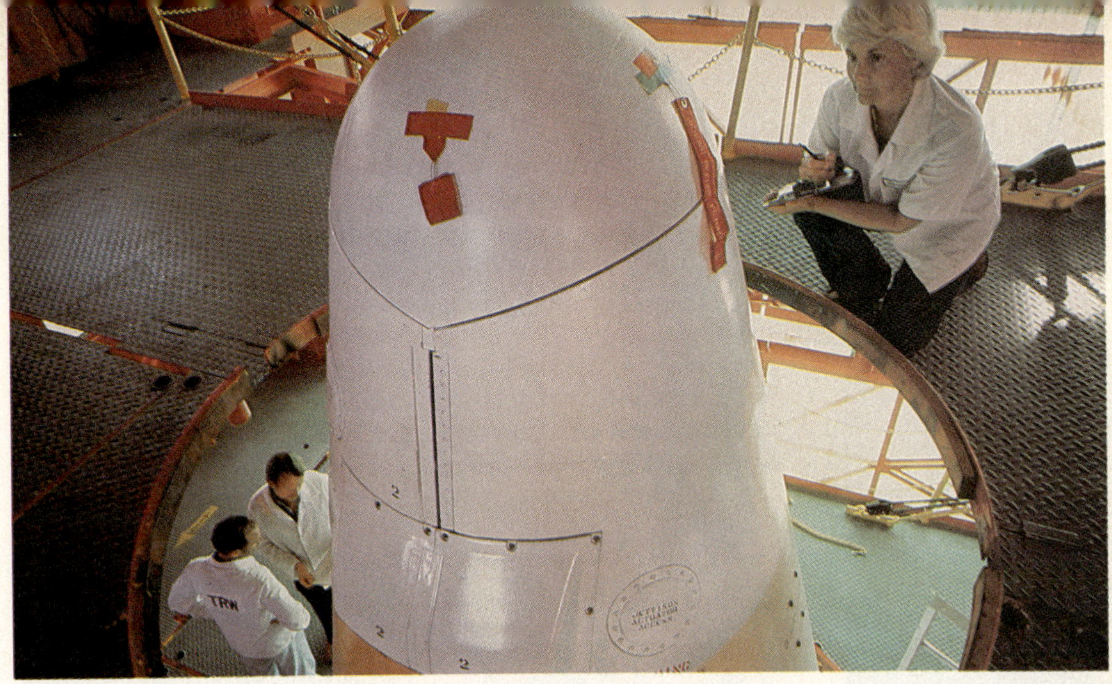

Records on the use of electricity jump from 220 B.C. to 1200 A.D. At about this time, the lodestone came into use. Lodestone is a natural magnetic mineral. Pieces of lodestone were used to build the first compasses. A piece of lodestone was tied to a stick and floated in a bowl of water. This permitted the lodestone to seek a north-south orientation within the earth's magnetic field. The people of that time didn't understand why it worked. But they realized the stone always pointed toward the north.

Broadcast technicians are a specialized group. Their responsibilities include the monitoring of schedules and quality of transmissions. (Courtesy TRW, Inc.)

The first published study on electricity and magnetism was produced by a British scientist, William Gilbert, in the late sixteenth century. As you will learn later in this book, Gilbert's name was given to a basic unit of measure for magnetism.

The connection between electricity and the forces of nature was established by Benjamin Franklin, the American statesman-scientist. In the late eighteenth century, Franklin performed his famous experiment that involved flying a kite in the rain. He used wire as a kite string. Electricity from the lightning of a storm was conducted along the wire, causing an electric arc, or spark, to a key hung from the wire. Franklin also originated the concept of using plus (+) and minus (−) signs to indicate positive and negative electrical polarity.

Development of electricity owes much to the experiments of Benjamin Franklin.

A major improvement in the design of batteries was contributed, in about 1800, by a scientist from northern Italy, Alessandro Volta. His name was given to a basic unit of measure for electrical pressure, the volt.

The first machine to derive electricity from mechanical energy was developed in 1831 by a British scientist, Michael Faraday. Faraday built a device that used the movement of a magnet to generate electricity.

The first commercial use of electricity for a working product was patented by an American, artist-inventor Samuel F. B. Morse. Morse was awarded a patent on an electric telegraph device in 1840. Very quickly, the electric telegraph became the main means of communication in the United States and much of Europe. Morse also developed a system for encoding letters and numbers by means of dots and dashes.

Invention of the telegraph was the beginning of modern long-distance communication. An early telegraph key is shown in the illustration to the left. Soon telegraph lines became a common part of the country's landscape. Today, telephone and power lines, as shown in the photo above, form vital network links for modern living and communication. (Photo courtesy TRW, Inc.)

During the same general era, a British scientist who was a contemporary of Faraday, James Maxwell, studied Faraday's work. From these studies, he developed a theory about electromagnetic disturbances and their function in the production of electricity. Maxwell's major findings were published in 1864. The names of both Faraday and Maxwell are used, today, for basic units of electrical measure.

A few decades after Maxwell's publication, a German scientist, Heinrich Rudolf Hertz, actually created the electromagnetic disturbances that Maxwell had theorized. Hertz's work led to the later introduction of alternating electric current. Hertz's name is used as a basic measure for alternating current.

George Westinghouse, whose career began just after the Civil War and continued into the early 1900s, began as a mechanical engineer and later turned to what was then the new field of applied electricity. He founded the Westinghouse Electric Company and helped establish alternating current as the United States standard for electrical energy.

Charles Proteus Steinmetz lived at about the same time as George Westinghouse. Steinmetz was born in Germany in 1865. He received his formal education in Germany and came to the United States in 1893 to join the engineering staff of the General Electric Company. During his work for General Electric, Steinmetz developed techniques that led to more than 100 inventions. These included the magnetite arc lamp, the induction regulator, and the transformer. The transformer is a device for changing voltage levels of electric current. Its use has made it possible to deliver electricity to homes, factories, and offices. In addition, Steinmetz, who died in 1923, is widely regarded as a scientist who made major contributions to the theoretical understanding of electricity.

One of the great inventors in modern technology, Thomas Edison, began working in his New Jersey laboratory in the 1870s and continued almost to the time of his death in 1931. Among Edison's contributions to the application and use of electricity are the light bulb, the electric generator, the phonograph, and the motion picture.

Another great American inventor, Alexander Graham Bell, lived in the same era as Edison. Bell was one of several inventors racing to patent the first practical telephone. He beat out the other inventors by a matter of days. Bell's telephone patent was awarded in 1876. He later helped found the Bell Telephone system.

Electricity and electronics took on international dimensions of service in 1901, when an Italian scientist, Guglielmo Marconi, succeeded in sending radio signals across the Atlantic Ocean. Early radio transmissions used dots and dashes, adapting the Morse code to this new form of communication. Worldwide radio communication grew from this beginning.

The big breakthrough that led to the development of modern electronics came in 1906, when Lee DeForest invented the triode vacuum tube. The DeForest triode, or audion tube, was the first device to use the principles of electronics to amplify, or increase, electrical signals. All modern electronic devices rely on the principle of signal amplification.

In 1912, at the age of 16, an amateur radio operator named David Sarnoff picked up the radio signal from the sinking *Titanic.* He stayed in continuous contact with the ship until it sank. He then began his career in electronics which eventually led him to the top position at RCA. Sarnoff's technical background, coupled with his genius for organization and business, contributed greatly to the growth of electronics in

Communication progress is dramatized by both compactness and capabilities of these telephone instruments. The early telephone above made it possible for people to talk to each other over electric wires—provided they were willing to shout over the noisy connections. The modern instrument, below, makes possible pushbutton dialing convenience—and is also fitted with a device (the flat, platform-like unit below the phone itself) that makes it possible to transmit data as well as conversation. (Photo by Liane Enkelis)

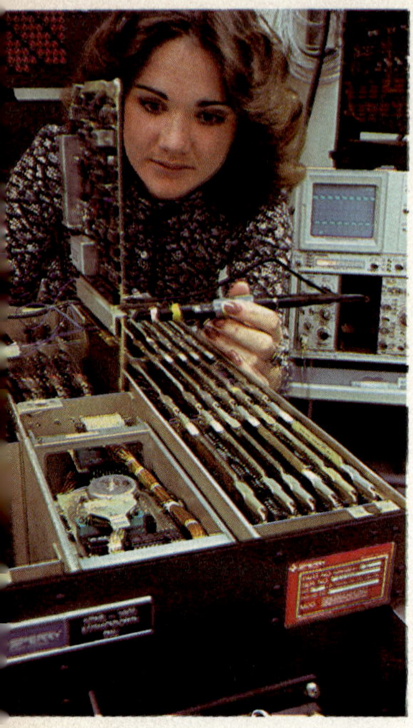

Packaged circuits hold the key to modern electronic devices. Entire circuit boards are replaced as units. The idea is to keep critically important units in service, leaving the repair of circuit boards to specialized shops. (Courtesy Sperry Corporation)

the United States. It was his vision that transformed radio and television from laboratory curiosities to everyday conveniences in the average American home.

Television, one of the most widely used forms of modern electronics, resulted from the work of many people over more than a generation. One of the more prolific inventors was Vladimir Zworykin, credited as the inventor of the standard television picture tube and the modern television camera tube. Zworykin is considered to be the father of television. He also had other research interests, including computer memory and the electron microscope.

A development that was to have a major impact in making possible commerical television in the United States came about in 1927. Philo T. Farnsworth, who was 21 years old at the time, received a patent for the first practical electronic camera tube. This was the dawn of television. Continuous research and investment were necessary before television systems could be demonstrated publicly in 1934. Commercial broadcasting began in the United States in 1941.

The modern-day revolution in electronics came in 1948. A team of three scientists working at the Bell Telephone Laboratories developed what they called a *crystal amplifier*, later known as a *transistor* because it combined the properties of transferring and resisting electrical current. The three scientists who performed this work were John Bardeen, William Shockley, and Walter Brattain. This group shared a Nobel Prize in physics for their work.

Modern developments in electronics, about which you will learn as you read this book, include integrated circuits (digital and linear) and the devices they made possible. These include the latest high-performance computers and miniature calculators.

Many of the inventors mentioned here formed businesses which later grew into giant corporations. They had the ability to combine technical genius with business skills. They not only created careers for themselves but also developed thousands of jobs for employees in their companies.

ELECTRONICS AND THE FUTURE

The entire field of electronics is perhaps 100 years old. In its century of existence, the electronics industry has affected every facet of life and work. However, the opportunities in

Small computers, large opportunities. Desktop microcomputers, like the one in the photo on the left, are being delivered to homes and offices at rates estimated in the range of 3 million annually. Maintenance of this equipment promises to become a big business.

Designing computer. A major area of electronics expansion in industry involves the use of computers for product design, as shown in the photo below. The process, known as computer-aided design (CAD), makes it possible for engineers to design and preview performance of complex products.

the field of electronics still have a long way to grow. The electronics field promises to continue to expand—with new opportunities opening as far into the future as anyone can predict.

During the next decade or so, much of the growth in electronics will come as the technologies of computers and communications are joined. This combination will lead to development of many new systems, new businesses, and new opportunities. Some of these developing areas are reviewed below.

The Office of the Future

Many electronic improvements will impact the activities of office workers and managers. Greatly increased sums will be spent in the future on equipment used in performing office jobs.

In this respect, office workers will find themselves in situations approaching those already established in manufacturing and other industries. For example, the typical industry worker already uses equipment valued at $50,000 to $100,000. Many are backed by far more capital. A single refinery worker may control an entire refinery worth hundreds of millions of dollars. An operator at a nuclear power station may be in control of a billion dollars worth of equipment.

Infinite varieties of information. can be presented on CRT displays, like the one shown above.

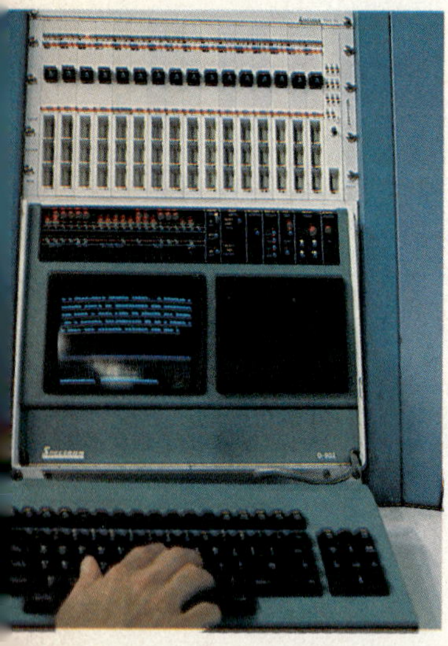

Communication network connections can be established and monitored electronically through use of computerized systems like this. Telephone and data transmission exchanges are now primarily electronic. (Photos by Liane Enkelis)

By contrast, a typical office worker uses only about $2,000 to $3,000 worth of equipment. One major development that has already begun to revolutionize office work and change the level of investment in equipment is the microcomputer.

The microcomputer uses components from the growing microelectronics industry which is based on the design and building of extremely small parts to handle complex electronics functions. For example, microelectronics chips now available are smaller than a fingernail. Yet a typical chip can store and retrieve up to 64,000 characters of information. Very soon, the capacity of such chips is expected to expand considerably.

Already, microcomputers are making possible sophisticated accounting and record-keeping systems in offices that could never have afforded older models of computers. These new miniature systems are sold for just a fraction of the price tags that computers carried through the 1970s. With microcomputers, small businesses—or small offices of large companies—are able to do computer processing work that only giant organizations could afford a few years ago.

Thousands of microcomputers are being manufactured and delivered every week. With these units, companies employing just a few people are able to buy and use full-scale computer systems. These systems make it possible for companies to send bills to their customers, to schedule production, to control inventories, and to process complete accounting systems.

In addition, the same principles that made the microcomputer possible have also started to change the way in which businesses produce letters and other business documents. The new development is known as *word processing*. The work of word processing is done on microcomputers, on larger computer systems, or on special electronic equipment that handles documents but not accounting jobs. The main features of word processing are electronic storage of information and high-speed production of documents.

An even more advanced use of electronics in the office of the future is known as *electronic mail*. Electronic mail is a technique that enables computers and word processing systems in different offices to talk to each other. An electronic mail system is thus a network of computers, microcomputers, word processing stations, and communication lines. The need to produce and transport documents is eliminated. Texts move from computer to computer over telephone or other communication lines.

All of these developments will continue to be translated into opportunities for persons with training in electronics. Opportunities will grow for persons who can work with computer equipment and computer programs, in designing systems, and in servicing customer equipment.

The Factory of the Future

Although the factory has been far ahead of the office in the use of electronics and other automatic devices to improve productivity, the future will see major new developments. In the past, emphasis was on improving the capacity and automatic capabilities of individual machines.

Today and into the future, factories are evolving as complete, continuous manufacturing systems. Machines and conveyors are being tied together so that the manufactured products can move continuously and automatically through a whole series of production steps.

Robots are being used to perform a multitude of manufacturing tasks, such as painting and welding. In addition, electronic equipment is being developed that inspects

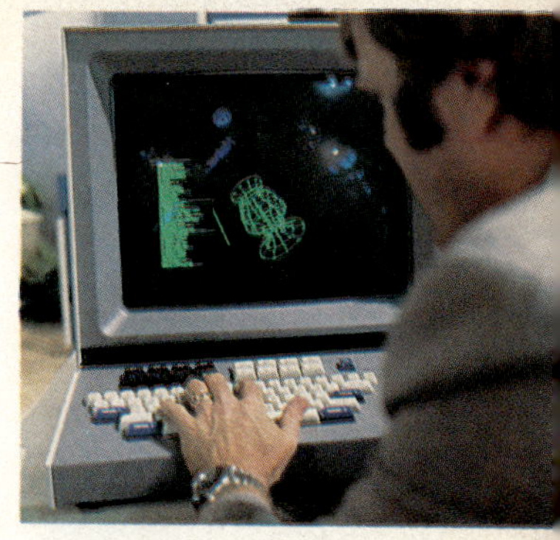

Computer-aided manufacturing (CAM) continues the streamlining that involves the use of computers for both design (photo above, by Liane Enkelis) and manufacturing of new products. Large complex machines are computer controlled for automated operations. (Photo below, courtesy McDonnell Douglas Automation, St. Louis)

Electronics for defense. Safety from military attack is being improved continuously through complex, computerized systems used for command and control (top photo, courtesy Sperry Corporation). More than half the costs of military aircraft go into electronic navigation and flight control equipment. The manufacture and maintenance of these ground and flight systems require many expert electronics technicians. The bottom photo shows navigation and flight control equipment being installed. (Courtesy TRW, Inc.)

the quality of products automatically as they are made. Products that do not meet specifications are set aside for rework or scrap.

Computer-aided design and computer-aided manufacturing (CAD/CAM) are helping to develop modern manufacturing systems. Computer-aided design (CAD) helps in the making of two- and three-dimensional geometric models used by manufacturers.

In the future, electronics technicians will make up an increasingly larger part of the work force of manufacturing organizations.

Transportation in the Future

Growing numbers of people will continue to travel; however, it is likely that business and governmental organizations may be able to substitute electronic communication for travel to business meetings. Already, many business executives are holding meetings through videophone, a system which connects people in distant places through a television-like network. Such services reduce the need to travel. As one result, people can be more productive because the time and fatigue of travel are reduced or eliminated.

Even with electronic "meetings," the amount of travel people undertake will continue. Travel operations of the future will be enhanced through the use of such innovations as advanced electronic systems for aircraft navigation, for routing traffic through cities, for safer and faster movement of trains, and for all-weather safety and navigation for ships.

Designing and producing these advanced electronic systems will offer continuing challenges and opportunities for persons with training in electronics.

Communication in the Future

In the past, expansion of communication systems has depended upon stringing wires to connect different points. Telephone calls have moved along wires. Even transcontinental television programming was carried by wires. Eventually, the wire-based communication networks became almost saturated.

The new horizon in communications is through satellites. Ground stations transmit signals to satellites in earth orbit. The signals are relayed to other ground stations. Worldwide installations of satellite ground stations are making it possible to multiply man's ability to communicate.

Space talking is made possible through use of antenna systems that send and receive microwave signals that are bounced off of satellites more than 100 miles up in space. Direct communication from vehicles, such as police cars (above) and fire trucks, to headquarters locations is possible through microwave relays (left) that are strategically located in most major cities. Building and maintaining such systems requires skilled people with electronics training. (Courtesy Glendale, CA, Police Department)

Satellite networks are electronic systems. The field of communications holds the potential for great expansion of opportunities for persons with backgrounds in electronics.

There is also a trend back to connecting points together by cable. However, the new cables are fine glass fibers. Electrical signals are converted to light signals and then sent over the fibers. To illustrate the breakthrough represented by this electronics development, consider that a single fiber can carry tens of thousands of telephone conversations.

The Home of the Future

Electronic devices—microwave ovens specifically—already represent the fastest growing method of preparing food for the family.

Electronic systems—burglar alarms—are being installed to protect increasing numbers of homes as the nation's crime problem spreads.

Electronic controls over heating and air conditioning systems are being used to conserve energy as costs for electricity and natural gas continue to mount.

Entertainment and learning facilities in the home continue to improve with the introduction of new electronic devices. Millions of homes are now equipped to play first-run movies or special programs from tape cassettes or discs. Closed circuit television systems now being installed can provide up to 250 channels of communication between homes and control centers.

In addition to traditional entertainment programming, these systems can provide two-way communication. You can

Personal computers (right) are used for work and play in hundreds of thousands of homes. Persons with trouble-shooting and maintenance skills will be needed in increasing numbers to keep them going.

communicate with the television control center right from your home. By using control devices linked to your television set, you can participate in study programs or can actually shop for and order merchandise displayed on your television screen. In addition, many television outlets are offering continuing displays of news and other important information through advanced communication systems.

These systems, like the others described above, are built with advanced electronics techniques. Thus, electronics can be seen as a field with a future.

Opportunities and Challenges

This brief overview should make clear that the field of electronics represents continuing opportunities and challenges. The opportunities occur because the field continues to grow. With growth, more jobs open. At the same time, there are challenges for persons who wish to maintain their positions in the field. Any body of specific, detailed knowledge acquired about electronics, at any given time, will become 50 percent obsolete in three to five years. This three-to-five-year period is known as the ''half-life'' of engineering or technology. To avoid becoming obsolete during this half-life cycle, persons in the field must continually read and study about new developments. On the other hand, a solid basis in the fundamentals of electricity and electronics, as covered in this book, will serve a person all through his or her career.

Electronics is a dynamic field that will continue to grow and to open new opportunities throughout your lifetime. Keeping up in the field of electronics also has its challenges. If you choose to work in electronics, the studies you begin now will continue throughout your working career.

Medical laboratory diagnosis (above) is completed in less time and with greater accuracy through use of computerized laboratory equipment. (Courtesy Glendale, CA, Memorial Hospital)

ELECTRICITY

1.1 SOURCES OF ELECTRICITY

Electricity is a form of energy that can produce heat, light, or force that causes motion. The energy that is electricity comes from these basic sources:

- Magnetism
- Chemical action
- Pressure applied to crystals
- Light
- Heat
- Friction/motion.

Electricity from Magnetism

A *magnet* is a type of metal or metal ore that attracts certain other metals. These include iron and related metals, such as steel.

When a magnet is in motion within a coil of wire with good electrical properties, electricity is generated. Magnetism is the main source used in generating electricity for commercial use. Therefore, magnetism is discussed in greater depth in Chapter 6.

To demonstrate the action of a magnet, Figure 1-1 shows a pattern of iron filings sprinkled on a paper placed over a magnetized iron bar. The filings distribute themselves along the lines of force of the magnet.

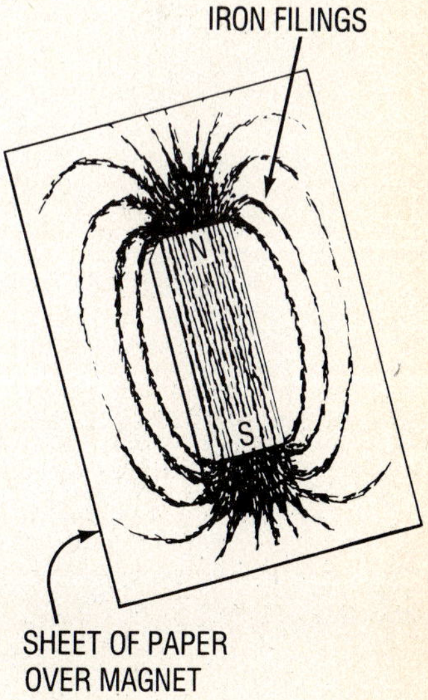

Figure 1-1. Lines of force of a magnet.

Figure 1-2 shows a simple method for generating electricity. A bar magnet is moved through the middle of a coil of wire. This causes electricity to flow through the wire. This is the basic principle behind the generation of electricity for commercial use.

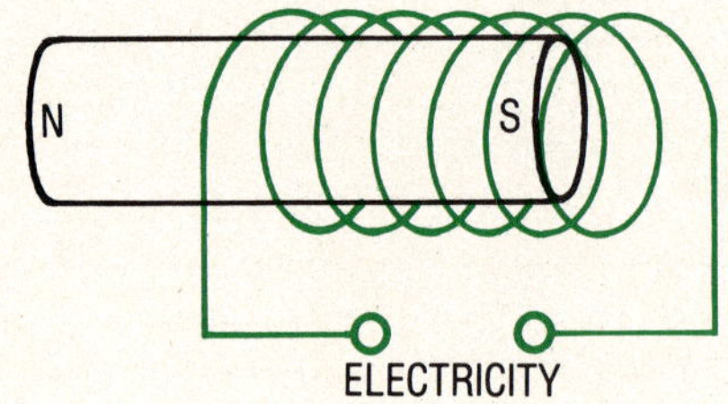

Figure 1-2. How electricity is generated with a magnet.

Electricity from Chemical Action

Chemical action is the principle applied to generate electricity in cells and batteries. A battery is made up of several interconnected cells. Typically, a cell contains two dissimilar solid materials. Each of these materials has different electrical properties.

The solid materials are separated by a chemical substance known as an *electrolyte*. The electrolyte has the ability to transfer electricity from one of the materials to the other. This is done through a chemical action of the electrolyte upon the basic materials of the cell. The electrolyte can be a liquid or it can be in the form of a paste.

Figure 1-3 is a diagram of a simple electrical cell from which batteries are assembled. Cells and batteries are described in greater depth in Chapter 5.

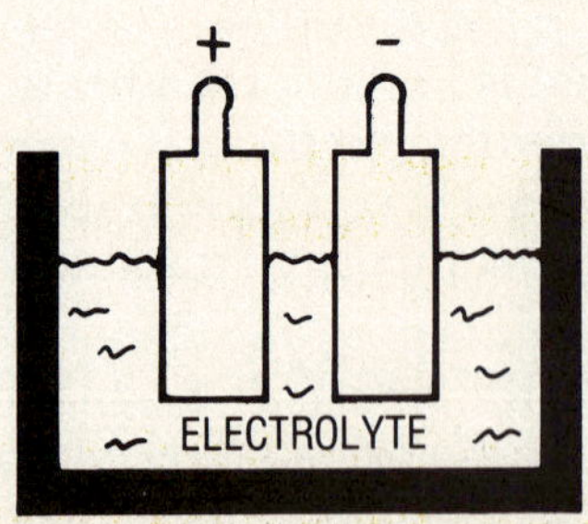

Figure 1-3. Simple electrical cell from which batteries are assembled.

Electricity from Pressure upon Crystals

A *crystal* is a material with a fine, regular structure. Pressure to a particular type of crystal by such actions as twisting or applying force causes electrical current. This property of the crystal is called *piezoelectricity.*

A phonograph or stereo record player is an example of how piezoelectricity can be used effectively. The needle of the player's pickup arm is made of a metal rod with a tip made from a hard substance, such as sapphire or diamond, which wears well. This needle comes under changes in pressure as it rides in the grooves of the record. The needle is connected to a quartz crystal. The changes in pressure which are transferred through the needle to the crystal cause varying electrical voltages. These voltages correspond to the recorded sounds. These electrical signals are transmitted from the quartz crystal into the electronic amplifier of the record player.

Electricity from Light

In some materials, electricity is generated by intense light. For practical purposes, the only source of light that can generate electricity economically is the sun. Materials that react to sunlight are called *photoelectric. Photo* means light sensitivity. A photoelectric material is one that generates electricity on exposure to light. One such material is selenium.

Possibly the best example of a practical use for photoelectric materials is the photographic light meter. This meter indicates light readings and can be hand-held or built into a camera. Within a camera, a meter can be attached to a mechanism that selects the correct exposure and may adjust the shutter or the aperture of the camera.

Electricity from Heat

Electricity can be generated from heat by using two dissimilar metals of differing electrical properties. Wires or strips of these metals are joined together mechanically. When the metals are heated, there is a difference in potential between them.

This principle is used in *thermocouples,* devices that generate electricity when heated. Thermocouples are frequently used in industrial ovens as temperature indicators.

Static Electricity

Static means still. In *static electricity* there is a lack of movement. Electricity exists, but it is not working. This is also called *electrical potential.*

Static electricity is found very commonly in nature. For example, static electricity builds up in clouds until it discharges. You see these discharges as lightning.

Static electricity also builds up when you rub certain materials together. The first recorded discovery of electricity by Thales of Miletus in approximately 600 B.C. came from rubbing a silk cloth on a piece of amber.

One application of static electricity is to control air pollution. The devices, used to trap dust and waste particles in smokestacks, often utilize static electricity. These electrical devices attract the dust particles, keeping them from getting into the environment.

Discussion Questions/Problems

1. How are magnetism and electricity related?
2. What is the name of the most common device using chemical action to produce electricity?
3. What kind of material is used to generate piezoelectricity?
4. What is the source of electricity produced by photoelectric materials?
5. What is the source for generating electricity from a thermocouple?
6. What kind of electricity forms in clouds to produce lightning?

1.2 NATURE OF ELECTRICITY

Electricity is one of the basic forces of nature. There is electricity in all matter. In this context, matter is any material that exists within the universe. Included are your body, water, earth, wood, and metal.

Matter is composed of a basic series of substances known as *elements.* Ninety-three elements occur naturally. A number of others have been produced in laboratories.

All matter is composed of tiny units of these 93 elements. These structural units are so small that they cannot be seen, even with powerful microscopes. These structures are called atoms. An *atom* is the smallest structural unit of an element that can retain the identity of that element.

Atomic Structure

All atoms have the same basic structural principle. Parts of atoms are interchangeable. Atoms of different elements are different from each other mainly because of the force that binds them together. These forces are electrical. Because it binds particles of atoms into a relationship with each other, this force is known as *binding energy.*

The most basic, simplest atomic structure is the hydrogen atom, diagrammed in Figure 1-4. Hydrogen is the simplest element because its atoms have the fewest number of parts. There is one electron. There is also a simple nucleus containing a single proton. The electron revolves around the nucleus in an orbit, or shell.

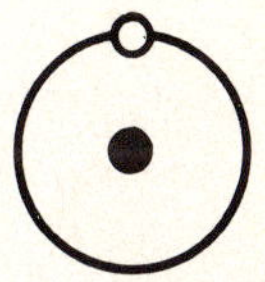

Figure 1-4. Diagram of hydrogen atom.

Natural forces provide the binding energy to attract these atomic parts to each other. This binding energy keeps the electron rotating around the nucleus. The forces that hold atoms together result from the electrical *charges.*

Electrical Charges

A charge is a type of electrical energy. There are two basic types of charges in electricity: *positive* and *negative.* Negative electrical charges are identified by a minus sign (−). Positive electrical charges are identified by a plus sign (+).

Electrons, the portions of atoms that are continuously in movement, have negative charges. Particles within the nucleus, called *protons,* have positive charges. A hydrogen nucleus has a single proton.

Virtually all of the mass, or weight, of an atom is in the nucleus. This is because protons and other particles in the nucleus are considerably larger and heavier than electrons. A proton weighs 1,837 *times* as much as an electron. However, the electrical charges of the protons and electrons are approximately equal in strength. It is this factor that helps to provide the binding energy to keep an electron orbiting around the nucleus of an atom.

The forces between electrons and protons also result from the basic nature of electrical charges. The rule is simple: *Unlike charges attract each other; like charges repel each other.*

In a hydrogen atom, the single negative electron has an equal, or balanced, attraction for the single positive proton.

This is basic for atomic structures: *In its normal, balanced state, each atom of a given material has the same number of protons and electrons. This electrical balance is basic to atomic structure. All atoms naturally seek a state of electrical balance.*

The attraction of negative and positive charges to each other is the basis for the flow of electricity. Electrical flow is the movement of electrons. Electrons are negative. Electrons flow from a negative source, or *pole,* to a positive sink, or pole.

Atomic Number

The hydrogen atom illustrated in Figure 1-4 has one proton. This atom is assigned an *atomic number* of 1. The atomic number is one method used for identifying atoms.

One type of particle found in the nucleus of an atom is the proton. Another is the *neutron.* The neutron has the same weight as a proton but has no electrical charge. As its name implies, it is electrically neutral. *The atomic number of an atom is equal to the total number of protons in its nucleus. The normal number of electrons orbiting in an atom is equal to the number of protons.*

Figure 1-5 is a diagram of a copper atom. Copper is one of the basic natural elements. It has good electrical properties. The copper atom has an atomic number of 29, meaning it has 29 protons and 29 electrons.

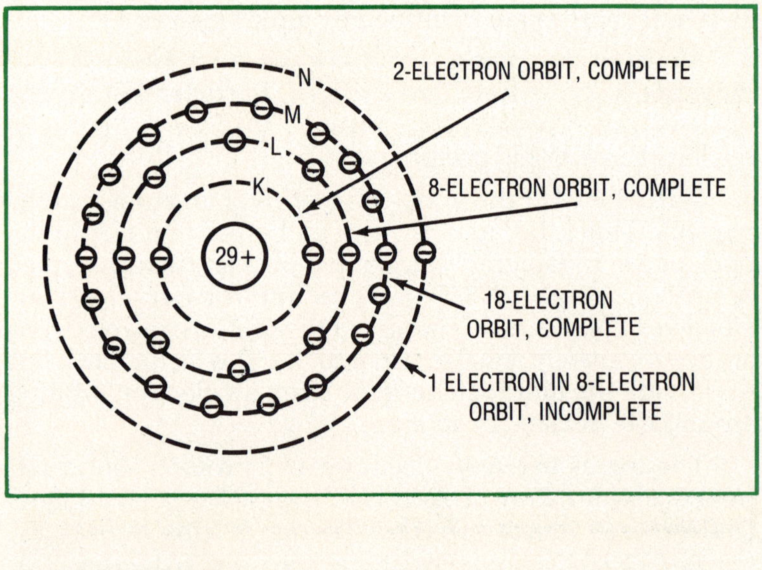

Figure 1-5. Diagram of copper atom.

Atomic Weight

Another method of identifying or describing atoms is by atomic weight. Most of the weight of an atom is in the nucleus, primarily in the protons and neutrons. Since each element has its own distinct number of protons and neutrons, an atom of each element will have its own distinctive weight. Of course, the weight of any one atom is extremely small and would be difficult to use or measure in a practical way. It is easier instead to establish a set, large number of atoms, enough to form a practical mass, such as a few pounds or kilograms. By taking the same number of atoms of every element, the differences in weight are easier to measure. Those elements made up of atoms with only a few protons and neutrons in the nucleus will be lighter than those elements made up of atoms containing many protons and neutrons. There is a very specific relationship to these differences in weight. For example, the weight of a set number of oxygen atoms would be four times that of the same number of helium atoms. The same number of bromine atoms would weigh just slightly less than five times that number of oxygen atoms.

As a convenience, scientists have decided to let the weight of this set number of oxygen atoms equal 16. Therefore, by taking equal numbers of atoms of all the other elements, the weights of these other elements will fall into a place order by weight. If a certain number of oxygen atoms weighs 16 kilograms, for instance, an equal number of helium atoms will weigh 4 kilograms, and that same number of bromine atoms will weigh 79.92 kilograms. If one atom of oxygen weighs four times as much as one atom of helium, then a trillion oxygen atoms will weigh four times as much as a trillion helium atoms. If the number of atoms of each element is equal, then their weights can be compared.

It has already been stated that copper has an atomic number of 29. If the atomic weight of an oxygen atom is 16, the atomic weight of a copper atom is 63.57. The atomic weight of an atom of a particular element is equal to the weight of a fixed number of those atoms, divided by the weight of the oxygen atoms, all multiplied by 16.

Periodic Table

Since all elements have a specific atomic number, specific atomic weight, and a specific number of electrons, it should be possible to list these characteristics in an orderly way. Appendix C shows the elements classified in a Periodic Table. This type of classification is credited to Dmitri Mendeleev, a Russian chemist who proposed it in 1869.

Electron Rings

The arrangement of electrons in an atom is important because it holds the key to the electrical properties of an element. Note the electrons in the copper atom in Figure 1-5. All electrons, in all atoms, are arrayed in *rings,* or *shells*. The basic structure, or arrangement, follows the same pattern for all atoms.

Electron rings, or shells, are identified by letters. The shell closest to the nucleus is always the K shell. Orbital rings beyond the K shell are lettered consecutively, beginning with L and running up through Q. Figure 1-6 lists the maximum number of electrons that may be contained in each shell. The shells fill up from the innermost ring to the outermost. The K shell can have two electrons; the L can have eight. The M shell may have either 8 or 18. The N shell may have 8, 18, or 32. The O and P shells may have either 8 or 18 electrons. The Q shell has a maximum of eight electrons.

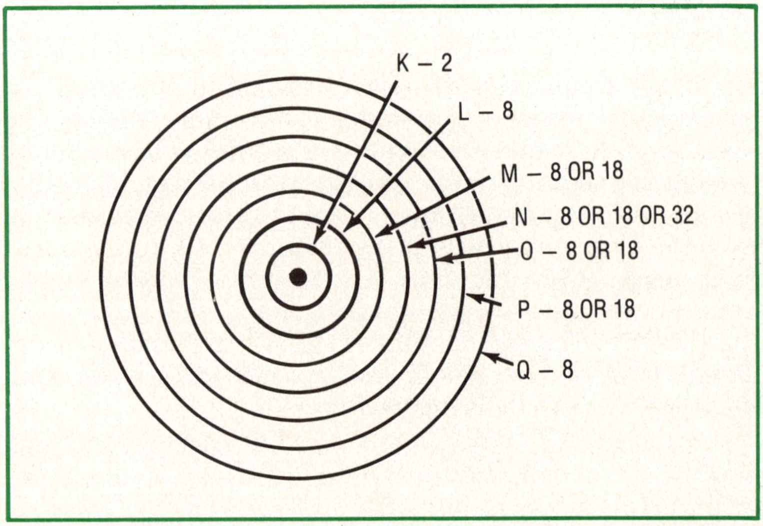

Figure 1-6. Maximum number of electrons that may be contained in each shell of an atom.

An important characteristic for any atom with two or more shells is that the outer ring, known as the *valence* ring, always has a capacity for a maximum of eight electrons. In terms of electrical properties, the valence holds the key.

To illustrate the importance of the valence ring, look at the diagram of the copper atom in Figure 1-5. Copper has a K ring with 2 electrons, an L ring with 8 electrons, and an M ring with 18. The N, or valence, ring has room for eight electrons but contains only one. A single electron orbiting in an outer ring is held less securely within its atomic structure than an electron in a complete, or full, valence shell. For example, the element neon is a gas. As diagrammed in Figure 1-7, neon

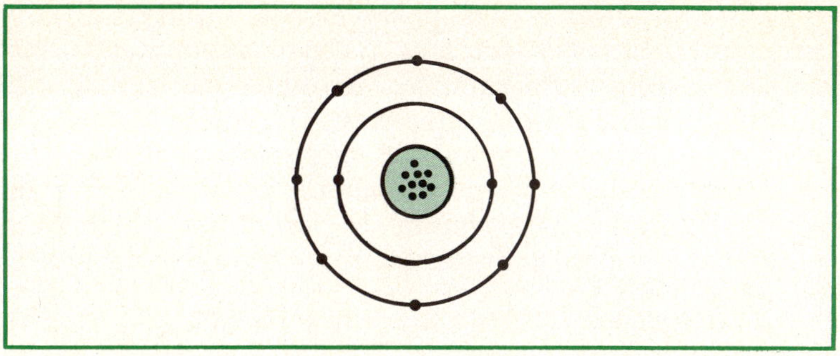

Figure 1-7. Diagram of neon atom.

has 10 protons and 10 electrons. This means that both the K and the L shells are complete, or filled. When the outer ring is complete, an atom is said to be *stable.* In the case of neon, this gas is said to be *inert.* That is, it does not react with the atoms of other elements.

Now refer to the diagram of the copper atom with its single valence electron. One of the basic properties of atoms is that incomplete valence shells can lose their electrons. These electrons can be dislodged from their atomic structures either through collisions with other atomic particles or through electrical attraction to positive forces. Once dislodged, they become *free* electrons.

Free electrons are the basis for electrical energy. Within the atomic makeup of the universe, there are many more electrons than atoms. Some elements attract more free electrons than others. These elements have incomplete valence shells. When an electron is dislodged from a valence shell, the atom takes on a positive charge. This is because the number of protons in the nucleus no longer balances the number of electrons. The excess proton causes a positive charge because it is positive. A positively charged atom attracts free electrons in its vicinity. This is the basis for the flow of electricity: Atoms that are missing valence electrons attract free electrons.

Within an ordinary strand of copper wire, there may be trillions of free electrons moving toward a positive pole. This movement forms *electric current.* Within the wire, copper atoms are continually losing and gaining free electrons in their valence shells. The positive charge, or polarity, of the atoms that are missing valence electrons facilitates the flow of current by attracting free electrons. This is a natural behavior, given the electrical properties of atoms.

When an atom is in its positive state, it still retains the basic properties of its elements. A copper atom that has lost a valence electron, for example, has only 28 electrons and 29 protons, but it is still a copper atom. The rule: *The number of protons in the nucleus determines what element an atom is.* This structural property of atoms makes it possible for certain elements, known as *conductors*, to permit free electrons to pass from one atom to the next without changing the type of element.

Discussion Questions/Problems

1. What is a natural element?
2. What is an atom?
3. Identify three particles found in atoms and describe their relationships?
4. What is binding energy and how does it function?
5. What are electrical charges and where do they originate within atoms?
6. What is atomic number and how is the atomic number of an element determined?
7. What particles of an atom determine the element to which it belongs?
8. What is atomic weight and how is it determined?
9. What are the rings or shells of an atom and what is their function?
10. What are valence rings and valence electrons?
11. What are free electrons?

1.3 ELECTRICAL CURRENT

Free electrons are the source of electrical current. Current can be defined as the *flow or movement of electrons from one point to another through materials known as conductors.* Current, then, is the flow of free electrons.

Current Flow and Electron Flow

Electric current is the result of electron flow. Over the years this has come to be described in two different, but not conflicting, ways.

Electron flow is from negative to positive. This is logical, because electrons are negative particles and can be expected to be attracted to and flow toward a positive point in a circuit.

Current flow is considered to be from positive to negative. This is a convention which developed over many years.

Both current flow and electron flow are correct. Just be sure that you are careful to notice which term is used in any given situation. Most of the discussions in this book, in cases where direction of flow is important, use electron flow.

Measuring Current

To use electricity productively, it is necessary to be able to measure it. There are a number of units of measure for electricity that must be learned. One of these is a measurement for the quantity, or amount, of current that flows through any given conductor or device.

The basic unit of measurement for current is the ampere. This unit is named after Andre M. Ampere (1775–1836), a French physicist who was one of the early experimenters with electricity. *An ampere is measured in terms of the number of electrons flowing past any given point in one second.*

Measuring charge. The unit of measurement for electrical charge is the coulomb. This unit is named after Charles A. Coulomb (1736–1806). Coulomb was a French physicist who conducted experiments in electricity and magnetism. *A coulomb is the charge that must flow for one second to produce 1 ampere of current.* One ampere is the flow of 6.28×10^{18} electrons per second. Because this is a large and somewhat inconvenient number, the ampere has also been defined as the flow rate of one coulomb per second. A coulomb of charge is 6.28×10^{18} electrons collected on a special surface suitable for storing electrons. This surface is sometimes called a *dielectric*.

Scientific notation. The figure, 10^{18}, is expressed in what is known as *scientific notation*. This is a form of shorthand used by scientists and mathematicians. The small, raised numbers alongside the 10 identify a multiplier to be used with the base number 10. This multiplier number is called an *exponent* or a superscript. In this example, the number 10 is raised to the 18th power. This means that the decimal point is moved 18 places to the right. Thus, in addition to the numerical values, zeros are added. If written out completely, the number of electrons that flow to equal 1 ampere is 6,280,000,000,000,000,000. This number is the equivalent of 6.28 billion-billion.

Each of the flowing electrons measured in an ampere has a charge of 0.16×10^{-18} coulombs. Notice the minus sign in front of the exponent. A minus sign in an exponent means that the value of the number derived is less than zero. In writing out the number represented by a negative mathematical notation, you add zeros to the left of the integer, rather than to the right, as you do with positive numbers. Thus, the numeric representation for the charge of one electron is 0.00000000000000000016.

Electrical Charge

For current to flow, there must be a source of free electrons. These free electrons gather and remain in a ready-to-use condition. Deposits of electrons are built up on materials with atomic properties suitable for storing free electrons. These are known as *dielectric* materials. A dielectric material is a nonmetal substance capable of storing, as electrical energy, a substantial number of free electrons. The electrons stored in this manner are known as an *electrical charge*.

To get the idea of building electrical charges on dielectrics, think of your own experiences. You have probably walked into a building that has thick carpets. After walking some distance, you remove a key from your pocket or purse and insert it in the lock of a door. As you do this, you notice an arc, or spark, between the key and the lock. This is an electrical discharge. A *discharge* is the release of a charge. Actually, as you walk, free electrons are building on your clothing and body. The charge that you carry is released as the key approaches the lock.

You may have noticed the same effect in combing your hair. This occurs when you use a comb made of a plastic material. Free electrons build on the comb as it passes through your hair. When the charge is built up sufficiently, the electrons are discharged as you bring the comb into contact with your hair. Your body acts as a conductor of those electrons.

Abbreviations

For convenience, many of the terms used often in electricity and electronics have been given standard abbreviations. Abbreviations that represent the electrical terms used in describing current include the following:

- Current is represented by I.
- Ampere is abbreviated as A.
- Coulomb is abbreviated as C.
- Charge is represented by Q.
- A period of time used in an electrical formula is represented by the Greek letter tau (τ).

Calculating Values of Current

The formula used for determining the value, or amount, of current is:

$$I = \frac{Q}{\tau}$$

To illustrate, suppose:

$Q = 20$ C

$\tau = 4$ sec.

$$I = \frac{20}{4} = 5 \text{ A}$$

The current is 5 A.

Discussion Questions/Problems

1. What is electrical current?
2. Name and describe the standard measure of electrical current.
3. Why is the unit of measure for current described through use of scientific notation?
4. Write out the number represented by the notation 23^{15}.
5. Write out the number represented by the notation 587^{-15}.

6. What is a coulomb and what does it measure?
7. What does the letter *I* represent?
8. What term is abbreviated with the letter A.
9. What is the abbreviation for coulomb?
10. What does the letter *Q* represent?
11. What term is represented by τ?
12. What is the current, in amperes, if a 50 C charge flows in two seconds?
13. What is the current, in amperes, if 4 seconds is required for the flow of a 64 C charge?

1.4 ELECTROMOTIVE FORCE (*EMF*)

To cause electron movement, a basic force has to be applied to a charge. This is known as *electromotive force*, or *emf*. A synonym frequently used for emf is *potential difference*. This term describes the excess electrons at a negative point that are available to flow toward a positive point on a conductor.

In equations, this book represents electromotive force with the letter *E*. The value of *E* is measured in *volts*, abbreviated V. The volt gets its name from Alessandro Volta (1745–1827), Italian physicist and inventor of the electric battery. One volt is equal to the amount of work (pressure or energy) needed to move a charge of 1 coulomb (C) along a conductor. It has been determined that the amount of work needed for this movement, or to create 1 V of electrical pressure, is 0.7376 foot pounds (English), which is equivalent to the metric measurement unit, 1 joule. Joule is abbreviated J and is named after the English physicist, James Prescott Joule (1818–1889), who formulated the first law of thermodynamics.

The total number of volts applied to a conductor is known as *voltage*. Voltage is the same as emf. Voltage measures potential difference between negative and positive points in a circuit. Thus, if, as stated above, 1 C requires 1 J to move from one point to another, the potential difference is 1 V. The formula for determining voltage or emf, represented by *E*, is:

$$E = \frac{\omega}{Q}$$

In this formula:

E = voltage or emf, expressed in volts

ω = work, expressed in joules

Q = the electrical charge, expressed in coulombs.

An example:

$$E = \frac{\omega}{Q} = \frac{30\ \text{J}}{3\ \text{C}} = 10\ \text{V}$$

In this example, a charge of 3 coulombs is divided into a work potential of 30 joules. The result is 10 volts.

Electrical Symbols

Electrical or electronic devices or networks through which electricity flows can be drawn or represented through certain standard symbols. Use of diagrams is common in electricity and electronics. Throughout this book, symbols are presented as new electrical and electronic devices or properties are discussed. To start, two very basic symbols are presented here.

Figure 1-8 is the symbol for an electrical cell. Notice that the negative pole, or terminal, of the cell is represented by a shorter vertical line. The negative side is indicated by a minus sign. The positive pole is represented by a longer vertical line and a plus sign.

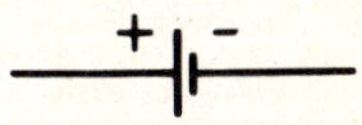

Figure 1-8. Symbol for an electrical cell.

Figure 1-9 is the symbol for a battery. This shows that a battery is composed of several electrical cells.

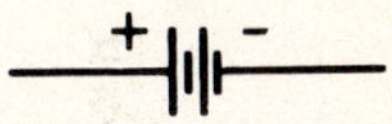

Figure 1-9. Symbol for a battery.

Discussion Questions/Problems

1. What is electromotive force?
2. What is the meaning of potential difference in describing emf?
3. What is represented by the letter E in an equation?
4. What is a volt and how is this term abbreviated?
5. What is a joule and how is this term abbreviated?
6. What is meant when the joule is described as a unit of work?

7. What voltage is present when 150 J of work is applied to move a 30 C charge?
8. How many volts are needed when a 10 C charge is driven by 90 J?

1.5 RESISTANCE

The amount of voltage needed to drive an electrical current along a conductor varies with the properties of the conductor. Some materials are considered good conductors of electricity. Other materials resist the flow of electrons. The property of resisting the flow of electrons is known as resistance, abbreviated *R*.

Resistance is an electrical property that plays a key role in many devices. For example, think about your toaster or electric oven. Wires or rods glow red hot when electricity is turned on. The heat is caused by resistance. The resistive properties of the materials used in toaster or oven heating elements require that extra energy be applied to move electrons. Resistance causes friction. The friction then causes heat. Thus, resistance is a necessary, basic property of materials used in building practical electrical devices.

The unit of measure for resistance is the *ohm*. This unit is named for George Simon Ohm (1787–1854), a German physicist who first published information on electrical resistance in 1827. The ohm is represented by the Greek letter *omega* (Ω). One ohm is the resistance present in a conductor when 1 ampere of current flows under 1 volt of pressure.

Figure 1-10. Symbol for a resistor.

The symbol for a resistor is shown in Figure 1-10.

Conductance

In evaluating conductor materials, resistance to electron flow is one property considered. Another important property is the ease with which electrons can flow, or conduct, through the material. This property is conductance, abbreviated *G*.

The unit for measuring conductance is the siemens, named after Ernst von Siemens (1816–1892), a German manufacturer and inventor who pioneered the use of telegraphy. Siemens is abbreviated S.

An older term used to describe a unit of conductance is *mho*, which is *ohm* spelled backwards. This term was chosen because it portrayed the reciprocal relationship between resistance and conductance. This is expressed mathematically as follows:

$$G = \frac{I}{R}$$

In this formula, G is equal to conductance, R is resistance. To express a reciprocal of a number, the value of the number is used as a denominator and 1 is the numerator. Thus, if a conductor had a resistance of 10 ohms (Ω), conductance would be calculated as follows:

$$G = \frac{1}{10\ \Omega} = 0.1 \text{ siemens}$$

Discussion Questions/Problems

1. What is resistance?
2. What is the unit of measure for resistance and what is its value?
3. How is resistance used to produce heat in an electrical device?
4. What is conductance?
5. What is the unit of measure for conductance?
6. What is the relationship between resistance and conductance?
7. What is the abbreviation for resistance?
8. What is the abbreviation for conductance?
9. If a conductor has a resistance of 100 Ω, what is its conductance?

1.6 CONDUCTORS, SEMICONDUCTORS, AND INSULATORS

Materials used in electrical or electronic devices and systems are classified according to their conductive properties.

Materials that conduct current easily are called *conductors*. Examples of conductors include copper, aluminum, and

MATERIAL	SYMBOL	RESISTANCE (ohms per cir mil foot)
Aluminum	Al	16.06
Arsenic	As	214
Chromium	Cr	15.87
Copper	Cu	10.37
Gold	Au	14.55
Iron	Fe	60.14
Lead	Pb	134
Manganese-Nickel	Ni-Mn	85
Nichrome	Ni-Fe-Cr	675
Nickel, Pure	Ni	60
Nickel-Silver	Cu-Zn-Ni	171
Silver	Ag	9.706
Steel	C-Fe	103
Steel, Manganese	Mn-C-Fe	427
Steel, Stainless	C-Cr-Ni-Fe	549
Tantalum	Ta	79.9
Tin	Sn	69.5
Titanium	Ti	292
Tungsten	W	33.22

Figure 1-11. Resistive properties of metals and alloys.

silver. Figure 1-11 is a table listing and rating the resistive properties of several metals. Notice that conductivity is rated according to resistance. The materials with the lowest ratings in ohms (for a given length and cross section) are the best conductors.

Several factors are considered in the choice of conductors for electrical or electronic jobs. For example, from the table in Figure 1-11, it is apparent that silver is a better conductor than copper. However, copper is the most-used conductor material because silver is far more costly.

Materials that offer some resistance to electrical current are called *semiconductors.* Semiconductors are selected for specific resistivities. Semiconductor materials are available in a wide range of resistive properties. The two materials used most as semiconductors are germanium and silicon. For specific jobs, these materials are mixed with other chemicals to impart exactly the desired resistive and other electronic properties.

Materials which offer strong resistance to the flow of current—or which block or impede electron flow almost entirely—are called *insulators.* Insulators are used to block the flow of electric current or to protect against the danger of electrical shock.

As one example, consider the power transmission lines that deliver electricity to homes and offices. At the points where these lines are connected to poles or other supports, insulators are necessary. These insulators prevent electricity from flowing through poles or other supports, which could include roof beams of homes. The insulators also help to assure that electricity will flow where it is wanted. All lighting fixtures use insulators. These permit electricity to flow to lamps but prevent the flow of current into the fixtures themselves. Similarly, insulators protect you from electrical shock.

Molecules and Compounds

Special materials are utilized to develop desired properties for electrical and electronic devices. These materials can result from chemical combinations of the 93 basic, natural elements.

Atoms combine to form *molecules*. In a molecule, or molecular structure, two or more atoms share valence electrons. This is easily illustrated by diagramming the structure of the world's most common molecule, water. See Figure 1-12. A molecule of water consists of two atoms of hydrogen and one of oxygen. The hydrogen atom, as you know, has a single electron in the K ring. The K ring has room for two electrons. Thus, hydrogen atoms have electron spaces to share in their outer rings.

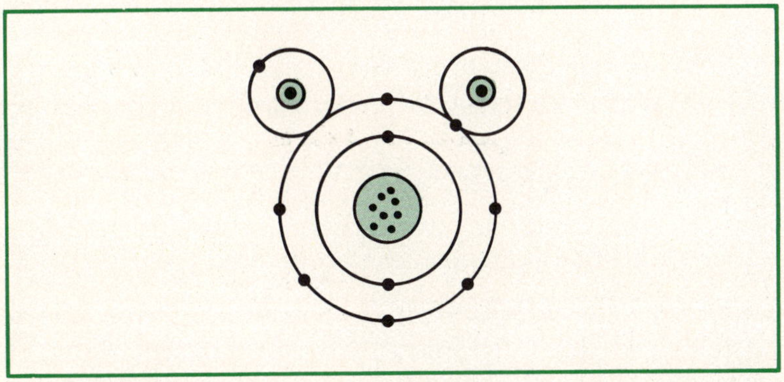

Figure 1-12. Molecular stucture of water (H_2O).

The oxygen atom has a total of eight electrons, six in the L, or valence, shell. The L ring has a capacity for eight electrons. This means that the oxygen atom has room to share its valence ring with the electrons from two hydrogen atoms.

The chemical formula for water is H_2O, indicating that the molecule has two parts of hydrogen and one of oxygen. When combined in this way, a water molecule is highly stable. That is, it resists being separated back into hydrogen and oxygen. This is largely because all valence rings are essentially complete.

Another interesting aspect of molecular chemistry is that a change in form or substance often takes place. Hydrogen and oxygen are both gases. But when combined, they form a liquid. As a liquid, water can also pick up and carry other materials in forms known as *solutions* and *suspensions.* The water delivered to your home contains a number of traces of minerals and chemicals suspended among H_2O molecules. Because of these suspended and dissolved materials, water can be an excellent conductor. This is why it is good advice *never to handle or touch electrical devices when your body is wet.*

Other chemical combinations are also used for conductors, semiconductors, and insulators. These other materials are known as *compounds.* Compounds are formed by combining structures of two or more molecules. Many electrical and electronic devices use chemical compounds.

Discussion Questions/Problems

1. What is a conductor?
2. Why is more copper used in conductors than silver?
3. What are the properties of semiconductors?
4. What are the properties of insulators?
5. What are some important applications of insulators?
6. What is a molecule?
7. What condition must exist in the orbital rings of an atom that becomes part of a molecule?
8. What is a compound?
9. How is a compound structurally different from a molecule?

1.7 TYPES OF CURRENT

There are two basic types of electric current. The type used depends largely on the job at hand. To illustrate, most portable devices, such as transistor radios, portable tape recorders, small television sets, photograph flash guns, and electronic watches all operate on *direct current,* abbreviated DC. A common characteristic for all of these devices is that they are powered by batteries. Batteries are a source of DC.

The term *direct current*, or *DC*, means that the flow of electrons is in a constant direction. This is shown in the diagram in Figure 1-13.

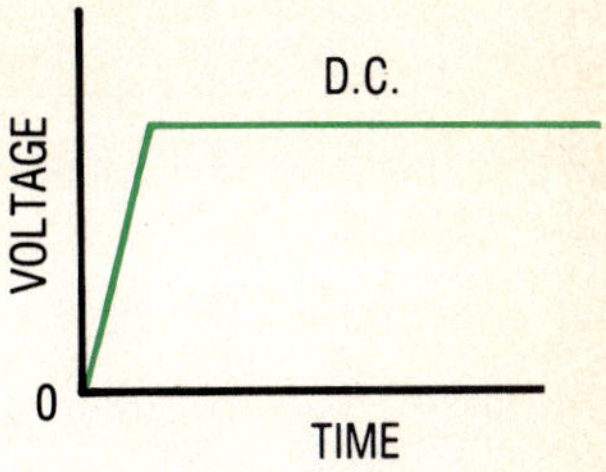

Figure 1-13. Flow of electrons in direct current (DC).

The other type is *alternating current*, abbreviated AC. See Figure 1-14. Where DC is represented by a straight line for continuous flow, AC is diagrammed as a wave form. The wave diagram shows that, at regular intervals, the direction of current flow changes. The wave of current flow alternates continuously from positive to negative. AC is a more economical method for transmitting electrical current. All convenience outlets for electricity, and all appliances or electrical devices you use around your home (except those that are battery operated) use AC.

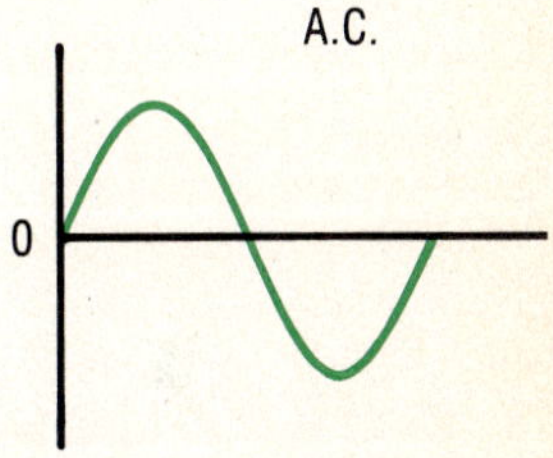

Figure 1-14. Flow of electrons in alternating current (AC).

Discussion Questions/Problems

1. Describe the flow of direct current.
2. Describe the flow of alternating current.
3. What is the main source of direct current?
4. What property of AC led to the name, alternating current?
5. What are the directions of alternation in AC?
6. Why is a diagram of AC called a wave form?
7. What is an example of an electrical device that uses DC?
8. What is an example of an electric device that uses AC?
9. Is AC or DC more economical to generate?

1.8 CIRCUITS

To make electricity usable, it is necessary to construct a circuit. A *circuit* is a complete, controlled path for electrical current. Every circuit needs the basic parts listed below:

- Control device
- Power source
- Load
- Conductive path.

Control Device

The function of a circuit control device is to permit or stop the flow of electrons. This is done by connecting or disconnecting a break in a conductor. The control device makes or breaks a circuit connection. The most common type of control device is a switch. The symbol for the simplest type of electrical switch—the single-pole, single-throw switch—is shown in Figure 1-15.

Figure 1-15. Symbol for a single-pole, single-throw switch.

An electrical switch has contact points that make or break electrical connections. When a switch is closed, electrons can flow. When it is open, the path for the electrons is broken.

One of the important functions of electrical switches is safety. Laws throughout the country require switches or other control devices in all circuits. Circuits must have a way to turn electrical current on or off. Switches are just one method. For example, plugging an appliance into the wall or unplugging it is a method of controlling the flow of current. Unscrewing a light bulb is another technique for current control. Every circuit must have some control method or device.

Power Source

Electrons flow from a negative source to a positive sink. This is basic to electrical circuits.

Figure 1-16. Symbol for an AC source.

In the majority of DC circuits, the power source is a battery. You already know the symbol for a battery.

Figure 1-16 shows the symbol that indicates an AC source.

Load

In an electrical circuit, the *load* is the user of electricity. A load consumes electrical energy. As a simple illustration, the bulb in your flashlight is the load placed upon a circuit for which the batteries provide a power source.

Figure 1-17. Symbol for a lamp.

In diagramming circuits, the symbol for a resistor is often used to indicate a load. Another symbol frequently used is the one for a lamp, shown in Figure 1-17. It indicates that there is a light bulb in the circuit.

A resistor may be used to represent the load of a circuit. For example, a general-purpose circuit such as those that deliver current to convenience outlets in your home can carry a wide variety of electrical loads. The load in such a general-purpose circuit is indicated with a resistor symbol.

Conductive Path

For a circuit to exist, there must be a complete path for electrons from the negative pole (source) to the positive pole (sink). Without this complete path from negative to positive, there is no circuit. For example, a wire connected to a negative source without a termination at a positive pole is not a circuit. No electricity flows. However, if you were to touch the positive pole and then pick up the end of such a wire, your body could complete the circuit, and you could get a shock.

To illustrate a circuit, consider the example of a flashlight. This circuit is diagrammed in Figure 1-18. The source of current is a battery, represented by two electrical cells. The circuit is controlled by a switch. When the switch completes the circuit, current is delivered to the load, a lamp.

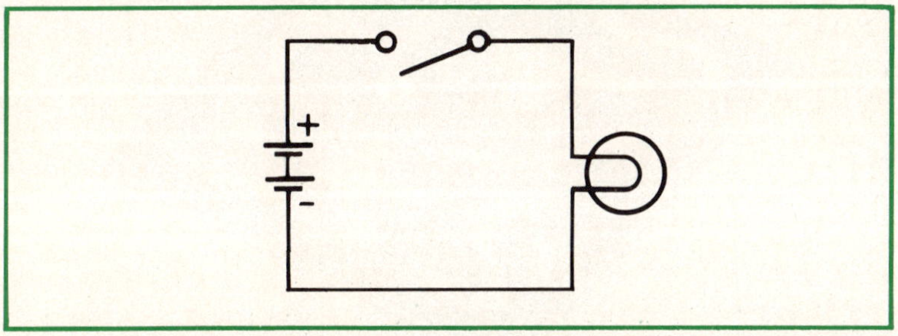

Figure 1-18. Flashlight circuit.

Notice the position of the switch in the circuit in Figure 1-18. It is a design convention for many electrical circuits that the switch is placed on the positive side, between the source and the load. This is done for safety reasons. When the switch to the positive lead is open, no electricity reaches the load.

Figure 1-19 is a diagram of a lamp circuit with an AC source. Notice that this is identical to the flashlight circuit, except for a difference in source. There will, of course, be other differences between AC and DC circuits. Those are covered beginning in Chapter 7.

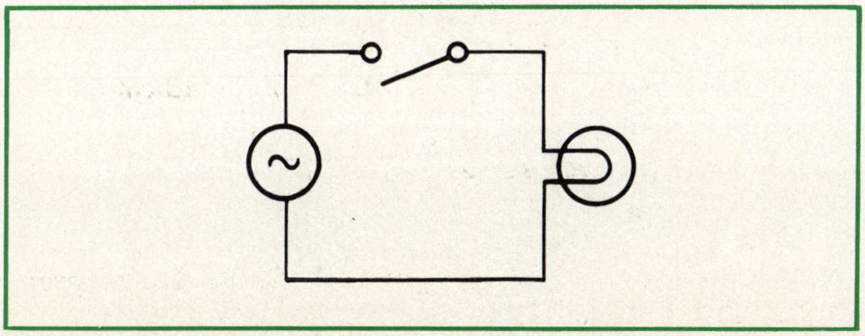

Figure 1-19. Lamp circuit with AC power source.

Most conductors, in most circuits, consist of wires. Wires are diagrammed with straight lines, as illustrated in Figures 1-18 and 1-19.

There are, however, other types of conductors used in electrical and electronic circuits. Possibly the most common alternative to wire is the bonded copper strip on a printed circuit board. In manufacture, areas of copper are laminated to (joined with) or deposited upon sheets of insulator materials. Unwanted areas of copper are then etched away, leaving copper circuit paths to carry electricity. A printed circuit board is illustrated in Figure 1-20.

Figure 1-20. Printed circuit board.

Discussion Questions/Problems

1. What is a circuit?
2. What parts must every circuit have?
3. What circuit requirement can be met by an electric lamp?
4. What circuit requirement can be met by a strand of copper wire?
5. What circuit requirement can be met by a switch?
6. What circuit requirement can be met by a convenience outlet in your home?
7. What type of circuit is powered by a battery?

8. What type of circuit is powered by a convenience outlet?
9. What circuit requirement can be met by a printed circuit board?
10. Draw the symbols for an AC source, a battery, a simple switch, a resistor, and an electric lamp.

1.9 OHM'S LAW

Electrical circuits are described in terms of these basic properties:

- Current (I)
- Voltage or electromotive force (E)
- Resistance (R).

George Simon Ohm, after whom the basic measure of resistance is named, developed a series of equations that describe relationships among current, voltage, and resistance. These electrical properties are always consistent in their relationships to each other. Thus, with Ohm's equations, you can always determine the value of a third property if you know the other two. The basic equations of Ohm's law for these three properties are as follows:

To determine current:

$$I = \frac{E}{R}$$

To determine voltage:

$$E = IR = I \times R = I \cdot R$$

To determine resistance:

$$R = \frac{E}{I}$$

Note that the letters *I* and *R* are shown alongside each other in the equation to determine voltage. This form of notation means that the values represented by the two letters are multiplied by each other. Alternate forms, as shown, include separating these values by a dot or by inserting a multiplication symbol (×).

A circuit diagram to be used as a basis for illustrating how to use these three equations is shown in Figure 1-21. Note that the source of electricity in this circuit is a battery. Ohm's law applies to direct current (DC) only. Somewhat different formulas are necessary to use these equations for the more complex AC circuits.

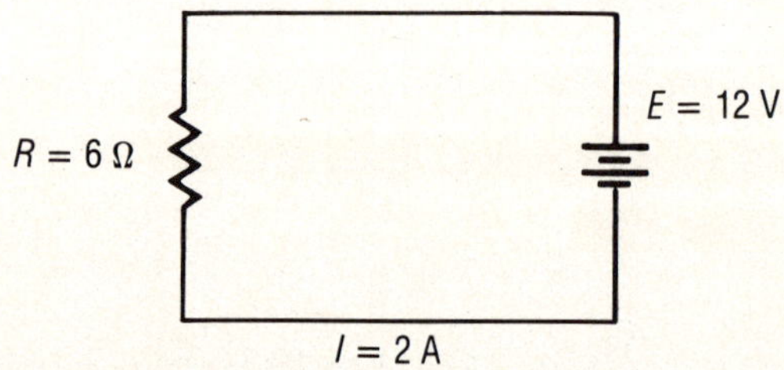

Figure 1-21. Circuit diagram to use for Ohm's law equations.

To illustrate use of Ohm's law to determine current, assume that the illustrated circuit has an emf value of 12 V and a resistance of 6 Ω.

$$I = \frac{E}{R} = \frac{12\ \text{V}}{6\ \Omega} = 2\ \text{A}$$

Now assume that the given values are 2 A of current and 6 Ω of resistance. Ohm's law is used to determine voltage.

$$E = IR = 2\ \text{A} \times 6\ \Omega = 12\ \text{V}$$

Using the same values, assume that you know the circuit has properties of 12 V and 2 A. To determine resistance, you use a third form of Ohm's law.

$$R = \frac{E}{I} = \frac{12\ \text{V}}{2\ \text{A}} = 6\ \Omega$$

Remembering Ohm's Law

Ohm's law is used frequently in work with electrical circuits. Therefore, you should remember these equations for ready use. A handy method for jogging your memory is shown in Figure 1-22. This is a circle in which the property E takes the upper half. The bottom half is divided between I and R.

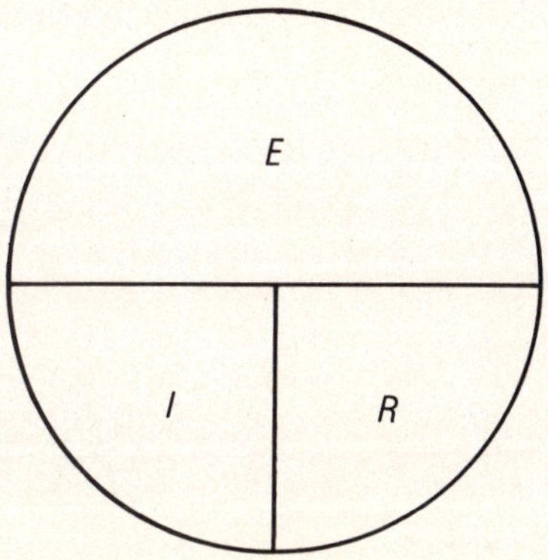

Figure 1-22. Memory aid for Ohm's law: $I = E/R$, $R = E/I$, and $E = IR$.

To use this reminder, just cover up the property that you want to figure. If you cover I, you see that it is equal to E over R. Cover R. This is equal to E over I. When you cover E, you can see that the value of this property is determined by multiplying I by R.

Determining Power (P) with Ohm's Law

Power is the relationship between the ability to do work and the time required for its completion. The relationship is as follows:

$$\text{Power} = \frac{\text{Work}}{\text{Time}}$$

The unit of measure for power is the *watt,* abbreviated W. This is named after James Watt (1736–1819), a Scottish inventor and engineer who developed, among other things, the modern steam engine. A series of equations can be used to state the relationships between power and other electrical properties, including current, voltage, and resistance. These equations are given and illustrated below. For the purposes of the examples, assume the same values as those given for Figure 1-21: $I = 2$ A, $E = 12$ V, and $R = 6\ \Omega$.

With *E* and *I* given:

$$P = EI = 12\text{ V} \times 2\text{ A} = 24\text{ W}$$

With *P* and *E* given:

$$I = \frac{P}{E} = \frac{24\text{ W}}{12\text{ V}} = 2\text{ A}$$

With *P* and *I* given:

$$E = \frac{P}{I} = \frac{24\text{ W}}{2\text{ A}} = 12\text{ V}$$

With *I* and *R* given:

$$P = EI = (IR)I = I^2R = (2)^2\text{ A} \times 6\,\Omega = 24\text{ W}$$

With *E* and *R* given:

$$P = EI = E\left(\frac{E}{R}\right) = \frac{E^2}{R} = \frac{(12\text{ V})^2}{6\,\Omega} = \frac{144}{6}\text{W} = 24\text{ W}$$

Notice the last two equations that are given and solved above. These use the squares of properties. When an exponent of 2 is used with a value, this is called the *square* of the number. The superscript, or exponent, 2 indicates that the number is multiplied by itself. Thus, I^2, when I is 2 A, is 4 A. E^2, when E is equal to 12 V, equals 144 V.

The formula: $P = I^2R$, is derived by substituting values. Note that in the first equation given above $P = EI$. In Ohm's law, $E = IR$. Thus, the value IR is substituted for E. This gives you the value of I multiplied by itself, then multiplied by R. A similar substitution was made in the last equation. That is, the value of I was exchanged for E over R.

Units of Electrical Measure

The particular numerical values used above have been chosen to give simple whole numbers as answers, just for the purposes of illustration. Electrical properties very rarely organize themselves into such neat, even relationships. Rather, they range widely from very large values to extremely small fractions. To keep pace with the realities of electrical measurements, therefore, scientists and engineers have devised a series of standard terms that are used to designate large or small values. These are illustrated in the chart in Figure 1-23.

	VALUE	NAME	SYMBOL		VALUE	NAME	SYMBOL
LARGE	10^{12}	tera	T	SMALL	10^{-12}	pico	p
	10^{9}	giga	G		10^{-9}	nano	n
	10^{6}	mega	M		10^{-6}	micro	μ
	10^{3}	kilo	K		10^{-3}	milli	m

Figure 1-23. Standard terms for large and small values of electrical measurements.

following equation:

$$I = \frac{E}{R} = \frac{117\ \text{V}}{1.2\text{K}\ \Omega} = 0.975\ \text{A} = 97.5\ \text{mA}$$

Discussion Questions/Problems

1. For what purposes is Ohm's law used?
2. State Ohm's law for determination of current.
3. State Ohm's law for determination of voltage.
4. State Ohm's law for determination of resistance.
5. If a circuit has 5 A of current and 10 Ω of resistance, what is the voltage?
6. If a circuit has 50 V of emf and 5 A of current, what is the resistance?
7. Give the basic formula for determining power in a circuit.
8. What is the unit of measure for electrical power?
9. What is the power of a circuit with ratings of 5 V and 20 A?
10. What is the term for 10^{9} V?
11. What is the term for 10^{-6} A?
12. What is the term for 10^{-6} seconds?

INTRODUCTION TO CIRCUITS

2.1 SERIES CIRCUITS—PRINCIPLES

A *series circuit* is one in which electrons move in a continuous path along a conductor that carries current in sequence through the devices within the circuit. Think of a clothesline with a number of items of clothing hanging in a row. If the line were an electric wire, you would have a picture of a series circuit in which current moves from one device to another, in order. Of course, the current originates from a negative pole (source) and moves toward a positive pole (sink).

Figure 2-1 is a schematic showing a number of different devices connected in series. You already know the symbols used to indicate resistors, cells, and electric lamps.

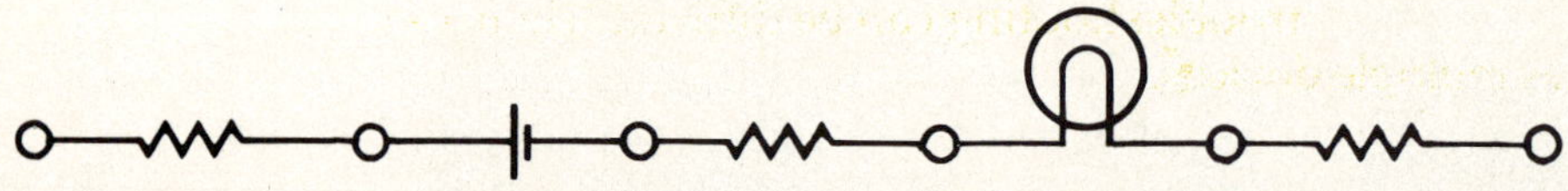

Figure 2-1. Schematic showing different devices connected in series.

A complete series circuit is shown pictorially in Figure 2-2. A schematic of this same circuit is presented in Figure 2-3. The illustrated circuit shows current flowing from the negative pole of a battery and returning to the positive pole of the same battery. Thus, there is a complete path for electrons. This meets one requirement for establishing a complete circuit. The current flows through two resistors, identified as R_1 and R_2. The current also moves through a lamp.

These illustrations demonstrate that a series circuit involves a continuous conducting path through each of the

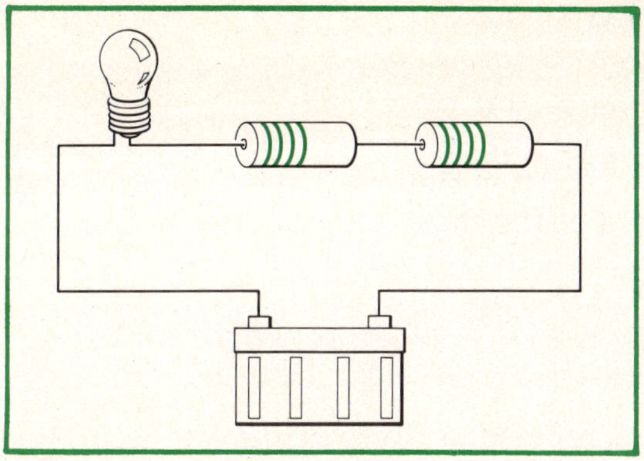

Figure 2-2. Pictorial representation of a complete series circuit.

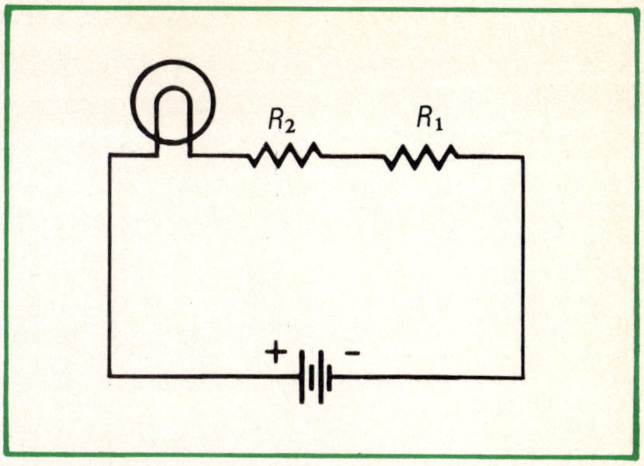

Figure 2-3. Schematic representation of the complete series circuit.

components. This points out another characteristic of a series circuit. If one of the components were to break the electron path by burning out or being removed, the circuit would no longer be complete—current would not be able to flow.

For an example of how series circuits work, consider the illustration of the string of decorative lights in Figure 2-4. This is typical of the lighting equipment sold for holiday decorations. Within the illustrated circuit, the sockets for all of the bulbs are connected in series. As long as all of the bulbs are functioning, as pictured in Figure 2-4, the circuit is complete and the lights are illuminated.

Figure 2-5 shows what happens when one of the bulbs burns out. None of the bulbs is lit.

This example shows the advantages and disadvantages of series circuits. Advantages include simplicity and low cost. On the other hand, troubleshooting can be difficult when a circuit has multiple devices. Suppose you have a dozen bulbs on a string of decorative lights and one bulb burns out. Most of the time, to find and replace the burned out bulb you would have to test all of the lights, one at a time.

Figure 2-4. String of decorative lights connected in series.

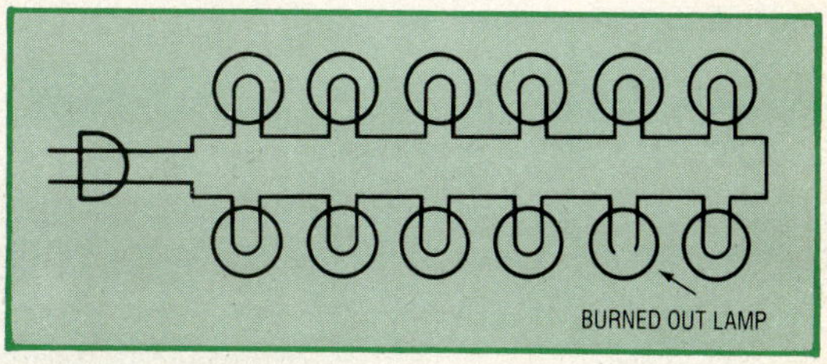

Figure 2-5. String of decorative lights with one bulb burned out.

Current in Series Circuits

An important characteristic of series circuits lies in current distribution: *Current is equal throughout a series circuit.* This means that the current through each of the components in a circuit is the same. Thus, in Figure 2-3, the current through R_1, R_2, and the lamp is the same for each. Mathematically, the statement can be made this way:

$$I_T = I_1 = I_2 = I_3 = \ldots = I_N$$

I_T is total current. I_N represents the total number of current-using devices in any given circuit.

Discussion Questions/Problems

1. Describe the electron path in a series circuit.
2. In what way can a series circuit be compared with a clothesline?
3. What are some advantages of series circuits?
4. What is the chief disadvantage common to series circuits?
5. What is the pattern of current distribution within series circuits?
6. If current in a series circuit is 1.6 A, what would be the value of I through each of three resistors? Assume resistors of 10 Ω, 20 Ω, and 30 Ω.

2.2 RESISTANCE IN SERIES CIRCUITS

Resistance in series circuits is cumulative. That is, *the total resistance is equal to the sum of all individual resistances.*

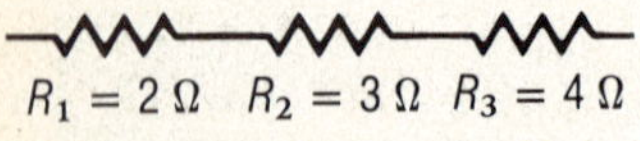

Figure 2-6. Three resistors connected in series.

As an example, consider the circuit in Figure 2-6. This schematic shows a simplified circuit that has three resistors with resistances of 2 Ω, 3 Ω, and 4 Ω. R_T, or total resistance, is developed by adding these three resistances:

$$R_T = 2\,\Omega + 3\,\Omega + 4\,\Omega = 9\,\Omega$$

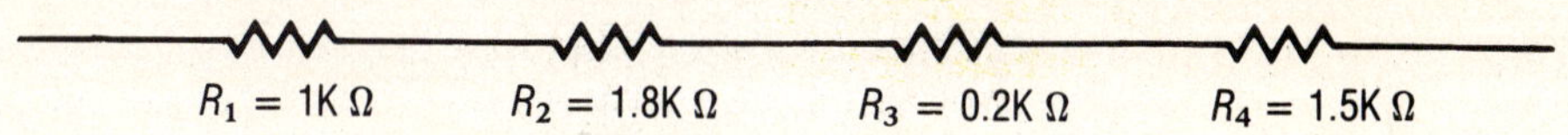

Figure 2-7. Four resistors connected in series.

For another example, look at the schematic in Figure 2-7. Total resistance for the four resistors is:

$$R_T = 1\text{K}\,\Omega + 1.8\text{K}\,\Omega + 0.2\text{K}\,\Omega + 1.5\text{K}\,\Omega = 4.5\text{K}\,\Omega$$

Again, to determine total resistance of a series circuit, just add the resistances of all individual resistors. The rule is:

$$R_T = R_1 + R_2 + R_3 + \ldots + R_N$$

It follows that, in a circuit with two or more resistors, the total resistance will always be *greater than* the resistance of any individual resistor. This can be a convenient way to check your work. Look at your total. Then look at the value of the largest resistor (the one with the highest resistance) within the circuit. The total must always be larger.

Discussion Questions/Problems

1. What is the rule for determining total resistance in a series circuit?
2. Write the formula used in determining total resistance in a series circuit.
3. If a series circuit has resistors of 1.5K Ω, 100 Ω, 300K Ω, and 200 Ω, what is the value of R_T?
4. If a series circuit has resistors of 1.5 Ω, 2 Ω, 1.5 Ω, and 20 Ω, what is the value of R_T?
5. What is the value of checking your work in computing resistance of series circuits by comparing R_T with the highest-value resistor within the circuit?

2.3 CURRENT IN SERIES CIRCUITS

You are already aware that in a series circuit current is the same through each component. To determine current in a series circuit, use Ohm's law. Within the equation for current, you use the totals for emf and R. The formula is:

$$I = \frac{E_T}{R_T}$$

As an example, consider the circuit in Figure 2-8. Voltage is given as 12 V. There are two resistors of 200 Ω and 2.2K Ω.

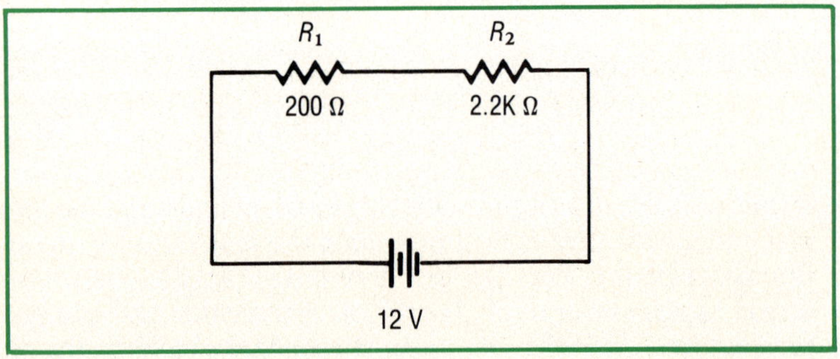

Figure 2-8. Circuit with two resistors in series.

From the figures given, you have the total voltage. Before you can find current, you need the total resistance. The equation for resistance in this circuit is:

$$R_T = 0.2\text{K}\,\Omega + 2.2\text{K}\,\Omega = 2.4\text{K}\,\Omega$$

Now you can use the Ohm's law formula to determine current:

$$I = \frac{E_T}{R_T} = \frac{12\text{ V}}{2.4\text{K}\,\Omega} = 0.005\text{ A} = 5\text{ mA}$$

The current delivered through this entire circuit would be 5 mA (milliamperes).

Discussion Questions/Problems

1. If a current of 2 A is applied to a series circuit that has two resistors, one of 200 Ω and one of 400 Ω, what current flows through each of the resistors?
2. If a series circuit has a source of 12 V and total resistance of 12 Ω, what is the current in the circuit?
3. If a series circuit has a source of 12 V and total resistance of 2 Ω, what is the current in the circuit?

2.4 VOLTAGE (*IR*) IN SERIES CIRCUITS

Voltage is the term usually used to describe the electromotive force of a circuit. The abbreviation E is generally used in formulas for circuits. Although you may see V used for voltage, this book uses E throughout. However, should you see equations using the abbreviation V, you should understand that it means voltage.

Under Ohm's law:

$$E = IR$$

Total voltage is determined by multiplying total current by total resistance.

There is no problem in determining current through any component of a series circuit. It will be the same for all components throughout the circuit.

Voltage levels are not the same for all resistors in series circuits. Rather, voltage is distributed in proportion to the resistances of individual resistors. The amount of the voltage across any resistor in a circuit is known as the *voltage drop,* or *IR drop.*

Remember the principle: Voltage is distributed among all resistors in a series circuit in proportion to their resistance. This means that *the sum of voltage drops within a circuit is equal to the source voltage.*

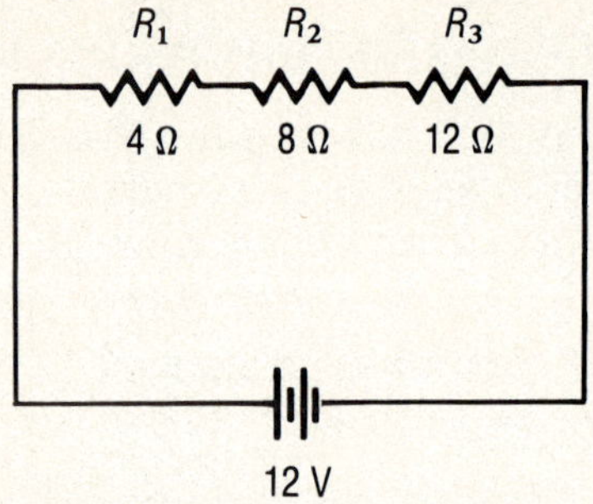

Figure 2-9. Circuit with three resistors in series.

For an illustration of these principles, consider the circuit in Figure 2-9. The value of the voltage source for this circuit is 12 V. Resistor values are 4 Ω, 8 Ω, and 12 Ω. Before you can determine voltage drops, you must find the current for the circuit. The formula is:

$$I = \frac{E_T}{R_T}$$

To solve this equation for *I*, you need to determine R_T.

$$R_T = R_1 + R_2 + R_3 = 4\ \Omega + 8\ \Omega + 12\ \Omega = 24\ \Omega$$

R_T is 24 Ω.

$$I = \frac{E_T}{R_T} = \frac{12}{24} = 0.5\text{ A}$$

The current is 0.5 A throughout the circuit. This figure can then be used to determine the voltage drops for each of the resistors through use of Ohm's law.

$$E = IR$$

The *IR* drop for R_1 is:

$$0.5\text{ A} \times 4\ \Omega = 2\text{ V}$$

The *IR* drop for R_2 is:

$$0.5\text{ A} \times 8\,\Omega = 4\text{ V}$$

The *IR* drop for R_3 is:

$$0.5\text{ A} \times 12\,\Omega = 6\text{ V}$$

To check the accuracy of these calculations, you can use the rule: *The sum of the voltage drops in a series circuit is equal to the source voltage*. Thus, in figures, the 2 + 4 + 6 = 12. The sum of the voltage drops does equal the source voltage.

Polarity of *IR* Drops

In connecting polarized devices into circuits, it is necessary that polarities correspond. That is, the negative connection for each polarized component must be set in the same current-flow direction as the negative lead of the circuit. Of course, the positive connector for each device must be lined up with a positive lead.

Failure to align polarity of devices within a circuit can have serious effects. For example, if you hook a DC motor to the wrong leads, it will turn in the wrong direction. If you connect other devices with opposite polarities, they may simply burn out. Therefore, it is important that you know how to determine polarity of voltage (*IR*) drops across series circuits. Figure 2-10 shows the polarity distribution when a battery is connected to a circuit.

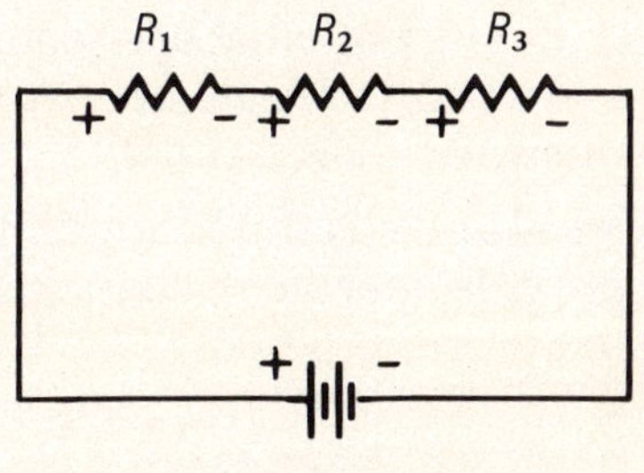

Figure 2-10. Polarity distribution when a battery is connected to a circuit with three resistors in series.

The rule is: *The polarity of all voltage drops in a series circuit will correspond with the polarity of current flow.* It is important that polarized components in a series circuit be connected correctly. This means that their polarities must correspond with those of the current flow. Thus, the negative lead from a battery goes to the negative lead for the first component.

Succeeding connections are from the positive lead of a component to the negative lead of the next one. The positive lead from the final component in a series circuit connects to the positive lead of the electrical source, or battery.

The Series Circuit as a Voltage Divider

Because voltage drops vary with resistance, it becomes possible to use series circuits as voltage dividers. This is done in many electronic devices and is an important principle.

A typical voltage-dividing circuit is illustrated in Figure 2-11. This circuit has 1 A of current and a source of 12 V. There are two resistors of 2 Ω and 10 Ω.

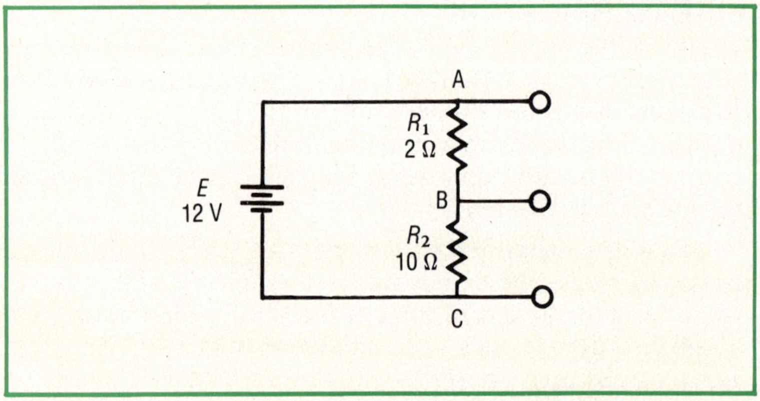

Figure 2-11. Voltage-dividing circuit.

If a branch circuit is established by connecting a conductor between these two resistors, voltage is divided in proportion to the resistances. In this example, the potential between points A and B is 2 V. The potential between points B and C is 10 V. These branch circuit voltages are equal to the voltage drops across the resistors they contain.

In many electrical and electronic applications, this technique is used to provide specific voltage to certain devices. Voltage dividers are an important tool for electrical and electronics designers.

Discussion Questions/Problems

1. How are the abbreviations *emf* and *V* related?
2. In a series circuit, are voltage drops across resistors the same or may they differ?

3. Which equation based on Ohm's law is used to find voltage across a resistor?
4. In a series circuit, a current of 15 A passes through a 100 Ω resistor. What is the voltage drop across this resistor?
5. In a series circuit of more than one resistor, how does the voltage drop across a resistor relate to its resistances?

2.5 POWER IN SERIES CIRCUITS

Power, as indicated in the last chapter, is a measure of the rate at which energy is used. In designing or assembling electrical circuits, it can be important, especially during times of increasing costs, to know how much power is consumed, or used.

In a series circuit, the total power consumption, or dissipation, (P_T) is equal to the sum of the power dissipated by all of the components. The formula is:

$$P_T = P_1 + P_2 + P_3 + \ldots + P_N$$

To determine power dissipated by individual components, you use the appropriate formula from those given in the last chapter:

$$P = EI = (IR)I = I^2R$$

$$P = EI = E\left(\frac{E}{R}\right) = \frac{E^2}{R}$$

Be aware that these equations apply to the calculation of power requirements for individual components. The values of E used in the equations should be the voltage drops across individual resistors. Do not use the total voltage for the circuit.

As an example of power calculations for individual components, see Figure 2-12. This is a series circuit with a 12 V battery as a source. There are three resistors of 8 Ω, 16 Ω, and 24 Ω. To determine power ratings for each of the resistors, you have to find current for the circuit and voltage drops across each of the resistors.

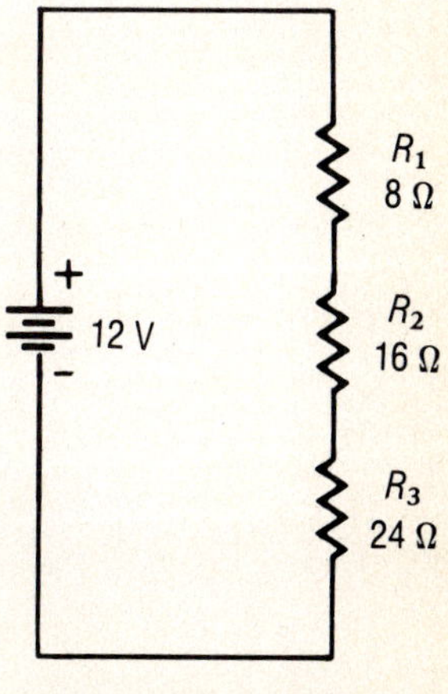

Figure 2-12. Series circuit with 12 V battery source.

To determine current for the circuit, you use the following formula:

$$I_T = \frac{E_T}{R_T}$$

For the circuit in Figure 2-12, E is 12 V. R_T is the sum of the resistances:

$$R_T = 8\ \Omega + 16\ \Omega + 24\ \Omega = 48\ \Omega$$

With these values, it is possible to determine current:

$$I_T = \frac{E}{R_T} = \frac{12\text{ V}}{48\ \Omega} = 0.25\text{ A}$$

With these values, it is possible to go ahead with calculations of power usages of the individual resistors:

$$P_1 = I^2R = 0.25^2\text{ A} \times 8\ \Omega = 0.5\text{ W}$$

$$P_2 = I^2R = 0.25^2\text{ A} \times 16\ \Omega = 1\text{ W}$$

$$P_3 = I^2R = 0.25^2\text{ A} \times 24\ \Omega = 1.5\text{ W}$$

To determine power for the entire circuit, simply add the power dissipated by the individual resistors.

$$P_T = P_1 + P_2 + P_3 = 0.5 + 1 + 1.5 = 3\text{ W}$$

To check this, note that the total power could be determined from the total voltage (12 V) and the total current (0.25 A).

$$P_T = E_T I_T = 12\text{ V} \times 0.25\text{ A} = 3\text{ W}$$

Note that power usage for each of the resistors could also have been determined by applying either of the other formulas:

$$P = \frac{E^2}{R}$$

$$P = I^2R$$

However, with either of these other approaches, it would have been necessary first to determine voltage drops across each individual resistor. Thus, the formula selected is a shortcut and is usually the easiest way to find the power used in a series circuit.

Discussion Questions/Problems

1. How is power usage of a series circuit determined if the resistance values are known for all resistors in the circuit?
2. What formula is used to determine power consumption in a series circuit?
3. What is the power usage of a resistor passing 10 A of current that has a resistance of 2 Ω?
4. What is the power usage of a resistor passing 20 A of current that has a resistance of 100 Ω?

2.6 BATTERIES IN SERIES CIRCUITS

Batteries can be connected into series circuits in either of two ways. First, they can be connected with matched polarity. That is, the batteries can be set up as though they were a group of matching cells connected in series, as illustrated in Figure 2-13. The positive pole of one battery is connected to the negative pole of the other, and current will flow in series through both. In the illustration, two 6 V batteries are connected with aiding polarity. Together, the batteries add up to 12 V and can operate a 12 V lamp.

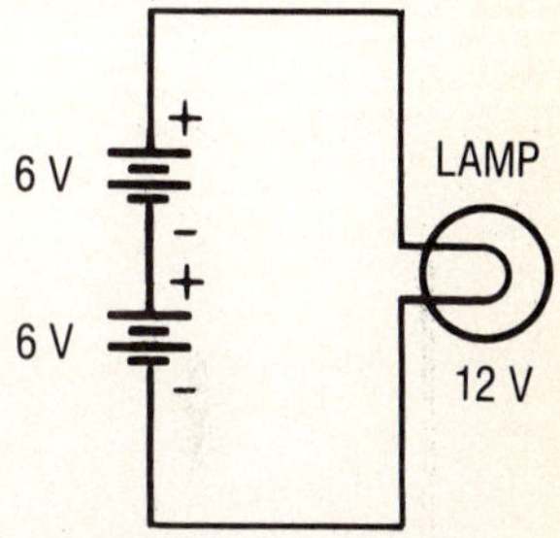

Figure 2-13. Batteries connected with matched polarities in series circuit.

The other method is to connect batteries in opposition to each other. This is illustrated in Figure 2-14. With this type of connection, the voltages of the two batteries oppose each other. The voltage delivered is the difference between the smallest voltage and the largest voltage. With 6 V and 9 V batteries connected in opposition, circuit voltage is 3 V. Batteries are not usually connected in this manner because it is more economical to use a single 3 V battery. Using a 6 V and a 9 V battery (for a total of 15 V) requires five times as much battery material as a single 3 V battery.

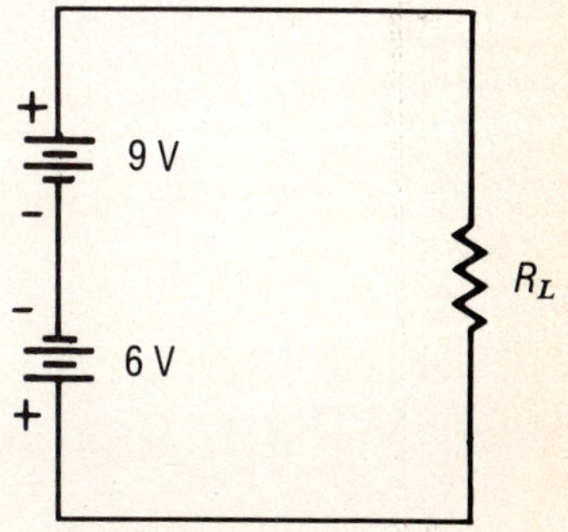

Figure 2-14. Batteries connected with opposite polarities in series circuit.

Note an abbreviation used in Figure 2-14. R_L is resistance for the entire circuit load. The L stands for load.

Discussion Questions/Problems

1. How can two batteries be connected in a series circuit so that their voltages are added to each other?
2. How can two batteries be connected in a series circuit so that the total voltage in the circuit is smaller than the voltage for either battery?
3. What is the rule for determining total circuit voltage when two or more batteries in a series circuit are connected with matched polarity?
4. What is the rule for determining total circuit voltage when two or more batteries in a series circuit are connected, but not all polarities are matched?

2.7 PARALLEL CIRCUITS—PRINCIPLES

A *parallel circuit* is one in which two or more components are connected across one voltage source.

A parallel circuit using three resistors is shown in Figure 2-15. To understand how parallel and series circuits differ, compare this drawing with the one in Figure 2-6, which shows three resistors connected in series.

A major difference between series and parallel circuits lies in the nature of the current path. In a series circuit, electricity must flow through all components in the circuit in order. Thus, if you have a string of decorative light bulbs and one bulb burns out, all of the bulbs go out. If the same bulbs were connected in parallel, only the one that was burned out would go out. The others would remain lit.

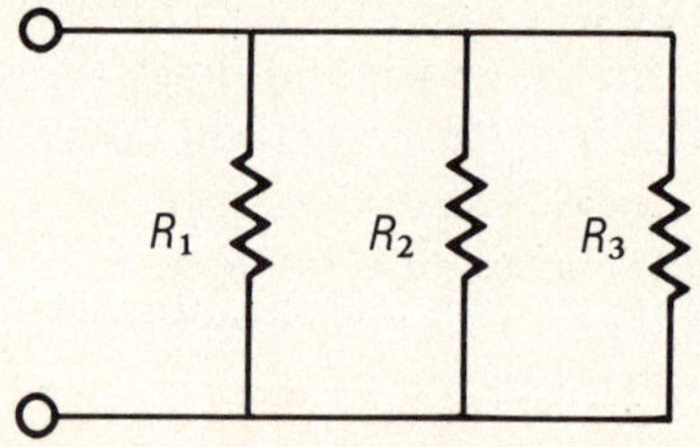

Figure 2-15. Circuit with three resistors connected in parallel.

Discussion Questions/Problems

1. How are components connected in a parallel circuit?
2. What are the differences in current flow between series and parallel circuits?
3. What are the differences in results when a component burns out in a series circuit, compared with a parallel circuit?

2.8 RESISTANCE IN PARALLEL CIRCUITS

In parallel circuits the reciprocal of total resistance is equal to the sum of the reciprocals of individual resistances. A reciprocal is a mathematical term in which a given value is used as a denominator under a numerator of one. As was illustrated in Chapter 1, the reciprocal of resistance is conductance. Conductance is measured in siemens. Thus, the reciprocal of a resistor of 120 Ω would be expressed:

$$\text{Reciprocal} = G = \frac{1}{120\ \Omega}\ \text{siemens}$$

Expressing this as a formula for a number of resistors in parallel:

$$\frac{1}{R_T} = \frac{1}{R_1} + \frac{1}{R_2} + \frac{1}{R_3} + \cdots + \frac{1}{R_N}$$

This can be solved to give the total resistance (rather than the reciprocal of total resistance):

$$R_T = \frac{1}{\dfrac{1}{R_1} + \dfrac{1}{R_2} + \dfrac{1}{R_3} + \cdots + \dfrac{1}{R_N}}$$

Consider the schematic in Figure 2-16. The calculation to determine total resistance for this circuit is as follows:

$$\frac{1}{R_T} = \frac{1}{20} + \frac{1}{50} + \frac{1}{25} + \frac{1}{200}$$

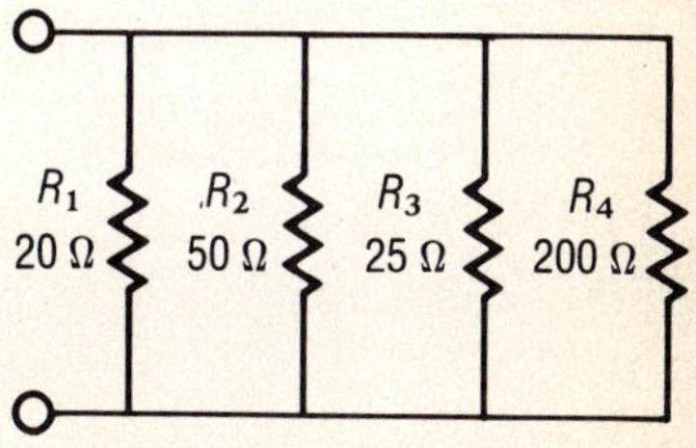

Figure 2-16. Parallel circuit with four resistors.

To solve this equation, you develop the lowest common denominator:

$$\frac{1}{R_T} = \frac{10}{200} + \frac{4}{200} + \frac{8}{200} + \frac{1}{200} = \frac{23}{200}$$

Now, to solve for the reciprocal, you invert the fraction:

$$R_T = \frac{200}{23} = 8.695\ \Omega \approx 8.7\ \Omega$$

Note that the decimal value has been rounded. Resistances rarely come out even. In this book, simple illustrations sometimes have even solutions for the purposes of explanation.

Another approach to solving the same problem is to develop decimal equivalents instead of building fractions with the lowest common denominator. The equation would be solved as follows:

$$\frac{1}{R_T} = 0.05 + 0.02 + 0.04 + 0.005 = 0.115 \text{ siemens}$$

$$R_T = \frac{1}{0.115} = 8.695\ \Omega \approx 8.7\ \Omega$$

Determining Resistance When Resistor Values Are Uniform

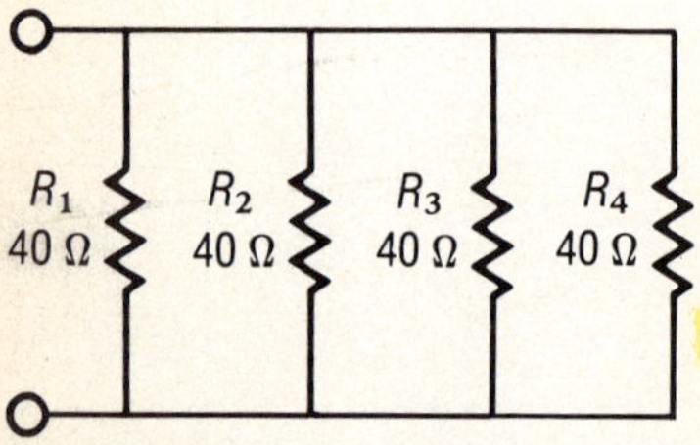

Figure 2-17. Parallel circuit with resistors of equal value.

The resistance formula presented and demonstrated above is universal. You can use this approach to determine resistance in any parallel circuit. However, in special cases, abbreviated, short-cut calculation techniques may be used. One of these special cases occurs when all resistors in a circuit have the same value. A circuit of this type is diagrammed in Figure 2-17. This circuit has four resistors, each with a value of 40 Ω. To find the total resistance of this circuit, you simply divide the common resistance value by the number of resistors:

$$R_T = \frac{40}{4} = 10\ \Omega$$

The total resistance of the circuit in Figure 2-17 is 10 Ω.

Determining Resistance for Two-Resistor Parallel Circuits

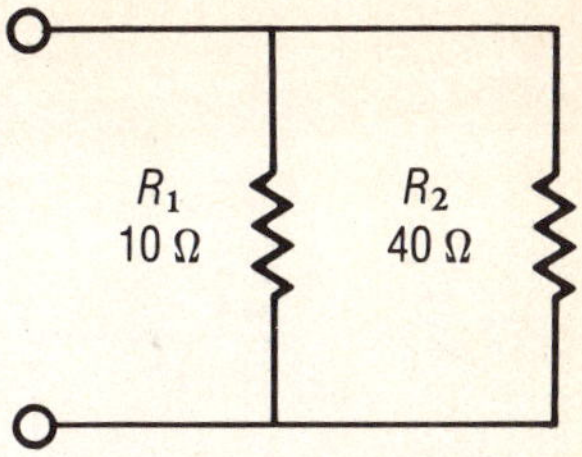

Figure 2-18. Parallel circuit with two resistors.

Another special case for the determining of total resistance in parallel circuits occurs when there are only two resistors in a circuit. A circuit of this type is illustrated in Figure 2-18. This circuit has resistors of 10 Ω and 40 Ω. The following formula can be applied:

$$R_T = \frac{R_1 \times R_2}{R_1 + R_2} = \frac{10 \times 40}{10 + 40} = \frac{400}{50} = 8\ \Omega$$

You can extend the two-resistor formula for application to the total resistance in any parallel circuit. All you do is compute two resistances at a time. After the first set of computations, you use the answers to compute a next set of answers. Then continue until all of the resistors in a given circuit have been included.

To illustrate, the resistance for the circuit in Figure 2-16 could be computed under the short-cut method as follows:

$$R_{T_1} = \frac{20 \times 50}{20 + 50} = \frac{1{,}000}{70} = 14.29\ \Omega$$

Next, the same calculation is performed for the values of R_3 and R_4:

$$R_{T_2} = \frac{25 \times 200}{25 + 200} = \frac{5{,}000}{225} = 22.22\ \Omega$$

Finally, the two answers are used to compute an overall result:

$$R_T = \frac{14.29 \times 22.22}{14.29 + 22.22} = \frac{317.52}{36.51} = 8.695\ \Omega \approx 8.7\ \Omega$$

This is the same answer that resulted from use of the full formula. You can select the technique that will do your job best, depending upon the makeup of individual circuits.

As a means of checking the reasonableness of your answers in determining resistance of parallel circuits, compare your answer with the smallest resistance value in any circuit. R_T should always be smaller in value than the smallest resistor.

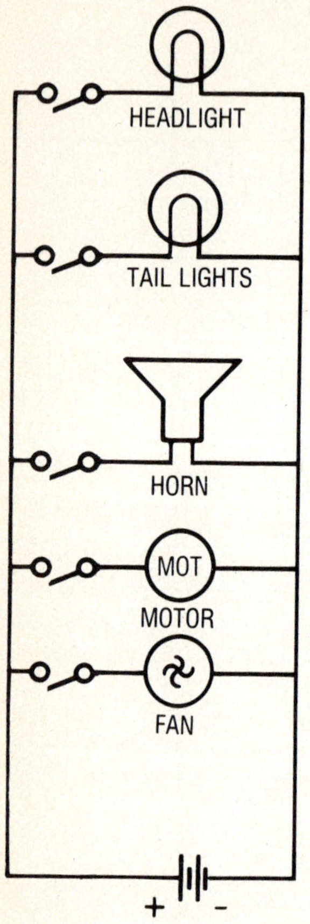

Figure 2-19. Automobile electrical devices connected in parallel.

Discussion Questions/Problems

1. What is the rule for determining total resistance of a parallel circuit?
2. What is the total resistance of a parallel circuit with three resistors of 20 Ω, 30 Ω, and 40 Ω?
3. What is the formula for determinig total resistance of a parallel circuit that has a group of resistors that are uniform in value?
4. What is total resistance of a circuit that has five resistors, all of 50 Ω?
5. What is the formula that can be used to determine total resistance of a parallel circuit that has only two resistors?
6. What is the total resistance of a circuit that has two resistors of 10 Ω and 15 Ω? Work this problem in your head.

2.9 VOLTAGE IN PARALLEL CIRCUITS

In parallel circuits, applied voltage is the same for all components. The rule for voltage in parallel circuits is the same as the rule for current in series circuits.

Figure 2-19 illustrates this principle. This diagram shows a number of electrical devices connected in parallel on an automobile. The electrical source is a 12 V battery. The same 12 V is applied across the headlights, the starter motor, the heater fan, horn, tail lights, and any other electrical devices on the car.

Discussion Questions/Problems

1. What is the rule for determining voltage in parallel circuits?
2. If a parallel circuit has a 12 V source and resistors of 10 Ω and 50 Ω, what is the voltage drop across each resistor?
3. What rule covering series circuits is similar to the rule for voltage in parallel circuits?

2.10 CURRENT IN PARALLEL CIRCUITS

Total current in a parallel circuit is equal to the sum of the current in all branch circuits. To find total current, add the current flowing through each component in a circuit. Current in a parallel circuit is determined by the same type of formula as is used for voltage in a series circuit. The formula is:

$$I_T = \frac{E_1}{R_1} + \frac{E_2}{R_2} + \frac{E_3}{R_3} + \cdots + \frac{E_N}{R_N}$$

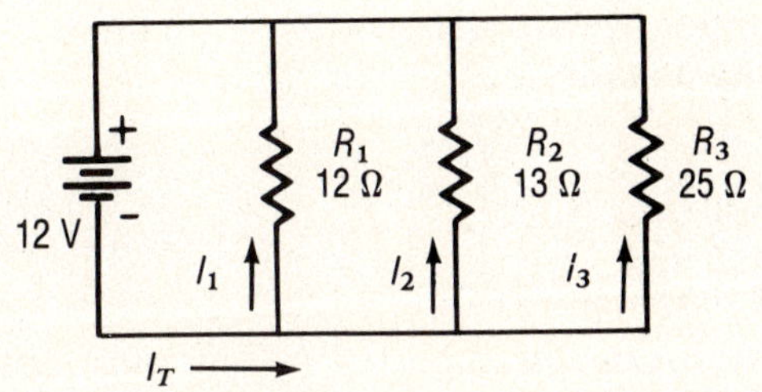

Figure 2-20. Parallel circuit with three resistors.

In practice, it is usually best to compute the current for individual components, then add the total. As an example, consider the schematic in Figure 2-20. This is a parallel circuit with resistors of 12 Ω, 13 Ω, and 25 Ω. Current for these components is calculated as follows:

$$I_1 = \frac{E_1}{R_1} = \frac{12}{12} = 1\ \text{A}$$

$$I_2 = \frac{E_2}{R_2} = \frac{12}{13} = 0.92\ \text{A}$$

$$I_3 = \frac{E_3}{R_3} = \frac{12}{25} = 0.48\ \text{A}$$

To determine total current for the circuit, add these values:

$$I_T = 1 + 0.92 + 0.48 = 2.4\ \text{A}$$

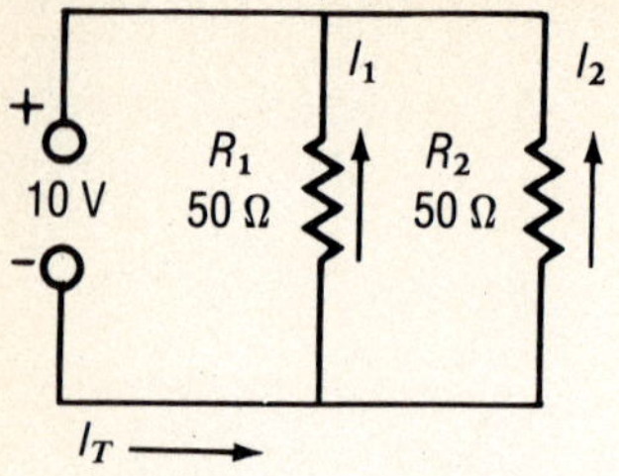

Figure 2-21. Parallel circuit with two equal resistors.

Circuits with Equal Resistances in Parallel

When resistances in a parallel circuit are all equal, current for each branch circuit will also be equal. If you determine the current through one resistor, you know the current through all resistors. For an illustration, see Figure 2-21. This shows a circuit with a 10 V source and two resistors of 50 Ω each.

To determine circuit resistance, you use the special formula presented earlier:

$$R_T = \frac{50}{2} = 25\ \Omega$$

Total resistance for the circuit is 25 Ω. To determine current, you use Ohm's law:

$$I = \frac{E}{R} = \frac{10}{25} = 0.4\ \text{A}$$

Total current for the circuit is 0.4 A. Since you know that the current in each of the branch circuits is equal, you simply divide 0.4 A by 2:

$$I = \frac{0.4\ \text{A}}{2} = 0.2\ \text{A}$$

The current through each resistor is 0.2 A.

Discussion Questions/Problems

1. What is the rule for determining current in parallel circuits?
2. How is the current through each resistor in a parallel circuit determined?
3. In a parallel circuit with a 12 V source, how much current passes through a resistor of 2 Ω?
4. If a parallel circuit has a 12 V source and resistors of 2 Ω, 6 Ω, and 12 Ω, what is the total current flowing through the circuit?
5. What is the rule for determining current in a parallel circuit in which all resistors are equal?

2.11 POWER IN PARALLEL CIRCUITS

Total power consumed in a parallel circuit is equal to the sum of the power usage of the individual components. The formula for this relationship is:

$$P_T = P_1 + P_2 + P_3 + \ldots + P_N$$

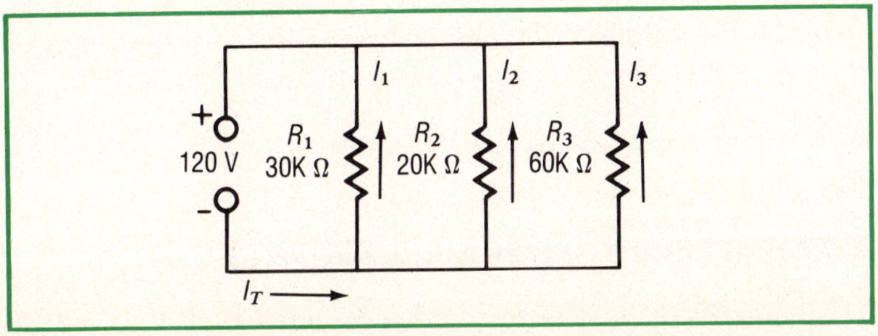

Figure 2-22. Parallel circuit with three resistors and a 120 V source.

For an illustration of this principle, see Figure 2-22. This drawing shows a parallel circuit with a 120 V source and three resistors. The resistors are 30K Ω, 20K Ω, and 60K Ω. To determine power usage, you can follow any of three methods:

$$P_T = \frac{E^2}{R_T} \qquad P_T = I^2 R_T \qquad P_T = EI_T$$

Note that, to use two of these three equations, you have to compute the value of I first, since this is not given. Similarly, for two of these equations you have to compute the value of R_T first. The fastest method is to use a combination of the first equation and an adaptation of the second.

$$P_T = P_1 + P_2 + P_3$$

$$P_T = \frac{120^2}{30K} + \frac{120^2}{20K} + \frac{120^2}{60K}$$

$$P_T = 0.48 + 0.72 + 0.24 = 1.44 \text{ W}$$

Total power usage for the circuit in Figure 2-22 is 1.44 W. The decimal values for watts in the equation above can be stated in milliwatts.

$$P_T = 480 \text{ mW} + 720 \text{ mW} + 240 \text{ mW} = 1.44 \text{ W}$$

Discussion Questions/Problems

1. What is the rule for determining power in parallel circuits?
2. If you know voltage and resistance for a resistor in a parallel circuit, how do you determine power usage?
3. If a parallel circuit has a source of 10 V and three resistors of 10 Ω each, what is the power usage?

2.12 BATTERIES IN PARALLEL CIRCUITS

When batteries with the same voltage rating are connected in parallel, the entire circuit has the same voltage as each individual battery. Figure 2-23 shows a circuit in which two 12 V batteries are connected in parallel. The entire circuit has 12 V.

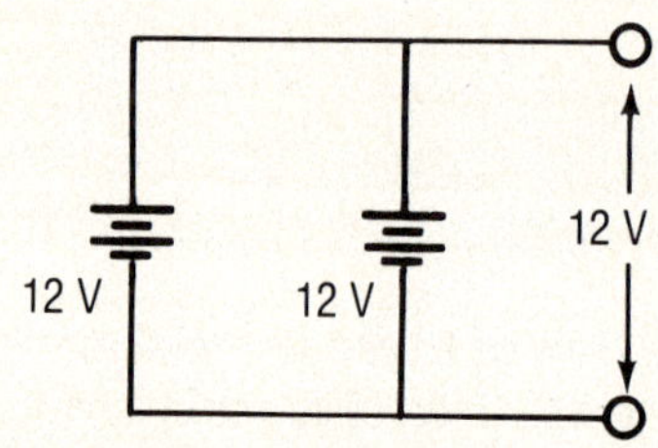

Figure 2-23. Circuit with two batteries of equal voltage connected in parallel.

When batteries are connected in parallel, some precautions must be followed:

- All of the batteries must have the same voltage. If they do not, the battery with the lowest voltage will drain current from the others.
- All batteries must be in good condition. If a single battery has one weak cell, it will drain all of the batteries in the circuit.

The value of linking batteries in parallel is that the batteries can supply more current. The amount of current that can be delivered is equal to the sum of the current ratings for all of the individual batteries.

Discussion Questions/Problems

1. What is the rule for determining the total voltage when two or more batteries are connected in a parallel circuit?
2. Why is it not recommended that batteries of unequal voltage be connected together in parallel?
3. Why is it necessary that all batteries in a parallel circuit with two or more batteries be in good condition?
4. What happens if a battery with a weak cell is connected into a parallel circuit that has two or more batteries?
5. What is the main value that results from connecting two or more batteries in parallel?
6. What is the rule for determining the total current in a parallel circuit with two or more batteries connected in parallel?

2.13 COMPARISON OF SERIES AND PARALLEL CIRCUITS

Series and parallel circuits have different electrical properties. These properties of circuits can be valuable when it comes to using electricity. It is important to understand the basic nature and properties of these two types of circuits.

Current

In *series* circuits, current is the same throughout.

In *parallel* circuits, current can be different in each branch, depending upon resistance.

Voltage

In *series* circuits, the voltage drop is different across each resistor.

In *parallel* circuits, the voltage drop is the same across all branches.

Resistance

In *series* circuits, total resistance is the sum of all individual resistors.

In *parallel* circuits, total resistance is found using reciprocals in this manner:

$$\frac{1}{R_T} = \frac{1}{R_1} + \frac{1}{R_2} + \frac{1}{R_3} + \cdots + \frac{1}{R_N}$$

Current flow through resistors. In *series* circuits, current flows in sequence through all resistors. If one resistor fails, the circuit is broken.

In *parallel* circuits, resistors are connected independently. If a resistor fails, only that branch of the circuit stops functioning.

Relationship of total resistance to individual resistances. In *series* circuits, total resistance must be greater than the largest individual resistance.

In *parallel* circuits, total resistance must be smaller than the smallest individual resistance.

Power

In a *series* circuit, total power use, or consumption, is equal to the total power consumed in all of the devices. The rule is the same for *parallel* circuits. In both types of circuits, the power used is the sum of the individual power consumptions.

Discussion Questions/Problems

1. How is current determined in series and in parallel circuits?
2. How is voltage determined in series and in parallel circuits?
3. How is resistance determined in series and in parallel circuits?
4. What are the differences in current flow patterns through resistors in series and in parallel circuits?
5. What are the differences in relationships between the largest or smallest resistors in series and in parallel circuits to total resistance?
6. How is power consumption determined in series and in parallel circuits?

2.14 SERIES-PARALLEL CIRCUITS

One reason why it is important to understand the electrical properties of series and parallel circuits is that these types of circuits are often combined in electrical applications. To diagnose and design complex circuits, it is often necessary to work with both series and parallel branches in the same circuit.

To illustrate, examine the circuit in Figure 2-24. This circuit has two parallel branches. Within each of the branches, two resistors are connected in series. Thus, within the individual branch, resistances are cumulative. For the circuit as a whole, the resistances of the two branches are combined as reciprocals, following the rules for parallel circuits.

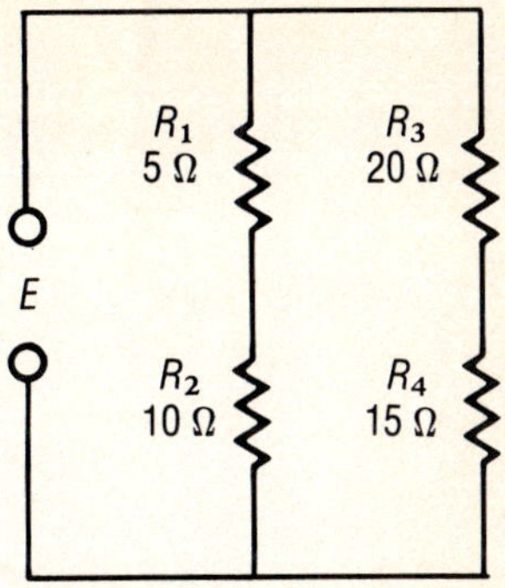

Figure 2-24. Circuit with two parallel branches, each with two resistors connected in series.

The resistance for this circuit is determined as follows:

For the left branch:

$$R_1 + R_2 = 5\ \Omega + 10\ \Omega = 15\ \Omega$$

For the right branch:

$$R_3 + R_4 = 20\ \Omega + 15\ \Omega = 35\ \Omega$$

For the entire circuit:

$$R_T = \frac{15 \times 35}{15 + 35} = 10.5\ \Omega$$

Discussion Questions/Problems

1. How are electrical properties of components determined within series branches of series-parallel circuits?
2. How are electrical properties of components determined within parallel branches of series-parallel circuits?
3. When a circuit has resistors connected in series within branches of a parallel circuit, how is total resistance determined?

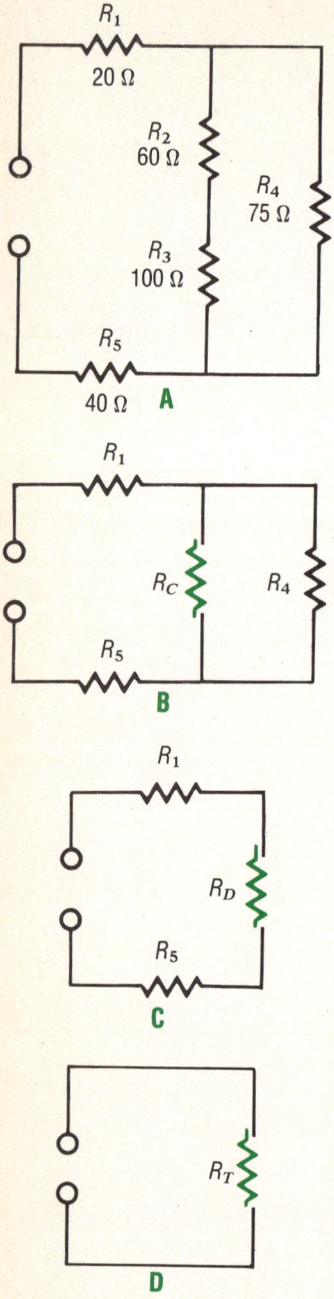

Figure 2-25. Diagrams used to determine resistance for a series-parallel circuit.

2.15 DETERMINING RESISTANCE IN SERIES-PARALLEL CIRCUITS

The technique used in determining resistance of the circuit in Figure 2-24 can be extended to determine the resistance of any series-parallel circuit. The principle is simple: *Simplify the circuit as much as possible, determining combined resistances as you are solving the equation.*

For a more extensive example of this technique, look at the drawings in Figure 2-25. Figure 2-25A diagrams a series-parallel circuit that includes five resistors.

In Figure 2-25B, the values of resistors R_2 and R_3 have been combined. These resistors were connected in series. So, their resistances are added (60 Ω + 100 Ω = 160 Ω). This combined value is shown as a newly drawn resistor, R_C.

In Figure 2-25C, the value of R_C has been combined with R_4. This computation is:

$$R_D = \frac{R_C \times R_4}{R_C + R_4} = \frac{160 \times 75}{160 + 75} = 51.06\ \Omega$$

With the determination that R_D has a resistance of 51.06 Ω, the circuit has been reduced to a point where there are, effectively, three resistors left. These are connected in series, so their values can be added to determine total resistance for the circuit. Figure 2-25D shows a simplified circuit with a single resistor, R_T. The value of R_T is determined by adding the three resistances from the circuit in Figure 2-25C:

$$R_T = R_1 + R_D + R_5 = 20 + 51.06 + 90 = 161.06\ \Omega$$

Total resistance for the circuit in Figure 2-25 is 161.06 Ω.

Discussion Questions/Problems

1. What is the best approach to follow in determining resistance in a series-parallel circuit?
2. When a series-parallel circuit has been simplified to represent resistors in series, how is total resistance determined?
3. What is the formula for determining resistance in series circuits?
4. What is the formula for determining resistance in parallel circuits?

2.16 DETERMINING CURRENT AND VOLTAGE IN SERIES-PARALLEL CIRCUITS

The same approach—simplifying circuits by combining branches—is used to determine current and voltage for series-parallel circuits. In many cases, it will be necessary to determine resistance before the values of current or voltage can be calculated. This is the case in the circuit in Figure 2-26.

The complete circuit is shown in Figure 2-26A. There are six resistors in three branches of a parallel circuit. Values of all the resistors are shown. However, there is no figure for

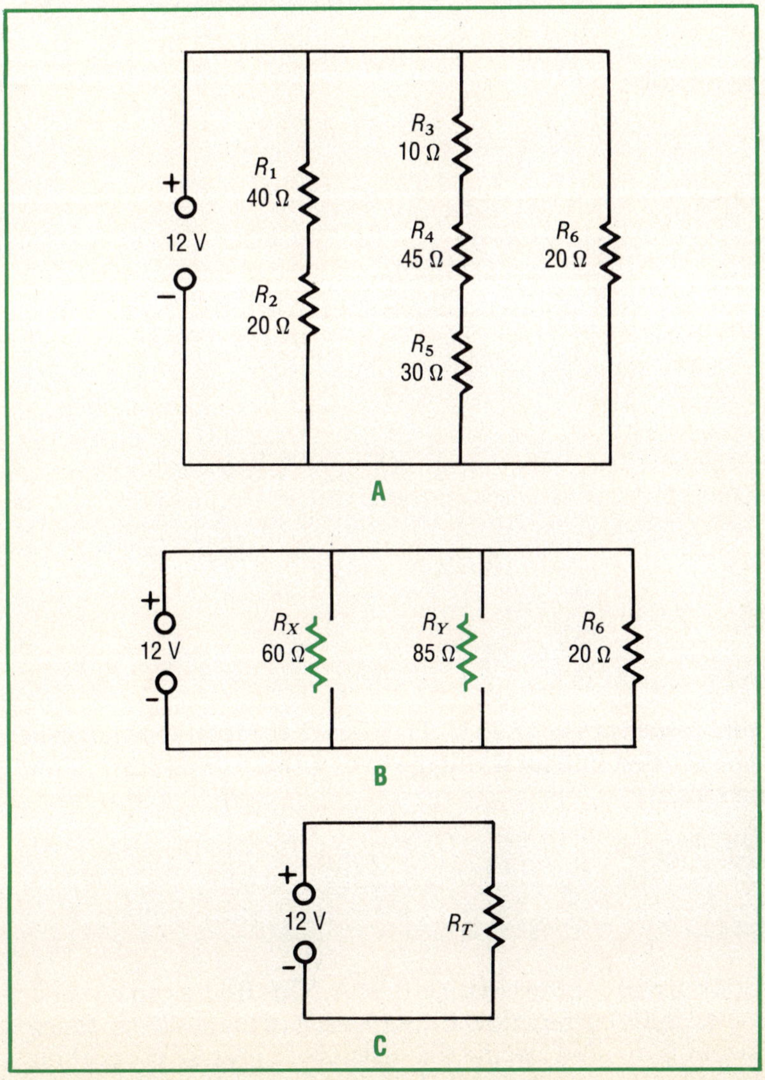

Figure 2-26. Diagrams used to determine current and voltage for a series-parallel circuit.

total resistance. This figure is needed for use, along with the given value of voltage, to determine total current.

Figure 2-26B shows a simplified version of the same circuit. The resistances have simply been added for each branch. They show that the three branches have resistances of 60 Ω, 85 Ω, and 20 Ω. The arithmetic is relatively simple:

$$R_X = R_1 + R_2 = 40 + 20 = 60\ \Omega$$

$$R_Y = R_3 + R_4 + R_5 = 10 + 45 + 30 = 85\ \Omega$$

$$R_6 = 20\ \Omega$$

To determine total resistance for the circuit, you add reciprocals:

$$R_T = \frac{1}{\frac{1}{R_X} + \frac{1}{R_Y} + \frac{1}{R_6}}$$

$$= \frac{1}{\frac{1}{60} + \frac{1}{85} + \frac{1}{20}}$$

$$= \frac{1}{0.016 + 0.0118 + 0.050}$$

$$= \frac{1}{0.0784}$$

$$R_T = 12.75\ \Omega$$

Total resistance is 12.75 Ω. This figure for total resistance is then divided into total voltage to determine total current for the circuit.

$$I_T = \frac{E_T}{12.75} = \frac{12}{12.75} = 0.94\ \text{A}$$

Total current for the circuit is 0.94 A. This figure can be used to determine current flow through each of the resistors. Since current is divided among the resistors in inverse proportion to their resistances, the simplest way to find the current through each resistor is to use the equation:

$$I = \frac{E}{R}$$

Figure 2-26C shows that E is the same 12 V for each resistor. Thus:

$$I_1 = \frac{E}{R_X} = \frac{12}{60} = 0.2 \text{ A}$$

$$I_2 = \frac{E}{R_Y} = \frac{12}{85} = 0.14 \text{ A}$$

$$I_3 = \frac{E}{R_6} = \frac{12}{20} = 0.6 \text{ A}$$

Note that the sum of these is 0.94 A, as determined above for the total current.

Discussion Questions/Problems

1. How is current determined in a series-parallel circuit?
2. How is voltage determined in a series-parallel circuit?
3. How is total resistance determined in a series-parallel circuit?
4. How is current through individual resistors determined in a series-parallel circuit?

2.17 VOLTAGE DIVIDERS

The use of a series circuit as a voltage divider is introduced earlier in this chapter. The principle is simple: *In a series circuit, voltage across resistors divides in proportion to the resistances of the resistors.*

In addition: *When a load is added, the resistance of the load must be combined with that of the resistor it parallels.*

These characteristics of voltage divider circuits are illustrated in Figure 2-27. In Figure 2-27A, two equal resistors are

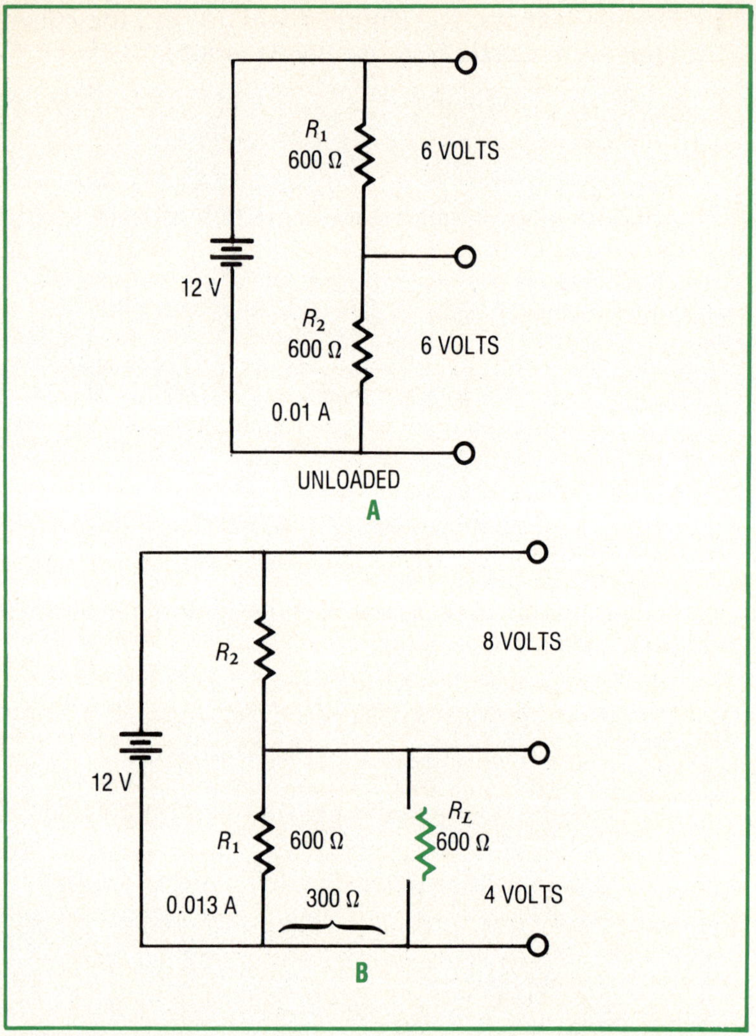

Figure 2-27. A load added to a circuit with equal resistors connected in series creates two parallel resistors in a branch circuit.

connected in series in a 12 V circuit. Each of the resistors has a voltage drop of half the total. Two 6 V circuits are created.

However, Figure 2-27B shows what happens when a load is introduced. A 600 Ω load is connected in parallel with the 600 Ω divider resistor, R_1. As a result, there are two parallel resistances in this branch circuit. Following the rule for parallel resistors, the resistance of the lower portion of the circuit is as follows:

$$R_1 = \frac{600\ \Omega \times 600\ \Omega}{600\ \Omega + 600\ \Omega} = 300\ \Omega$$

The addition of the load in the branch then changes the voltage drop for the basic series circuit. With the resistance across the parallel circuit decreased to 300 Ω, the voltage drop is also decreased to 4 V. At the same time, the voltage drop across the other circuit is increased to 8 V.

Remember: In designing voltage divider circuits, the effects of all loads must be considered.

Designing a Voltage Divider

A hypothetical example demonstrating how to design a voltage divider circuit is shown in Figure 2-28. This circuit has three loads with known requirements, as follows:

- Load 1 needs 24 V and 1 A.
- Load 2 needs 12 V and 0.4 A.
- Load 3 needs 6 V and 0.4 A.
- Total current requirement for the loads is 1.8 A.

As part of the design, it is necessary to determine the resistance and power requirements of each of the resistors in the circuit.

Figure 2-28. Voltage divider circuit with three loads and three resistors.

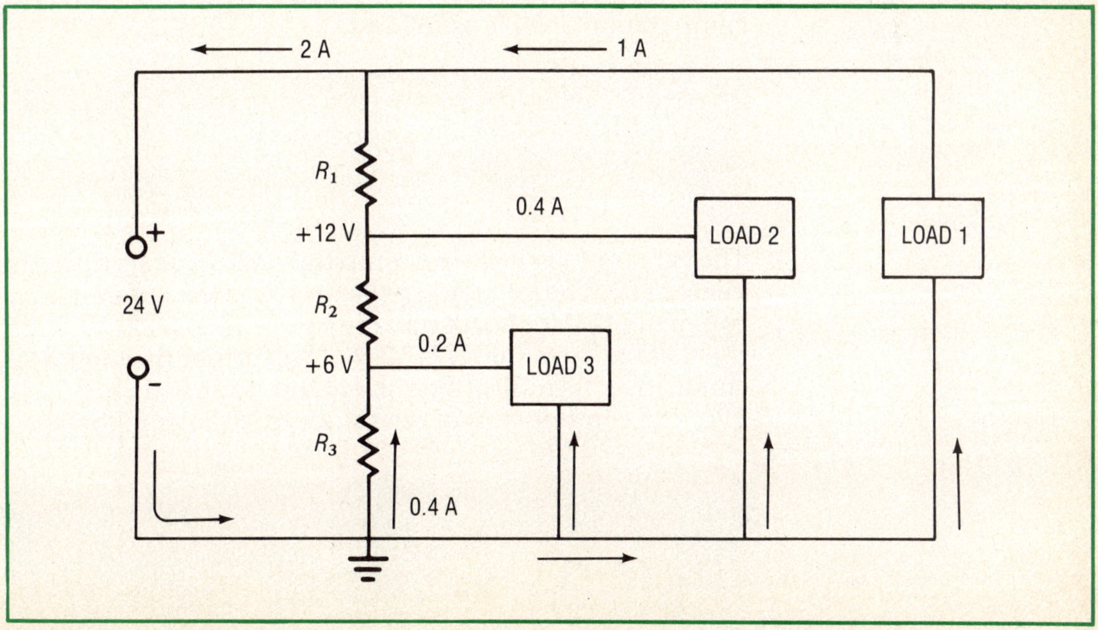

Resistance required by R_3 is determined by choosing a current to flow through it. You can choose almost any value you wish, within the ability of the power source to supply current. If $Load_3$ varies in resistance by perhaps 10 percent, the current through $Load_3$ will also vary by 10 percent. To minimize the effect of such a variation in $Load_3$, you could choose to pass a current through R_3 which is significantly different compared with the current through $Load_3$. Thus, a change in the resistance of $Load_3$ will have less effect on the overall circuit. If you choose to pass 0.4 A through R_3, which is more than the current through $Load_3$, then:

$$R_3 = \frac{6\ \text{V}}{0.4\ \text{A}} = 15\ \Omega$$

The 6 V figure is the voltage drop across resistor R_3. This is divided by the value of the current selected, 0.4 A. Note that this computation uses the figure for the voltage drop across the resistor, which is the same as the voltage across $Load_3$.

For resistor R_2, the voltage drop is also 6 V. This represents the difference between the 12 V across $Load_2$ and the 6 V voltage drop of R_3 and $Load_3$, in combination. To compute the value of R_2, the 6 V voltage drop is divided by the cumulative current flow through $Load_3$ and R_3. Current for $Load_3$ is 0.2 A, for R_3 0.4 A, producing a total of 0.6 A. The computation of R_2 is as follows:

$$R_2 = \frac{6\ \text{V}}{0.6\ \text{A}} = 10\ \Omega$$

The value of R_2 is 10 Ω. The computation of R_2 bears upon the value of R_1. The voltage drop across R_1 is the difference between the 12 V for $Load_2$ and the source voltage, 24 V. Thus, the voltage drop for R_1 is 12 V. The current through R_1 is cumulative, including the current through $Load_2$ and R_2, or 1 A. The computation of resistance for R_1 is as follows:

$$R_1 = \frac{12\ \text{V}}{1\ \text{A}} = 12\ \Omega$$

The power requirement for each of the resistors is computed with the formula:

$$P = \frac{E^2}{R}$$

$$P_3 = \frac{6^2}{15} = 2.4\text{ W}$$

$$P_2 = \frac{6^2}{10} = 3.6\text{ W}$$

$$P_1 = \frac{12^2}{12} = 12\text{ W}$$

The total power dissipated by all three resistors is:

$$P_T = 2.4 + 3.6 + 12 = 18\text{ W}$$

In practice, these figures would be used as a guide only. The circuit would be built with resistors that had the closest available power rating. P_3 would probably use a 3 W resistor. P_2 would use a 4 W resistor. Information on resistors and their ratings is presented in Chapter 4.

The three loads of this example could also be supplied by another circuit, as shown in Figure 2-29. Since Load$_2$ and Load$_3$ draw the same current, they can be connected in series. Since together the voltage drops represent only 18 V, the other 6 V (24 − 18 = 6) must be represented by resistance R_1. Since R_1 has 6 V across it and 0.4 A current through it:

$$R_1 = \frac{E}{I} = \frac{6}{0.4} = 15\ \Omega$$

In this circuit, the total current drawn from the power supply is only 1.4 A rather than the 2.0 A of the circuit in Figure 2-28. This circuit requires only one resistor. The power dissipated in R_1 is:

$$P = \frac{E^2}{R} = \frac{36}{15} = 2.4\text{ W}$$

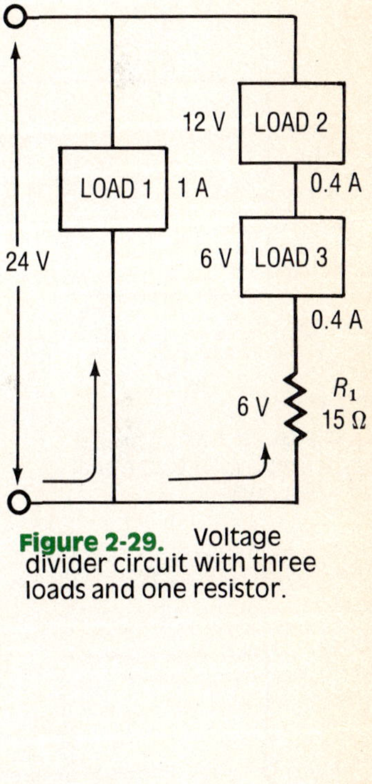

Figure 2-29. Voltage divider circuit with three loads and one resistor.

This circuit then dissipates (uses) 2.4 W rather than the 18 W of the circuit in Figure 2-28. The circuit in Figure 2-28 wastes 5.6 watts more than the circuit in Figure 2-29.

Discussion Questions/Problems

1. What is the relationship between resistors and voltage in series circuits?
2. What is the role of the load in a voltage divider circuit?
3. How are resistances computed in voltage divider circuits?
4. How are power requirements computed for resistors in voltage divider circuits?
5. How can loads be combined in voltage divider circuits so as to reduce the number of resistors and the power wasted in resistors?

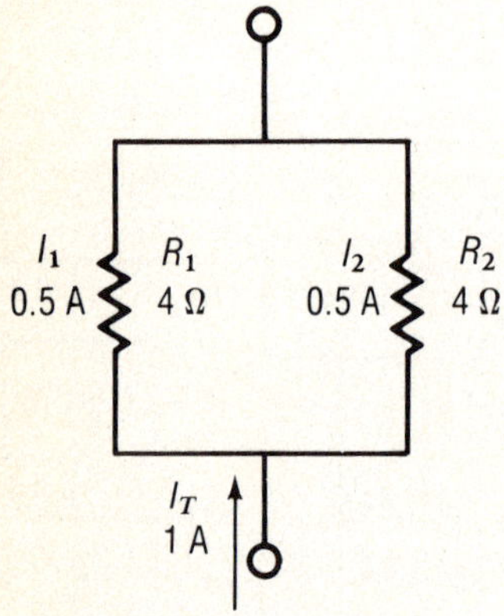

Figure 2-30. Parallel circuit with two equal resistors.

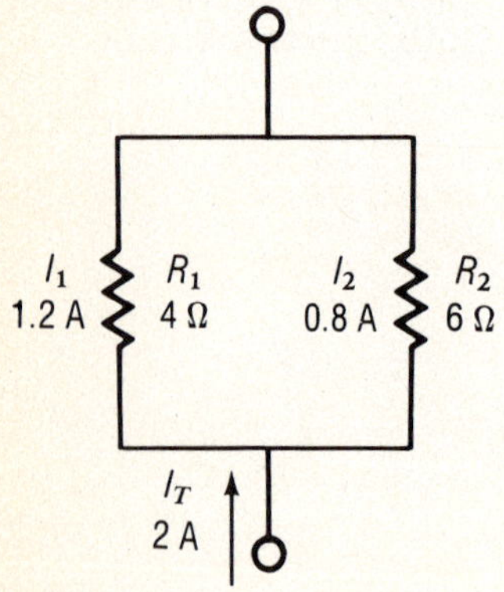

Figure 2-31. Parallel circuit with two unequal resistors.

2.18 CURRENT DIVIDERS

Current does not divide in a series circuit. Remember that current is the same throughout a series circuit.

However, current does divide in a parallel circuit. This division is in inverse proportion to the resistances in circuit branches. This means the higher the resistance, the lower the current flow. The reverse also holds true. This behavior of current in parallel circuits can be valuable in the design of electrical and electronic circuits.

To illustrate the principle of current division, see Figure 2-30. In this circuit, there are two equal resistors. So, the current divides evenly. Source current is 1 A. Current through each of the resistors is 0.5 A.

Remember, current divides inversely in proportion to the resistance. For an example of current division when resistors are unequal, see Figure 2-31. A parallel circuit with a current of 2 A has resistors of 4 Ω and 6 Ω.

The formula used to find the value of current through each of the resistors is derived from Ohm's law. The basic formula is:

$$I = \frac{E}{R}$$

Since the voltage is constant across a parallel circuit, E is the total voltage, E_T. It is possible to substitute the values of current and resistance which correspond to this. Therefore, since: $E_T = I_T R_T$, the formula can be restated as follows:

$$I_1 = \frac{I_T R_T}{R_1}$$

In this formula, I_1 is the value for R_1.

The circuit can then be solved as follows:

$$R_T = \frac{4 \times 6}{4 + 6} = 2.4\ \Omega$$

$$I_1 = \frac{I_T R_T}{R_1}$$

$$I_1 = \frac{2 \times 2.4}{4} = 1.2\ \text{A}$$

$$I_2 = \frac{2 \times 2.4}{6} = 0.8\ \text{A}$$

The current through R_1 is 1.2 A, and the current through R_2 is 0.8 A.

To determine current in parallel circuits with more than two resistances, use the reciprocal approach. Consider the circuit in Figure 2-32.

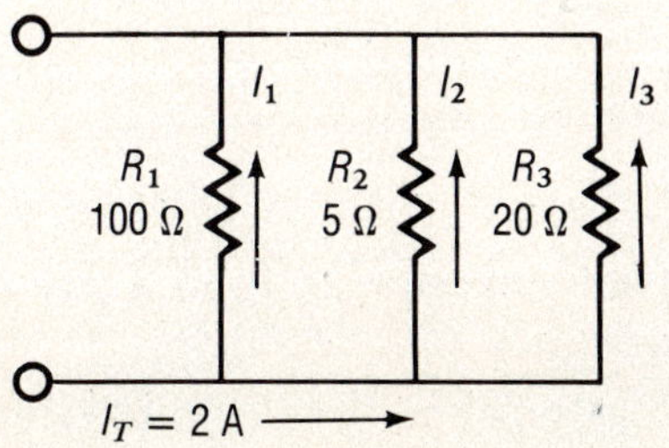

Figure 2-32. Parallel circuit with three resistors.

This has three parallel resistors, of 100 Ω, 5 Ω, and 20 Ω. With the reciprocal method, the current values for the circuit are determined as follows:

$$R_T = \frac{1}{\frac{1}{100} + \frac{1}{5} + \frac{1}{20}} = 3.85\ \Omega$$

Total resistance for the circuit is 3.85 Ω. This figure can then be used as a factor to determine current through each resistor.

$$I_1 = \frac{R_T}{R_1} \times I_T$$

Note that I_T is given in Figure 2-32 as 2 A. All of the needed data are available to compute current values for the three resistors:

$$I_1 = \frac{3.85}{100} \times 2\text{ A} = 0.077\text{ A}$$

$$I_2 = \frac{3.85}{5} \times 2\text{ A} = 1.54\text{ A}$$

$$I_3 = \frac{3.85}{20} \times 2\text{ A} = 0.38\text{ A}$$

As a check for accuracy, add the current values for the three resistors. The sum should equal the total current for the circuit (0.077 A + 1.54 A + 0.38 A = 2 A).

Discussion Questions/Problems

1. What is the rule for current distribution in series circuits?
2. How does current divide in parallel circuits?
3. How does current divide in a parallel circuit with two equal resistors?
4. How does current divide in a parallel circuit with unequal resistors?

2.19 COMMON

A complex electrical circuit includes a number of voltages, currents, and loads. The relationships among these voltages, currents, and loads are important to the way the circuit operates.

Figure 2-33 shows two batteries, one of 9 V, the other of 6 V. In this configuration, it is possible to measure the voltages across terminals A–B and C–D. However, voltages across A–C and B–D cannot be measured. Neither can the voltages across A–D or B–C be measured. Measurements cannot be made because there is no connection between Batteries 1 and 2. Therefore, there are no circuits.

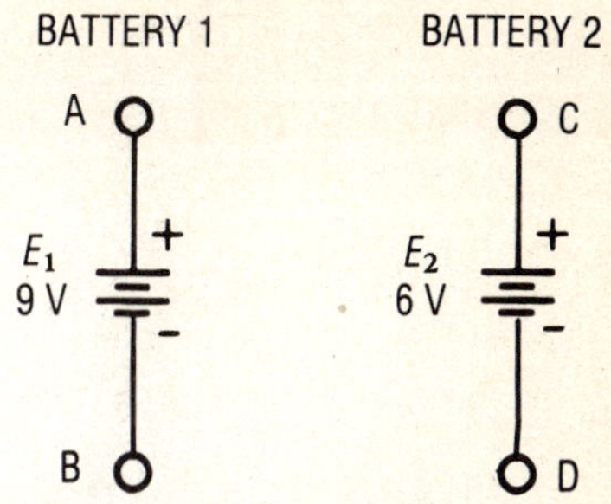

Figure 2-33. Two unconnected batteries.

Figure 2-34 shows a reconfiguration in which points B and D have been connected. In this new circuit, it is possible to measure voltages A–C (3 V) or B–D (0 V). These measurements are possible because the circuits have been connected. They have a common set of terminals, B–D, to which voltages can be referenced. Complex circuits always have a common reference point for various voltages. This is usually referred to simply as the *common*.

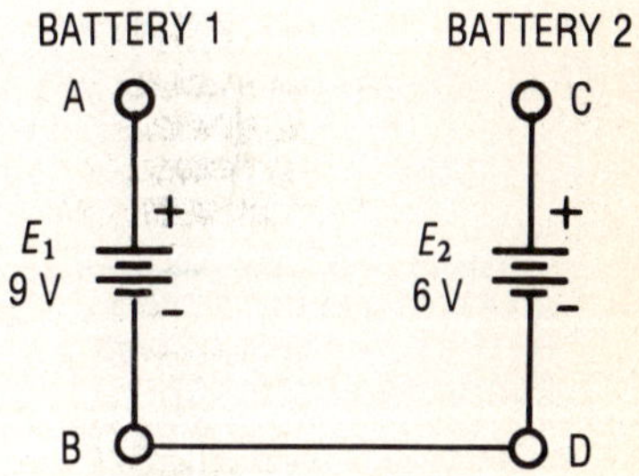

Figure 2-34. Two batteries connected across terminals B and D.

Now, suppose connection B–D was with a small resistance, R_c, as shown in Figure 2-35. In this case, a load across C–B would cause a current to flow through Battery 2 and R_c. The voltage across the load would be less than E_2 by the amount of the *IR* drop across R_c. Even though a voltage of 6 V might be expected across the load, it would actually be less than 6 V. In most circuits, it is important to make this voltage difference as small as possible—and to assure that the reference is the same for all points of the circuit. For this reason, the common is made of a very good conductor, such as a heavy copper wire or bar, or a metal chassis, so that R_c will be very small.

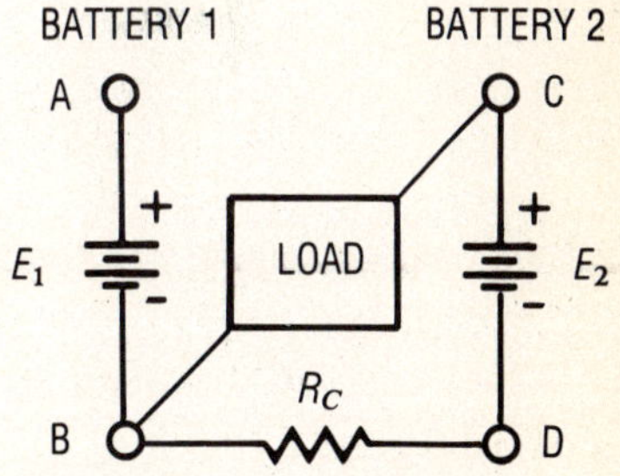

Figure 2-35. Load connected across terminals B and C of two batteries.

The symbol for a common chassis connection is shown in Figure 2-36.

Figure 2-36. Symbol for a common chassis connection.

Discussion Questions/Problems

1. What would happen if a connection were made between A and C in Figure 2-35?
2. Look at Figure 2-35. Can the load be placed across terminals A and D instead of being connected as shown?

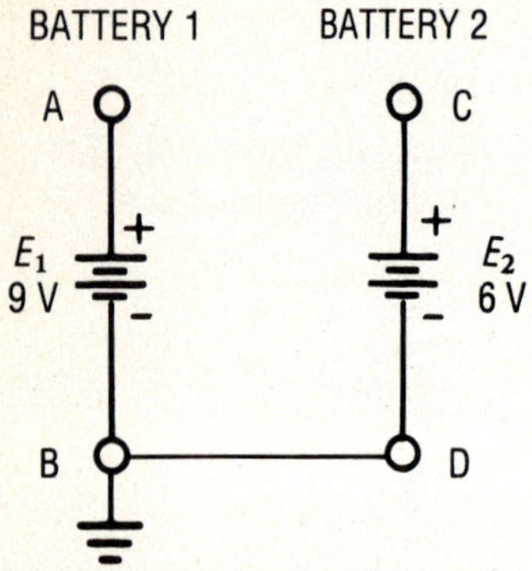

Figure 2-37. Symbol for an earth ground added to the schematic of two connected batteries.

2.20 GROUND

Suppose a large circuit is spread out over a large area. For a common, it would be necessary to have a long copper bar. Rather than use such an expensive method, engineers have learned that the earth can be a good common point of reference. Many large circuits, such as electrical power systems, create a common reference by connecting certain points of the system to a metal rod placed in damp soil fairly deep in the ground.

The ground is not expected to be a conductor for the circuit's currents. That is, the ground is not used instead of connection B–D in Figure 2-37. Rather, the ground is a point of reference. In effect, Figure 2-37 is the same as Figure 2-34, with a ground added. Notice the symbol for an earth ground.

An earth ground is important for certain types of radio receiving and transmitting antennas, particularly those which operate at lower frequencies. Standard AM radio stations usually have an elaborate earth ground. In situations where the conductivity of the earth is not adequate, the ground includes a large number of buried cables.

Discussion Questions/Problems

1. In Figure 2-37, could the ground be connected to point D?
2. In Figure 2-37, would it be possible to connect A–C and install the ground at point A?

3

ANALYZING CIRCUIT NETWORKS

3.1 NETWORK THEOREMS— TOOLS FOR CIRCUIT ANALYSIS

Series-parallel circuits can become highly complex. A circuit that includes multiple electrical or electronic functions is known as an electrical or electronic *network*.

The formulas and equations with which you worked in Chapter 2 are adequate to analyze and design relatively simple series or parallel circuits. However, as you deal with networks, you need more sophisticated methods. The most common techniques to analyze electrical and electronic networks use *network theorems*. A theorem is a rule or set of rules that can be proved mathematically. In the course of mathematical proof, it is necessary to work through a series of related, sequential equations.

There are literally scores of theorems that can be applied to network analysis. In practice, however, you will probably rely on one or two theorems as your major network analysis methods. But you should also be familiar with several different theorems and their associated equation sets. It would be impossible and impractical to try to cover all available network theorems in this book. Instead, the most commonly used theorems have been selected. These are:

- Kirchhoff's law for current
- Kirchhoff's law for voltage
- Superposition theorem
- Thevenin's theorem
- Norton's theorem
- Thevenin–Norton conversions
- Millman's theorem.

Discussion Questions/Problems

1. What is an electrical network?
2. What is a theorem?
3. What is a network theorem?
4. What steps are followed in proving a theorem?

3.2 KIRCHHOFF'S LAW FOR CURRENT

There are two sets of theorems known as Kirchhoff's laws. One set deals with current, the other with voltage. These laws were formulated by a German physicist, Gustav R. Kirchhoff (1824–1877).

Kirchhoff's law for current states that *the amount of current delivered to any point (or node) in a circuit will be equal to the current leaving that point (or node)*. This law can also be explained algebraically: *The algebraic sum of currents entering and leaving a point in a circuit must equal zero.*

A simple illustration of the application of Kirchhoff's law for current is seen in Figure 3-1. In this illustration, 2 A of current is delivered to point A. At this point, the current divides.

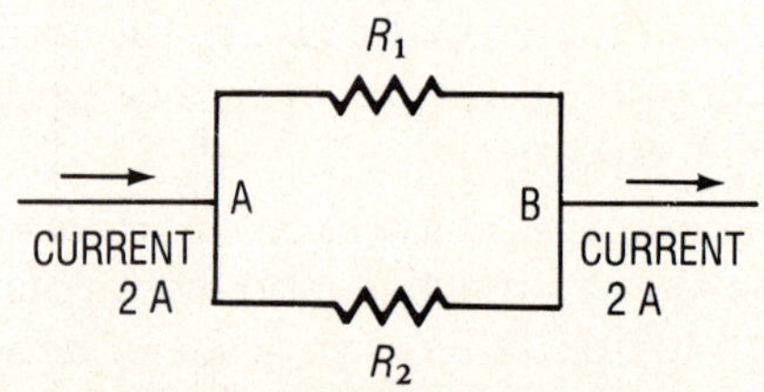

Figure 3-1. Current delivered to point A divides in parallel circuit with two resistors.

Current through resistor R_1 is 1.5 A. Through R_2, the current is 0.5 A. The total current for the two resistors adds up to 2 A. Also, the current flowing through point B is 2 A. In terms of algebraic addition, this circuit can be described as follows:

$$+2\text{ A} - 1.5\text{ A} - 0.5\text{ A} = 0\text{ A}$$

Algebraic addition is a means of adding numbers (addends) that have plus or minus values. Plus values are added to each other. Minus values are also combined. Where plus and minus values occur in the same addition problem, the minus numbers are subtracted from the plus numbers. The sum can be either plus or minus, depending on the values of the addends. Some examples:

$$-3 + 4 = +1$$
$$-4 + 3 = -1$$

Applying Kirchhoff's Law for Current

In using Kirchhoff's law for current to analyze an electrical network, you follow the flow of current from its source through each circuit branch. As you do, you make algebraic notations in sequence. Values of current entering a branch are assigned plus values. Currents leaving the dividing point are given minus values.

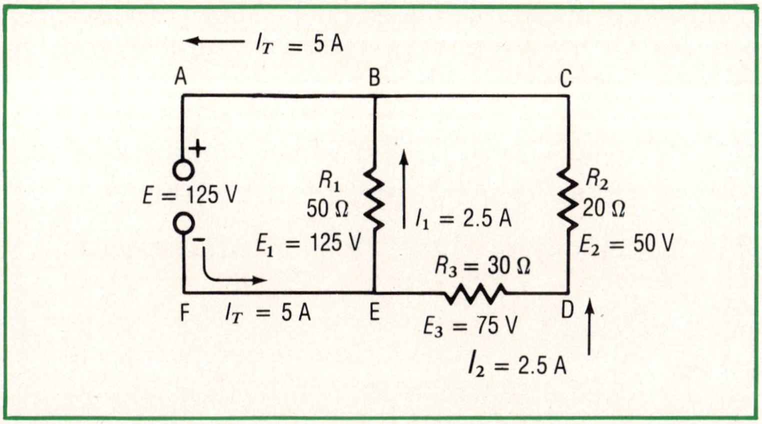

Figure 3-2. Series-parallel circuit to illustrate Kirchhoff's law for current.

As an example, consider the circuit in Figure 3-2. In this illustration, 5 A leaves point F and flows to point E. There, it divides, with 2.5 A flowing through R_1 and 2.5 A flowing through R_3 and R_2. The algebraic notation for the flow at point E is:

$$+5\ \text{A} - 2.5\ \text{A} - 2.5\ \text{A} = 0\ \text{A}$$

Now consider current flow through point D. The circuit branch E–D–C–B (containing resistors R_3 and R_2) is connected in series. Therefore, current is equal throughout this circuit loop. The current leaving point D will thus be equal to the current entering point D.

$$+2.5\text{ A} - 2.5\text{ A} = 0\text{ A}$$

Now consider point B. It has been established that current flowing from C to B is 2.5 A. It has also been shown that the current through resistor R_1, moving to point B, is also 2.5 A. The algebraic addition covering point B is as follows:

$$+2.5\text{ A} + 2.5\text{ A} - 5\text{ A} = 0\text{ A}$$

This indicates that the current out of point B must be 5 A, rather than 2.5 A or some other value. The same algebraic addition works for total circuit current, I_T. The current leaving point F is 5 A and the current returning to the source through point A is 5 A. Thus:

$$+5\text{ A} - 5\text{ A} = 0\text{ A}$$

The concept of current through a point applies also to circuit sources: *Current leaving the source must be equal to current returning to the source.*

Discussion Questions/Problems

1. How does Kirchhoff's law for current apply to a single current dividing point in an electrical network?
2. How does Kirchhoff's law for current apply to the electrical source in a network?
3. What is algebraic addition?
4. How do you develop an algebraic notation that applies Kirchhoff's law for current by using a diagram of an electrical network?

3.3 KIRCHHOFF'S LAW FOR VOLTAGE

Kirchhoff's law for voltage can be stated in two separate ways:

- *The algebraic sum of voltage drops and voltage rises around any closed series loop (or path) is equal to zero.*
- *The sum of voltage drops around any closed series path (or loop) is equal to the sum of voltage rises.*

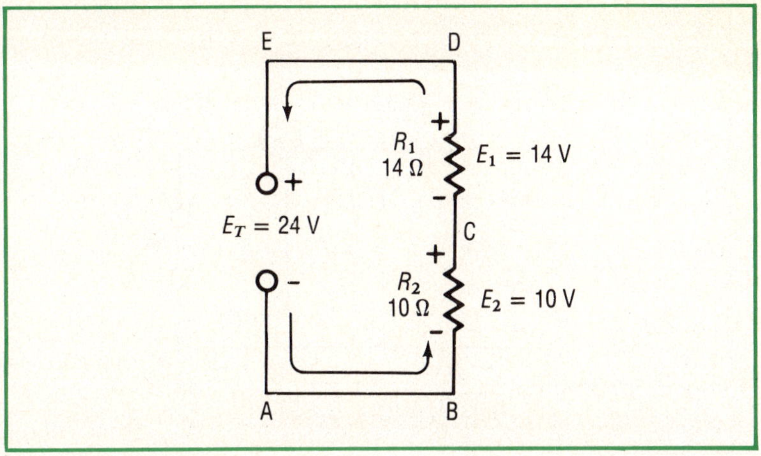

Figure 3-3. Circuit with two resistors in series.

For an illustration of how these two approaches are applied, look at the circuit in Figure 3-3. This shows a source of 24 V and a circuit with two resistors in series. Under the first approach, the source voltage is considered to be the reverse of a drop and is assigned a plus sign. The loads are considered drops and are assigned a negative sign.

$$+24\ \text{V} - 10\ \text{V} - 14\ \text{V} = 0\ \text{V}$$

The algebraic formula for this first statement of applying Kirchhoff's law for voltage is:

$$E_T - E_1 - E_2 - \ldots - E_N = 0$$

Using the second statement, algebraic addition is not necessary. Instead, you think of the source voltage as a rise, or increase. The resistors in the circuit are treated as voltage drops. The two amounts should then be equal. When an equation is used, the values on both sides of the equal sign should be the same. The result is not zero.

$$24\ \text{V} = 10\ \text{V} + 14\ \text{V}$$

Applying Kirchhoff's Law for Voltage

Figure 3-4 shows an example of a circuit that lends itself well to solution through Kirchhoff's law for voltage. Note that there are two voltage sources in a single circuit. Analyzing this type of circuit with Ohm's law would be complex. Kirchhoff's law for voltage offers a better approach.

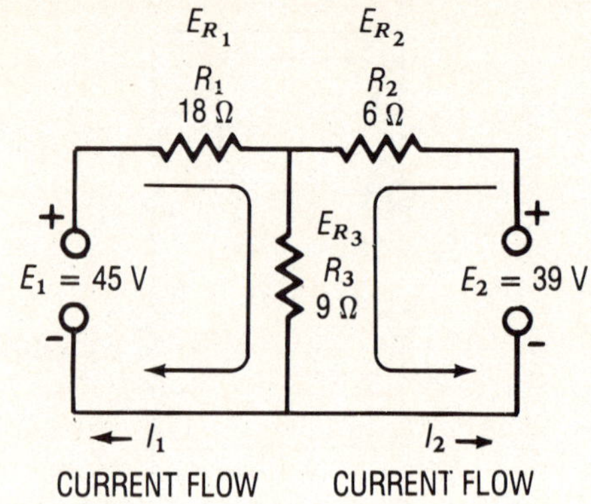

Figure 3-4. Circuit with two voltage sources.

Voltages for the two electrical sources are given. Also given are resistances for the three resistors in this circuit. Current levels and voltage drops across the three resistors are unknown and will be derived. Kirchhoff's law for voltage is used in the example below to determine the voltage drops. To do this, it is necessary to determine current levels. The basic equations for finding voltage drops are as follows:

$$+E_1 - E_{R_1} - E_{R_3} = 0$$
$$+E_2 - E_{R_2} - E_{R_3} = 0$$

Applying known values to these equations, you get the following:

$$+45 - E_{R_1} - E_{R_3} = 0$$
$$+39 - E_{R_2} - E_{R_3} = 0$$

Both equations are used to solve the problem. In each equation the value of E is replaced by IR. In the situation of R_3, both I_1 and I_2 are used in the replacement formula since both currents flow through this resistor.

A basic mathematical principle is that values can be added to or subtracted from both sides of any equation. As long as the same function is performed on both sides of the equal sign, the equation remains true.

The following set of equations establishes initial values for the E_1 loop.

$$+45 - R_1 I_1 - R_3 (I_1 + I_2) = 0$$

$$+45 - 18\, I_1 - 9\, (I_1 + I_2) = 0$$

$$+45 - 18\, I_1 - 9\, I_1 - 9\, I_2 = 0$$

Note that the signs for the value of I_1 and I_2 are both minus. This is because both values have been multiplied by -9. When factors with opposite signs, $-$ and $+$, are muliplied by each other, the result is negative. When like signs, either negative or positive, are multiplied by each other, the result is always positive.

The next step in solving this equation is to combine like terms. The values, $9\, I_1$ and $18\, I_1$ are combined:

$$+45 - 27\, I_1 - 9\, I_2 = 0$$

The value of E_1 is subtracted from both sides of the equation. This is done by subtracting 45 from both sides. The effect is as follows:

$$-27\, I_1 - 9\, I_2 = -45$$

The same steps are followed for the E_2 loop of the circuit:

$$+39 - R_2 I_2 - R_3 (I_1 + I_2) = 0$$
$$+39 - 6\, I_2 - 9\, (I_1 + I_2) = 0$$
$$+39 - 6\, I_2 - 9\, I_1 - 9\, I_2 = 0$$
$$+39 - 9\, I_1 - 15\, I_2 = 0$$
$$-9\, I_1 - 15\, I_2 = -39$$

At this point two equations have been developed, each with two unknown quantities:

$$-27\, I_1 - 9\, I_2 = -45$$
$$-9\, I_1 - 15\, I_2 = -39$$

According to mathematical theory, it is possible to solve a problem of two unknowns when there are two equations. Since there are two unknowns, neither equation can be solved until at least one of the values is determined. The value of I_2 can be derived by multiplying the values on both sides of the equation for loop E_2 by a factor of -3:

$$\begin{array}{r} -9\,I_1 - 15\,I_2 = -39 \\ \times\,(-3) \\ \hline +27\,I_1 + 45\,I_2 = +117 \end{array}$$

Next, the equations for the E_1 loop and the E_2 loop are added:

$$\begin{array}{r} -27\,I_1 - 9\,I_2 = -45 \\ +27\,I_1 + 45\,I_2 = +117 \\ \hline 0 + 36\,I_2 = +72 \end{array}$$

$$36\,I_2 = 72 \qquad I_2 = \frac{72}{36} = 2\text{ A}$$

The value for the current in loop E_2 is 2 A.

The value of I_2, just established, can now be used to determine the value of I_1, as shown below:

$$-27\,I_1 - 9\,(2) = -45$$
$$-27\,I_1 - 18 = -45$$
$$-27\,I_1 = -27$$
$$I_1 = \frac{-27}{-27} = 1\text{ A}$$

The current in the E_1 loop is 1 A. With the completion of these calculations, the value is known for the current in each loop. These figures can then be used to calculate voltage drops, as shown below:

$$E_{R_3} = R_3\,(I_1 + I_2) = 9\,(1 + 2) = 27\text{ V}$$
$$E_{R_1} = R_1 I_1 = 18 \times 1 = 18\text{ V}$$
$$E_{R_2} = R_2 I_2 = 6 \times 2 = 12\text{ V}$$

The voltage drop across R_1 is 18 V; across R_2, 12 V; and across R_3, 27 V.

Finally, Kirchhoff's law for voltage can be used to check the accuracy of the computations:

$$E_1 = E_{R_1} + E_{R_3} = 18 + 27 = 45 \text{ V}$$
$$E_2 = E_{R_2} + E_{R_3} = 12 + 27 = 39 \text{ V}$$

The computations are correct because the total of voltage rises is equal to the total of voltage drops for each loop.

Discussion Questions/Problems

1. Using Kirchhoff's law for voltage, how do you add the voltage drops around a series circuit?
2. What should be the relationship between voltage drops and voltage rises in a series circuit?
3. A series circuit has three resistors and a supply of 18 V. Voltage drops across two of these circuits are 12 V and 4 V. What is the voltage drop across the third resistor?
4. A series circuit has four voltage drops with a total value of −24 V. What would be the voltage rise for this circuit?

3.4 SUPERPOSITION THEOREM

This is believed by many to be the most logical of the network theorems. The superposition theorem holds that *the effect of two or more sources of current or voltage for any part of a circuit is equal to the algebraic sum of the effect from each source, acting separately*.

The logic of this theorem is that it follows a basic arithmetic principle: *The whole is equal to the sum of its parts.*

Applying the Superposition Theorem

The superposition theorem is used principally to determine current for circuits or circuit loops which have two or more supply voltages that are known. Consider an example with two voltage sources. In using the theorem, the polarities of the applied voltages must be observed. As an illustration,

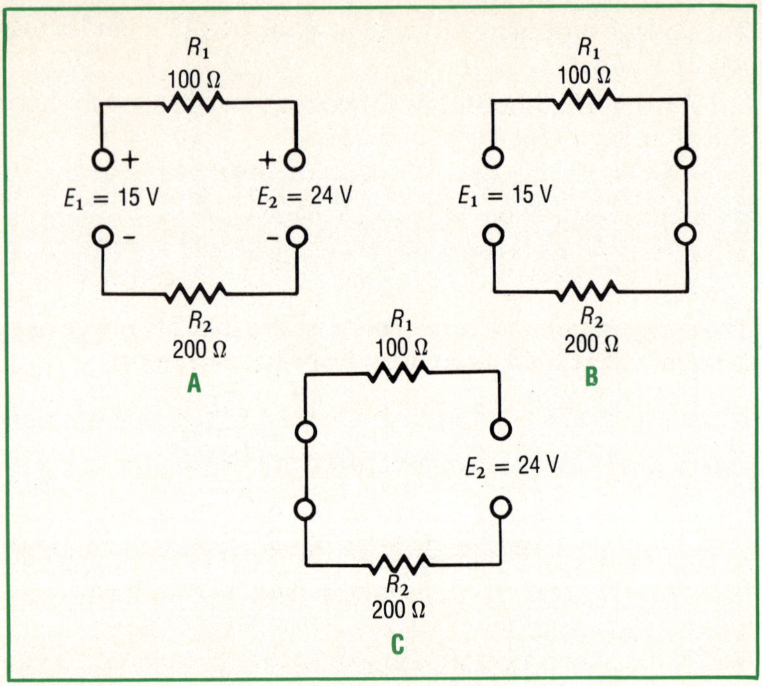

Figure 3-5. Superposition theorem is used to determine current for circuit with two known supply voltages.

consider the diagram in Figure 3-5A. There are two voltage sources, 15 V and 24 V. The sources oppose each other; their positive and negative terminals oppose, rather than aid. The smaller source is assigned a negative value and the larger source is assigned a positive value. (This is done as a matter of convenience. The same answer would result if the larger source was assigned a negative value and the smaller was assumed to be positive.) Algebraic addition is used to determine voltage. The net applied voltage of this circuit is 9 V.

To determine current through use of the superposition theorem, you have to assume that one of the voltage sources doesn't exist—that it has been grounded or shorted out. As an example, consider the configuration in Figure 3-5B. The larger voltage source has been shorted out, or bridged with a connecting wire. The current from the voltage source E_1 is determined as follows:

$$I_T = \frac{E_1}{R_T} = \frac{15}{300} = 0.05\ \text{A}$$

The current from the lower voltage source is 0.05 A. Since this is the current from the smaller source, it will be assigned a negative value in determining total current.

Next, the same procedure is followed for the other voltage source. This is illustrated in Figure 3-5C. Current for this hypothetical configuration is determined as follows:

$$I_T = \frac{E_2}{R_T} = \frac{24}{300} = 0.08 \text{ A}$$

The larger voltage source delivers a current of 0.08 A. To determine total circuit current, this current is assigned a positive value. The net current in the circuit is then determined as follows:

$$0.08 \text{ A} - 0.05 \text{ A} = 0.03 \text{ A}$$

Applied current in this circuit is 0.03 A.

Discussion Questions/Problems

1. If a circuit has voltage sources of 24 V and 36 V applied so as to aid each other, what is the total applied voltage?
2. If a circuit has current sources of 0.5 A and 4.5 A, what is the total applied current?
3. In using the superposition theorem, can there be more than two voltage sources?
4. Calculate the answer to the problem in Figure 3-5 when the smaller source is assigned a positive value and the larger source is assigned a negative value. Review your answer.

3.5 THEVENIN'S THEOREM

This theorem takes its name from M. L. Thevenin, a French engineer. Its main application is in simplifying the relationship between voltage sources and drops in a network, making it easier to determine applied voltage. Its basic concept is that any circuit can be reduced, or simplified, to an equivalent circuit consisting of a single supply voltage, a related single supply resistance, and a single load resistance. Thus, the complicated circuit in Figure 3-6A can be reduced to the Thevenin equivalent in Figure 3-6B.

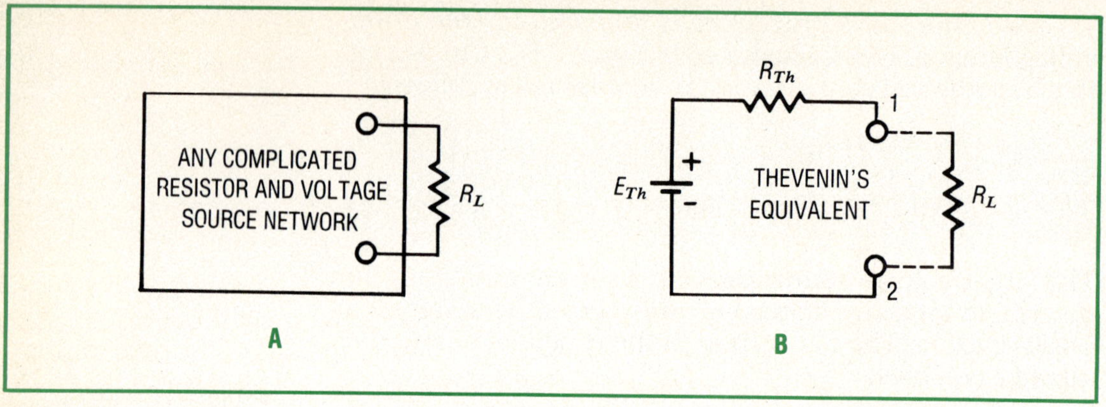

Figure 3-6. Any complicated network circuit can be reduced to a Thevenin equivalent.

Thevenin's theorem holds that *the current in any load resistance, R_L, connected to two terminals of a network consisting of any number of resistors and/or voltage sources is the same as though R_L were connected to a single source whose open circuited voltage is the open-circuit voltage of the two terminals and whose resistance is the equivalent resistance looking into the two terminals.*

Algebraically, Thevenin's theorem can be expressed this way:

$$I_T = \frac{E_{Th}}{R_{Th}}$$

Applying Thevenin's Theorem

In applying Thevenin's theorem to determine either voltage or current, you begin by simplifying a circuit. As an example, consider the circuit in Figure 3-7. There are three resistors across a 12 V battery source. The idea of Thevenin's theorem is that this circuit can be represented as shown in Figure 3-8. All resistances are considered to have been consolidated and are represented as a single resistance, R_{Th}. All voltages (in this case only one) are represented as a single voltage, E_{Th}.

To arrive at a conceptual result like the one shown in Figure 3-8, you select any starting point within the circuit, as long as it is logical for you. The most usual starting point is the output terminal. Mentally remove one selected component from the circuit: Resistor R_L is removed, as indicated in Figure 3-9. Note that resistor R_L has been positioned beyond the circuit's terminals, removing it effectively from the circuit. With the circuit reconfigured in this way, Thevenin's

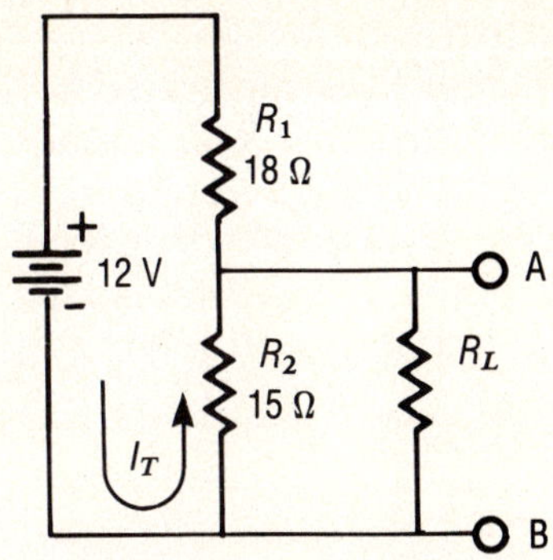

Figure 3-7. Circuit with three resistors across 12 V battery source.

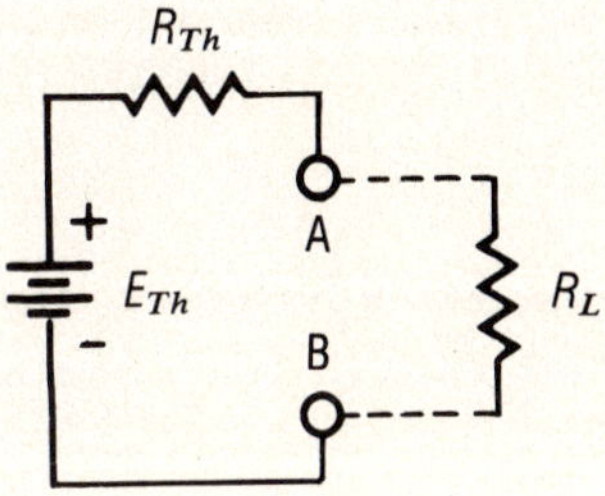

Figure 3-8. In this reconfiguration, E_{Th} and R_{Th} represent all the voltage and resistance of the circuit.

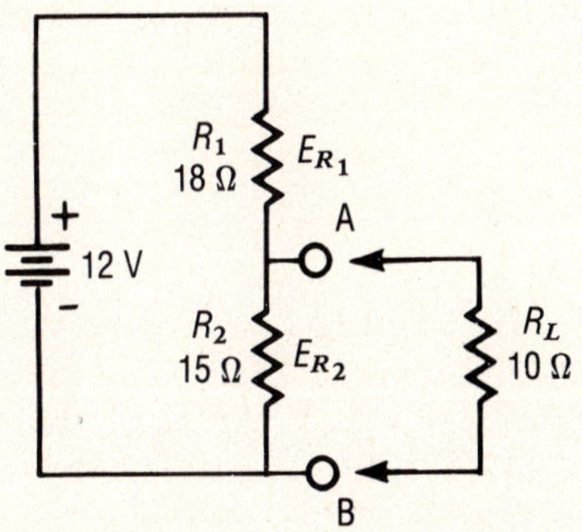

Figure 3-9. To apply Thevenin's theorem, resistor R_L is positioned beyond the circuit's terminals.

theorem can now be applied to determine total current and voltage drops for the remaining series circuit. To find voltage across points A and B, you use Ohm's law. First you determine the total current, I_T:

$$I_T = \frac{E_T}{R_1 + R_2} = \frac{12}{33\ \Omega} = 0.36\ \text{A}$$

The total current through this circuit loop is 0.36 A. This value can then be used to determine the voltage drop across resistor R_2. Note that resistor R_2 is connected between terminals A and B. The voltage drop for R_2 is equal to the voltage that would exist across A–B if R_L were not connected—because R_L isn't connected! The voltage drop for R_2 meets the stated definition and is equal to E_{Th}. It is also necessary to find the value of R_{Th} to determine I_{Th} in solving the equation above.

The calculation for the voltage drop across R_2 is as follows:

$$\begin{aligned} E_{R_2} &= I_{T_1} \times R_2 \\ &= 0.36 \times 15 \\ &= 5.45 \text{ V} \end{aligned}$$

The voltage drop across R_2 is 5.45 V. This is the voltage drop between open-circuit terminals A and B. Thus, E_{Th} also equals 5.45 V.

The next step in applying Thevenin's theorem is to determine an equivalent series resistance, R_{Th}. To do this, the basic circuit is simplified further. The voltage source is eliminated. The resistor R_L still remains out of the circuit. The result is the circuit shown in Figure 3-10. This remaining circuit loop

Figure 3-10. To determine equivalent series resistance, the circuit of Figure 3-7 is further simplified.

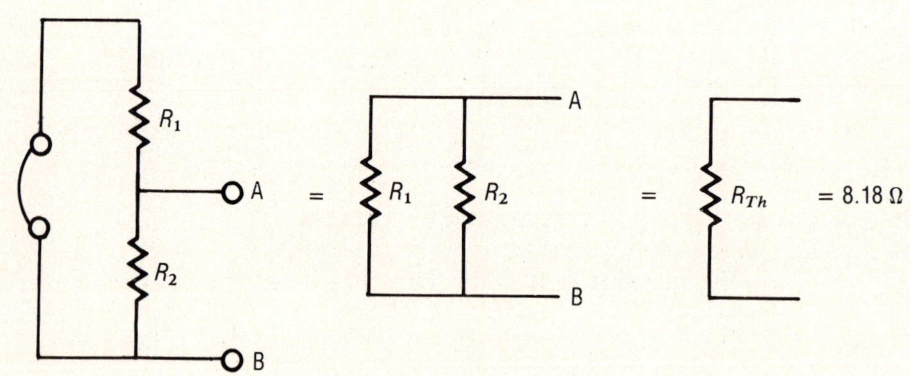

has two resistors of known value in parallel. The value for R_{Th} for this circuit is as follows:

$$R_{Th} = \frac{R_1 \times R_2}{R_1 + R_2} = \frac{18 \times 15}{18 + 15} = \frac{270}{33} = 8.18\ \Omega$$

R_{Th} for the circuit in Figure 3-10 is 8.18 Ω.

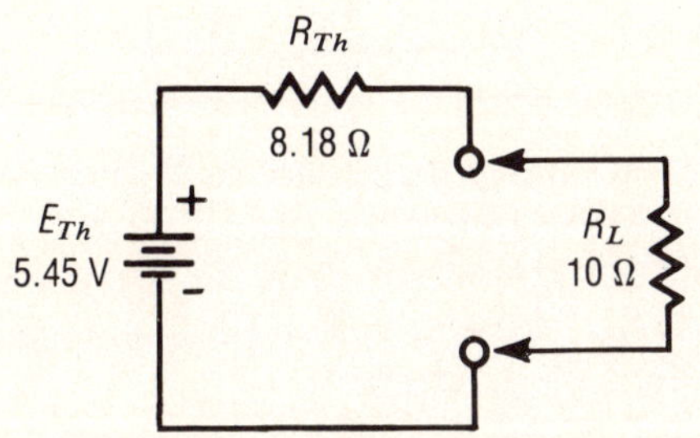

Figure 3-11. Composite circuit used to determine current through R_L.

Now you are ready to deal with the resistor that was deleted from the circuit initially. A composite circuit that includes the values of E_{Th} and R_{Th}, as well as R_L, is shown in Figure 3-11. The only unknown within the circuit now is the current through R_L. This is determined as follows:

$$I_L = \frac{E_{Th}}{R_{Th} + R_L} = \frac{5.45}{10 + 8.18} = 0.30\ \text{A}$$

The current through resistor R_L is 0.30 A. In this example, the circuit in Figure 3-7 has been reduced to that in Figure 3-11 with an equivalent voltage of 5.45 V and an equivalent Thevenin resistance of 8.18 Ω according to Thevenin's theorem. From this, the current through resistor R_L is determined as 0.3 A.

Networks encountered in actual electrical or electronic design situations would be a lot more complex than this example. The illustration is purposely simplified to introduce and demonstrate the principles of the network theorems.

When the example is a relatively simple network, it is possible to check the work by applying Ohm's law. This is difficult on extremely complex networks. To apply Ohm's law to determine the current through R_L, you would begin by establishing circuit values for resistance and current. Refer

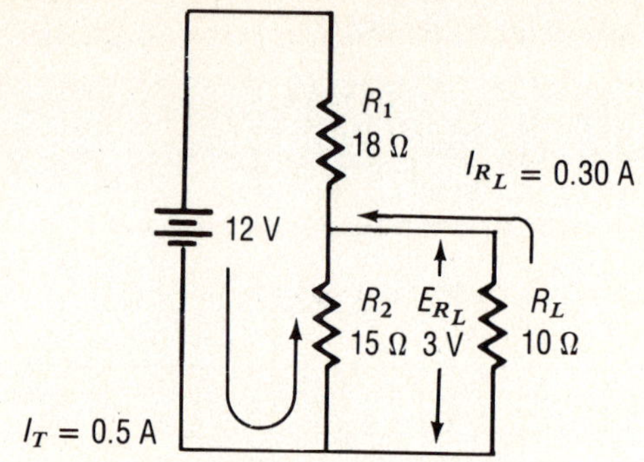

Figure 3-12. To check application of Thevenin's theorem, Ohm's law is applied to the circuit of Figure 3-7.

to Figure 3-12, which is a duplicate of Figure 3-7. To determine total resistance, begin by simplifying the circuit and finding the combined voltage drop across the two parallel resistors, R_2 and R_L. The equivalent parallel resistance across R_2 and R_L is called R_{2+L}. Then:

$$R_{2+L} = \frac{R_2 \times R_L}{R_2 + R_L} = \frac{15 \times 10}{15 + 10} = \frac{150}{25} = 6\ \Omega$$

The resistance of R_{2+L} is 6 Ω. Next, this value is added to the value of R_1, which is given as 18 Ω:

$$R_1 + R_{2+L} = 18 + 6 = 24\ \Omega$$

Total resistance for the circuit is 24 Ω. This figure can then be used to determine total current in the circuit:

$$I_T = \frac{E_T}{R_T} = \frac{12}{24} = 0.5\text{ A}$$

Total current in the circuit is 0.5 A. With this figure, it is now possible to determine the voltage drop across R_L. In determining the voltage, E_{R_L}, across resistor R_L, you have to remember that there are two resistors in parallel. Therefore, you use the calculated value for resistance. This calculation, shown above, establishes that the resistance of R_{2+L} is 6 Ω. This is the value used. The calculation, then, is:

$$E_{R_L} = I_T \times R_{2+L} = 0.5 \times 6 = 3\text{ V}$$

The voltage across resistor R_L is 3 V. This figure can now be used to determine the current through resistor R_L:

$$I_{R_L} = \frac{E_{R_L}}{R_L} = \frac{3}{10} = 0.30 \text{ A}$$

The current through resistor R_L is 0.3 A. This checks against the result reached through use of Thevenin's theorem.

Thevenin's Theorem for Two Voltage Sources

The same approach of simplifying circuits applies when there are two voltage sources. The circuit has to be reduced in stages. To illustrate, this discussion uses the same circuit, now shown in Figure 3-13, that was used to demonstrate the application of Kirchhoff's law for voltage.

The selected starting point is the resistor R_3. This is removed from the circuit. Then, total voltage and resistance can be determined for the circuit.

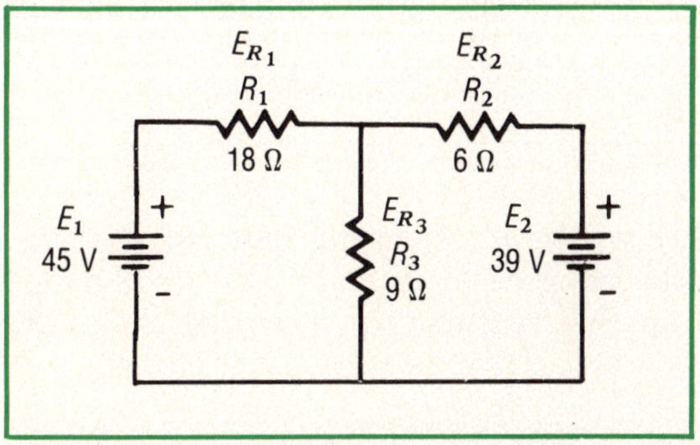

Figure 3-13. Circuit with two voltage sources to be analyzed with Thevenin's theorem.

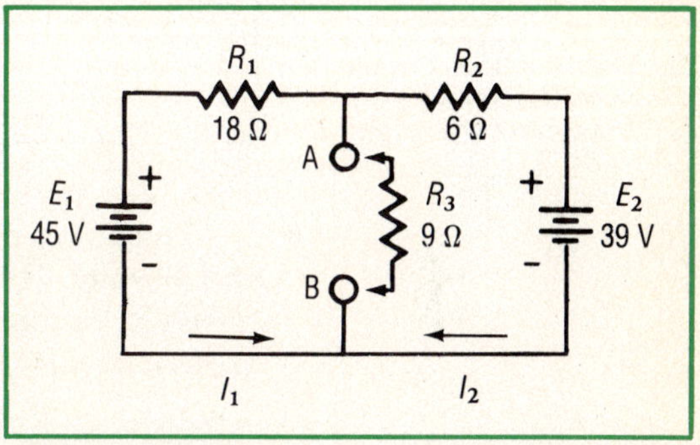

Figure 3-14. A resistor is removed from circuit at terminals A–B.

Figure 3-14 is a drawing showing R_3 removed from the circuit at terminals A–B. Since the revised circuit has two resistors in series, their values can be added:

$$R_T = R_1 + R_2 = 18 + 6 = 24\ \Omega$$

Total resistance for this reduced circuit is 24 Ω. The net voltage across this resistance of 24 Ω is the larger voltage value minus the smaller one:

$$E_1 - E_2 = 45 - 39 = 6 \text{ V}$$

With these figures, the current through R_1 and R_2 can be determined:

$$I_T = \frac{E_T}{R_T} = \frac{6}{24} = 0.25 \text{ A}$$

The current through R_1 and R_2 is 0.25 A. The revised circuit is redrawn as in Figure 3-15 to relate it to the definition of Thevenin's theorem.

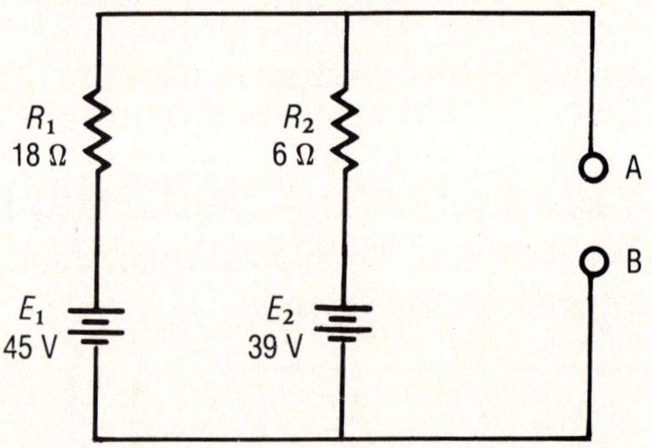

Figure 3-15. Circuit redrawn to relate to Thevenin's theorem.

The voltage at A–B will be at some value between 45 volts and 39 volts. This voltage will be less than 45 V by the *IR* drop across R_1 or more than 39 V by the *IR* drop across R_2. The *IR* drop across R_1 is:

$$E_{R_1} = I_T R_1 = 0.25 \text{ A} \times 18\ \Omega = 4.5 \text{ V}$$

And so the voltage at A–B is:

$$E_{A\text{-}B} = E_1 - E_R = 45 \text{ V} - 4.5 \text{ V} = 40.5 \text{ V}$$

Similarly, the *IR* drop across R_2 is:

$$E_{R_2} = I_T R_2 = 0.25 \text{ A} \times 6\ \Omega = 1.5 \text{ V}$$

So the voltage at A–B is:

$$E_{A\text{-}B} = E_2 + E_{R_2} = 39 \text{ V} + 1.5 \text{ V} = 40.5 \text{ V}$$

Of course, E_{A-B} should be the same whether it is calculated from E_1 or from E_2. Following the definition of Thevenin's theorem, E_{A-B} is the open-circuit voltage, and so:

$$E_{A-B} = E_{Th} = 40.5 \text{ V}$$

Next, reduce the circuit of Figure 3-15 by removing both voltage sources and replacing each with a simple connection. The result is Figure 3-16.

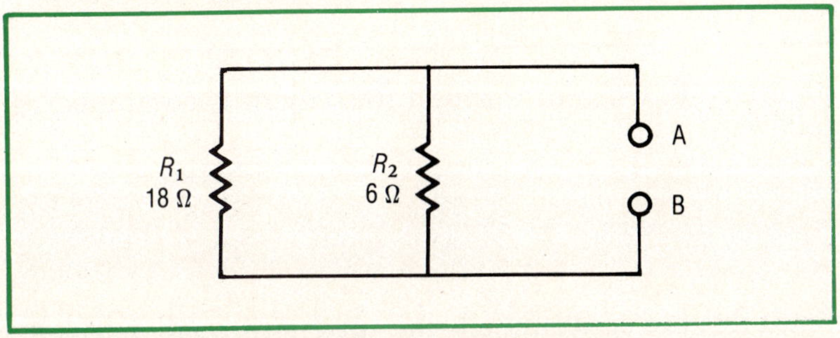

Figure 3-16. Voltage sources are removed and replaced with a simple connection to reduce circuit of Figure 3-15.

The equivalent resistance of R_1 and R_2 is:

$$R_{1+2} = \frac{18 \times 6}{18 + 6} = \frac{108}{24} = 4.5 \, \Omega$$

Since this is the equivalent resistance looking into the terminals A–B, this must be the equivalent Thevenin resistance. Thus:

$$R_{1+2} = R_{Th} = 4.5 \, \Omega$$

The Thevenin equivalent of Figure 3-13 now is as shown in Figure 3-17.

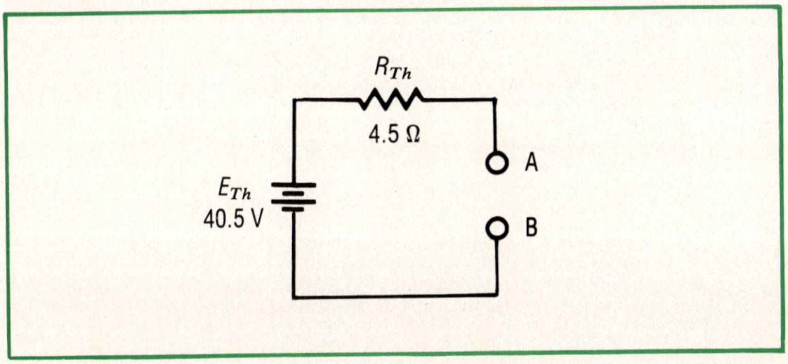

Figure 3-17. Thevenin equivalent of Figure 3-13.

The originally removed resistor, R_3, is now placed across the terminal of the Thevenin equivalent circuit as shown in

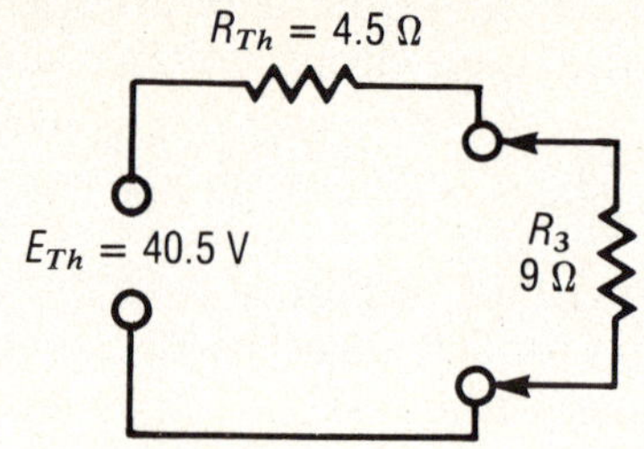

Figure 3-18. Originally removed resistor placed across the terminal of the Thevenin equivalent circuit.

Figure 3-18. Current through this circuit is computed as follows:

$$I = \frac{E_{Th}}{R_{Th} + R_3} = \frac{40.5\ \text{V}}{4.5\ \Omega + 9\ \Omega} = \frac{40.5\ \text{V}}{13.5\ \Omega} = 3\ \text{A}$$

Since this is a series circuit, the current through R_3 is 3 A. Reviewing the calculations covering the circuit in Figure 3-4, you will find that the same result was realized by applying Kirchhoff's laws to this same network. With either calculation, 3 A of current flows through R_3.

Discussion Questions/Problems

1. What is the basic concept of Thevenin's theorem?
2. What is the main application for Thevenin's theorem?
3. What is the meaning of the expression R_{Th}?
4. Is I_T in Figure 3-7 different from I_T in Figure 3-12?
5. Why is E_{Th} in Figure 3-10 different from E_{R_L} in Figure 3-12?
6. Is E_{R_L} in Figure 3-7 different from E_{R_L} in Figure 3-12?
7. If a circuit has two voltage sources, how is total applied voltage determined?
8. A circuit has two resistors of 10 Ω and 30 Ω, one of which is the load. What is the R_{Th} for the circuit?

3.6 NORTON'S THEOREM

Norton's theorem is another approach that determines electrical values by simplifying networks. The theorem is named after E. L. Norton (1898–1983), a scientist with Bell Telephone Laboratories. Whereas Thevenin's theorem simplifies circuits in terms of resistance and voltage, Norton's theorem simplifies circuits in terms of resistance and current. The basic idea of Norton's theorem is that any circuit can be reduced, or simplified, to an equivalent circuit consisting of a single supply current; a related, single parallel internal resistance; and a single load resistance. In many ways, as demonstrated below, Norton's theorem may be easier to apply than Thevenin's.

Norton's theorem states that *the current in any load resistance,* R_L*, connected to two terminals of a network consisting of any number of resistors and/or current sources is the same as though* R_L *were connected to a constant current source whose current is equal to the current that flows through the two terminals when they are short-circuited*. This assumes that the constant current source is across a resistance equal to the resistance of the network—looking back from the two terminals.

Figure 3-19 demonstrates how a parallel circuit is simplified under the Norton approach. Figure 3-19A shows a three-resistor circuit in which two resistors are in parallel. The Norton equivalent to this circuit is shown in Figure 3-19B. This consists of a single, composite resistor, identified as R_N, and a single current source, identified as I_N.

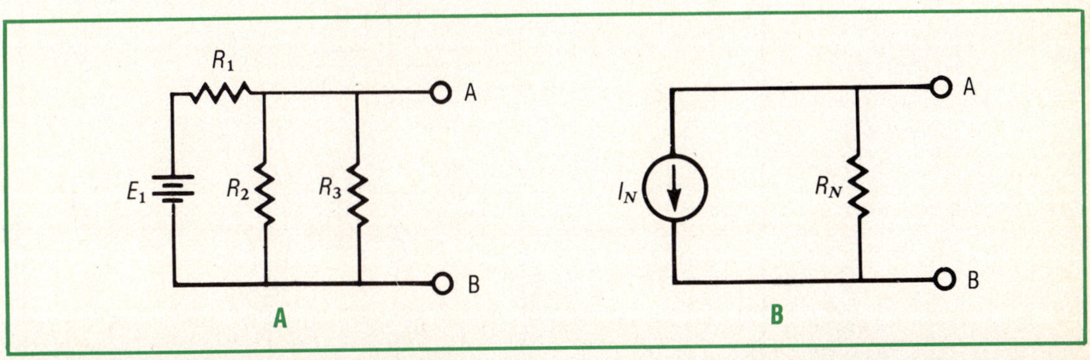

Figure 3-19. Three-resistor circuit with two resistors in parallel is simplified by applying Norton's theorem.

Applying Norton's Theorem

For demonstration purposes, the same basic circuit which has been used twice before is again illustrated in Figure 3-20. This circuit will now be used to determine current through R_3, using Norton's theorem. The first step is to simplify the circuit

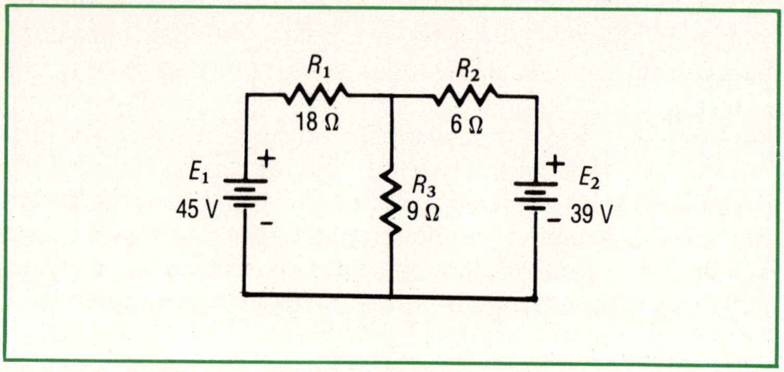

Figure 3-20. Circuit of Figure 3-7 to be analyzed with Norton's theorem.

by removing one portion. Again, R_3 is removed. Figure 3-21A shows R_3 removed and the load terminals A–B shorted out. It also indicates current flow directions for the remaining, consolidated circuit. These flow directions are a matter of choice. If the current flow directions were chosen as in Figure 3-21B, the answer would still work out the same.

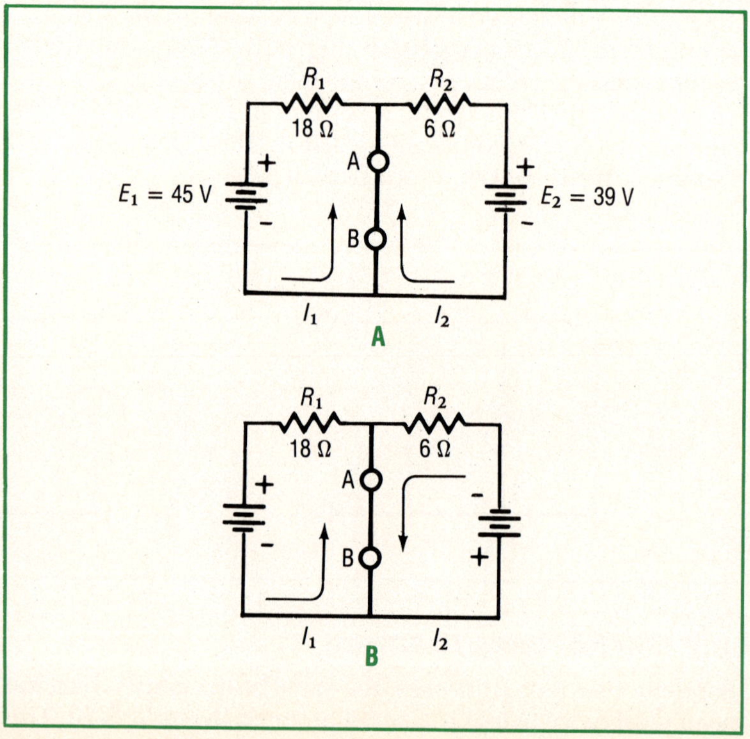

Figure 3-21. Circuit simplified by removing one portion.

For the purposes of determining current flow, the circuit must be simplified further. The circuit is considered as though both voltage sources deliver current independently of each other. Then, current flows through each portion of the circuit as though the other were not there. Such a circuit is shown in Figure 3-22. On the basis of this revised circuit, current is determined as follows:

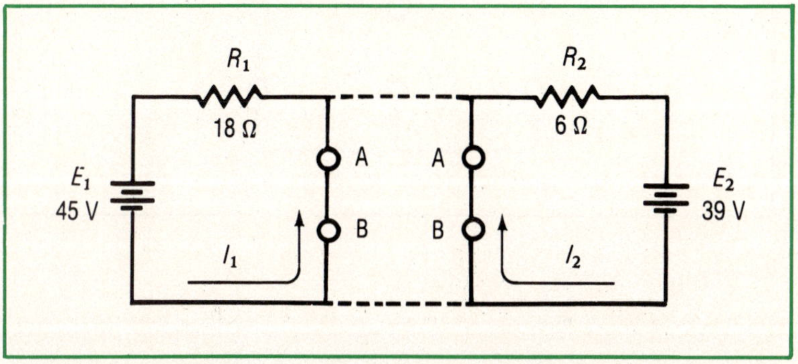

Figure 3-22. Circuit simplified further to determine current flow.

$$I_1 = \frac{E_1}{R_1} = \frac{45}{18} = 2.5 \text{ A}$$

Current through resistor R_1 due to E_1 is 2.5 A.

$$I_2 = \frac{E_2}{R_2} = \frac{39}{6} = 6.5 \text{ A}$$

Current through resistor R_2 due to E_2 is 6.5 A. At this point, because both I_1 and I_2 flow through terminals A–B, and both in the same direction, current values can be added to get the total current flowing through the terminals A–B.

$$I_T = I_1 + I_2 = 2.5 + 6.5 = 9 \text{ A}$$

Total current flowing through the terminals A–B is 9 A. This is the current that would flow through the two terminals when they are shorted out. This is I_N according to the definition of Norton's theorem. So, I_N is equal to 9 A.

With I_N determined, the next task is to determine the resistance looking back from A–B. To do this, E_1 and E_2 are shorted out; R_3 is left open. The result is shown in Figure 3-23. R_1 and R_2 are in parallel. On this basis, the value of R_N can be determined:

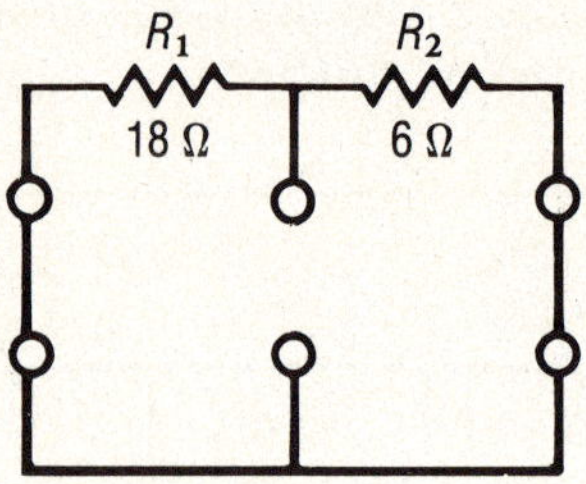

Figure 3-23. Circuit redrawn to determine resistance looking back from A–B.

$$R_N = \frac{R_1 R_2}{R_1 + R_2} = \frac{18 \times 6}{18 + 6} = \frac{108}{24} = 4.5\ \Omega$$

R_N, or the overall resistance for the circuit diagrammed in Figure 3-23, is 4.5 Ω. The Norton equivalent of Figure 3-20 is as shown in Figure 3-24.

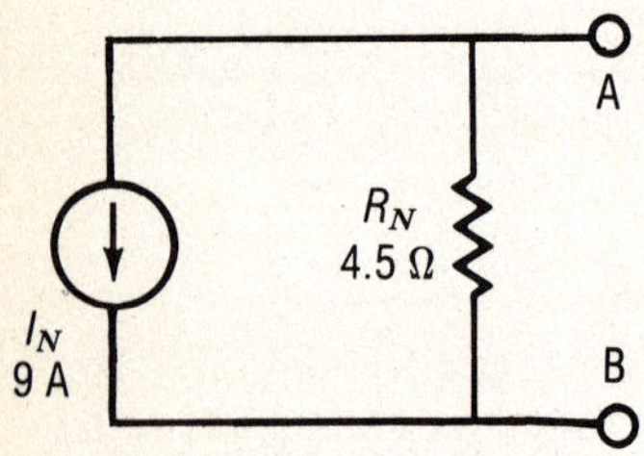

Figure 3-24. Norton equivalent of Figure 3-20.

R_3 can be added to the circuit, and the values just determined for I_N and R_N can be used to find the current flow through R_3. The circuit used as a basis for this calculation is shown in Figure 3-25. Since R_N and R_3 are in parallel, their parallel equivalent resistance can be determined:

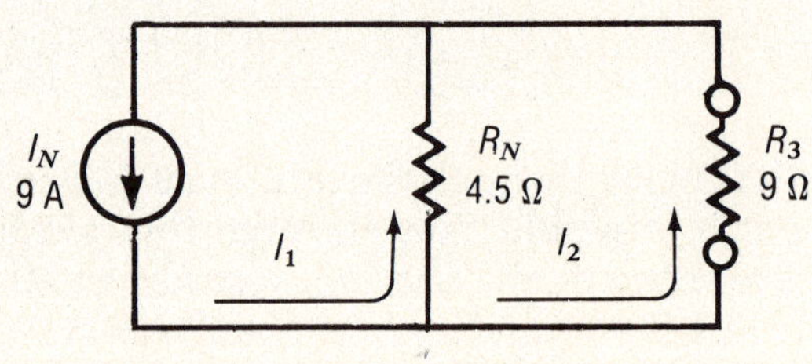

Figure 3-25. Circuit used as basis for determining current through R_3.

$$R_T = \frac{R_N R_3}{R_N + R_3} = \frac{4.5 \times 9}{4.5 + 9} = \frac{40.5}{13.5} = 3\ \Omega$$

The value of R_T is 3 Ω. This value can next be used to find the current flowing through each branch of the original circuit. Refer to Figures 3-20 and 3-4. First, since 9 A flows through R_T, the voltage drop across R_T is:

$$E_T = IR = 9\text{ A} + 3\ \Omega = 27\text{ V}$$

The same 27 V appears across R_3 alone, so the current through R_3 is:

$$I_{R_3} = \frac{E_T}{R_3} = \frac{27\text{ V}}{9\ \Omega} = 3\text{ A}$$

Also, since there is 27 V across R_1, the voltage across R_1 must be:

$$E_{R_1} = 45\text{ V} - 27\text{ V} = 18\text{ V}$$

And:

$$I_1 = \frac{E_{R_1}}{R_1} = \frac{18\text{ V}}{18\ \Omega} = 1\text{ A}$$

Similarly, the voltage across R_2 must be:

$$E_{R_2} = 39\text{ V} - 27\text{ V} = 12\text{ V}$$

And:

$$I_2 = \frac{E_{R_2}}{R_2} = \frac{12\text{ V}}{6\ \Omega} = 2\text{ A}$$

If you check back, you will see that these results are the same as those derived by applying Kirchhoff's laws and Thevenin's theorem.

Discussion Questions/Problems

1. Which two electrical properties of a circuit are used to represent the circuit under Norton's theorem?
2. What is the meaning of the notation R_N?
3. Which components of a network are simplified first under the Norton approach?
4. Which electrical property of a network is most frequently computed through application of Norton's theorem?
5. If a 24 V circuit has parallel resistors of 10 Ω and 24 Ω, what is the current?

3.7 THEVENIN–NORTON CONVERSIONS

In network analysis, there may sometimes be situations in which it is desirable to convert values between the Thevenin and Norton theorems. This makes it possible for each theorem to be used more widely.

- With Thevenin's theorem, a network can be represented by a voltage source and a series resistance.
- With Norton's theorem, a network can be represented by a current source and a parallel resistance.

The ability to convert values from one theorem to the other makes it easier to deal with voltage, current, and resistance in both series and parallel branches of networks.

Figure 3-26 shows equivalent representations for the same network under the Thevenin and Norton theorems. Figure 3-26A is a Thevenin representation. Figure 3-26B is a Norton version.

Figure 3-26. Thevenin and Norton equivalents for the same networks.

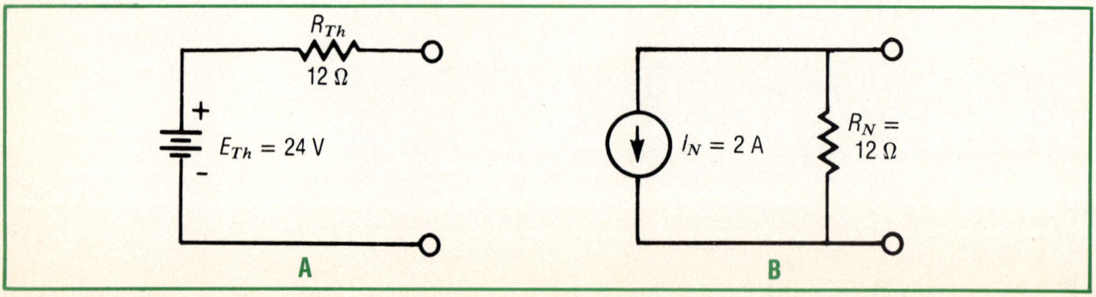

To convert from Thevenin values to Norton values, use these formulas:

$$R_N = R_{Th}$$
$$I_N = \frac{E_{Th}}{R_{Th}}$$

To convert from Norton values to Thevenin values, use these formulas:

$$R_{Th} = R_N$$
$$E_{Th} = I_N R_N$$

Theory Versus Practice

Norton's and Thevenin's theorems assume theoretical current and voltage sources. In Thevenin's theorem, no allowance is made for internal resistances of batteries or other voltage sources. In practice, voltage sources have internal resistances that affect the performance of the circuit. In Norton's theorem, it is assumed that all current from the source goes into the circuit elements. In practice, however, there is always current lost in the equivalent parallel resistance of the source itself. Figure 3-27A shows the effect of the internal resistance on a voltage source, and Figure 3-27B shows the effect of internal resistance on a current source.

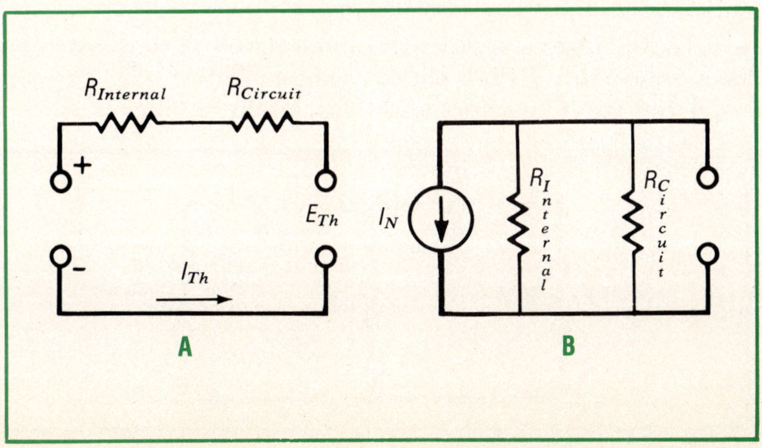

Figure 3-27. Effect of internal resistance on (A) voltage source and (B) current source.

Discussion Questions/Problems

1. What electrical properties should be used to represent a Thevenin circuit in a conversion to Norton values?
2. What electrical properties should be used to represent a Norton circuit in a conversion to Thevenin values?
3. If $R_N = 12.5\ \Omega$, what is the value of R_{Th}?
4. If $I_N = 24$ V and $R_N = 10\ \Omega$, what is the value of E_{Th}?

3.8 MILLMAN'S THEOREM

Millman's theorem is appropriate for use with networks that have many sources, making application of either the Thevenin or Norton techniques tedious.

Millman's theorem states that *each voltage source branch within a circuit may be changed to its Norton-current source equivalent; that these current sources may be analyzed along with whatever parallel current sources are originally part of the circuit and with whatever resistances result from the Norton equivalent along with whatever resistances are originally part of the circuit.*

The concept of Millman's theorem is to change each voltage-source branch within a circuit to its Norton-current source equivalent. These are then analyzed along with whatever parallel current sources are originally part of the circuit. The original resistances plus those developed from the Norton equivalent are also used.

To illustrate, consider the same network that has been used repeatedly. This is shown again in Figure 3-28. The network has three resistors and two voltage sources.

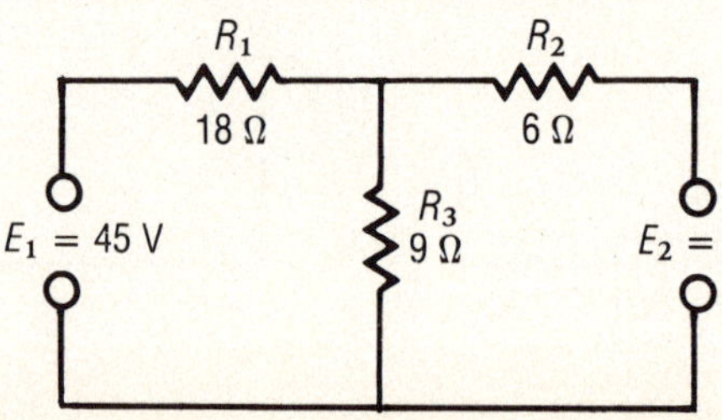

Figure 3-28. Circuit with three resistors and two voltage sources to be analyzed with Millman's theorem.

To apply Millman's theorem, reconfigure the circuit as a ladder of parallel current sources. This makes it easier to solve the network because, in a parallel circuit of this type, currents can be added simply. The contrived circuit in which this is done is shown in Figure 3-29. In this illustration, each voltage-source branch has been converted to its Norton equivalent. Note that resistor R_3 is included in this composite, since it is in parallel with the remainder of the circuit.

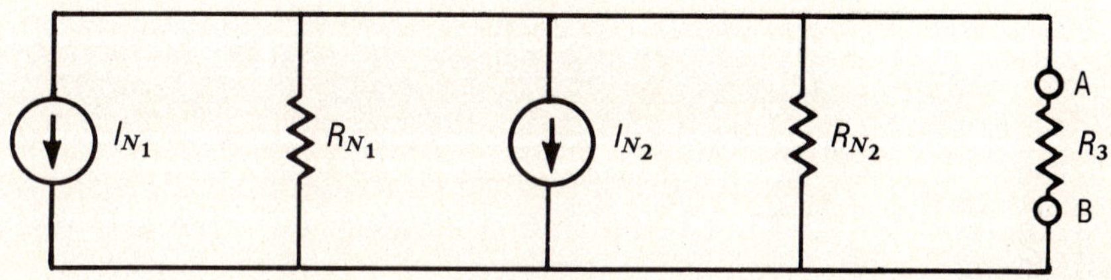

Figure 3-29. Circuit reconfigured as a ladder of parallel current sources and resistances.

In solving this circuit, convert the left half of Figure 3-28 to its Norton equivalent. The value of I_{N_1} is determined first, using Norton's theorem. Assume R_3 is shorted and solve for I_{N_1}:

$$I_{N_1} = \frac{E_{Th}}{R_{Th}} = \frac{E_1}{R_1} = \frac{45\ \text{V}}{18\ \Omega} = 2.5\ \text{A}$$

I_{N_1} is a current of 2.5 A. Next, make the same computation for I_{N_2}:

$$I_{N_2} = \frac{E_2}{R_2} = \frac{39\ \text{V}}{6\ \Omega} = 6.5\ \text{A}$$

The current through R_2 is 6.5 A. With these figures, simple algebraic addition provides the total current for the circuit:

$$I_{N_T} = I_{N_1} + I_{N_2} = 2.5\ \text{A} + 6.5\ \text{A} = 9\ \text{A}$$

Total current is 9 A. This 9 A is the total current through the load, R_3, and through the other resistances of the network.

Next determine the Norton equivalent resistances R_{N_1} and R_{N_2}. Since the Norton equivalent resistance is the resistance looking back into the network, the Norton equivalent resistance for the left half of Figure 3-28 is:

$$R_{N_1} = 18\ \Omega$$

Similarly, for the right half of Figure 3-28, the Norton equivalent resistance is:

$$R_{N_2} = 6\ \Omega$$

The resulting Norton equivalent circuit is shown in Figure 3-30. The Norton equivalent currents and resistances are given in the drawing.

Figure 3-30. Each voltage-source branch converted to its Norton equivalent in applying Millman's theorem.

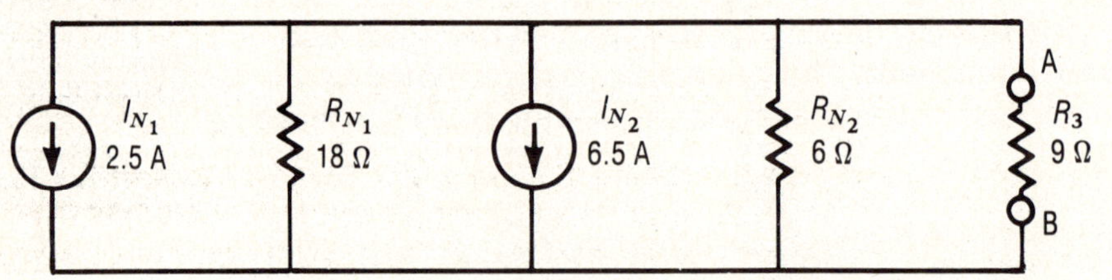

It is now necessary to determine what portion of the 9 A total current flows through R_3 and what portion flows elsewhere. To do this, the network is redrawn as shown in Figure 3-31A. Since R_{N_1} and R_{N_2} are in parallel, determine

Figure 3-31. Circuit redrawn to determine current flow through R_3.

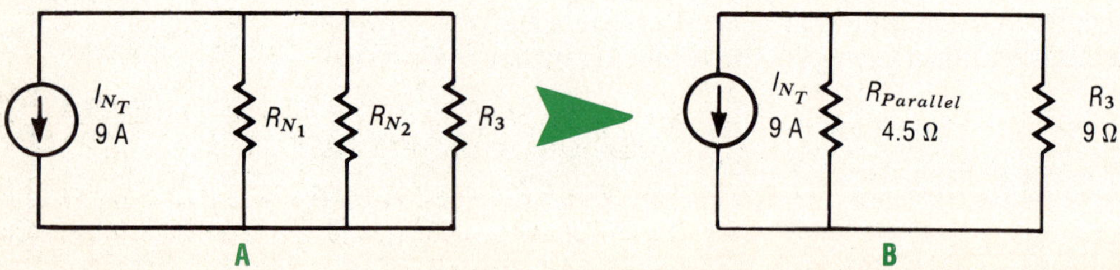

their equivalent resistance. This can be done by using reciprocals. In doing this, the conductances are in siemens (S). Convert the resistances of R_{N_1} and R_{N_2} to conductances with the following calculations:

$$G_1 = \frac{1}{R_{N_1}} = \frac{1}{18} = 0.056\ \text{S}$$

$$G_2 = \frac{1}{R_{N_2}} = \frac{1}{6} = 0.167\ \text{S}$$

Then add the two conductances:

$$G_1 + G_2 = 0.056\ \text{S} + 0.167\ \text{S} = 0.223\ \text{S}$$

Take the reciprocal to find their combined resistance:

$$R_{\text{parallel}} = \frac{1}{0.223} = 4.5\ \Omega$$

Figure 3-31A then becomes Figure 3-31B, and the 9 A current is divided as follows:

$$I_3 = 9\ \text{A} \times \frac{4.5}{4.5 + 9} = 9 \times \frac{1}{3} = 3\ \text{A}$$

The current through R_3 is 3 A. The remainder of the current (9 A − 3 A = 6 A) flows through R_{N_1} and R_{N_2}.

Again, the results are the same. These network theorems make it possible to analyze network parameters in several ways.

Discussion Questions/Problems

1. To what types of circuits would Millman's theorem be applied?
2. In applying Millman's theorem, what is the conversion for each current-source branch in a network?
3. How do you reconfigure a network to apply Millman's theorem?
4. Using Figure 3-28: Given that R_1 is 9 Ω, R_2 is 3 Ω, and R_3 is 3 Ω, to what would the current through each branch of the circuit be equivalent?

REVIEW

CHAPTERS 1, 2, AND 3

SUMMARY

1. Electricity is a form of energy that can produce heat, light, or force that causes motion. (1.1)

2. A magnet is a type of metal or metal ore that attracts certain other metals. When a magnet is in motion near a coil of wire with good electrical properties, electricity is generated. (1.1)

3. Chemical action is the principle applied to generate electricity in batteries. A battery contains two dissimilar solid materials, each of which has different electrical properties. (1.1)

4. A crystal is a material with an extremely fine structure. Pressure, such as twisting or applying force, to a crystal causes electrical current. This characteristic of crystal is called *piezoelectricity*. (1.1)

5. In some materials, electricity is generated by intense light. Materials that produce electricity as a result of sunlight are called *photoelectric*. (1.1)

6. Electricity can be generated from heat by using two dissimilar metals of different electrical properties. Wires, or strips, of these metals are joined together mechanically. (1.1)

7. *Static* means still. In static electricity, there is a lack of movement. Electricity exists, but it is not working. (1.1)

8. A charge is a type of electrical energy. There are two basic types of charges in electricity: positive and negative. Unlike charges attract each other; like charges repel each other. (1.2)

9. Electrons are the portions of atoms that are continually in movement. They have negative charges. Protons, particles within the nucleus of atoms, have positive charges. (1.2)

10. *Current* can be defined as the flow or movement of electrons from one point to another through materials known as *conductors*. Electron flow is from negative to positive. The basic unit of measurement for current is the ampere, abbreviated A. (1.3)

11. The unit of measurement for electrical charge is the coulomb, abbreviated C. (1.3)

12. To cause electron movement, a basic force has to be applied to a charge. This is known as *electromotive force*, or *emf*. (1.4)

13. The property of resisting the flow of electrons is known as *resistance*, abbreviated *R*. (1.5)

14. In evaluating conductor materials, resistance to electron flow is one property considered. Another important value lies in the ease, or facility, permitted for the flow of the electrons. This property, which is the inverse of resistance, is conductance, abbreviated *G*. (1.5)

15. Materials that conduct current easily are called *conductors*. (1.6)

16. Materials that offer some resistance to electrical current are called *semiconductors*. (1.6)

17. Materials which offer strong resistance to the flow of current—or which block or impede electron flow almost entirely—are called *insulators*. (1.6)

18. A current flow that is a constant value and that flows in only one direction is called a *direct current*. (1.7)

19. Alternating current is a type of current that changes both magnitude and direction. (1.7)

20. A circuit is a complete, controlled path for electrical current. Every circuit needs four basic parts: control device, power source, load, and conductive path. (1.8)

21. Electrical circuits are described in terms of their three basic properties: current (*I*), voltage (*E*), and resistance (*R*). Ohm's law describes the relationships among current, voltage, and resistance. (1.9)

22. Power is the relationship between the ability to do work and the time required to complete it. (1.9)

23. A series circuit is one in which electrons move in a continuous path along a conductor that carries current in sequence through the devices within the circuit. Resistance in series circuits is cumulative. The total resistance is equal to the sum of all individual resistances. In a series circuit, current is the same through each component. However, voltage levels are not the same for all resistors in series circuits. Rather, voltage is distributed in proportion to the resistances of individual resistors. (2.1, 2.2, 2.3, 2.4, and 2.5)

24. A parallel circuit is one in which two or more components are connected across one voltage source. In parallel circuits, the reciprocal of total resistance is equal to the sum of the reciprocals of individual resistances. In parallel circuits, applied voltage is the same for all components. Total current in a parallel circuit is equal to the sum of the currents in all branch circuits. (2.7, 2.8, 2.9, and 2.10)

25. The most common techniques to analyze electrical and electronic networks use network theorems. (3.1)

26. Kirchhoff's law for current states that the amount of current delivered to any point (or node) in a circuit will be equal to the current leaving that point (or node). (3.2)

27. Kirchhoff's law for voltage can be stated in two separate ways:

    1. *The algebraic sum of voltage drops and voltage rises around any closed series loop (or path) is equal to zero.*
    2. *The sum of voltage drops around any closed series path (or loop) is equal to the sum of voltage rises.* (3.3)

28. The superposition theorem holds that the effect of two or more sources of current or voltage for any part of a circuit is equal to the algebraic sum of the effect from each source, acting separately. The superposition theorem is

used principally to determine current for circuits or circuit loops which have two or more supply voltages that are known. (3.4)

29. Thevenin's theorem states that the current in any load resistance, R_L, connected to two terminals of a network consisting of any number of resistors and/or voltage sources, is the same as though R_L were connected to a single source whose open-circuited voltage is the open-circuit voltage of the two terminals and whose resistance is the equivalent resistance looking into the two terminals. (3.5)

30. Norton's theorem says that the current in any load resistance, R_L, connected to two terminals of a network consisting of any number of resistors and/or current sources, is the same as though R_L were connected to a constant current source whose current is equal to the current that flows through the two terminals when they are short-circuited. (3.6)

31. Millman's theorems are appropriate for use with networks that have many sources. In such situations, applying either the Thevenin or Norton techniques can become tedious. (3.8)

PROBLEMS

1. Write out the number that is represented by the notation 2^{14}. (1.3)
2. What is the current, in amperes, if a 25 C charge flows in 5 seconds? (1.3)
3. Draw the electrical symbol for a cell. (1.4)
4. What voltage is present when 100 J of work is applied to move a 25 C charge? (1.4)
5. If a conductor has a resistance of 50 Ω, what is its conductance? (1.5)
6. Draw a basic circuit having a light bulb, cell, wire, and switch as its parts. (1.8)
7. If a circuit has 1 A of current and 5 Ω of resistance, what is the voltage? (1.9)

8. If a circuit has a supply of 25 V and a current of 5 A, what is the resistance? (1.9)

9. If a given circuit has 2 A of current and 10 Ω of resistance, what is the voltage? (1.9)

10. What is the power dissipated in a circuit with ratings of 120 V and 2 A? (1.9)

11. If a circuit has a supply of 50 V and a resistance of 25 Ω, what is the current? (1.9)

12. Using the circuit in Figure R1-1, find R_T, I_T, E_{R_1}, E_{R_2}, E_{R_3}, and E_{R_4}. (2.2, 2.3, and 2.4)

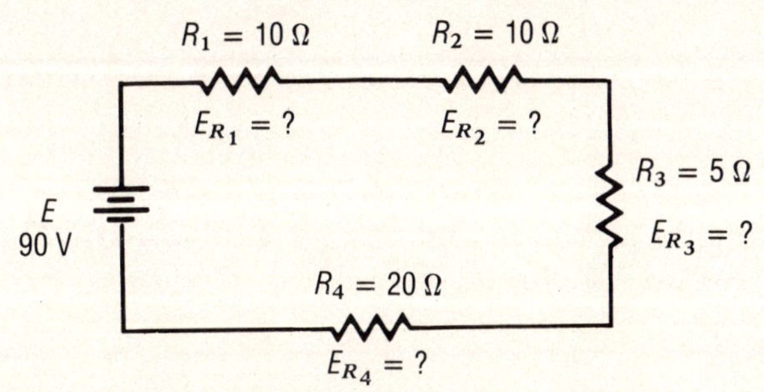

Figure R1-1. Series circuit with four resistors to use in solving Problem 12.

13. Using the circuit in Figure R1-2, find R_T, E, I_{R_1}, I_{R_2}, and I_{R_3}. (2.8, 2.9, and 2.10)

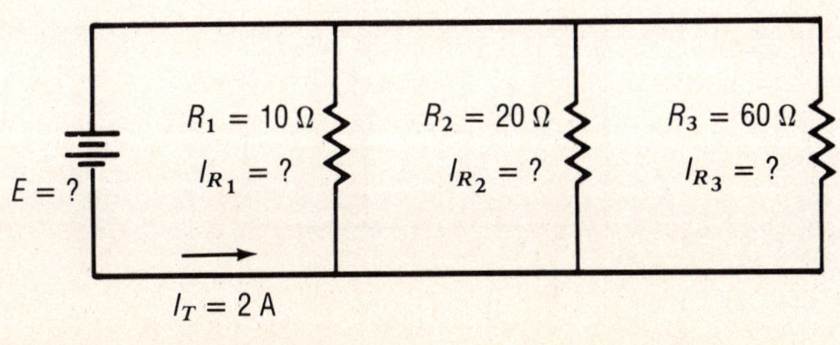

Figure R1-2. Parallel circuit with three resistors for Problem 13.

14. Using the circuit in Figure R1-3, find I_T, R_T, I_{R_1}, I_{R_2}, I_{R_3}, I_{R_4}, E_{R_1}, E_{R_2}, E_{R_3}, and E_{R_4}. (2.14, 2.15, and 2.16)

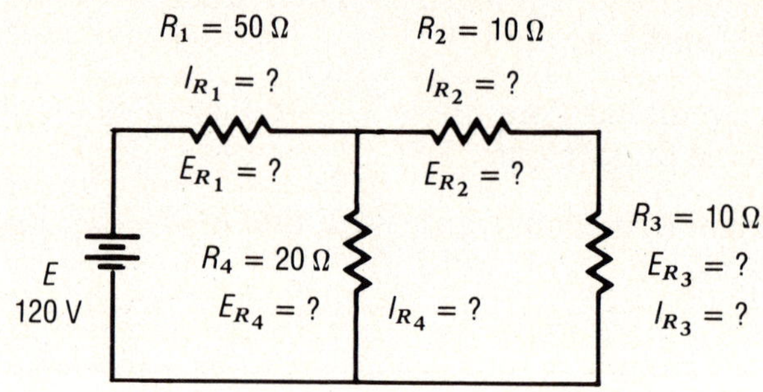

Figure R1-3. Series-parallel circuit for Problem 14.

15. Using the circuit in Figure R1-4, find R_{Th} and E_{Th} using Thevenin's theorem. (3.5)

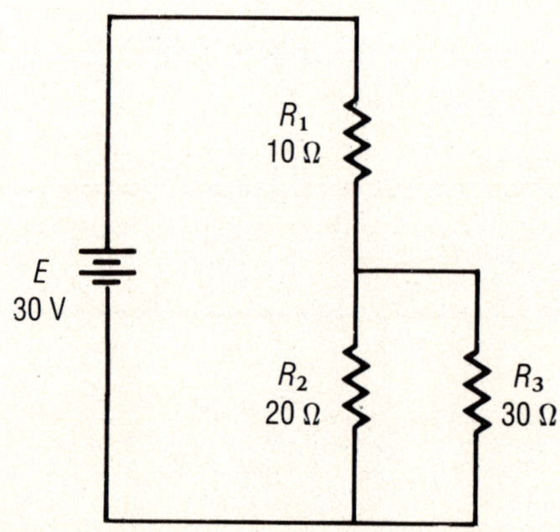

Figure R1-4. Circuit for Problem 15.

16. Using the circuit in Figure R1-5, find the Norton R_N and I_N. (3.6)

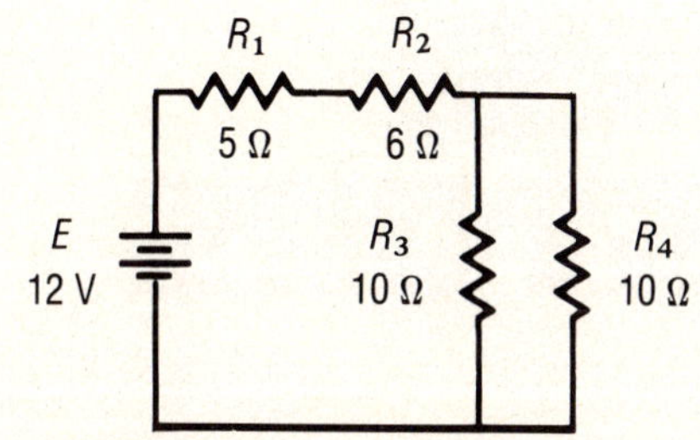

Figure R1-5. Circuit for Problem 16.

17. Using Kirchhoff's law for current, draw the path of current in Figure R1-6. (3.2)

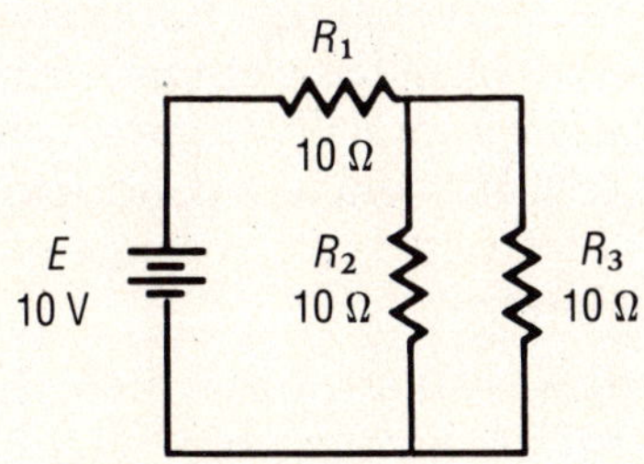

Figure R1-6. Circuit for Problem 17.

(Answers to the odd-numbered problems can be found at the back of the book.)

CONDUCTORS, INSULATORS, AND RESISTORS

4.1 CONDUCTORS

Any carrier of electricity is called a *conductor*. The term is, however, usually applied to materials that are relatively good, or efficient, carriers of electricity.

Most conductors used in electrical systems or electronic devices are made of metal. There are a few exceptions. For example, carbon, though not a metal, is used in important applications. These include brushes that make contact in electric motors, electrodes in batteries, and also electrodes in high-intensity arc lamps.

In selecting metals as commercial conductors, some practical trade-offs must be considered. For example, silver is the best known conductor of all the elements. However, the cost of this mineral makes it impractical for many uses. Therefore, copper, although less efficient as a conductor, is more practical as wiring for electrical transmission and service or for appliances or machinery.

Another example is aluminum which, though not as efficient a conductor as copper, has weight advantages. Because of this weight advantage, aluminum is used for long-distance transmission lines.

The table in Figure 4-1 lists a number of commonly used conductors, along with their relative resistance ratings, more correctly known as their resistivities, and common applications.

Resistances of Conductors

As indicated in Figure 4-1, one of the factors that affects how materials conduct electricity is the conductivity of the materials

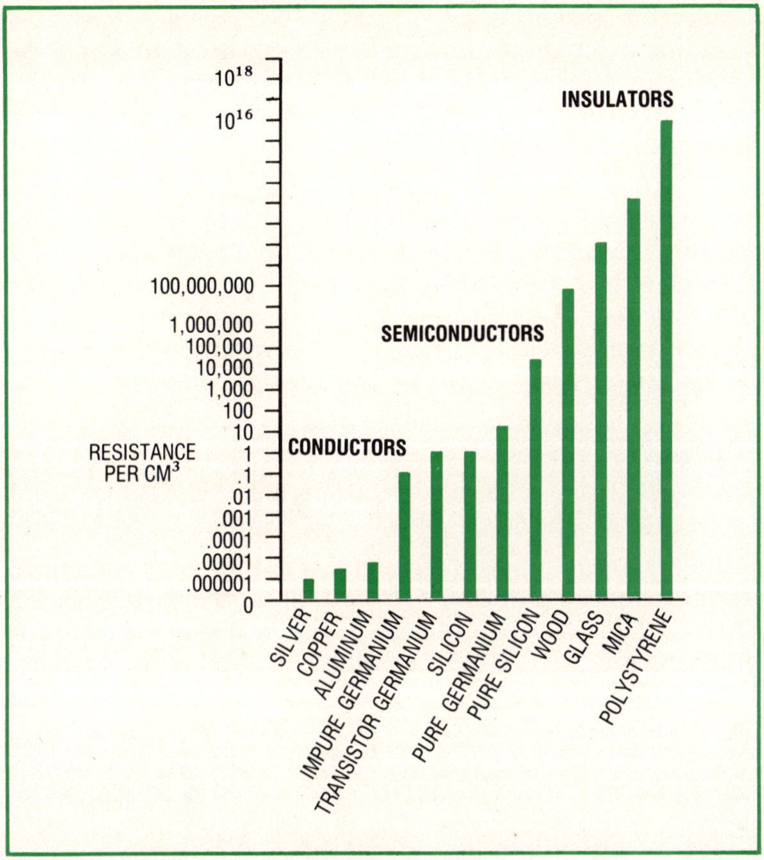

Figure 4-1. Conductivity of materials.

used. Different conductors have different resistivities and different levels of efficiency in conducting current. Other factors are discussed below.

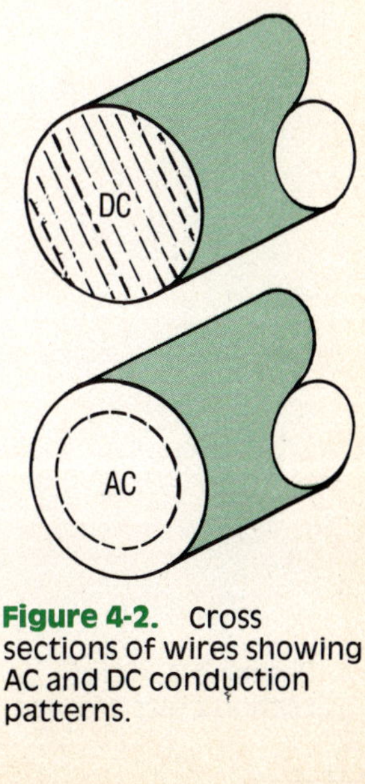

Figure 4-2. Cross sections of wires showing AC and DC conduction patterns.

Type of current. The type of current carried has an effect on the performance of conductors. This is because of the different ways in which AC and DC travel through conducting wires. DC moves through the entire cross section of the conductor. On the other hand, AC travels along the exterior of the conductor, penetrating into the conductor only slightly. These conduction patterns are illustrated in Figure 4-2. Because of these patterns, transmission of a large AC current usually requires wires with larger diameters than would be needed for the same current in DC.

Length of conductor. Another factor is the length of the conductor itself. Quite simply, the longer the wire, the higher the resistance to current flow.

Conductor cross section. The cross section is the area across a conducting material or wire. The greater the diameter of the conductor, the less resistance it will offer to the flow of electrical current. Conversely, then, the smaller the diameter, the greater the resistance.

A safety precaution: Using wires too small for an application can lead to overheating. Many fires are caused by overloading electrical wires. There have been many tragic results with window air conditioners that were installed in older homes. The smaller gauge wires used in these homes became overloaded. The heat generated started many fires.

The cross section area of any wire would be:

$$\frac{\pi d^2}{4}$$

Rather than using the factor $\pi/4$, it is more convenient to define a special unit, the circular mill, as a measure for the cross section of a wire. The cross section area of a wire in circular mills is the diameter in mills, squared. A mill is one-thousandth of an inch.

$$\text{Cross section area (circular mills)} = [\text{diameter (mills)}]^2$$

Figure 4-3 illustrates application of this formula.

Length-diameter relationship. In combination, the length and cross section of a conductor can affect resistance to

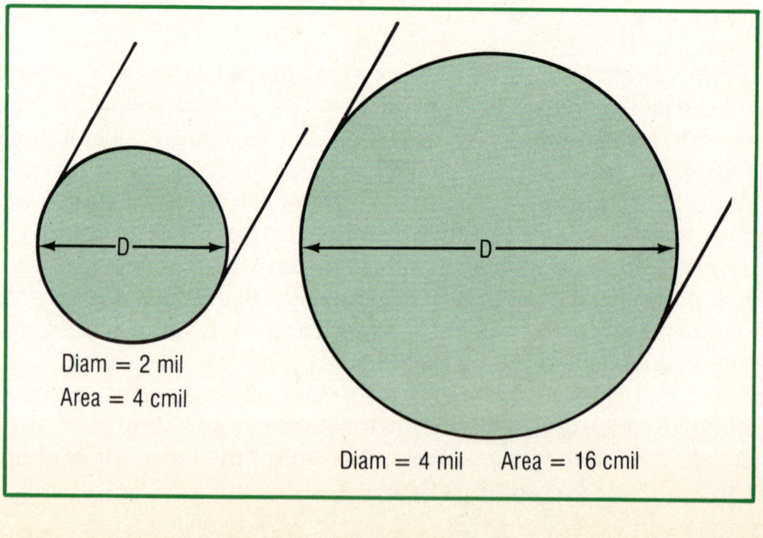

Figure 4-3. Cross section areas of wires.

current. The relationship of these factors is illustrated in Figure 4-4.

American Wire Gauge No.	Diameter In Inches	Area Circular Mils	Ohms at 68° Fahrenheit Per 1,000 Ft.
0000	0.460	211600.	0.04906
000	0.40964	167805.	0.06186
00	0.3648	133079.	0.07801
0	0.32486	105534.	0.09831
1	0.2893	83694.	0.12404
2	0.25763	66373.	0.1563
3	0.22942	52634.	0.19723
4	0.20431	41743.	0.24869
5	0.18194	33102.	0.31361
6	0.16202	26251.	0.39546
7	0.14428	20817.	0.49871
8	0.12849	16510.	0.6529
9	0.11443	13094.	0.7892
10	0.10189	10382.	0.8441
11	0.090742	8234.	1.254
12	0.080808	6530.	1.580
13	0.071961	5178.	1.995
14	0.064084	4107.	2.504
15	0.057068	3257.	3.172
16	0.05082	2583.	4.001
17	0.045257	2048.	5.04
18	0.040303	1624.	6.36
19	0.03589	1288.	8.25
20	0.031961	1021.	10.12
21	0.028462	810.	12.76
22	0.025347	642.	16.25
23	0.022571	509.	20.30
24	0.0201	404.	25.60
25	0.0179	326.	32.2
26	0.01594	254.	40.7
27	0.014195	201.	51.3
28	0.012641	159.8	64.8
29	0.011257	126.7	81.6
30	0.010025	100.5	103.
31	0.008928	79.7	130.
32	0.00795	63.	164.
33	0.00708	50.1	206.
34	0.006304	39.74	260.
35	0.005614	31.5	328.
36	0.005	25.	414.
37	0.004453	19.8	523.
38	0.003965	15.72	660.
39	0.003531	12.47	832.
40	0.003144	9.88	1,049.

Figure 4-4. Length, diameter, and current resistance of wires.

Effect of temperature. The resistance of a metallic conductor usually increases as the temperature is raised. The resistance at a higher temperature, R_2, will be equal to the resistance at a lower reference temperature, R_1, multiplied by a factor that includes the temperature coefficient of resistance, α, and the increase in temperature, ΔT.

$$R_2 = R_{20^\circ}(1 + \alpha_{20^\circ}\,\Delta T)$$

α_{20° = temperature coefficient of resistance at 20° C.
R_{20° = resistance at 20° C.

Types of Conductors

Many types of commercial conductors are made. Each has its own characteristics and applications. The construction of a number of common types of wires and conductors is illustrated in Figure 4-5. These conductors and their applications are discussed below.

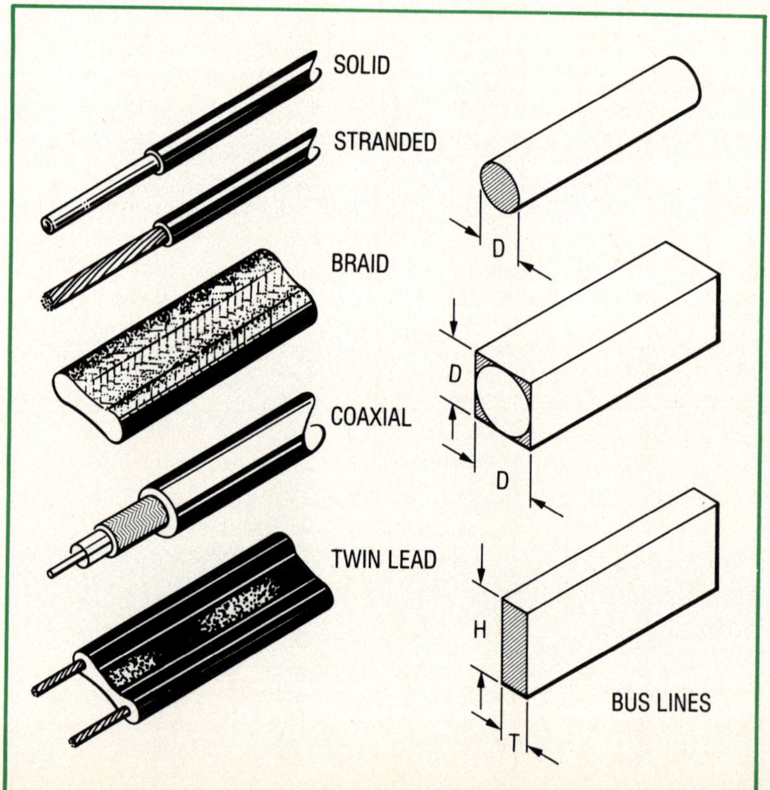

Figure 4-5. Construction of wires and conductors.

Solid wire. These wires consist of strips of solid metal, usually copper, covered by an insulation material. Solid strands of wire are used for service distribution in homes, offices, and factories. They also are used as parts of chassis for devices such as television sets.

Stranded wire. Stranded wires are groups of narrow wires packaged together. Stranded wire has an advantage over solid wire: flexibility. It is used for applications in which wire has to be bent frequently. An excellent example is the cord for the average electric lamp. On electronic instruments, stranded wire is typically used for connector cords. In addition, test leads usually use stranded wire.

Braided wire. Braided wire is constructed from a series of small wires braided like rope. Braiding serves a packaging function; it keeps strands of wire together. Braided wire can carry large currents and yet not require an insulation coating to hold it together. Braided wire is typically used in applications requiring high current-carrying capacities. Examples include grounding wires for television-set chassis, test equipment, or audio equipment.

Coaxial cable. Coaxial cable consists of a pair of conductors within a single package. A central, insulated wire is surrounded by a braided exterior set of wires which are insulated as a package. The central element of the coaxial cable carries the main electrical signal. The exterior braided wire serves as the ground, providing electrical and electromagnetic insulation for the interior wire. An application for coaxial cable is to carry television signals. This type of cable is used to connect television cameras with control consoles and to link video tape recorders with television sets that play their signals.

Twin-lead wires. A twin-lead wire has two separated strands of wire encased in insulation materials. This insulation casing avoids certain types of electrical interference in applications involving two or more currents in a single wire. The most common application is the antenna lead for a television set.

Bus lines. These are strips, or rods, of metal used to distribute current within electronic or electrical circuits or devices. Bus lines are usually square or rectangular, though they are sometimes round. The square or rectangular shape is used most often because it has the advantage of holding wires wrapped around the bus line with less slippage than round carriers. Bus lines provide common sources of electricity to which components of electronic devices can be wired. Bus lines that carry large currents are often made of heavy bars of copper.

Printed circuit boards. A flat insulating material, usually some form of plastic, serves as the base of a printed circuit board. To this base, a sheet of copper is bonded. Then, in manufacturing, parts of the copper are etched away, exposing the insulation material and leaving in place a series of conductive strips, or paths. Printed circuit boards are used as circuit networks delivering current to electronic components. A printed circuit board is illustrated in Figure 4-6.

Figure 4-6. A printed circuit board with circuits in place after etching. (Courtesy TRW, Inc.)

Wire Gauges

Gauge is a unit of measure for wire. This term describes the diameter of a wire conductor. For example, common references are to 14-gauge or 18-gauge wire.

A *wire gauge* is also an instrument for measuring the diameter, or gauge, of wires. A wire gauge is illustrated in Figure 4-7.

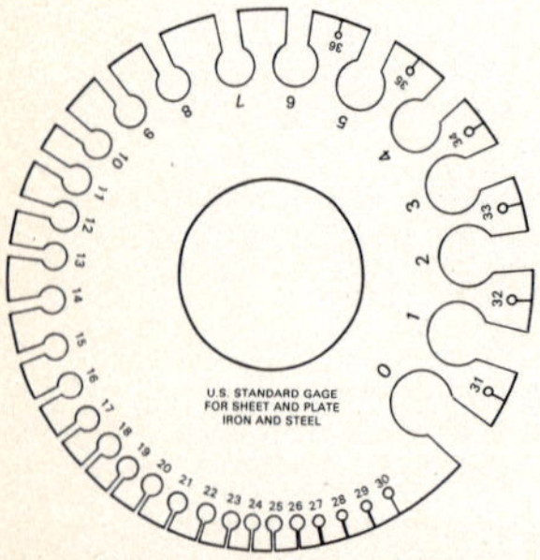

Figure 4-7. Wire gauge.

Standard sizes for wires range from 0000 through 40, according to the system of American Wire Gauge. Figure 4-4 is a table giving American Wire Gauges and their dimensions.

To use a wire gauge, wire is drawn through the slots around the circumference of the instrument. When the wire passes through a slot, touching both sides with only a slight drag, this is the correct gauge. Most wire gauges measure from 0 to 30 AWG.

Note that the wires with the higher gauge numbers are smaller in diameter. Thus, a number 14-gauge wire is larger than a number 18-gauge wire. A wire with a rating of 0000 has a diameter of 0.460 inches (in.) or 1.1684 centimetres (cm). A number 40 wire has a diameter of 0.003144 in. or 0.00798576 cm. For special purposes, wire is also fabricated in other sizes.

Discussion Questions/Problems

1. Why are most conductors made from metal?
2. Describe the trade-offs between conducting efficiency and cost in the selection of materials for commercial conductors?
3. What are the differences in conductive patterns between AC and DC?
4. How do length and cross section of the conductor relate in determining the resistance of an electrical carrier?
5. What are some relative advantages and disadvantages of solid and stranded wires?
6. What are typical applications for braided wires and what are the characteristics of this conductor that adapt it for those applications?
7. What are the main features of coaxial cable and how do these adapt the cable for its application as a television signal carrier?
8. What is the relationship between gauge numbers and wire diameter measurements?
9. What is an AWG and how is it used?

4.2 INSULATORS

Materials that obstruct, or prevent, the flow of current are called *insulators*. Some common types and their resistance ratings are listed in Figure 4-8.

MATERIAL	DIELECTRIC STRENGTH (Volts per Mil)
Air	31
Bakelite (paper base)	250–585
Glass (window)	760
Mica	2030–5080
Nylon	285–470
Polystyrene	300–710
Porcelain	200–400
Rubber (hard)	470

Figure 4-8. Resistances of insulating materials.

In addition to the materials covered in the table, a number of other materials or types of insulators are in common use.

Insulated coverings for wires. The majority of wires sold for use in electrical or electronic applications are coated with insulating materials. The insulation is what makes it possible to handle wires without getting a shock when they are carrying current. Insulation materials for wires are usually made from plastic or rubber. The most common plastic materials are polyvinyl chloride (PVC) and Teflon®. Many other specialized materials are used for individual applications.

Figure 4-9. Electrical tape.

Electrical tape. This is a plastic material cut into strips and packaged in rolls. The tape has an adhesive backing. It is used for insulation at points where two or more wires are connected. Electrical tape (shown in Figure 4-9) replaces insulation that has been stripped away from wires to form the connection.

Shrink tubing. This is a plastic material used to create a tight coating over electrical connectors or joints. Shrink tubing is made from plastic material that contracts when heated and retains its shrunken shape. A length of shrink tubing is slid over a wire to be connected to an electrical device. The connection is usually done by soldering, a method for joining wires with melted metal. After the solder joint has been completed, the shrink tubing is slipped over the connection point. The tubing material is then heated with a heat gun which is similar to a hair dryer, except that temperature levels are much higher. The heat shrinks the tubing, providing permanent, insulated protection for the connected joint. Figure 4-10 illustrates how shrink tubing is applied.

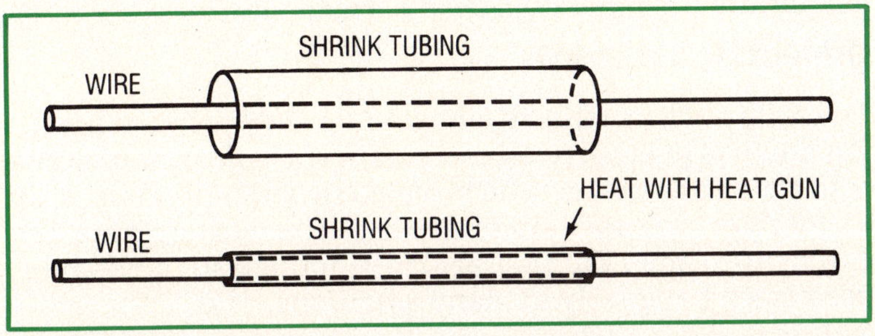

Figure 4-10. Application of shrink tubing.

Stand-off insulators. These are insulating fixtures that hold current-carrying wires away from surfaces that could ground the current. Stand-off insulators are used on poles or towers holding electrical transmission wires. Smaller stand-off insulators are used inside chassis of electrical and electronic equipment. Their function is to provide insulated separation between components. Stand-off insulators are frequently used to assure separation between circuit boards within electronic instruments. Several types of stand-off insulators are shown in Figure 4-11.

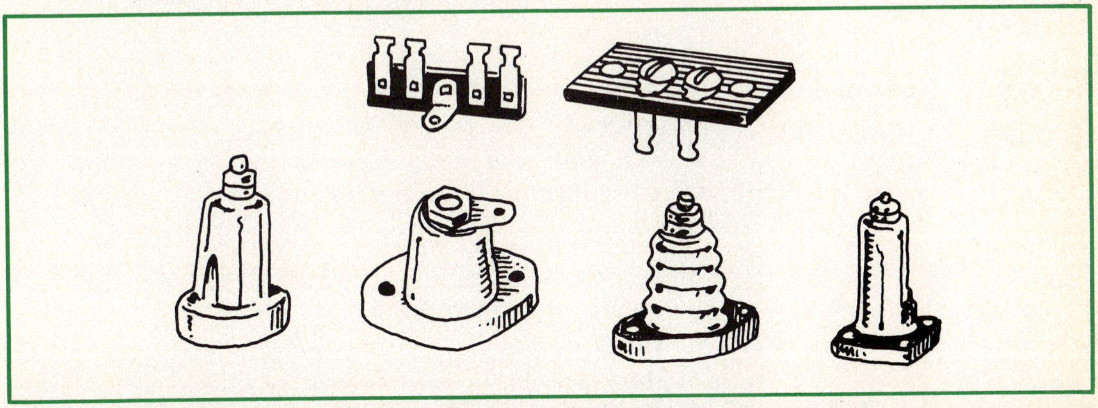

Figure 4-11. Stand-off insulators.

Circuit boards. The base materials for circuit boards are usually made from insulating plastics with high structural strength. Typical circuit boards are constructed from fiberglass and plastic resins.

Porcelain insulators. Porcelain is a ceramic material prepared from a clay-like substance fired in furnaces at high temperatures after formation. By their nature, porcelain materials can withstand high temperatures. Thus, porcelain insulators tend to be used for heavy loads or in heated areas. Many porcelain insulators are used for high-power electrical transmission. Another application is in toasters or ovens where surrounding temperatures are normally high.

Dielectric Strength

The ability of a material to withstand voltage without breaking down is referred to as its *dielectric strength*. You may encounter references in which dielectric strength is used as a synonym for insulation properties. The two terms are not exactly the same. But they are close enough so they can be used interchangeably in many situations.

Dielectric strength is stated in terms of volts per mil of thickness for an insulator. For example, hard rubber has a dielectric strength of 450. This means that it can withstand 450 V per mil of thickness without breakdown.

Dielectric strength is progressive. If you double the thickness of a dielectric material, you can also double the voltage that can be withstood. If a sheet of hard rubber that is 1 mil thick can withstand 450 V, a sheet of the same material that is 2 mils thick can withstand 900 V, and so on.

Discussion Questions/Problems

1. Which material among the following has the best insulation properties: copper, paper, ceramic, rubber, aluminum?
2. What are two of the most common insulation coverings used for electrical wires?
3. What is the main application for shrink tubing?
4. What are the main purposes and functions of stand-off insulators?
5. What are the insulation materials that provide the bases for printed circuit boards?
6. Porcelain has a dielectric strength of 150. How much voltage could a porcelain insulator that is two mils thick withstand?

4.3 RESISTORS

Materials that impede the flow of current predictably are called *resistors*. The properties of resistors lie between the extremes of conductors and insulators. Conductors, it might be said, allow electrons to flow. Insulators block the flow of current. Resistors permit current to flow, but with some difficulty. The degree of difficulty, or the amount of resistance, is predictable and can be measured. This predictable performance becomes critical to the design and construction of electrical circuits.

Resistance is measured in ohms. The symbol for ohms is the Greek letter, *omega* (Ω). As explained in Chapter 1, 1 Ω of resistance is present in a conductor when 1 A of current flows under 1 V of potential.

Figure 4-12 shows a resistor and also repeats the symbol for a resistor: a zig-zag line.

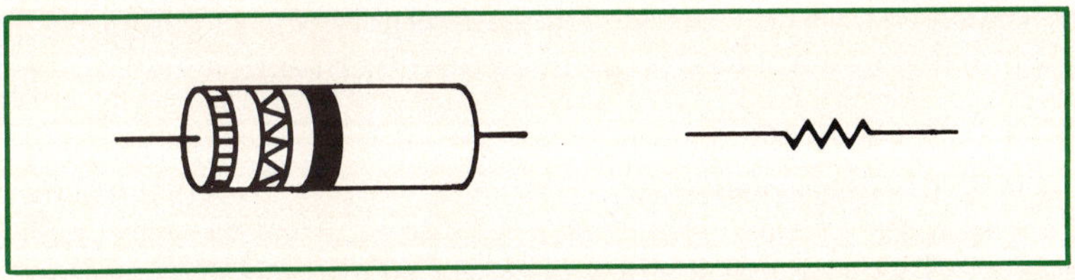

Figure 4-12. Resistor and symbol.

An important property of resistors, in terms of their usefulness in circuits, lies in their power ratings. The power rating is the ability of a resistor to handle electrical current through it at the voltage across it. Power for resistors is measured in watts.

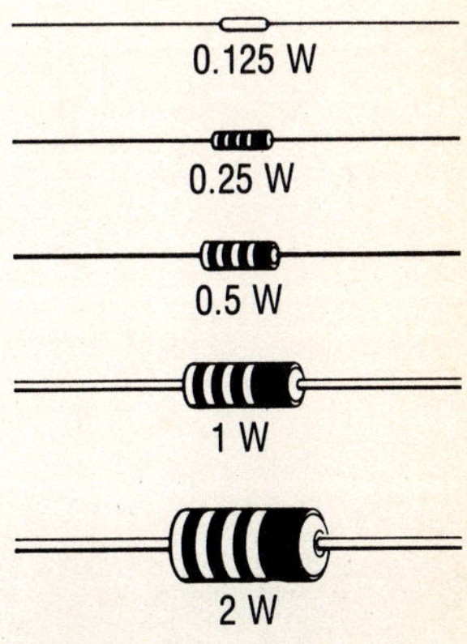

Figure 4-13. Resistor sizes and power ratings. (Courtesy Sprague Electric)

Power ratings are important because of the way resistive conductors behave when current flows through them. Resistive materials impede the flow of electrons. Because of this restriction on the flow of current, the electrons that do flow must force their way down the conductor. In the process, electrons rub against each other. Friction results. This friction gives off heat.

If a resistance wire—or any wire, for that matter—is overloaded with current, it burns out. If resistors are overloaded —subjected to currents greater than their rated capacity—they can burn out or crack.

Power ratings for a particular type of resistor are directly proportionate to their size. Figure 4-13 demonstrates this. This figure contains drawings of five resistors, with power ratings from 0.125 W to 2 W. The illustrated resistors are the most commonly used. However, resistors do come in a much wider range of power ratings. At higher power ratings, resistors are sometimes built with aluminum fins that dissipate the heat, as shown in Figure 4-14.

Figure 4-14. Resistor with aluminum fins.

Types of Resistors

Resistors can also be described in terms of the way they are constructed, or fabricated. The common types, or categories, of resistors based upon construction techniques are:

- Wire wound
- Carbon composition
- Film.

Wire wound resistors. A wire wound resistor is made from resistance wire wrapped around an insulating core. The wire is of a resistive material. The core is either ceramic or plastic. The amount and type of wire used depend upon both the power rating of the resistor and the resistance desired.

The wound core is encapsulated, or packaged, in a round, tubular case. The case materials are also plastic or ceramic. Within the case, the wire wound core is further enclosed, or *potted,* with a plastic material that provides both insulation and protection. Metal caps are placed over the ends of the tubular case. Connecting wires are brought through holes in the caps.

Wire wound resistors are frequently referred to as *power resistors* because they are usually used for high-power applications.

Special wire wound resistors are used in applications where *precision* is needed in a circuit. These are precision resistors because the length and gauge of resistive wires can be measured more exactly than other materials used in resistors.

Carbon composition. The resistance elements of carbon resistors are nonmetallic. This is one of the major types of resistors in electricity and electronics where nonmetallic materials are current carriers.

The resistive elements of carbon-composition resistors are made by compressing granules of carbon or graphite. These granules are suspended in insulating materials such as plastic or ceramic. The resistive cores are round or oblong in shape. At the ends of the cores are metal caps that carry current to and from the resistive elements. Wires are connected to the metal caps.

Carbon-composition resistors are packaged in plastic tubes, as shown in Figure 4-15. Note the circular bands on the outside of the resistor package. These bands are applied in different colors. The colors represent the resistance values of the resistors. Color coding of resistors is discussed later in this chapter. Note that color coding is used primarily for carbon-composition resistors. Other types of resistors generally have their value printed on the outside of their cases.

Figure 4-15. Carbon-composition resistor.

Carbon-composition units are the least expensive and most widely used resistors. They are made in the most popular sizes and are used by the millions. Carbon-composition resistors are generally fabricated in standard resistances and a range of power handling capability from 0.25 W to 2 W.

Resistance values for carbon-composition units vary over an extremely broad range. Standard units can be purchased with resistance values in fractions of an ohm up to as high as 22 million Ω.

Film resistors. Film resistors are formed by depositing films of vaporized material on different carriers. These carriers are then wrapped to form resistors. In general, the basic types of film resistors are:

- Carbon-film resistors
- Metal-film resistors
- Cermet-film resistors.

All three of these types of resistors are fabricated through similar techniques. Granules of resistive materials are compressed and suspended in insulators such as plastic or ceramic. The resistive cores are then wrapped in a film. The film makes possible improved control or precision. In part, this is because the film provides additional control and direction for the flow of current. Film-wrapped resistors tend to have lower levels of electrical *noise* than ordinary carbon-composition resistors.

Carbon-film resistors are made with carbon-composition cores that are film wrapped.

Metal-film resistors use a film that is made from a metallic material. This metallic film is wrapped around a glass core, or *substrate*. Metal-film resistors, because of the properties of their construction, have extremely low electrical noise levels.

Cermet-film resistors use a core of cermet material. Cermet is a combination of ceramic and metal. Granules of ceramic and metal materials are compressed, then fired to their melting point. The fused materials become the cores for the resistors. These cores are then coated with deposits of carbon. This type of resistor is highly stable. Performance is extremely reliable over wide temperature ranges. These resistors tend to be used for applications in which performance and quality are critical.

Variable Resistors

Sometimes, electrical circuits and the devices they operate require the ability to vary the current delivered. This need for variation cannot always be predicted. Sometimes the need to vary current is a basic part of the application itself.

One common example is the need to vary the speed of electrical motors. Consider the case of a photographic print dryer. Wet photographs are pressed against the surface of a rotating, heated drum. In one full revolution of the drum, the prints are dried and dropped into a basket. Usually, when operation of a dryer starts, the drum is rotated at slow speeds. As it becomes apparent that the prints are drying before they complete the cycle, the speed of the drum can be increased. A variable resistor is used to do this. By adjusting the variable resistor, the amount of current fed to the motor is increased. The rotation rate of the motor, and of the drum it drives, increases.

Variable resistors are often called *rheostats*. The symbol used to represent variable resistors is shown in Figure 4-16. The variation comes from moving one of the contacts along the resistive element within the circuit. When current travels a shorter distance through the resistive element, the amount of current delivered increases. As increased resistive area is added to the circuit, the amount of current drops.

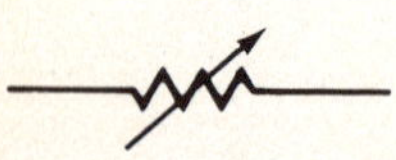

Figure 4-16. Symbol for variable resistor.

Variable resistors are made with both wire wound and carbon-composition elements. Wire wound resistors are usually used for higher power applications. With either type, there must be an adjustable control that makes contact at different points along the resistive core. Sometimes wire wound resistors use circular elements with a knob that can be rotated to make the contact. This is shown in Figure 4-17.

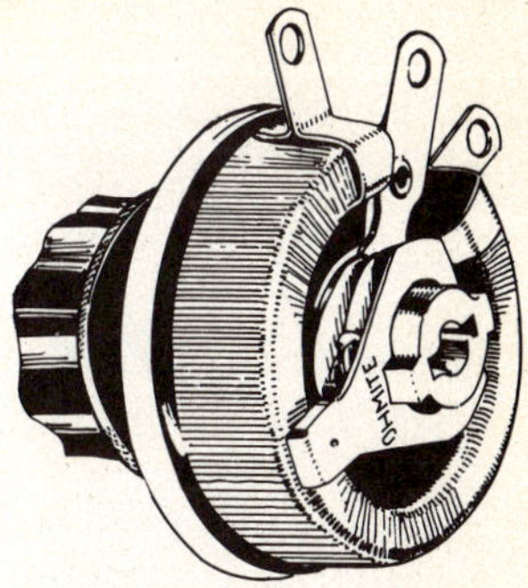

Figure 4-17. Adjustable control with rotating element for wire wound resistor.

Alternatively, there can be a slide-type control, like the one illustrated in Figure 4-18.

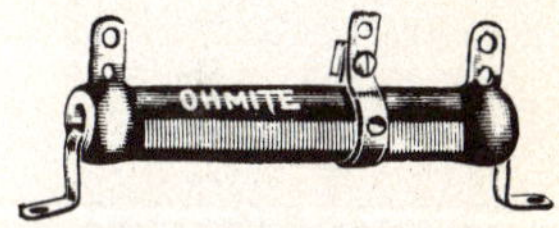

Figure 4-18. Resistor slide-type control.

Variable control resistors. The most common designs of variable control resistors are *log* and *linear*. Both of these types of controls are referred to as *tapered* controls. Tapered means that the output of a circuit through the resistor varies (tapers) in a specific manner.

A log taper, or log control, means that output varies in a logarithmic relationship with the position of the control device, as shown in Figure 4-19. By contrast, a linear control varies uniformly through the entire range of adjustments, as represented by the graph in Figure 4-20.

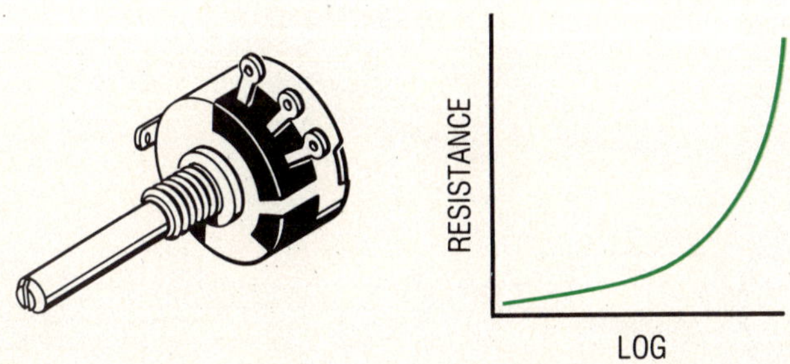

Figure 4-19. Log tapered variable control resistor and graph showing relationship of control positions to output variances

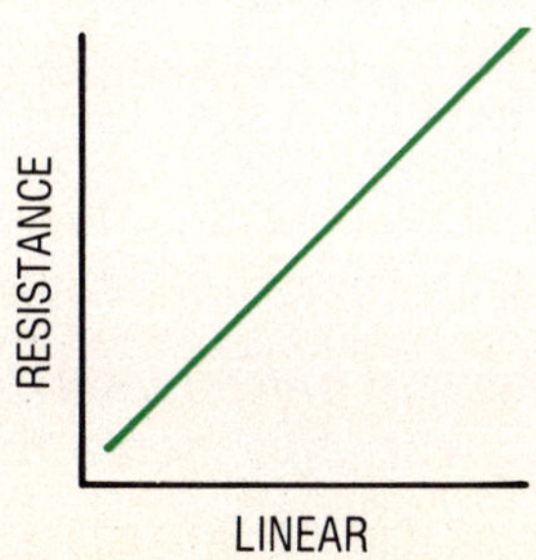

Figure 4-20. Relationship of linear control postions to output adjustments.

It can be important to know whether you are using a log-type or linear-type control. With a linear control, you know you are getting half the resistance if the control is set at a halfway point. With a log control, the percentage or degree of output variance is not proportional to the percentage of rotation.

Potentiometers

A *potentiometer* is a device that uses variable resistance to change or adjust the voltage within a circuit. The principle is similar to that of a variable resistor except that potentiometers divide voltage rather than limit current. The potentiometer makes the circuit adjustable. The symbol for a potentiometer is shown in Figure 4-21. What makes the potentiometer different from the variable resistor is that the output connection is varied along the resistor by the third lead connector. This connector, which moves along the resistor, is called a *wiper*. The name comes from its function: The wiper connection actually moves across the face of the circular-shaped resistor. The control is through a rotating knob. Figure 4-22 is a cutaway drawing showing the functioning of a typical potentiometer.

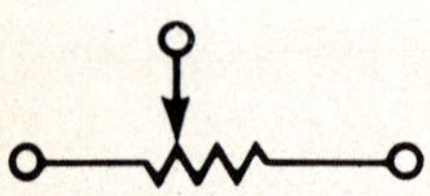

Figure 4-21. Symbol for potentiometer.

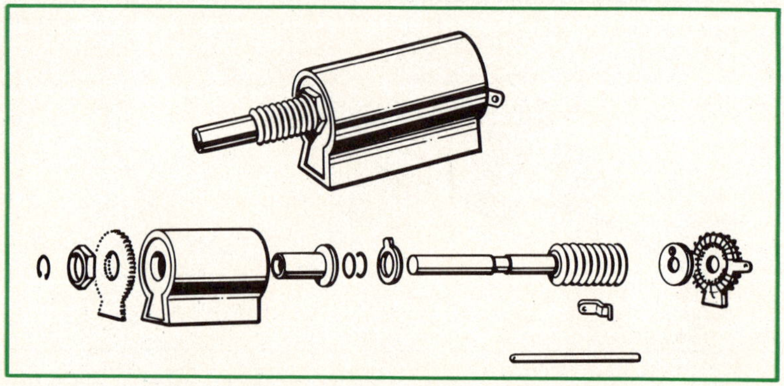

Figure 4-22. Cutaway of potentiometer.

Trim potentiometers. A special form of low-power potentiometer is encountered frequently in electronic circuits. You find them in radios, television sets, computers, and a wide range of electronic instruments. The term *trim potentiometer* describes a small unit that is designed to be soldered onto a circular control board.

Trim potentiometers, or *trim pots*, are used in applications that do not require frequent adjustments. Rather they are used for settings that will be unchanged for some time. Therefore, trim potentiometers, like that in Figure 4-23, are usually set with screwdrivers rather than with adjustable knobs.

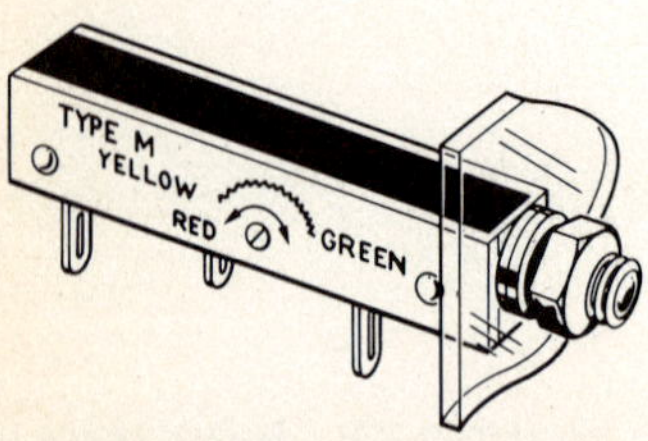

Figure 4-23. Trim potentiometer.

Color Coding for Resistors

Because it is vital that resistors be matched to specific resistance requirements in a circuit, a clear method is needed for marking and identifying these devices. For larger units, especially the high-power wire wound resistors, it is possible simply to print identification information right on the cases. However, low-wattage carbon-compostion resistors are so small that there would be no room to print all of the information needed. This is particularly true when resistors get down to ratings of 0.25 W or 0.5 W.

In addition, it would be impossible to print a mark on a resistor where it could be seen when the resistor was wired into a circuit or placed on a board. At least some of the printing would be covered up when the resistor was in use.

To overcome these problems, the electrical industry standardized with a system of color bands that provide visible identification of resistance values. These color bands comprise what is known as the *resistor color code*.

Every resistor is identified by at least three color bands. Many resistors have a fourth band; some have a fifth band. The location, color codings, and meanings of these bands are given in Figure 4-24.

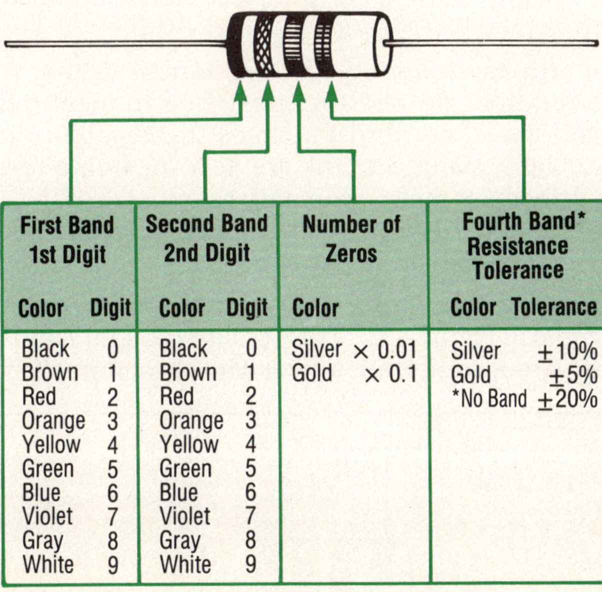

First Band 1st Digit		Second Band 2nd Digit		Number of Zeros	Fourth Band* Resistance Tolerance	
Color	Digit	Color	Digit	Color	Color	Tolerance
Black	0	Black	0	Silver × 0.01	Silver	±10%
Brown	1	Brown	1	Gold × 0.1	Gold	±5%
Red	2	Red	2		*No Band	±20%
Orange	3	Orange	3			
Yellow	4	Yellow	4			
Green	5	Green	5			
Blue	6	Blue	6			
Violet	7	Violet	7			
Gray	8	Gray	8			
White	9	White	9			

Figure 4-24. Color coding for resistors.

The color coding on the bands must be read in a specific order. *Always start reading from the band that is closest to an end of the resistor.* The first band may be on the left, as in Figure 4-24. However, depending upon how a resistor is placed in a circuit, you may find the starting band on your right. Note that the fifth band is always set in slightly from the edge. Thus, if you start right at the edge and read in you will get a proper reading.

The first two bands in a color code represent a two-digit number. The same set of colors is used to represent values from 0 through 9 in these two bands.

The third band is a multiplier. This indicates either the number of zeros that should be added after the first two digits or a fractional factor that should be used to reduce the value of the first two digits.

The fourth band (or the absence of this band) indicates the tolerance of the resistor. Tolerance is the degree to which a resistor is permitted to vary from its value. For example, assume a resistor has a value of 100 Ω. If the resistor has a silver band in the fourth position, this will indicate that the tolerance is ±10 percent. This means that, if you use a 100 Ω resistor, it may not be exactly 100 Ω; the actual resistance in your circuit will range anywhere from 90 Ω to 110 Ω. If there is a gold band, the tolerance will be ±5 percent. If the fourth band is missing entirely, the resistance can be assumed to be ±20 percent.

The fifth band is used only on some resistors. A fifth band is required on resistors designed to meet military specifications. This band indicates the reliability of the resistor. Many manufacturers are now including this fifth band routinely on their top-quality resistors. Reliability is a measurement of the expected failure rate of these resistors per each one thousand hours of use.

To illustrate how resistor color coding works, consider the example in Figure 4-25. This resistor has four bands. The first one is red, indicating a two. The second one is black,

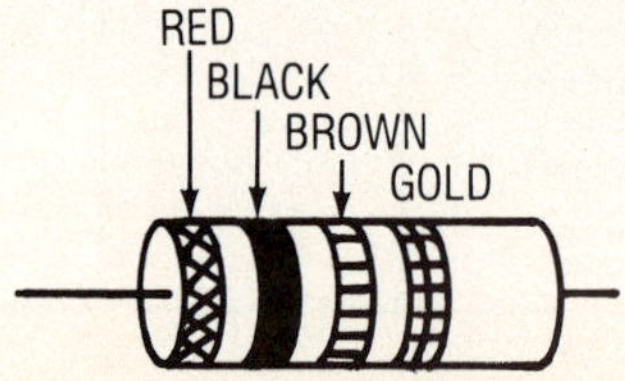

Figure 4-25. Color-coded resistor.

representing a zero. The third band is brown, indicating a multiplier of ten. The resistance is 20 × 10 or 200 Ω. The final band is gold, meaning that the tolerance is ±5 percent. The tolerance means that actual resistance of this device may be anywhere between 190 Ω and 210 Ω.

Now look at the color-coded resistor in Figure 4-26. The first band is yellow, indicating a value of four. The second

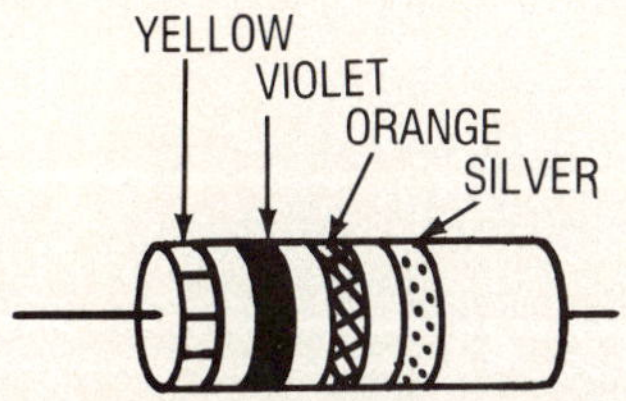

Figure 4-26. Color-coded resistor.

band, violet, stands for a seven. The third band is orange, giving it a value of three zeros. Thus, the resistor is rated at 47,000 Ω. This can also be written as 47K Ω. The fourth band is silver, indicating a tolerance of ±10 percent. This tolerance figure is 4,700 Ω. Thus, actual resistance for this unit can be anywhere from 42,300 Ω to 51,700 Ω. Note that there is a fifth band (brown) on this resistor. This means that up to 1 percent of these resistors may fail per each 1,000 hours of use.

Resistor-Related Problems

Eventually, the understanding you are building about circuits will be put to work in diagnosing and troubleshooting electrical or electronic devices. You should be aware that, even though it happens rarely, electrical or electronic problems may be due to resistor problems. Because resistor failures happen rarely in a properly designed circuit, there can be a tendency to forget about the possibility of resistor failure in troubleshooting a circuit.

However, resistors can crack, burn out, or change values. They can do this in response to power overloads, to vibration, or to environmental temperature levels.

When a resistor fails, the signs are usually apparent. Therefore, it is always a good practice, when you are troubleshooting, to look at the resistors. Look for cracks. Also check for discolored or deformed cases. A burned-out resistor may have bubbles in the case.

On checking your resistors, look for anything unusual. As mentioned, resistor failures are infrequent, and there is a tendency to overlook them as a possible source of trouble. After you have done trouble-shooting for a while, you will realize that it pays to check everything.

Discussion Questions/Problems

1. What are the bases used for rating resistors in terms of both power and resistance?
2. What are the three basic types of resistors?
3. What are the basic differences between rheostats and potentiometers?
4. What are the differences between log and linear tapers in variable resistors?
5. What is the primary application area for trim potentiometers?
6. A resistor has bands with the following colors: brown, orange, yellow, and no fourth or fifth bands. What is its resistance? What is its power rating?
7. A resistor has the following bands: green, grey, gold, and silver. What is its resistance and what are the ranges of tolerance?

4.4 OTHER DEVICES THAT CONDUCT, INSULATE, AND RESIST ELECTRICITY

The properties of conductors, insulators, and resistors are the basis for many practical devices used in electrical circuits. Reviewing just a few of these will help build your appreciation for the design of safe, practical electrical devices and circuits. The types of devices that are particularly worth noting are:

- Fuses
- Circuit breakers
- Lamps
- Switches.

Fuses

A fuse uses the properties of resistive wire to build safety features into a circuit. The function of a fuse is to break the circuit should it become overloaded. Fuses are represented in circuits by the symbol in Figure 4-27.

Figure 4-27. Symbol for a fuse.

The key to protecting a circuit with fuses is in a strip of wire known as a *link*. A link is made of resistive foil. The foil can be made of any of these materials:

- Aluminum
- Tin-coated copper
- Nickel.

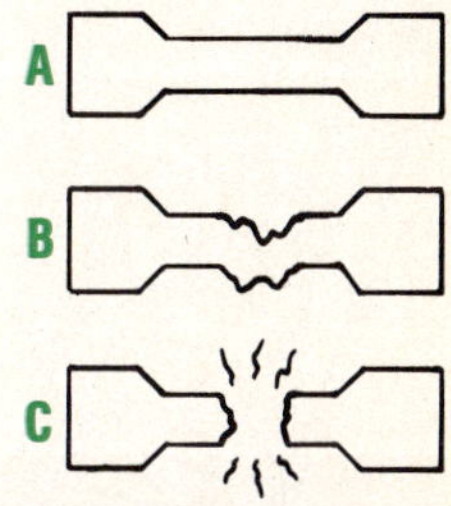

Figure 4-28. Three states of a fuse link.

In use, fuses are wired into circuits such as those that provide power to lighting fixtures or convenience outlets. The fuse link is designed to carry a rated current. The thickness and width of the material are designed for specific current levels. When a circuit becomes overloaded, the fuse link melts. When the link melts through, the circuit is broken and the devices it serves are saved from burnout or overload. Figure 4-28 shows three states of a fuse link. In Figure 4-28A, a fuse link is intact. In Figure 4-28B, the foil is distorted as melting begins. In Figure 4-28C, melting has reached the point at which the fuse link breaks.

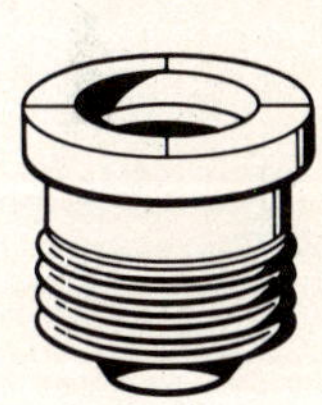

Figure 4-29. Old-syle household fuse.

Figure 4-29 illustrates an old-style household fuse. (Most modern homes have circuit breakers.) Figure 4-30 shows a cartridge fuse of the type used in many appliances and in automotive applications. Figure 4-31 shows another type of fuse cartridge. These fuses are economical since only the cartridges need be replaced when a fuse burns out. The cartridges fit into fuse holders like those shown in Figure 4-32.

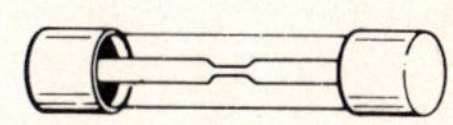

Figure 4-30. Cartridge fuse.

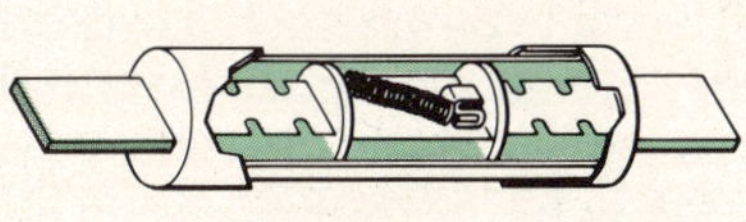

Figure 4-31. Fuse cartridge.

Figure 4-32. Fuse holders for cartridge fuses.

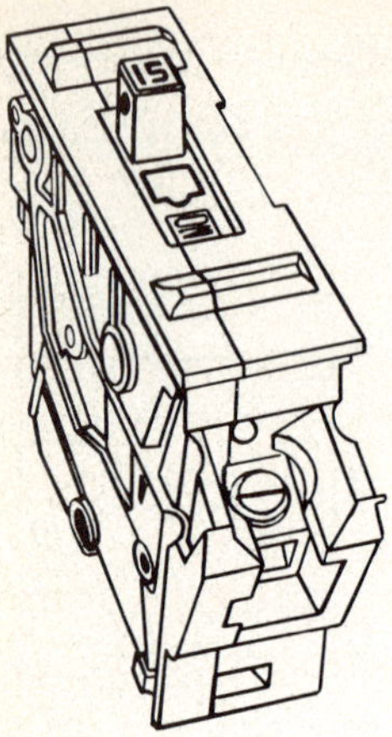

Figure 4-33. A circuit breaker.

Circuit Breakers

Circuit breakers serve the same basic function as fuses. They break the circuit when it becomes overloaded. Where fuses use melting links, circuit breakers have electromechanical devices that are forced open to interrupt the flow of electricity when an overload occurs.

Figure 4-33 is a drawing of a circuit breaker. A circuit breaker is a mechanical device that passes current through a spring made of a material that expands when it heats. Overheated, the spring forces apart a mechanical set of connectors. After the circuit is broken, the heat-sensitive element cools rapidly. Thus, in a matter of seconds, a breaker can be reset. This makes it possible to re-establish the circuit conveniently after an overload condition has been corrected. For example, if a toaster plugged into a convenience outlet causes a circuit to break, all you have to do is unplug the toaster, then reset the circuit. Before plugging the toaster in again, remove some of the other load from the circuit. Or, if the toaster is beyond the rating of the circuit, try another circuit.

Figure 4-34. Threaded circuit breaker.

Circuit breakers also come in threaded packages like the one shown in Figure 4-34. These circuit breakers are designed to replace fuses. Designed for older homes or other buildings initially built with fuse boxes, these units screw into the same sockets used for fuses.

Lamps

Lamps are devices used for illumination and also for information about the status of circuits. Illuminating lamps include those used in homes, schools, and other buildings. They include incandescent lamps that give off light from glowing wire and fluorescent lamps that give off light from a glowing, fluorescent phosphor on the inside surface of the lamp. Other common types of illuminating lamps include units that trap gaseous vapors such as sodium or mercury. These gases glow when current is applied. Illuminating lamps are parts of electrical circuits. They are of little direct interest in this book, where emphasis is on electronic circuits and devices.

Electronic circuits use many types of lamps including:

- Indicator lamps
- Neon lamps
- Display devices that may not be actual lamps, such as those using light-emitting diodes (LED).

Indicator lamps. Some typical indicator lamps are illustrated in Figure 4-35. There are several types of mounts that fit a number of different sockets. The main function of these lamps, when they are used individually, is to indicate an on-off status for a circuit. For example, when you turn an electronic device on, a small indicator light is often illuminated to show this. When the light is on, the power is being delivered to the circuit. When the light is off, the power has been interrupted.

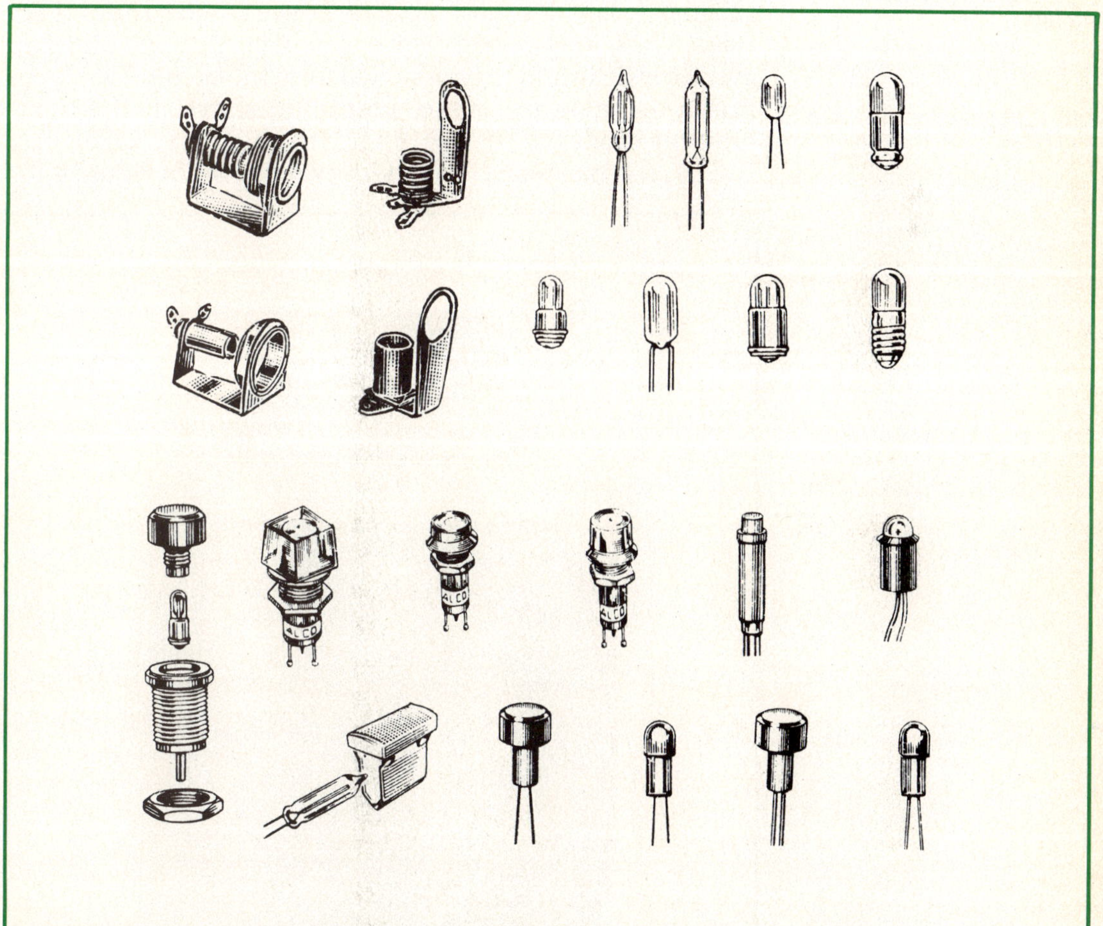

Figure 4-35. Indicator lamps.

Neon lamps. Neon lamps are used for indicators as well as for other purposes. Some of the units shown in Figure 4-35 are neon indicator lamps. These, generally, are the only type of neon lamp you will encounter when working with electronic circuits. The neon lamps used for advertising signs or displays use gas-filled tubes and are of a different type than those discussed here. The neon indicator lamp is usually a miniature device with a pair of wires that carry current to and from neon gas enclosed (encapsulated) in a glass envelope. When power is applied, the neon gas glows. The most common neon lamp used in electronic circuits is the NE-2. This is illustrated in Figure 4-36.

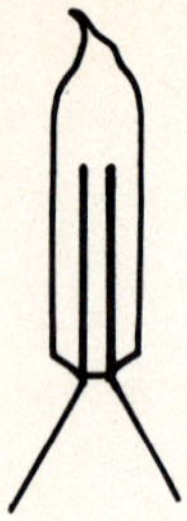

Figure 4-36. Neon lamp (NE-2) used in electronic circuits.

Display devices. These are solid-state devices that glow when current is applied. Their advantages in circuits lie in reliability, long life, and rapid response when current is applied. Thus, a solid-state display device goes on and off when current is delivered to or removed from the circuit. Two types are in common use. One, identified above, is the light-emitting diode (LED). The other is the liquid crystal display (LCD). A typical solid-state display is shown in Figure 4-37.

Figure 4-37. Wristwatch with typical LCD display.

Switches

In both electrical and electronic circuits, switches provide safety and convenience. Switches provide a means of controlling current flow. With switches, people have the ability to complete or break circuits—instantly and as required. Electronic circuits use several types of switches, including:

- Toggle
- Rotary
- Pushbutton
- Knife
- Slide switches.

Toggle switches. Toggle switches are operated by pivoting a lever. Internally, toggle switches are spring loaded. When the lever is operated, the contact pivots, completing or breaking the circuit in a positive manner. Toggle switches are illustrated in Figure 4-38.

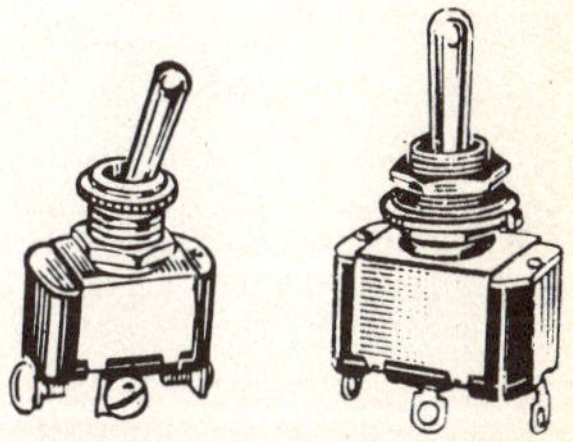

Figure 4-38. Toggle switches.

Rotary switches. Rotary switches are also spring loaded. Contact is made or broken by matching points on flat surfaces, or wafers, within the switches. These switches are operated simply by rotating the shaft. A rotary switch is illustrated in Figure 4-39.

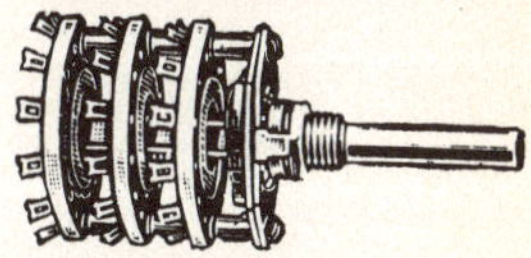

Figure 4-39. Rotary switch.

Pushbutton switches. Pushbutton switches are also spring loaded. Contact is made by pressing inward. Pushbutton switches may be either locking or temporary. On the locking type, contact is retained when the button is pushed inward. Pushing the button a second time releases the switch and breaks the circuit. On temporary pushbutton switches, the circuit is completed only while pressure is retained on the button. When you release your finger, the circuit breaks. A pushbutton switch is shown in Figure 4-40.

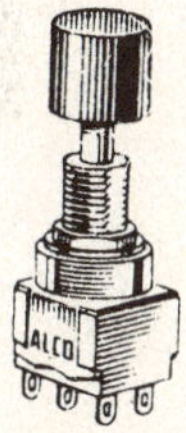

Figure 4-40. Pushbutton switch.

Knife switches. Knife switches have mechanisms similar to penknives or pocketknives. A knife switch is illustrated in Figure 4-41. By comparison with the others described above, knife switches are relatively large, cumbersome, and slow to operate. Because of these characteristics, knife switches are usually used in circuits that carry large amounts of current and do not have to be completed or broken very often. The main power switch on an elevator power system might be a knife switch. In this situation, the switch is used infrequently —possibly to turn the system on in the morning and off in the

Figure 4-41. Knife switch.

evening. The advantage of knife switches is that they can handle large amounts of current reliably. A person operating a knife switch can see clearly when contacts are made or broken. Therefore, they are mechanically very reliable.

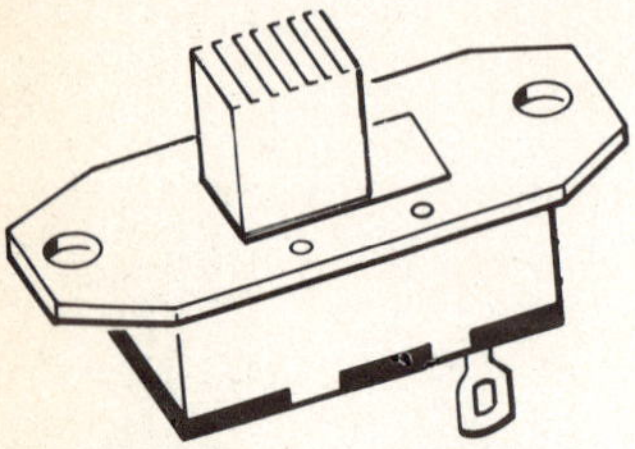

Figure 4-42. Slide switch.

Slide switches. These also are manually operated switches. The switch mechanisms slide back and forth along tracks. Wafers within the switches make and break contact. Slide switches might be used in applications where there is not enough clearance to install a toggle or pushbutton unit. A slide switch is illustrated in Figure 4-42.

Switching Contact Methods

Each of the types of switches identified above can have different configurations of contacts for making or breaking circuits. The requirements of the circuit control the type of contact used. Some contact configurations include:

A *single-pole, single-throw* switch makes contact along one conductor. The *single-pole* designation indicates that there is one conductor. The *single-throw* descriptor explains that the switch connects a lead with only one other lead. In effect, this is a binary-type switch. The switch within a circuit is either on or off.

Double-pole, as part of a switch description, indicates that circuits are completed or broken along two conductors in a single-switch operation.

Double-throw indicates that a switch can complete one of two circuits by setting it in one of two positions. When this switch is in position 1, circuit 1 has power. When this double-throw switch is in its second, or alternate, position, circuit 2 has the power.

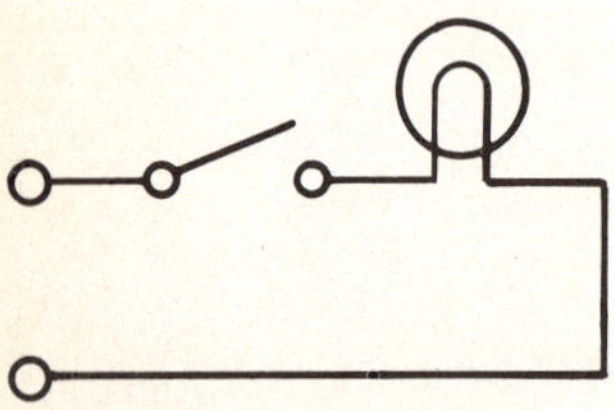

Figure 4-43. Circuit with single-pole, single-throw switch.

Many combinations of contacts can be built into switches. A single-pole switch can have a double-throw capability. On the other hand, double-pole switches that can have a double-throw capability may have only a single-throw capability in a particular circuit. Other switches may have a greater number of poles.

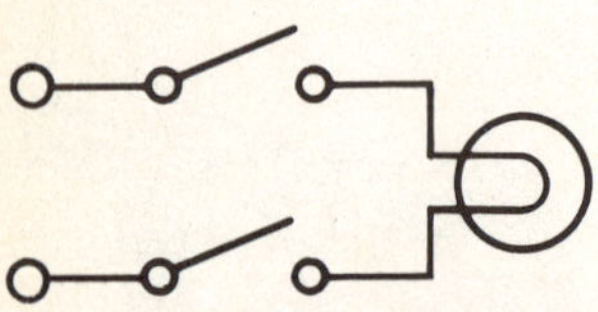

Figure 4-44. Circuit with double-pole, single-throw switch.

Figure 4-43 illustrates these concepts by showing a circuit with a single-pole, single-throw switch.

Figure 4-44 shows a circuit using a double-pole, single-throw switch.

Discussion Questions/Problems

1. What is the function of a fuse?
2. How does a fuse operate?
3. Describe the similarities and differences between circuit breakers and fuses.
4. What is the main function of lamps in electronic circuits?
5. What are the differences in light emission between incandescent and neon lamps?
6. What is the main function of a switch within an electronic circuit?
7. What is the meaning of single-pole, double-throw?

4.5 COPPER-CLAD BOARDS

Copper-clad boards are also known as *printed-circuit boards.* They are used as the primary bases for electronic circuits. Most modern electronic devices are assembled on printed-circuit boards.

A printed-circuit board is a single unit based upon the principles of both insulation and conductivity. The base material for a board is an insulator. The most common materials used are fiberglass impregnated with plastic resins. These are formed into flat surfaces of many different thicknesses and dimensions, depending on use. The thickness of the board is the distance between its top and bottom flat surfaces. This can be as thin as paper or as thick as perhaps three-eighths of an inch. Dimensions of the board can range from the miniature to more than a foot across.

The surface of the board is then coated with copper. This is usually done by bonding (under pressure) copper foil to the plastic base, or substrate. Copper can be bonded either to a single side or to both sides of a board. When bonding is done, the entire surface—top, bottom, or both—of a board is covered. A circuit board with a full surface of copper bonding is shown in Figure 4-45.

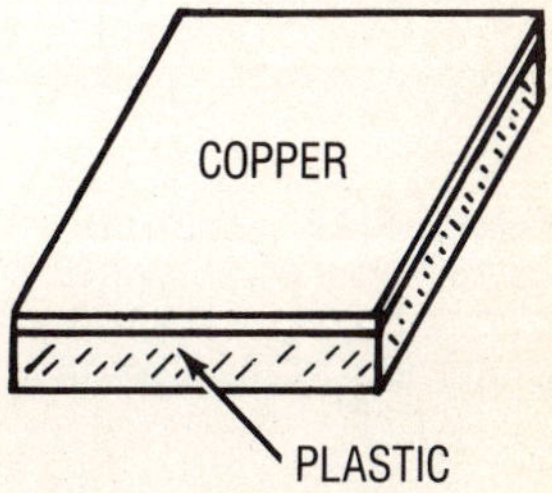

Figure 4-45. Copper bonding on full surface of a circuit board.

Obviously, a full-surface coating of copper does not provide the basis for circuits. If electricity were applied to a full-coated board, it would simply distribute throughout the surface. Instead, the idea is to create circuit-type carriers on the insulated substrate, or base, by removing portions of the bonded copper. This is usually done by protecting the areas to be used as circuit elements. Once the protection has been applied, the unwanted copper surfaces are etched away. The boards are immersed in acid baths that eat away the unprotected copper but do not affect the plastic base.

The covered portions are protected. After etching, the protective coatings are removed and the circuit configurations are in place, as shown in Figure 4-46.

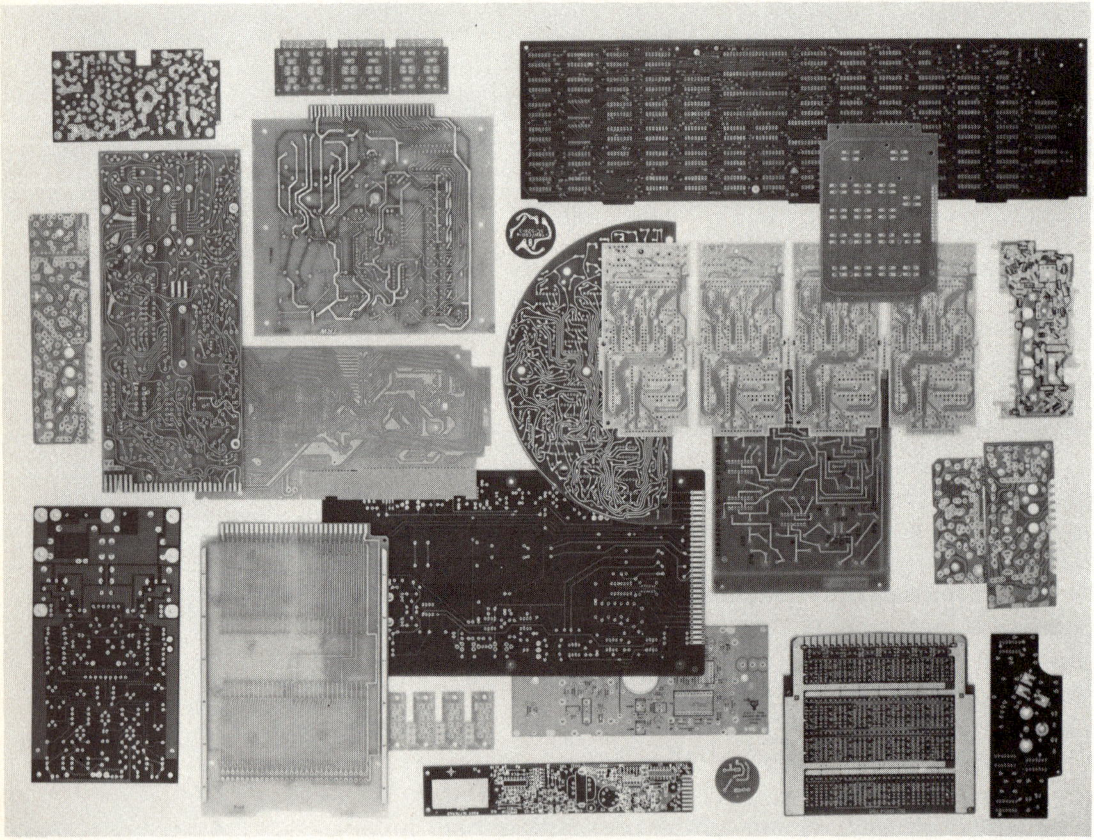

Figure 4-46. Printed circuit boards showing circuit configurations in place. (Courtesy TRW, Inc.)

For advanced electronic applications, copper-clad boards can be sandwiched, or laminated, in multiple layers. Individual circuits are etched on boards with extremely thin insulating bases. Several layers of boards are then bonded together, or laminated, under pressure. In use, the boards are drilled at the points where contacts or devices are to be installed. Where circuits on boards at different levels meet, or coincide, the devices or contacts become parts of multiple circuits. These methods are known as high-density fabrication techniques.

Fabricating Circuit Boards

Many career opportunities in electronics deal directly with designing and fabricating printed-circuit boards. Therefore, you should understand how these boards are produced in commercial quantities.

As with any circuit, a printed circuit begins with a schematic design. An engineer or advanced technician must decide what the circuit is to achieve and what components are necessary. Part of this design effort determines the configuration and connection points on the circuit.

Once a circuit is designed, engineers and technicians develop designs for the layouts of the printed-circuit boards themselves. This is done in a series of steps beginning with rough sketches and ending with exact drawings. Full precision circuit-layout drawings are made many times the size of the actual boards themselves. A skilled drafter working on a circuit layout is shown in Figure 4-47.

Figure 4-47. Drafter working on circuit layout. (Courtesy TRW, Inc.)

The large drawings are then reduced photographically to the size of the actual boards. Large process cameras, like the one shown in Figure 4-48, are used. The result is a reduced-size negative of the original drawing. Figure 4-49 shows a negative being inspected.

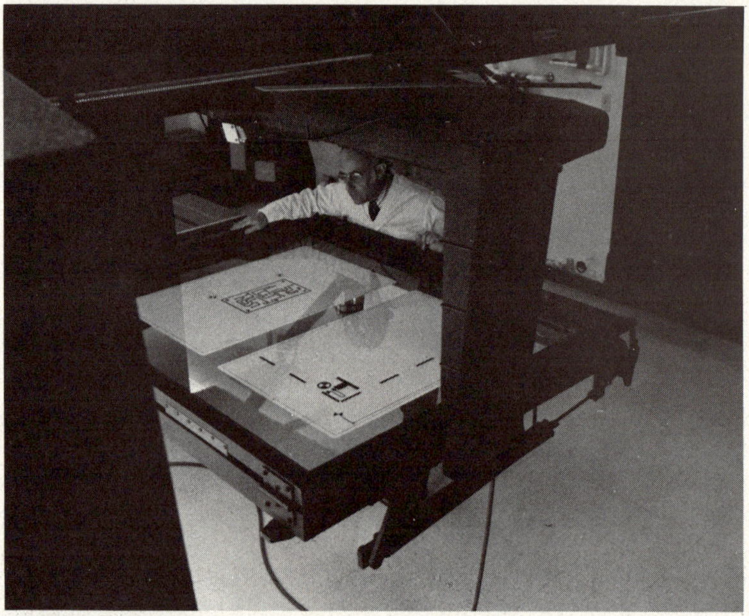

Figure 4-48. Camera being used to process printed circuit boards. (Courtesy TRW, Inc.)

Figure 4-49. Inspecting negative of circuits. (Courtesy TRW, Inc.)

The circuit-sized photographic negatives can then be used to produce boards by either of two techniques. One method is the screening process. The other is photo-etching.

With the screening process, the negative is used to produce a printing screen that is mounted on a frame. The reduced-size negative pattern appears on the screen as opened and closed patterns. A screen is shown in Figure 4-50. The screen frame is placed above a circuit board with a fully coated copper surface. An acid-resistant, thick liquid, known as a *resist*, is placed in the screen frame. The resist used for screening usually has a lacquer base which dries quickly and will form a hard surface. Other chemicals are used in resists, including water-based materials.

Figure 4-50. Printing screen and negative being inspected. (Courtesy TRW, Inc.)

The resist is forced through the screen with a squeegee, a device with a rubber blade. The squeegee is passed over the screen and a pool of resist material. The resist is forced through the uncovered portions of the screen, coating the board. The resist covers the board in an exact image of the photo negative. Resist can be screened onto boards either manually or with special screen-printing machines.

Figure 4-51. Cart carrying resist-covered boards to vats for emersion in heated etching solution. (Courtesy TRW, Inc.)

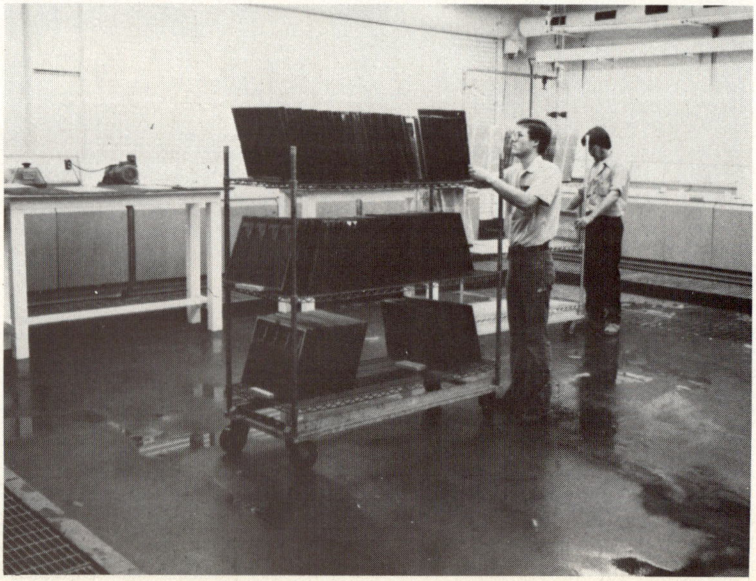

Figure 4-52. Boards being prepared for exposure during photo-etching. (Courtesy TRW, Inc.)

The resist-covered boards are then dried and etched. The most common etching materials used commercially are ferric chloride and ammonium persulfate. As shown in Figure 4-51, groups of boards are being prepared for lowering into vats with heated etching solutions. The unwanted copper coatings are quickly etched away, leaving the circuit elements in place.

Under photo-etching methods, the surfaces of the boards themselves are covered with a resist which is sensitive to light. Photo negatives are placed directly over the boards and exposed to intense sources of ultraviolet light. The light hardens the photo-sensitive resist. The exposed boards are then developed photographically, similar to film or photographic prints. During this process, the unexposed portions of resist are washed away, leaving a hardened resist over the circuit areas. These boards are then etched in the same manner as described above.

Figure 4-52 illustrates boards being prepared for exposure as part of the photo-etching technique.

Discussion Questions/Problems

1. What is the primary electrical property of the base of the copper-clad board?
2. How is electricity carried on printed-circuit boards?
3. What is the purpose of creating sandwiches, or laminates, in printed-circuit boards?
4. How is unwanted copper removed from printed-circuit boards?
5. What two techniques are used most commonly in applying etching-resist materials to the surface of copper-clad boards?

ELECTRICITY PRODUCING DEVICES

5.1 BATTERIES

A battery is a device composed of two or more *voltaic cells.* In dealing with electricity and electronics, it is important to understand the difference between batteries and cells. These terms are frequently confused. A cell or a battery can be used as a source of electricity. A cell is one unit; a battery is two or more cells which are assembled as a single working entity. The ordinary flashlight battery is actually a voltaic cell.

The voltaic cell is a basic source of electricity. Electricity is generated through chemical action among the components within the cell. Because of the effect they produce, cells and batteries are frequently described as devices that store electricity. This description, however, is incorrect. A cell contains the materials needed to generate electricity upon demand. A voltaic cell is a generator of electricity, not a storage device.

One of the problems facing the power industry is that there are few practical and cost-effective methods of storing electricity. Energy can be stored in a number of forms and converted to electricity. But electricity itself cannot usefully be stored and discharged on demand.

The Voltaic Cell

The voltaic cell is a device that contains three interactive materials as well as structural and conducting components. A basic cell is diagrammed in Figure 5-1. The materials within the cell that interact to generate electricity are:

- Positive electrode
- Negative electrode
- Electrolyte.

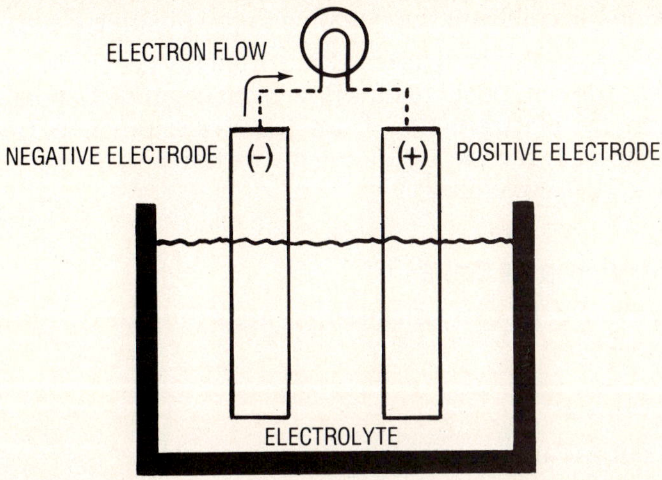

Figure 5-1. Basic voltaic cell.

Positive electrode. The positive electrode of a voltaic cell collects electrons. When a cell is linked to a circuit, the flow of electrons is from the negative electrode through the circuit and its devices, to the positive electrode.

Within the cell, positive ions collect on the positive electrode. The negative electrode, in fulfilling its function, discharges electrons into the circuits it serves. The positive electrode continually collects electrons.

Negative electrode. The negative electrode receives electrons provided by the chemical action within the cell. The negative electrode is also a source of electrons to a circuit that is powered by the cell. The voltage difference between the positive and negative electrodes is called the *potential difference* of the cell.

Electrolyte. The electrolyte both promotes chemical action between the electrodes and carries electrical energy. Electrons must flow through the electrolyte to complete the flow around the loop from negative electrode, to load, to positive electrode, then back to negative electrode. At the same time, positive ions within the electrolyte are attracted to the positive electrode. The positive ions flow (inside the cell) from the negative electrode to the positive electrode, keeping the terminal positive. Each positive ion flowing inside the cell from negative to positive electrode is the same as one or several—depending on the electrolyte—electrons flowing outside the cell from the negative to the positive electrode.

Thus, a voltaic cell forms a kind of energy generating loop. Within the cell, positive ions flow from negative to positive. Externally, electrons flow from negative to positive. (Refer to Chapter 1, section 1.3 for the distinction between current flow and electron flow.)

Categories of Cells

All voltaic cells can be classified according to two broad categories:

- Primary
- Secondary.

Primary cells. Primary cells are those that cannot be recharged. All chemical generating potential is built into the cell when it is manufactured. Electrons flow in a single direction—from negative electrode through the load to the positive electrode. Reversing flow will not recharge the cell. When the last available electrons have flowed through and out of the negative electrode, the cell or battery is "dead." A flashlight cell is a primary cell.

Secondary cells. Secondary cells can be recharged. Another way of saying this is that the chemical action within the cell is reversible. It is possible to redeposit electrons on the positive electrode, providing a source for the continuing chemical generation of current. Probably the most common example is the automotive battery. Electrical current from the battery is used to start the automobile's engine. When the engine is running, it has a generator that creates its own electricity. Electricity not used for the operation of the vehicle is routed to the terminals of the battery. The chemical action is reversed, the original materials are reconstituted, and the battery is recharged and ready again to supply current. Automotive batteries use secondary cells, which are rechargeable.

The Carbon-Zinc Dry Cell

The carbon-zinc dry cell is the most commonly used cell. It is relatively small. Carbon-zinc cells are used to power flashlights, radios, cameras, and toys.

A diagram of a carbon-zinc cell is shown in Figure 5-2. The symbol for showing a dry cell in an electronic circuit is shown in Figure 5-3.

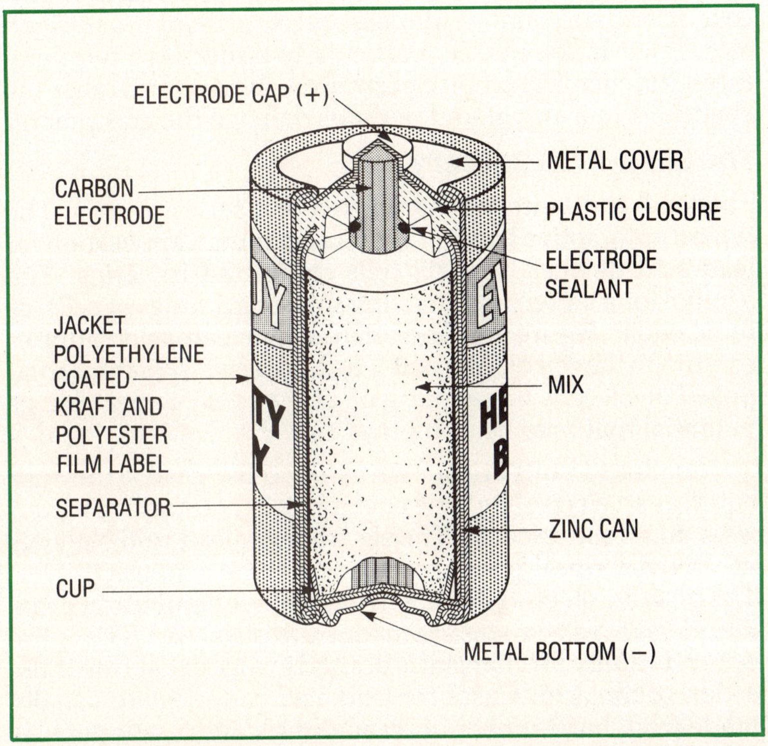

Figure 5-2. Cutaway of carbon-zinc cell.

In dry cells, the positive electrode is known as the *anode.* In carbon-zinc cells, the anode is a carbon rod placed in the center of the cell.

Negative electrodes of dry cells are known as *cathodes.* In a carbon-zinc cell, the negative electrode is made of zinc. The negative electrode is actually the outer case of the cell. The entire housing of the cell is its cathode.

Within a carbon-zinc cell, the electrolyte is a paste material containing the chemical *ammonium chloride.* Another name for this chemical is *sal ammoniac.*

Carbon-zinc cells are primary. That is, they cannot be recharged. Electrons flow from the carbon anode through the electrolyte and are delivered to circuits or devices from the cathode that forms the case. The life of the cell depends upon current drain by the device or circuit supported.

The voltage range of carbon-zinc cells is usually 1.4 V to 1.6 V.

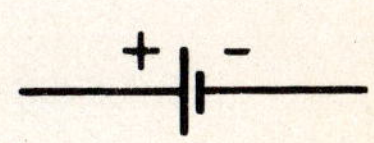

Figure 5-3. Symbol for dry cell.

A note of caution and economy about the use of carbon-zinc cells: Chemical action within the cells is continuous. Chemical action also accelerates at higher temperatures. If you are not going to use a device powered by carbon-zinc cells, it is best to remove the cells and store them in a refrigerator. If left in the device, particularly under high temperatures, the chemical action can eat away the case and cause the chemicals to leak out and perhaps damage the equipment.

The Lead-Acid Battery

The lead-acid battery is composed of a series of cells. The typical automotive battery, the most common application for lead-acid devices, has six cells connected in series. The symbol for a battery is a drawing showing a series of cells, as indicated in Figure 5-4. The cells used in lead-acid batteries commonly have voltages of 2 V to 2.2 V each. A battery composed of six cells has a rating of 12 V and an actual voltage potential that may be as high as 13.2 V.

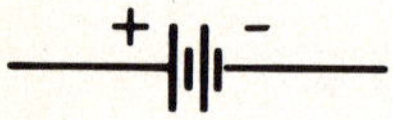

Figure 5-4. Symbol for battery.

Lead-acid units are known as wet cells because they use liquid as an electrolyte. The electrolyte in automotive batteries is dilute sulfuric acid. The electrodes of lead-acid batteries are in the form of plates suspended in the acid electrolyte. The positive electrolytes are usually made from lead peroxide. Negative electrodes are made from plain lead. The entire cell and battery mechanism is housed in a case that resists the action of the acid. Cases for lead-acid batteries are usually made from hard rubber or an acid-resistant plastic such as polypropylene. A drawing of a lead-acid battery is shown in Figure 5-5.

Figure 5-5. Lead-acid battery.

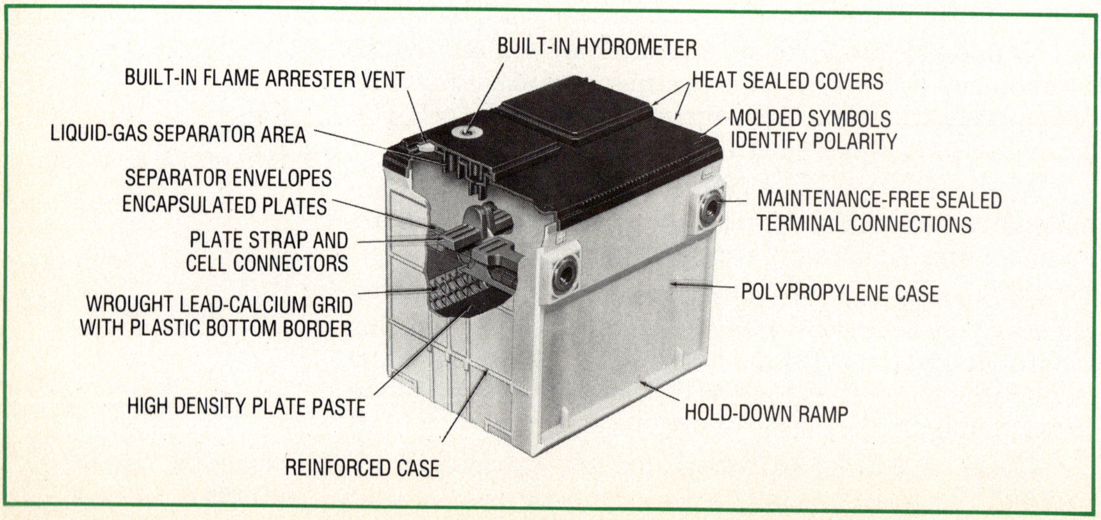

Lead-acid cells are secondary, rechargeable. The cycles for charging and discharging lead-acid batteries are diagrammed in Figure 5-6.

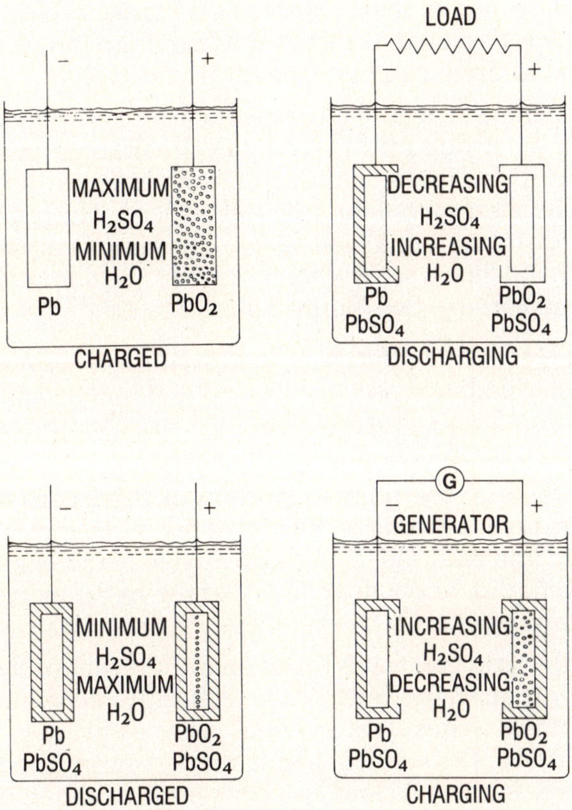

Figure 5-6. Charging and discharging cycles for lead-acid batteries.

Referring to Figure 5-6, note that the symbol for lead is Pb. The symbol for water, used to fill the liquid-electrolyte batteries, is H_2O. The symbol for sulfuric acid is H_2SO_4. The symbol for lead peroxide is PbO_2. Lead sulfide is $PbSO_4$.

When the battery is fully charged, the electrolyte has its strongest concentration of sulfuric acid. Deposits of lead peroxide are greatest on the positive electrode.

When a load is placed across the electrodes, the lead peroxide moves through the electrolyte—traveling from the

Figure 5-7. Recharger for lead-acid batteries. (Courtesy Delco Remy)

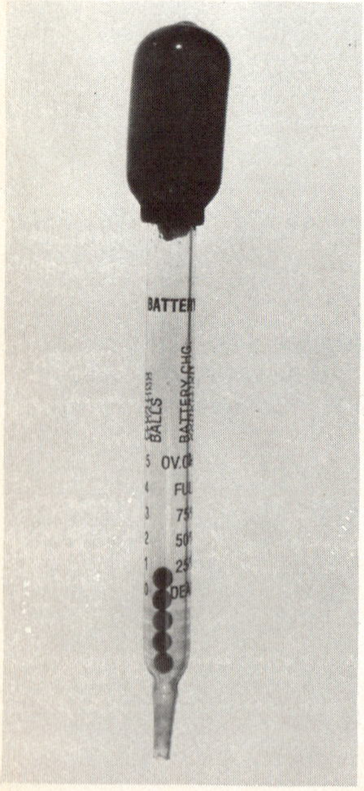

Figure 5-8. Hydrometer.

positive to the negative electrodes. Lead peroxide then converts at the negative electrode to lead sulfide, and this chemical process supplies current to the load. When the battery is fully discharged, lead sulfide builds up on the negative electrode, which has exhausted its ability to absorb and convert lead peroxide. The sulfuric acid level in the electrolyte is at its lowest level while the water content of the electrolyte has built up and its acid content has decreased. The positive electrode is depleted to a level at which it no longer has the potential to drive positive ions to the electrolyte.

To recharge a lead-acid battery, current is applied to the positive electrode and the chemical/electrical process is reversed. This causes a buildup of lead peroxide on the electrode. The chemistry of the cell is rebalanced. The cell is ready to provide current to a load again. Charging is done with a device that delivers somewhat more than the nominal 12 volt DC to the battery. See Figure 5-7.

Testing lead-acid batteries. To determine the state of the charge of a lead-acid cell, a device is used that measures the mix of water and sulfuric acid in the cell. The device, illustrated in Figure 5-8, is called a *hydrometer*.

For testing, the pointed portion of the hydrometer is placed in a cell. The bulb at the top of the hydrometer is then squeezed and released. The glass capsule of the instrument fills with liquid, and a float device in the hydrometer floats at a certain level.

The hydrometer test is based on the fact that sulfuric acid is heavier than water. Specifically, concentrated sulfuric acid is 1.835 times as heavy as an equal volume of water. As discussed above, the sulfuric acid is more concentrated when the battery is charged and less concentrated when the battery is discharged. The hydrometer measures the concentration of the acid. In the hydrometer is a float device which is a sealed capsule containing both air and metal weights. The weights are carefully selected so the device, or capsule, will rest suspended partially above and partially below the level of the liquid. The float device will not submerge quite as deeply in a dense liquid as it will in a liquid that is dilute. Thus, the float indicates the density of the liquid, and thereby indicates the amount of charge on the battery.

Other Types of Cells and Batteries

The cells and batteries described are the most widely used devices for generating electricity on demand. There are also other materials that can interact chemically to generate electricity. There is not time—or a need—to cover all possible types of cells and batteries in this book. However, the four additional cells discussed below are in sufficiently wide use so that you should be familiar with them.

Mercury cells. A main feature of mercury cells is their ability to provide a constant output voltage. Each mercury cell can be counted on to deliver 1.35 V, even when it is almost depleted.

Figure 5-9 is a cutaway drawing showing the construction of a mercury cell. The cathode is made from mercuric oxide. The anode is made of zinc. The electrolyte is an alkaline material.

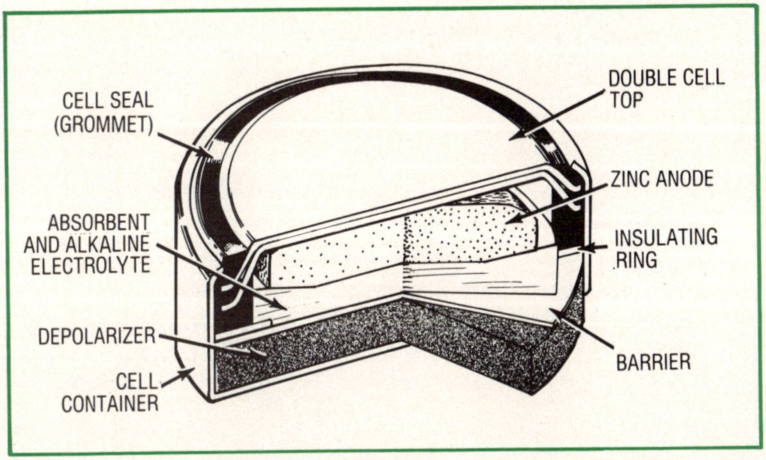

Figure 5-9. Cutaway showing construction of mercury cell.

Mercury cells come in a variety of shapes and sizes, some of which are shown in Figure 5-10.

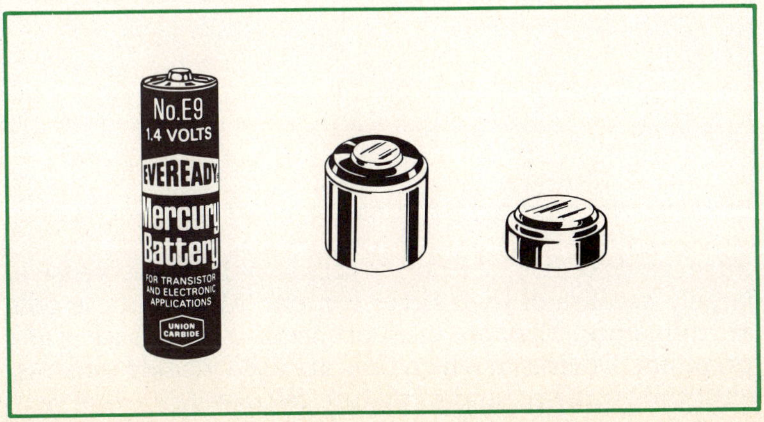

Figure 5-10. Mercury cells.

Nickel-cadmium cells. The main feature of nickel-cadmium, or "NiCad," cells is that they are secondary cells. This means they are rechargeable. They also produce a relatively stable output voltage—1.25 V.

The cathode of a nickel-cadmium cell is made of metallic cadmium. The anode is nickel hydroxide. The electrolyte is potassium hydroxide. An alkaline electrolyte material is used in these cells. A diagram of a nickel-cadmium cell is shown in Figure 5-11.

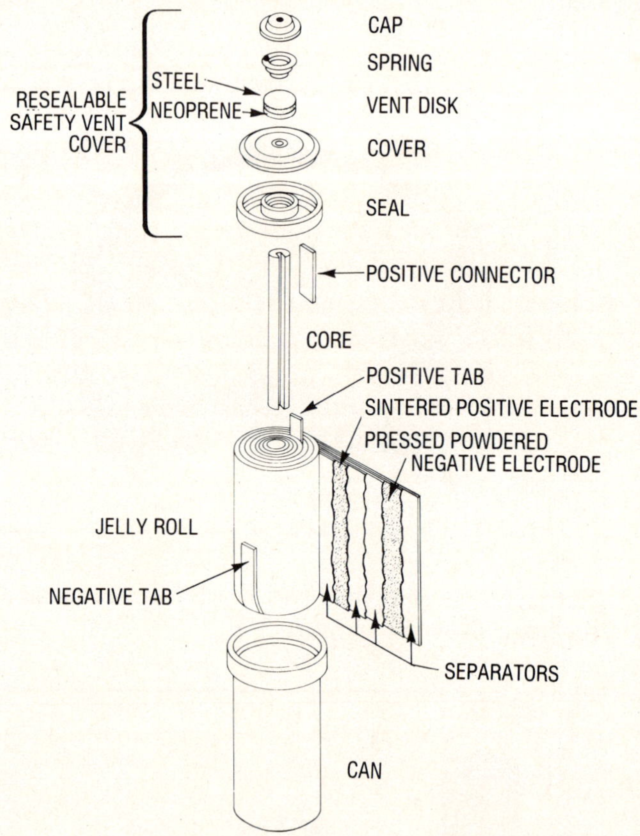

Figure 5-11. Construction of nickel-cadmium cell.

Silver oxide cells. Silver oxide cells are relatively expensive, largely because of their silver content. Most of these cells are miniature. They are used primarily where space is at a premium. Devices in which they are used include cameras, electronic watches, and miniature calculators.

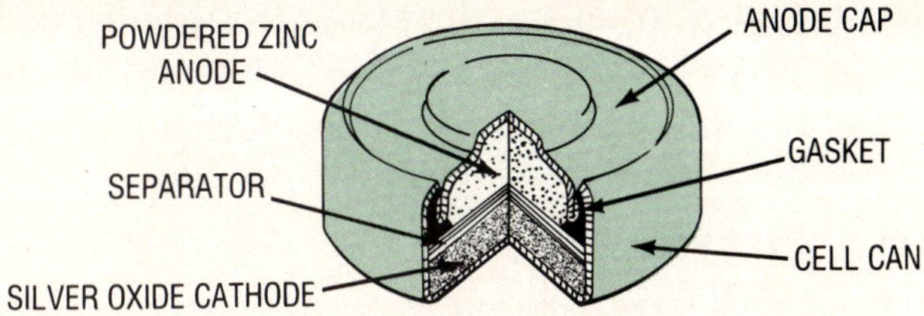

Figure 5-12. Cutaway of silver oxide cell.

Silver oxide cells produce a constant output of 1.5 V. Their cathodes are made of silver oxide. Anodes are made from powdered zinc. The electrolyte is alkaline. A silver oxide cell is illustrated in Figure 5-12.

Manganese-alkaline cells. These are primary cells which produce an output of 1.5 V. Their main features are stability and reliability of output. They tend to be used in devices that require long running life and constant voltage. For example, many portable dictating machines use maganese-alkaline cells.

The cathodes are manganese dioxide. The anodes are zinc. Either of two electrolytes may be used: potassium hydroxide or sodium hydroxide. A manganese-alkaline cell is illustrated in Figure 5-13.

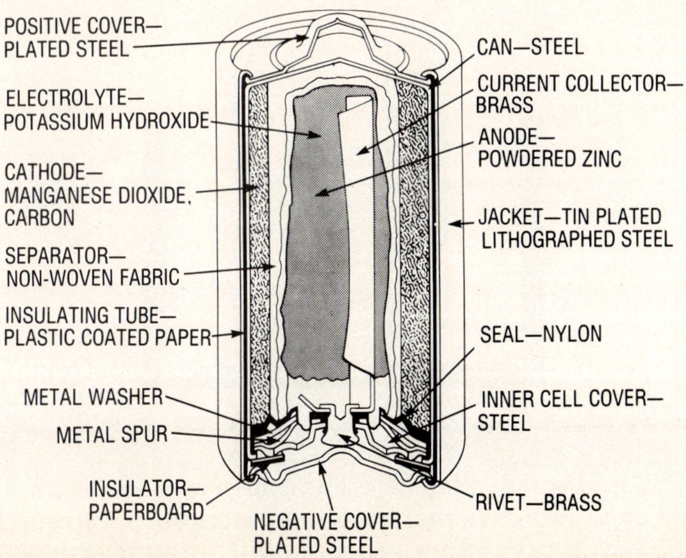

Figure 5-13. Cutaway of manganese-alkaline cell.

Some other types of cells. Most of the other types of cells you will encounter are variations of those described above. Included are lead-acid cells with gelled electrolytes, nickel-iron cells, silver-cadmium cells, and silver-zinc cells.

In addition, there are some primary cells that use lithium in their cathodes. Still another type is zinc-chloride.

Internal Resistance of Cells

All generators have some internal resistance. In some circuits with close tolerances, it may be necessary to make allowances for the resistance within the generator itself. *There are no perfect generators.* This applies to cells and batteries, as well as to other types of generators.

The resistance within a generator, including a cell or battery, is always in series with the load. Within a cell, the electrolyte is the resistance. Current traveling between the cathode and anode meets resistance from the electrolyte. Since the cathode and anode are linked into circuits in series, they add series-type resistance to external circuits.

Figure 5-14 illustrates the series-type resistance of a cell. In this illustration, a lamp is used as the load. The resistance internal to the cell or other generator is identified as R_i. (See also Figure 3-20 for an example.)

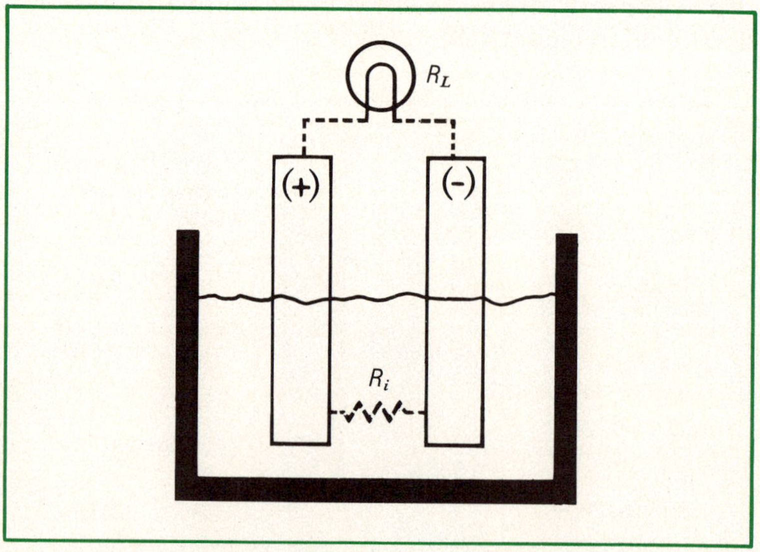

Figure 5-14. Series-type resistance within cell.

Should you need to determine the internal resistance of a cell or battery, use the following procedure. Connect the cell (or battery) to an external load R_{L1} and measure the voltage E_1 and the current I_1 drawn by the load. Then connect a different load R_{L2} and measure the voltage E_2 and the current I_2.

With the first load:

$$I_1 = \frac{e}{R_{L1} + R_i}$$

And:

$$E_1 = I_1 R_{L1}$$

With the second load:

$$I_2 = \frac{e}{R_{L2} + R_i}$$

And:

$$E_2 = I_2 R_{L2}$$

After solving for I_1 and I_2 and E_1 and E_2:

$$\begin{aligned} I_1(R_{L1} + R_i) &= I_2(R_{L2} + R_i) \\ I_1 R_{L1} + I_1 R_i &= I_2 R_{L2} + I_2 R_i \\ I_1 R_{L1} &= I_2 R_{L2} + I_2 R_i - I_1 R_i \\ &= I_2 R_{L2} + R_i(I_2 - I_1) \\ \frac{I_1 R_{L1} - I_2 R_{L2}}{(I_2 - I_1)} &= R_i \end{aligned}$$

Using E_1 and E_2:

$$R_i = \frac{E_1 - E_2}{(I_2 - I_1)}$$

Thus, the value for R_i, the internal resistance of the cell (or battery), can be derived from the two load currents I_2 and I_1

and the measured voltages E_1 and E_2. Notice that it is not necessary to determine the resistances R_{L1} and R_{L2} to obtain this result.

Cells and Batteries in Series and Parallel

When cells or batteries are connected in series, their voltage is added. Current through cells or batteries in series is the same through each. This principle is illustrated in Figure 5-15. The circuit shown is typical for a multi-cell flashlight battery. Series connection of three 1.5 V batteries delivers a total voltage of 4.5 V. Current remains the same through this entire circuit.

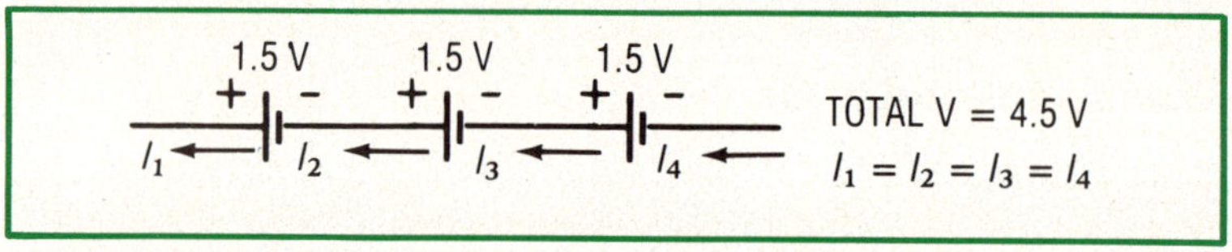

Figure 5-15. Circuit for multi-cell flashlight battery.

When cells or batteries are connected in parallel, the current capability of each unit is added. Voltage remains the same, as shown in Figure 5-16. This type of connection is used frequently in portable audio devices, such as radios or tape recorders. The effect is to increase the capacity to provide current while maintaining the same voltage.

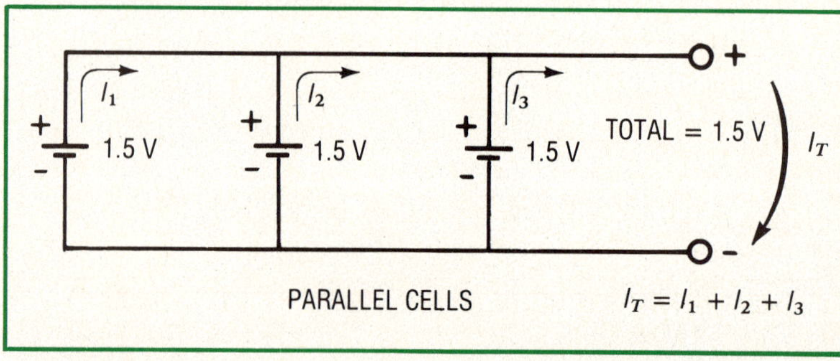

Figure 5-16. Circuit for cells connected in parallel.

Cells and batteries are also connected in series-parallel circuits for many applications. This is illustrated in Figure 5-17, which shows the type of circuit that might be used for a portable cassette player. The circuit in Figure 5-17 has two branches. Each branch has two 1.5 V cells. Thus, the total voltage output for this circuit is 3 V. Because the two branches of the circuit are in parallel, the current is cumulative. If each of the batteries in the circuit could deliver 1 A of current, the overall circuit could deliver 2 A.

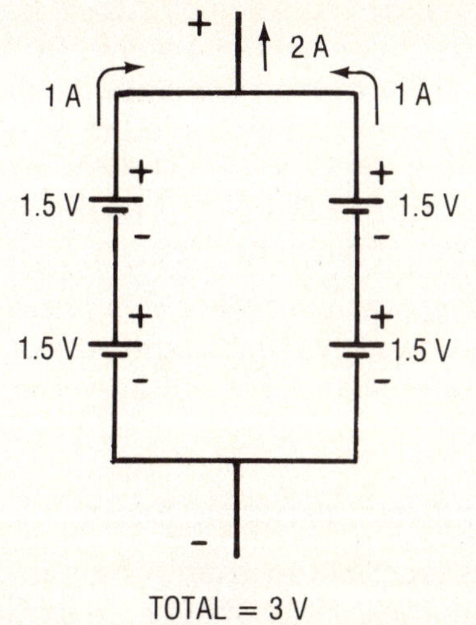

Figure 5-17. Circuit with two parallel branches, each with cells connected in series.

Discussion Questions/Problems

1. How does a voltaic cell generate electricity?
2. What is the function of the electrolyte in a cell?
3. What is a primary cell?
4. What is a secondary cell?
5. What are the differences between dry and wet cells?
6. What is the result when cells are connected in series?
7. What is the result when cells are connected in parallel?
8. A battery is rated at 9 V. When it carries a load, the battery delivers 8 V. Current is 0.5 A. What is the internal resistance of this battery?

5.2 PHOTOVOLTAIC (SOLAR) CELLS

Photovoltaic cells generate an emf upon exposure to light, particularly bright sunlight. The voltage is created as semiconductor materials react to energy contained in a beam of light.

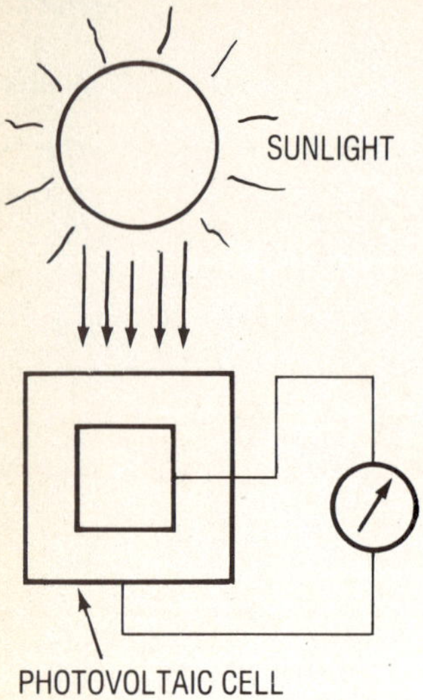

Figure 5-18. Light source (sun) generates a measurable voltage when it shines on a photovoltaic cell.

Applications for photovoltaic energy have been devised around a theory stated initially by Albert Einstein. Einstein theorized that light consists of energy. Building upon Einstein's work, J. J. Thompson developed experiments which showed that electrons can be forced to flow out of metallic materials when they are exposed to beams of light. From these initial findings, further study led to the compounding of specialized photovoltaic materials.

A circuit diagram illustrating this principle is shown in Figure 5-18. This indicates that a light source, in this case the sun, generates a measureable voltage when it shines on a photovoltaic cell.

Materials for Photovoltaic Cells

Some materials react better than others in generating emf when exposed to the energy of light. In particular, semiconductor materials are ideal for use in photovoltaic applications. This is because semiconductors contain substantial numbers of surplus electrons (or their counterparts—holes in atomic structures which attract electrons).

Specific materials that provide excellent photovoltaic properties include:

- Germanium
- Silicon
- Selenium.

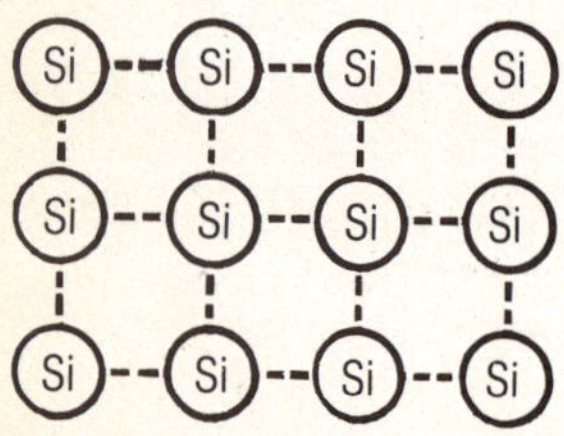

Figure 5-19. Molecular structure of silicon.

Most modern photovoltaic or solar cells use silicon as a base. Silicon is used because it is abundant and available. Further, silicon is easiest to process among the suitable materials. It is easily melted and grown into the rods needed to produce the thin, flat, round wafers that are the basis for light-sensitive cells. Silicon is also ideal because, when processed, this material can be brought to a high level of purity. Silicon assumes a lattice-like molecular structure, as illustrated in Figure 5-19, and can be processed to a chemical purity of 99.99 percent. Initial processing delivers pure silicon rods shaped like the one illustrated in Figure 5-20. These rods are the source material for the fabrication of solar and photovoltaic cells.

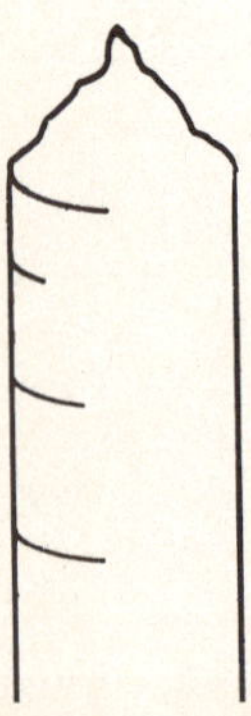

Figure 5-20. Pure silicon rod.

Pure silicon, by itself, does not have the properties necessary to form operational semiconductors. In its pure state, silicon functions as a resistor. Its properties are similar to those of glass. For use in semiconductors, silicon requires an additive material that imparts the proper conductive properties. The additives used are called *dopants*, or *doping elements*. Dopants used commonly to form silicon semiconductors include antimony, arsenic phosphorus, aluminum, boron, gallium, and indium.

The dopant used depends upon the desired properties of the specific semiconductor to be fabricated. The chemical symbols, atomic numbers, and numbers of valence electrons for commonly used doping elements are listed in the table in Figure 5-21.

ELEMENT	SYMBOL	ATOMIC NUMBER	VALANCE ELECTRONS	USES
(*DONOR* IMPURITY ELEMENTS—GIVE ELECTRONS TO FORM N-TYPE SEMICONDUCTOR)				
ANTIMONY	Sb	51	5	FOR GERMANIUM
ARSENIC	As	33	5	FOR GERMANIUM
PHOSPHORUS	P	15	5	FOR SILICON
(*ACCEPTOR* IMPURITY ELEMENTS—TAKE ELECTRONS TO FORM P-TYPE SEMICONDUCTOR)				
ALUMINUM	Al	13	3	FOR SILICON
BORON	B	5	3	FOR SILICON
GALLIUM	Ga	31	3	FOR GERMANIUM
INDIUM	In	49	3	FOR GERMANIUM

Figure 5-21. Commonly used dopants.

Depending upon the impurity of the dopant added in doping, silicon can be formed into a semiconductor material that conducts either negative or positive electrical charges. This is because some of the doping materials contain free electrons while others contain free holes.

For example, as illustrated in Figure 5-22, adding a small amount of phosphorus to a lattice structure of relatively pure silicon introduces free electrons. The resulting material has an excess of electrons, as shown in the drawing. Therefore, a silicon material doped with phosphorus produces a negative-type, or N-type, semiconductor material.

If aluminum is used as the doping element, a positive hole is added to the basic silicon material. Thus, a positive-type, or P-type, semiconductor material is formed. A P-type semiconductor material is diagrammed in Figure 5-23.

These, then, are the two basic types of semiconductor materials—N-type and P-type.

In fabrication, the silicon rod is sliced into extremely thin wafers. This is done by a process of cleaving. A sharp instru-

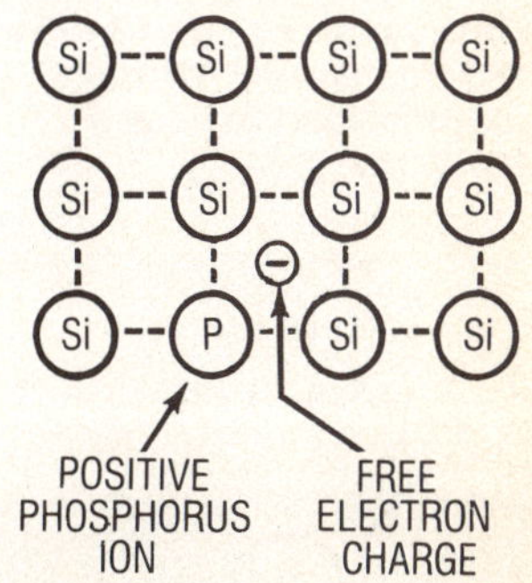

Figure 5-22. Phosphorus added to silicon produces negative-type semiconductor material.

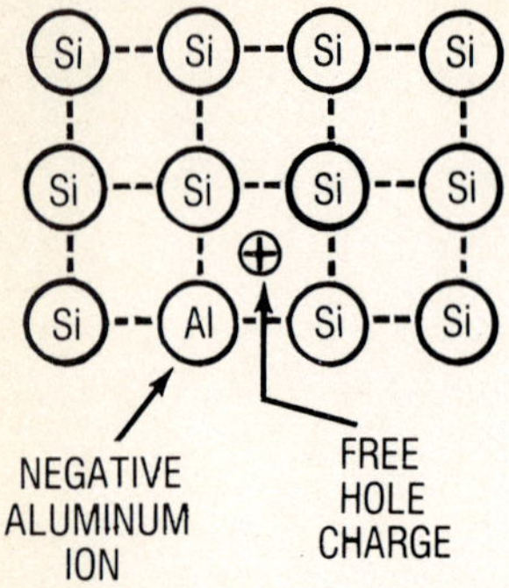

Figure 5-23. Aluminum added to silicon produces positive-type semiconductor material.

ment is placed at the slice point, then struck forcefully. Wafers fall away from the rod. Impurities are added when the wafers are processed. A common technique is to diffuse small amounts of impurities onto the surface of the wafers, where they are absorbed into the lattice structures of the silicon.

Fabricating Photovoltaic Cells

One of the most common methods of forming photovoltaic or solar cells is to create devices known as *P-N junction diodes.* A typical photovoltaic cell of this type is shown in Figure 5-24. Wafers of P-type and N-type semiconductors are brought together and fused under intense heat. When the resulting diode is exposed to an intense light source, a stream of electrons is forced to move out of the N-type material and toward the P-type material. The electrons must travel through an external load, as indicated in Figure 5-24.

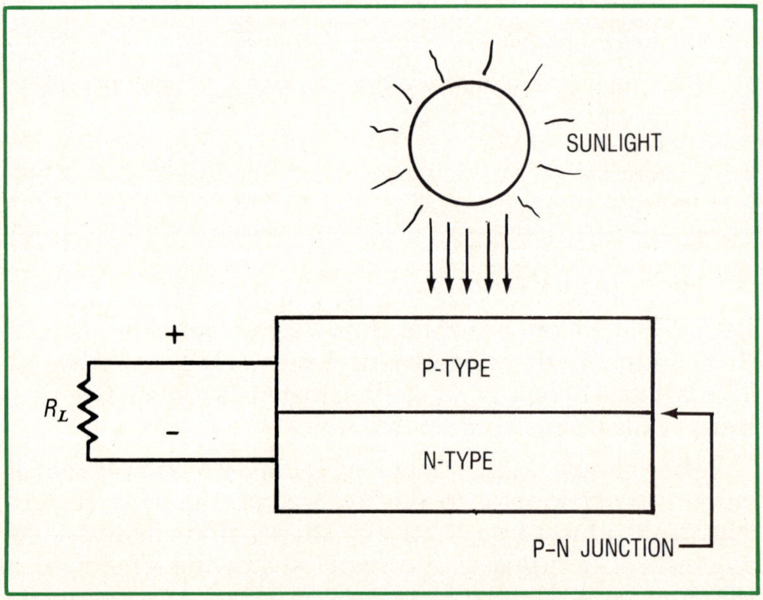

Figure 5-24. When photovoltaic cell is exposed to light, electrons move from N-type material through external load to P-type material.

A typical photovoltaic or solar cell of the type illustrated will produce from 0.35 V to 0.55 V of emf.

The symbol for a solar cell is shown in Figure 5-25.

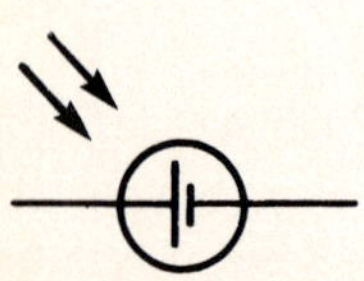

Figure 5-25. Symbol for photovoltaic or solar cell.

Applications for Photovoltaic and Solar Cells

To date, the solar cells that have been put to use have produced low-voltage, low-current outputs only. There is considerable research on ways to increase power output to potentially commercial levels. In the meantime, a number of important applications have been developed for photovoltaic cells. These include light meters in many automatic cameras. Photovoltaic cells are used for exposure readings or to generate signals to battery-operated devices that set exposures automatically.

Space vehicles and satellites use photovoltaic cells as their main source of generated power. This is possible because of the intense sunlight when there is no atmospheric interference. In space travel, exposure to sunlight is almost continuous.

Many industrial processes use photovoltaic sensors. For example, these units are part of systems that count pieces moving down assembly lines.

Governmental and private organizations throughout the world have experimental pilot projects involving photovoltaic cells. They are used to provide part or all of the power for remotely located water pumps, for air-conditioning systems, and for supplementary power to light office buildings.

Discussion Questions/Problems

1. What causes electrons to flow in metal or semiconductor materials when they are exposed to light?
2. Why is pure silicon not usable as a semiconductor material?
3. What is the function of dopants added to semiconductor materials?
4. Why do some dopants cause semiconductors to acquire negative charges while others create positively charged materials?
5. How is electricity generated when a P-N junction diode is exposed to sunlight?

5.3 THERMOCOUPLES

A thermocouple is a device that produces electricity from energy absorbed from heat. When the device is heated either by environmental temperatures or from external sources, a stream of electrons is released.

This process is known as the thermoelectric, or Seebeck effect. It was named after Thomas Johann Seebeck, who discovered this phenomenon in 1821.

A thermocouple is fabricated from two dissimilar metals, and includes two metal-to-metal junctions. For example, Figure 5-26 shows a thermocouple created by joining strips of iron and copper. If the low-temperature junction is kept cool while the high-temperature junction is heated, a thermoelectric potential will develop across the terminals A–B. Thus, an external circuit, such as a meter, can be connected to and driven by the potential difference at these terminals.

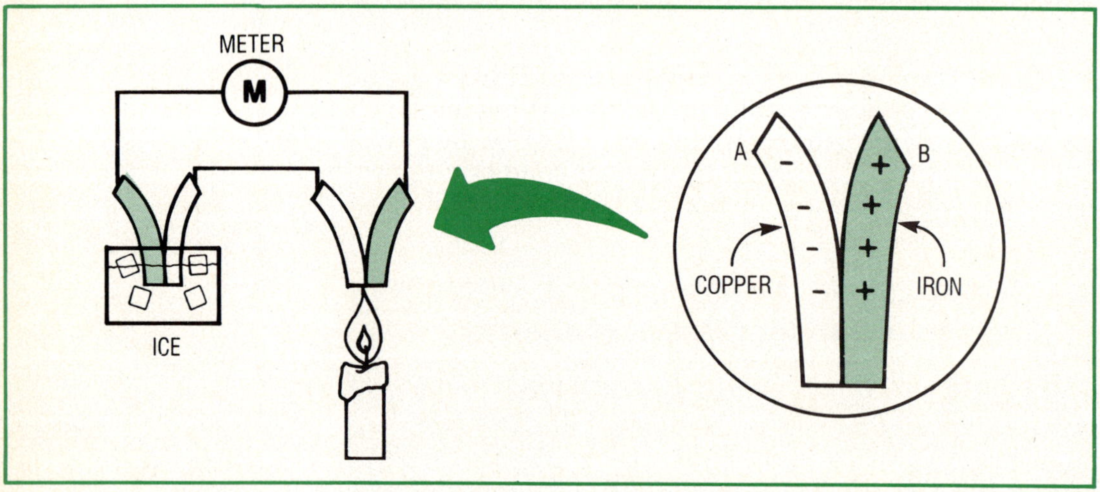

Figure 5-26. Creation of thermocouple by joining strips of copper and iron.

The low-temperature junction is kept at a known, fixed temperature such as the temperature of melting ice. Changes in the thermoelectric current will then be due only to changes in temperature at the high-temperature junction. The thermocouple can be calibrated so that a given change in current will correspond to a known change in temperature.

Thermocouples are low-powered devices. Their outputs are measured in millivolts. Also, they can be extremely sensitive to temperature change. Thermocouples are available that operate in ranges of −328 degrees F. (−200° C) to +500 degrees F (+260° C).

Thermocouples are usually used to measure temperatures. Examples include temperature indicators on heating chambers that must be closely controlled, such as kilns, used to cure pottery, or scientific ovens. Thermocouples are also used in pyrometers, devices that measure the temperatures of molten metals.

Discussion Questions/Problems

1. What is the source of energy for thermocouples?
2. Why is it necessary that one of the materials used in a thermocouple be more dense than the other?
3. In a thermocouple using copper and iron, which material is positive?

5.4 THE PIEZOELECTRIC EFFECT

Some materials generate electricity when they are put under mechanical stress. This stress can be caused by twisting or by applying pressure against the surface.

This effect takes place in crystalline substances, such as quartz, rochelle salts, or tourmaline. Other crystalline materials display the same effect. For example, diamonds and sapphires generate emf under stress.

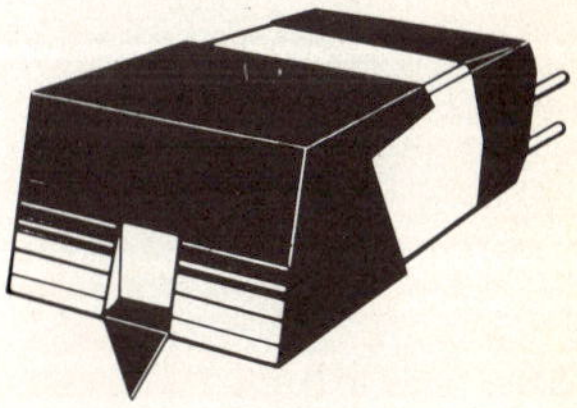

Figure 5-27. Piezoelectric effect created as phonograph needle vibrates in record grooves.

Figure 5-27 shows a common example of a device using the piezoelectric effect. A phonograph needle riding in record grooves vibrates in response to the shapes of the grooves. The twisting and pressure that result cause an electrical potential in the attached wires.

Other types of devices that utilize the piezoelectric effect include microphones, radio transmitters, and radio receivers.

Emf created by piezoelectric devices is extremely small, and is measured usually in millivolts or microvolts.

Discussion Questions/Problems

1. What type of force triggers the piezoelectric effect?
2. Are piezoelectric devices considered to be high- or low-power devices?
3. What are some examples of the use of piezoelectric devices?

REVIEW CHAPTERS 4 AND 5

SUMMARY

1. A carrier of electricity is called a *conductor*. (4.1)
2. One of the factors that affects how materials conduct electricity is the conductivity of the materials. (4.1)
3. DC moves through the entire cross section of a conductor. High frequency AC travels along the exterior of a conductor, penetrating the conductor only slightly. (4.1)
4. The cross section is the area across a conducting material or wire. Generally, the greater the diameter of a conductor, the less resistance it will offer to the flow of electrical current. (4.1)
5. There are many types of commercial conductors. Each has its own characteristics and applications. (4.1)
6. A wire gauge is an instrument for measuring the diameter, or gauge, of a wire. (4.1)
7. Materials that obstruct, or prevent, the flow of current are called *insulators*. (4.2)
8. Materials that impede the flow of current predictably are called *resistors*. (4.3)
9. A potentiometer is a device that uses variable resistance to change or adjust the voltage within a circuit. (4.3)
10. Resistors must be matched to the specific requirements of a circuit. A clear method is needed for marking and identifying these devices. Many years ago, the electrical industry standardized on a system of color bands that enable resistors to be identified visibly. These color bands comprise what is known as the *resistor color code*. (4.3)

11. The function of a fuse is to break a circuit when the circuit becomes overloaded. (4.4)

12. Circuit breakers serve the same basic function as fuses. Where fuses use meltable links, circuit breakers have electromechanical devices that are forced open to interrupt the flow of electricity when an overload occurs. (4.4)

13. Lamps are used for illumination and also for information about the status of circuits. (4.4)

14. In both electrical and electronic circuits, switches provide safety and convenience. Switches are a means of controlling current flow. (4.4)

15. Copper-clad boards are also known as *printed-circuit boards*. They are used as the primary bases for electronic circuits. Most modern electronic devices are assembled on printed-circuit boards. (4.5)

16. A battery is a device composed of two or more voltaic cells. A voltaic cell is a basic source of electricity. (5.1)

17. The positive electrode of a voltaic cell collects electrons from the external circuit. (5.1)

18. The negative electrode receives electrons provided by the chemical action within the cell and releases them to the external circuit. (5.1)

19. The electrolyte both promotes chemical action between the electrodes and carries electrical energy. (5.1)

20. All voltaic cells can be classified according to two broad catagories: primary and secondary. Primary cells are those that cannot be recharged. Secondary cells can be recharged up to a point. (5.1)

21. Photovoltaic cells generate an emf upon exposure to light, particularly bright sunlight. (5.2)

22. A thermocouple is a device that produces electricity from energy absorbed from heat. (5.3)

23. The piezoelectric effect is created by the ability of some materials to generate electricity when they are put under mechanical stress. This stress can be caused by twisting or by applying pressure against the surface. (5.4)

PROBLEMS

1. What is the wire gauge number of a wire that has a diameter of 0.0201 inches? (4.1)
2. What would be the resistance of a 500-foot length of No. 32? (4.1)
3. What is the dielectric strength of mica? (4.2)
4. What is the value of a resistor whose color code is brown, yellow, red, silver? (4.3)
5. What is the value of a resistor whose color code is violet, orange, black, gold? (4.3)
6. What is the value of a resistor whose color code is white, blue, brown, no color? (4.3)
7. Draw the symbol for a fuse. (4.4)
8. Draw the symbol for a lamp. (4.4)
9. What is the internal resistance of a battery which is rated at 12 V but, when loaded with a 2.5 A current, delivers 10 V? (5.1)
10. What voltage is delivered by three 1.5 V cells connected in series? (5.1)
11. What voltage is provided by three 1.5 V cells connected in parallel? (5.1)
12. If three 9 V batteries are connected in series with three 9 V batteries connected in parallel, what is the total voltage? (5.1)

(Answers to the odd-numbered problems can be found at the back of the book.)

MAGNETISM

6.1 FERROMAGNETISM

Since magnetism is the basis for most generated electricity, it is important to understand the nature of magnets and methods for applying and controlling magnetism.

Magnets occur naturally. As a matter of fact, the entire planet Earth is enveloped by an overall magnetic force. This force is both reliable and powerful. For centuries, the earth's magnetic field has been a basis for navigation.

Ferrous compounds are the primary substances, or materials, in which magnetism occurs. The term *ferrous* means that a substance contains iron. The atomic structure of iron is an ideal environment for magnetism. Iron-based materials facilitate the presence of magnetism.

Figure 6-1 illustrates both the effect and nature of magnetism on ferrous materials. Figure 6-1A shows the state of iron molecules in a metal bar which has not been magnetized. The molecules are oriented in random directions. Figure 6-1B shows the molecules of a magnetized iron bar.

Figure 6-1. Effect and nature of magnetism on ferrous materials.

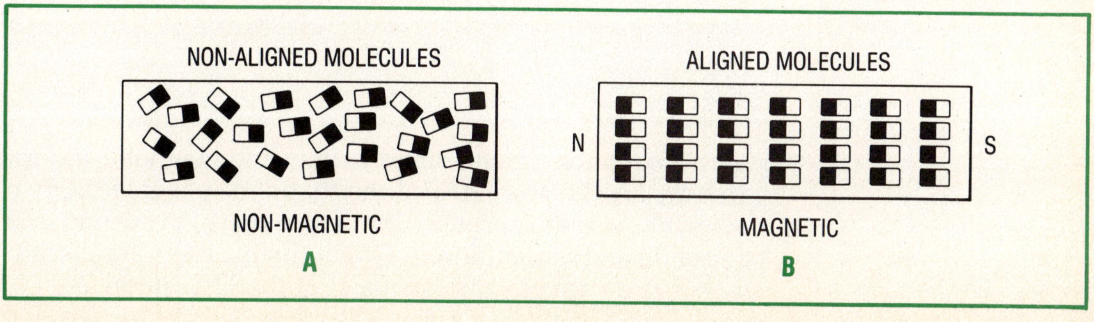

All of the ferrous molecules are aligned in the same direction. When molecules are aligned in this way within a piece of ferrous metal, the metal is a magnet. It is said to have a magnetic field.

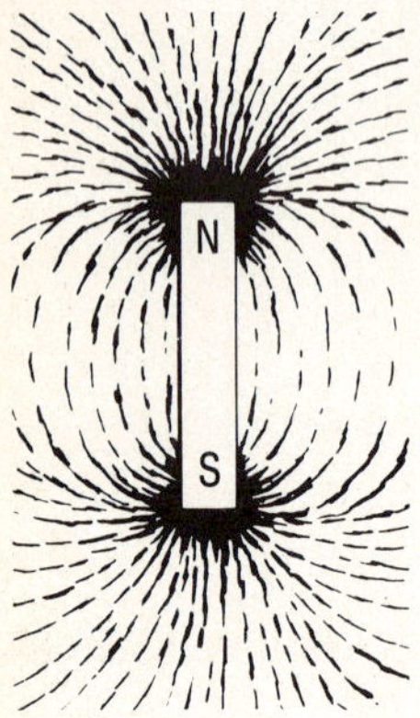

Figure 6-2. Magnetic field lines.

The magnetic field surrounds the aligned molecules in magnetized metal. Note, in Figure 6-1B, that the aligned molecules all point toward the ends of the bar. Given that these molecules are lined up, there must have been some force that caused the alignment. This force is greatest at the ends of the bars, in the direction toward which the molecules point, and less beyond the ends of the bars. In the air, with no metal to concentrate the lines of force, they spread out.

The field lines, or force lines, of a magnetic bar are illustrated in Figure 6-2. This shows that the field lines are strongest at the ends of the bar and that magnetic force falls off in relation to distance from these end points.

Note that the ends of the bar are identified with the letters *N* and *S* for *North* and *South*. The ends of magnets are called *poles*. For a basic understanding of how electricity is generated in commercial quantities, it is important to understand the concept of magnetic force.

When two magnets are placed near each other, one can cause the other to move. The force causing the movement is related to the strength of the magnetic field surrounding each magnet. The stronger the magnetic field of one magnet, the greater the force upon the other magnet will be. The closer the magnets are to one another, the greater the force will be. Conversely, as distance between the magnets increases, the magnetic force decreases. The decrease in magnetic force is inverse to the square of the distance between the magnets. This means that magnetic force would be one-fourth as great at 6 cm as at 3 cm. The strength of magnetic force also varies inversely with the permeability of the free space between the magnets, according to this formula:

$$F = \frac{m_1 m_2}{\mu_0 r^2}$$

In this equation r is the distance between magnets, and m_1 and m_2 are the strengths of the magnetic poles of the two magnets. μ_0 is the permeability of free space, a constant that depends on the medium in which the magnets are placed. $\mu_0 = 1$ in a vacuum and is only slightly different in air.

Figure 6-3 demonstrates the nature of polar magnetic attraction. A bar magnet is placed beneath a sheet of paper. Iron filings are sprinkled on top of the paper. These iron filings are attracted by the magnetic field. Greater numbers of iron filings are attracted to the poles of the magnet. The force lines of the magnetic fields are shown through the alignment patterns in which the iron filings distribute themselves.

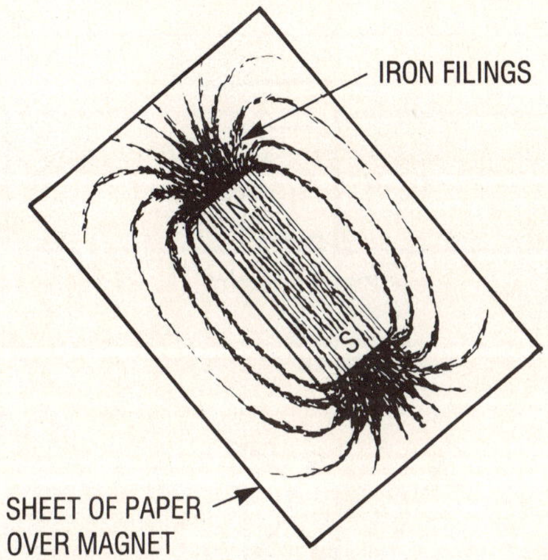

Figure 6-3. Iron filings attracted to poles of magnet.

Magnetic poles retain their force even when the form of a magnet changes. Figure 6-4A shows the magnetic lines of force for horseshoe-shaped magnets. In Figure 6-4B, iron filings are distributed around the poles of a horseshoe-shaped magnet.

Figure 6-4 highlights a basic characteristic of the magnetic field between the poles of a magnet. That is, lines of force are continuous between north and south poles. These are opposite poles. Opposite magnetic poles attract each other.

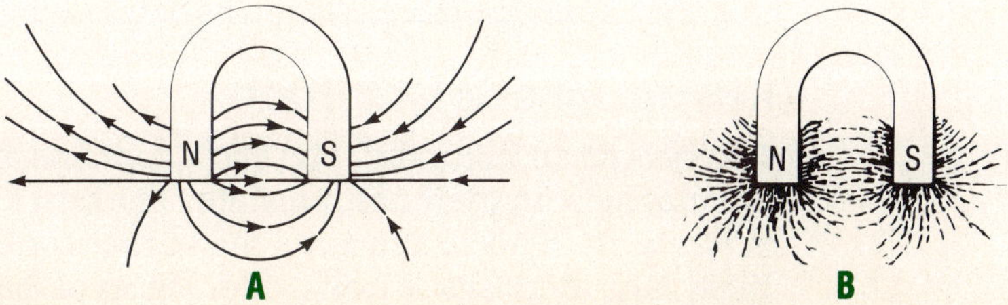

Figure 6-4. Magnetic lines of force.

It is also true that the magnetic forces of like poles are in opposition; the poles repel each other. This is illustrated in Figure 6-5. Figure 6-5A shows that the north and south poles of two bar magnets, when they face each other, are attracted to each other. Lines of force draw these poles together. In Figure 6-5B, two north poles are placed in proximity to each other. The force lines repel each other. The same is true for the two south poles that are placed close to each other, as in Figure 6-5C.

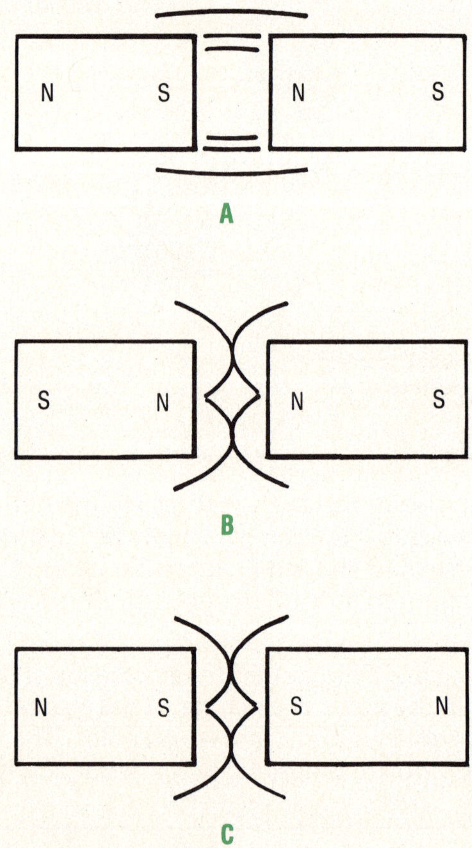

Figure 6-5. Like magnetic poles repel each other, opposite poles attract each other.

Ferromagnetic Materials

Magnets occur naturally. The earth contains materials which are magnetic in their natural state. The best known of these is *magnetite*. This is a ferrous metal that contains comparatively large deposits of iron oxide. Pieces of magnetite ore are called *lodestones*.

Within a lodestone, ferrous molecules are aligned; lodestones occur this way in nature. If iron particles are placed in the vicinity of a magnet such as a lodestone, they will line up along the fields of magnetic force. This means that magnets affect the molecules of all ferrous materials, whether they are naturally magnetized or not. Thus, a lodestone or other existing magnet can be used to create another magnet.

Figure 6-6 illustrates one way to create a magnet. This shows an existing magnet being rubbed along the length of a bar of ferrous material. As the magnet is moved along the iron bar, the molecules within the bar align themselves north and south. A new magnet is created.

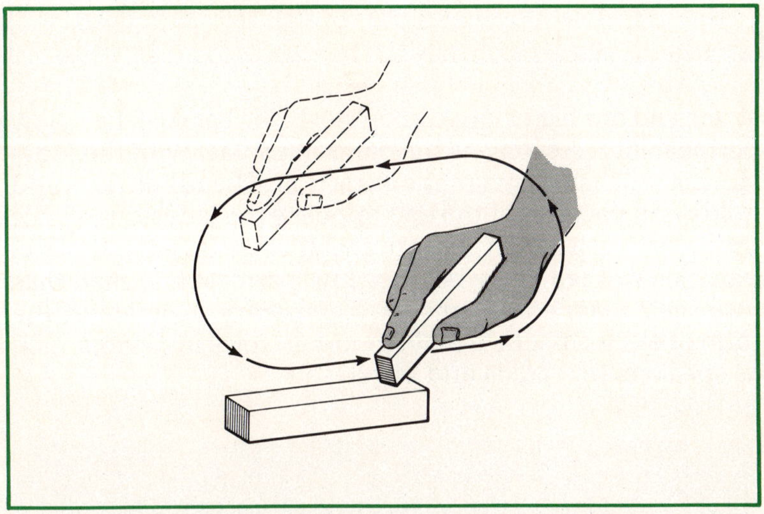

Figure 6-6. Magnet rubbed along bar of ferrous material creates another magnet.

Different types of ferrous materials have different properties for acquiring magnetism. The indication that a ferrous material has become magnetized is called *magnetic flux*. The ability to conduct magnetic flux is called the *permeability* of a material. The higher the permeability, the greater the ability of the ferrous material to conduct magnetic flux. Both high- and low-permeability materials have their places in electrical and electronic applications.

Retentivity is the ability to retain magnetism. A material of high retentivity, for example, will hold its magnetism for long periods of time, approaching permanence. In general, hard steel substances have high retentivity and high permeability ratings. High-carbon steel is a good example.

There are times when it is desirable to have a magnet that loses its field of force rapidly. This is known as a temporary magnet, since magnetism exists only when some external force is applied. Figure 6-7 illustrates the principle of temporary magnetic attraction. This shows a lodestone placed

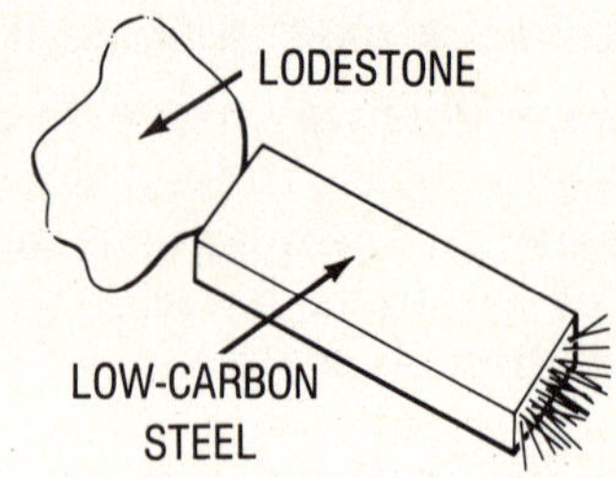

Figure 6-7. Temporary magnet created by placing lodestone at one end of bar of low-carbon steel.

at one end of a bar of low-carbon steel. The bar itself has a low permeability. As long as the bar is in contact with the lodestone it will have a magnetic attraction. The bar attracts and holds iron particles at the end opposite the lodestone. Once the lodestone is removed from a temporary magnet, the iron particles will fall away because low-permeability materials lose their magnetism quickly. Temporary magnets in the form of electromagnets play important roles in devices such as speakers for radios and stereo sets.

Discussion Questions/Problems

1. What are ferrous materials?
2. What happens to the molecules of a magnetized iron bar?
3. What are magnetic poles?
4. What are magnetic field lines?
5. What influence do like magnetic poles have upon each other?
6. What influence do opposite magnetic poles have upon each other?
7. What is permeability and how does it affect the magnetism of a ferrous substance?

6.2 PERMEABILITY TYPES AND INTENSITIES

The permeability of materials determines their value and potential use in magnetic applications. Because this factor is so important, systems for classifying and measuring permeability have been developed.

Permeability is based on a scale in which the value for a vacuum is 1. Other materials are measured on this scale. Individual materials, as seen in Figure 6-8, have permeability ratings ranging from less than 1 up to 74,000.

Permeability is classified into these broad categories:

- Ferromagnetic materials
- Paramagnetic materials
- Diamagnetic materials.

These three classifications are based upon the permeability ratings of materials. Ferromagnetic materials are those with the greatest permeability. Paramagnetic materials are those substances with permeability ratings of slightly more than 1. Materials with ratings below 1 are all classed as diamagnetic.

Classification	Material	Permeability
DIAMAGNETIC MATERIALS	BISMUTH	0.999833
	QUARTZ	0.999985
	WATER	0.999991
	COPPER	0.999995
PARAMAGNETIC MATERIALS	LIQUID OXYGEN	1.00346
	OXYGEN (S.T.P.)	1.0000018
	ALUMINUM	1.0000214
	AIR (S.T.P.)	1.0000004
FERROMAGNETIC MATERIALS	NICKEL	40.
	COBALT	50.
	IRON	7,000.
	PERMALLOY	74,000.

Figure 6-8. Permeability ratings of materials.

A general understanding of permeability and the groupings of materials according to permeability rating is necessary for learning about electromagnetic and electrical generating systems. These are discussed later in this chapter and in following chapters.

Discussion Questions/Problems

1. What is the significance of a low number on the permeability rating scale?
2. What is the significance of a high number on the permeability rating scale?
3. If a material has a permeability rating of 7,000, what is its permeability classification?

6.3 MAGNETIC CIRCUITS

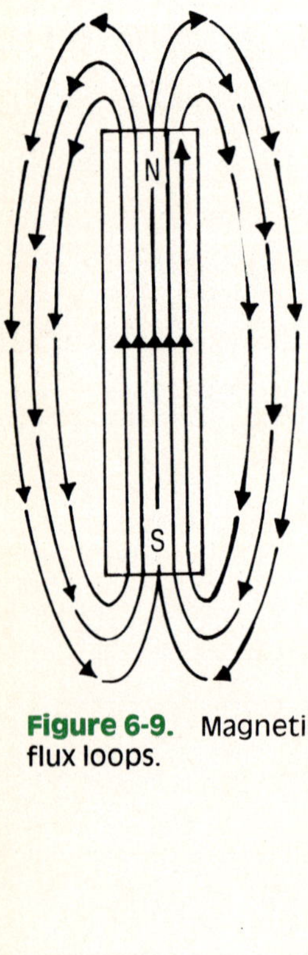

Figure 6-9. Magnetic flux loops.

Just as there are electrical circuits, there are also magnetic circuits. Magnetic circuit parameters are related by a law similar to Ohm's law for electric circuits. The formula for this is:

$$\phi = \frac{\text{mmf}}{\mathcal{R}}$$

In this equation, ϕ is magnetic flux, corresponding to current in Ohm's law; mmf is the magnetomotive force, which corresponds to the voltage in electrical circuits. The symbol $\mathcal{R}$ is the magnetic reluctance, corresponding to resistance in electrical circuits.

Magnetic flux (ϕ) is the group of magnetic field lines that surround a magnet. These lines form closed loops around a magnet, with the ends of the loops at the poles of the magnet. Flux loops are shown in Figure 6-9.

Magnetomotive force (mmf) is the force that creates magnetic flux.

Reluctance ($\mathcal{R}$) is the opposition to flux. The value of reluctance depends on the dimensions and material of the magnetic circuit. Reluctance may be combined in series or parallel like resistance in an electrical circuit. Series reluctances are added, just as are series resistances.

In addition, two other properties of magnetic circuits need to be understood:

- The *magnetic flux density* (B) is a measure of how concentrated the magnetomotive force is. A heavy concentration of magnetic flux lines results in a strong magnet. Conversely, the more dispersed, or spread out, the flux lines are, the less the magnetic flux density and the weaker the magnet. B is a measure of the magnetic flux per unit area (A), or:

$$B = \frac{\phi}{A}$$

- *Permeability* (μ) is the ability to conduct magnetic flux. A high permeability means that a material can conduct large amounts of highly concentrated magnetic flux. This relationship between permeability and magnetic flux density can be expressed as:

$$\mu = \frac{B}{H}$$

In this equation H is the *magnetizing force*. In air, H is numerically equal to the flux density (B). In iron, B can be many times larger than H because μ is large.

Discussion Questions/Problems

1. What is magnetic flux?
2. What is magnetomotive force?
3. What is a flux loop?
4. What is the formula for determining magnetic flux?
5. What is magnetic flux density?
6. What is the formula for describing magnetic flux density?
7. How does flux density vary according to the position on or near a magnet where it is measured?

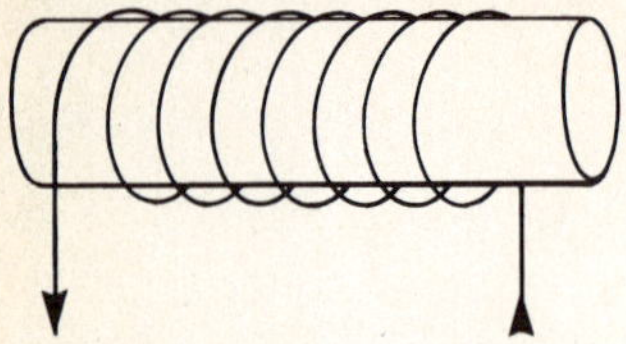

Figure 6-10. Temporary magnet created by surrounding a steel dowel with electrical coil and passing a current through coil.

Figure 6-11. Symbol for air-core coil.

Figure 6-12. Symbol for iron-core coil or electromagnet.

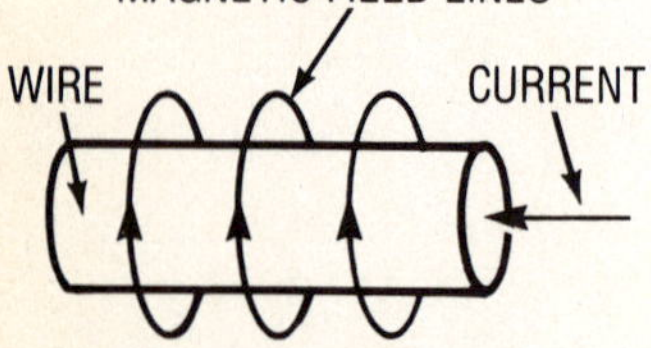

Figure 6-13. When current passes through length of wire, evenly spaced magnetic field lines encircle conductor at right angles to direction of current flow.

6.4 ELECTROMAGNETISM

An electromagnet is a temporary magnet. One way to create a temporary magnet is to place a piece of low-carbon steel in contact with a permanent magnet. The low-carbon steel acts as a magnet as long as the permanent magnet is in place. When the permanent magnet is removed, the magnetic flux goes with it.

Temporary magnets can also be created by placing pieces of low-carbon steel into magnetic fields. One configuration, shown in Figure 6-10, is to surround a round steel bar, a dowel, with an electrical coil. When current is applied to the coiled wire, the ferrous material becomes magnetized. If high-carbon steel is used, it is possible to create permanent magnets in this way. With low-carbon steel, the magnetism is temporary.

The magnetism transferred to ferrous materials in this way, either temporarily or permanently, comes from a property known as *electromagnetism*. Figure 6-11 is the symbol for an air-core coil. A coil of wire, with current flowing through it, becomes an electromagnet. The electromagnetic flux is increased when the electric coil is wrapped around a ferrous material, as shown in Figure 6-10. The symbol for an iron-core coil or electromagnet is shown in Figure 6-12.

Electromagnetism was discovered in 1820 by a Danish scientist, Hans C. Oersted (1777–1851). In his early experiments, Oersted placed a needle near an electric wire. He noted that when current was passed through the wire the needle would deflect.

Later research determined that current flowing through a conductor creates a magnetic field around the conductor. Magnetic field lines are formed around the electrical conductor. These magnetic field lines encircle the conductor at right angles to the direction of current flow. The field lines are uniformly spaced along the length of the conductor.

This principle is illustrated in Figure 6-13. The drawing shows a current passing through a length of wire. Evenly spaced magnetic field lines are formed around the conductor and at right angles to the direction in which the current is flowing.

When a single wire is used, the magnetic field generated is not strong. This explains why coils are commonly used for electromagnets. By winding wire tightly into coils, the electromagnetic field from each of the turns is added and concentrated in a relatively small volume. Magnets with large amounts of flux can be formed by coils of wire wound in such a way as to get a great many turns into a very small space.

An easy way to prove this principle through a simple experiment is to use any electric wire and a compass. Pass a current through the wire and place the compass near it. The compass needle will point at right angles to the flow of current through the wire, indicating that the needle is lining up with the circular lines of the magnetic field. The N pole will point in the direction of the magnetic lines. This is illustrated in Figure 6-14.

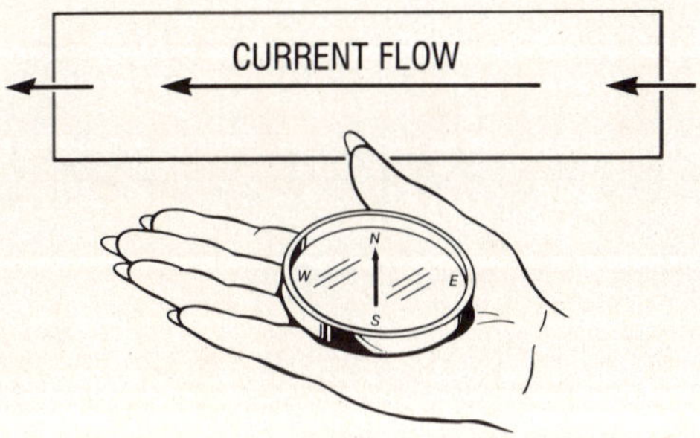

Figure 6-14. Compass needle points at right angles to flow of current through nearby electric wire, and N pole points in direction of magnetic lines.

A simple method can be used to determine the relationship between current direction and magnetic field lines: Suppose you have determined the magnetic field direction (the direction to which the N pole points), as shown in Figure 6-14, through use of a compass. You still don't know the direction in which the current is flowing. To determine the direction of current flow, use what is called the *Right-Hand rule,* which is illustrated in Figure 6-15. To do this, imagine that your right hand is wrapped around the electric wire with your fingers curled in the direction of the magnetic field lines. The direction in which your right thumb would point is the direction in which the electric current flows. (Similarly, if you

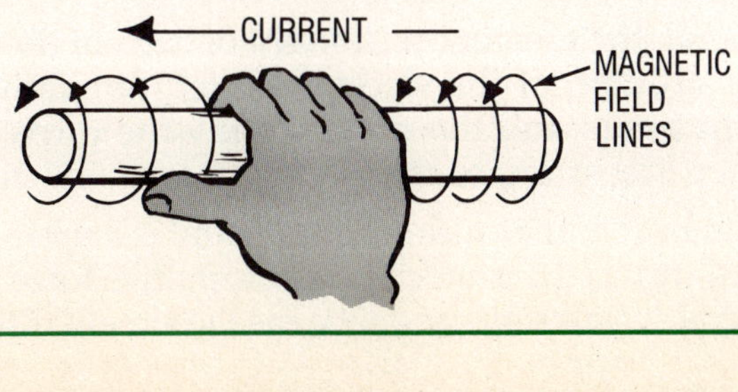

Figure 6-15. Using right-hand rule, thumb points in the direction of current flow.

were to imagine using your left hand, your left thumb would point in the direction in which the electrons flow.)

A word of caution: This test should never actually be done. It is a mental exercise only. Never wrap your hand around a bare wire!

Discussion Questions/Problems

1. What is the source of magnetic force in electromagnets?
2. Are all electromagnets temporary or can they be permanent as well?
3. What is the relationship, in terms of directions, between current flow and magnetic field lines?
4. If you know the direction of magnetic field lines around a conductor, how do you determine the direction of current flow?

6.5 PROPERTIES OF ELECTROMAGNETS

The performance of an electromagnet depends on a number of variables. These variables are the electrical and physical characteristics of an electromagnet's construction. Some of the key characteristics that give electromagnets their properties are discussed below.

Magnetomotive Force (mmf)

The strength of an electromagnetic field, which is the magnetomotive force (mmf), depends upon two factors:

- The amount of current that flows through a coil of wire
- The number of turns of wire in the coil.

In general, the greater the amount of current the greater will be the strength of the electromagnet. Similarly, the more turns in a coil, and the more closely the turns are wound together, the stronger the electromagnet will be.

The unit of measurement for mmf is *ampere-turns*. (The ampere-turn is the unit for magnetomotive force used in one of several systems for magnetic measurement discussed later in this chapter.) The symbol for ampere-turns in an electromagnet is *NI*. *N* stands for the number of turns. *I* is the amount of current in amperes. The value of *NI* is calculated by multiplying the number of turns by the current. As an example, suppose a coil has 100 turns and carries 2 A of current. The value of *NI* is determined as follows:

$$NI = 100 \text{ turns} \times 2 \text{ A} = 200 \text{ Ampere-turns}$$

Electromagnetic Field Intensity (*H*)

In an electromagnet, the closer the turns of wire are in a coil, the greater the field intensity will be. In the discussion above, ampere-turns were described without relationship to the closeness of wire turns to one another. Another dimension of electromagnetic properties lies in the radius of the coil and the length of the coil. As a principle, the smaller the radius, the greater the field intensity. Also, the longer the coil, the greater the field intensity.

In one of the measurement systems, field intensity (*H*) is measured in ampere-turns/metre. The formula used to calculate field intensity is as follows:

$$H = \frac{2\pi\, NI}{10\, r}$$

In this formula, the value of π is standard: 3.14. The symbol *r* represents the radius of the coil, in metres. An example of an application of this formula is given below. In this equation, *NI* is 200 ampere-turns. The value of *r* is 0.5.

$$H = \frac{2\pi\, NI}{10\, r} = \frac{6.28 \times 200}{10 \times 0.5} = \frac{1{,}256}{5} = 251.2$$

Measuring Electromagnetic Field Intensity

To calculate the magnetic field intensity, also known as *magnetizing force,* at a given point, measure the current (*I*) and the distance from the axis of the conducting wire. This distance is measured in metres. The symbol for distance in the formula for determining field intensity is *d*. The formula is:

$$H = \frac{I}{5d}$$

To illustrate, consider a circuit with 15 A of current. Assume field intensity is measured at a distance of 3 cm from the center or axis of the conductor. The calculation is as follows:

$$H = \frac{I}{5d} = \frac{15\ \text{A}}{5 \times 3\ \text{cm}} = 1\ \text{oersted}$$

Magnetic Polarity

The direction of the windings and the direction of the electron flow are the keys to determining polarity. To determine which end of an electromagnetic coil will be the north pole, imagine that you grasp the coil with your right hand. Your fingers would point in the same direction as the flow of electrons through the wire. Your thumb would point toward the north pole. See Figure 6-16. Figure 6-17 shows a series of electromagnets and illustrates that reversing the direction of either the winding or the current flow will reverse the polarity of the magnet. Reversing both will leave the polarity unchanged.

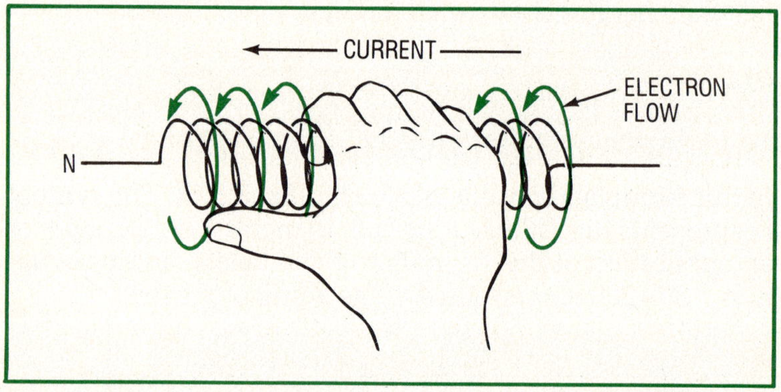

Figure 6-16. Directions of coil windings and electron flow determine magnetic polarity.

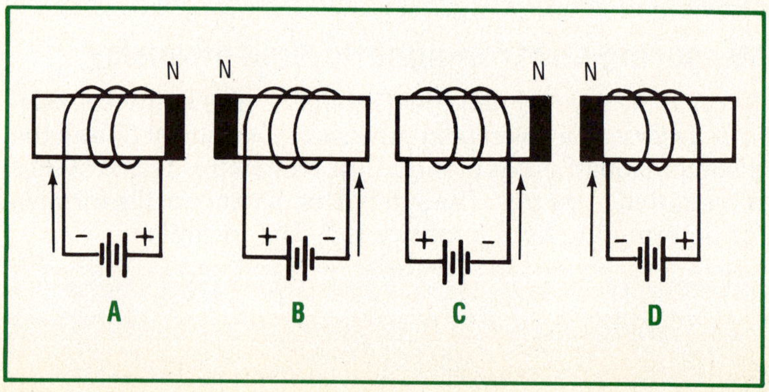

Figure 6-17. Reversing direction of winding or current flow reverses magnetic polarity while reversing both leaves polarity unchanged.

Residual Magnetism

When an electromagnet is de-energized, some degree of magnetism remains—if the electromagnet has an iron core. This residual magnetism is slight. But it must be taken into account in planning for applications using electromagnets.

Hysteresis

The term *hysteresis* means "lagging behind." When a magnetic material has been magnetized and the magnetic field around the material is then reduced to zero, there is still a residual magnetism. Hysteresis is a measure of the relationship between residual magnetism and an energy loss in the material. If an electromagnetic coil is continually energized and de-energized, there is considerable energy and heat produced in the iron core of the coil. This is due to the fact that the molecules in the material are constantly being aligned first in one direction, and then in the other.

The magnetic properties of a material are usually plotted in a curve called a *B/H curve* or *hysteresis loop*. As shown in Figure 6-18, a series of points are plotted showing the relationship of field intensity and flux density as the field intensity is increased, decreased to zero, reversed and increased,

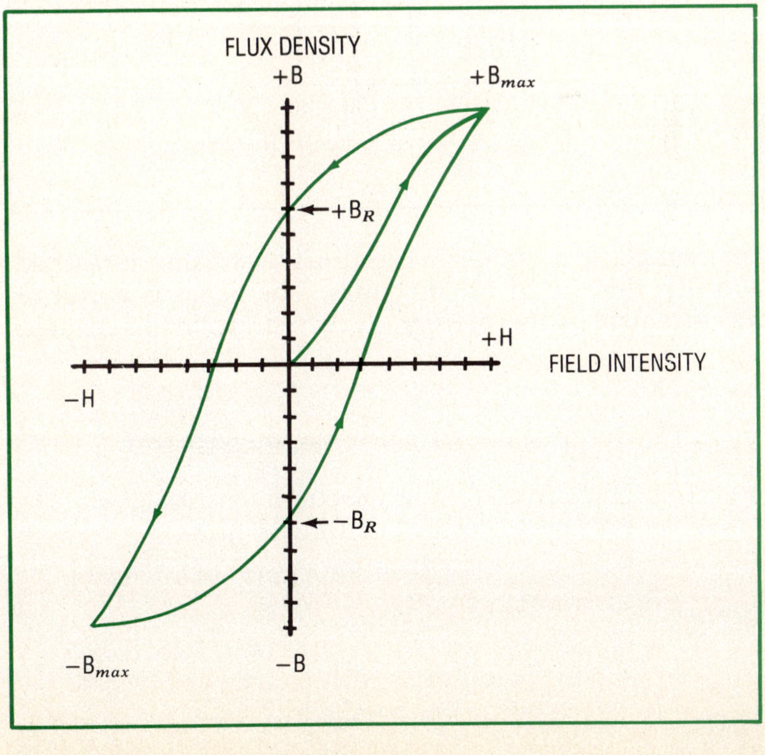

Figure 6-18. Hysteresis loop, or *B/H* curve, showing relationship of field intensity to flux density.

then reduced again to zero. The hysteresis loop shows the relationship of applied magnetism to residual magnetism for the particular magnetic material.

Saturation

A given material of a given size can build only a certain amount of magnetic force. After that, it reaches its limit and becomes saturated. Cores of electromagnets can become saturated. The levels of this saturation depend upon the permeability of the materials used.

When a coil material reaches saturation, there is nothing that can be done to increase the flux density (B) of a magnet.

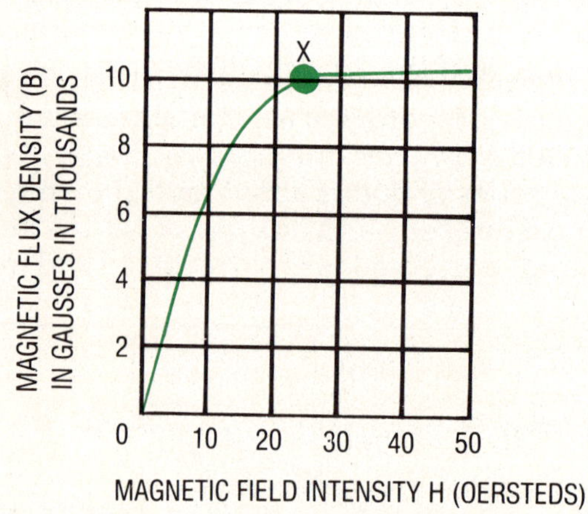

Figure 6-19. Saturation reached when material has reached its limit of flux density.

For example, in the graph in Figure 6-19, saturation is reached at a flux density of 10,000 gauss. The gauss is a unit of measurement for flux density, discussed later in this chapter.

Discussion Questions/Problems

1. What property of an electromagnet is determined by its number of ampere-turns?
2. What two factors determine the field intensity of an electromagnet?
3. If a conductor carries 20 A of current, what is the magnetic field intensity at a distance of 4 cm from the axis of the conductor?

4. What factor determines the polarity of an electromagnet?
5. What is residual magnetism?
6. What is hysteresis and how does it affect the performance of electromagnets?
7. What is saturation and how does it affect the performance of electromagnets?

6.6 USES OF ELECTROMAGNETISM

Electromagnetism provides many opportunities, products, and potential problems for the designer or engineer. Some of these are covered below.

The Hall Effect

The Hall effect is the small electrical potential that is created in a wire when a stationary magnetic field is applied perpendicular to the wire. The effect was discovered by and named after Edwin Herbert Hall (1855–1938), an American physicist.

The Hall effect is used to generate low-voltage electricity through magnetism. This principle is used in designing certain semiconductor components. These components, in turn, are used in instruments that measure magnetism.

Solenoids

A solenoid is a primary example of a practical use of electromagnetism. A *solenoid*, basically, is a tightly wound coil of wire in the shape of a cylinder. When current is applied, a magnetic field is created within and around the solenoid's coil.

An iron plunger rests loosely within the coil when there is no current. A relatively low-tension spring holds the iron core outside the center of the solenoid when no current is applied. When a current is applied, it creates a magnetic field. The magnetic attraction overcomes the holding power of the spring and the iron core is drawn into the solenoid.

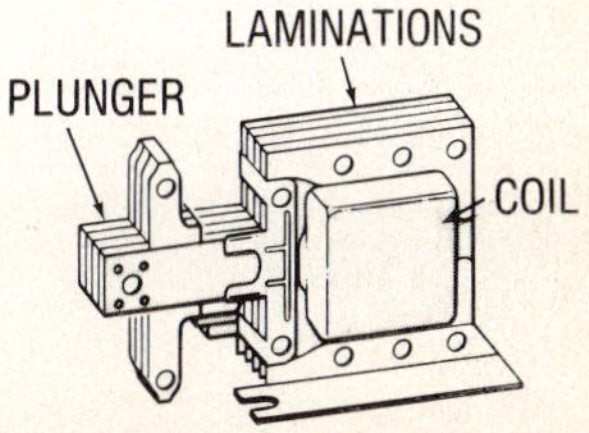

Figure 6-20. A solenoid.

A basic solenoid mechanism is illustrated in Figure 6-20. Applications for this type of device are widespread. For example, solenoids are commonly used as activators for relays, switches, circuit breakers, chimes, and motors.

Solenoids provide a factor of safety in the use of high-power electrical devices because they can be used as remotely operated switches. Thus, people do not have to be in the vicinity of high-power devices to turn them on and off. This greatly reduces the potential danger and increases the safety factor in the operation of high-power electrical and power transmission equipment.

Magnetic Shielding

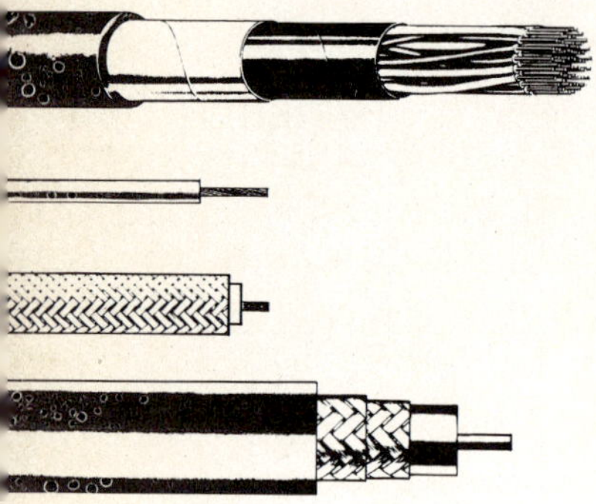

Figure 6-21. Shielding on electrical wires and cables.

Just as there are situations in which magnetism plays a valuable role, so also are there situations in which magnetic fields created by electrical wires could be a problem. For example, in sound (audio) systems, wires that pass close to each other can set up magnetic fields that interfere with the flow of current in adjacent carriers. These magnetic fields can distort electrical properties and the resulting sounds. The effect is often referred to as *crosstalk.*

To prevent the effects of magnetism in places where interference is unwanted, electrical carriers are shielded with magnetic materials. Figure 6-21 illustrates several types of shielding used on electrical wires and cables.

Discussion Questions/Problems

1. To create the Hall effect, what relationship must exist between a magnetic field and an electrical wire?
2. What component of a solenoid creates its magnetic field?
3. How is a solenoid plunger held in position when no current is applied to the unit?
4. What causes the solenoid plunger to move into the coil?
5. What is shielding and how is shielding used on electrical wires and cables?

6.7 INDUCTION THROUGH ELECTROMAGNETISM

If a magnet is moved across a wire, the movement of the flux field generates electricity. A current is created in the wire by the changing magnetic flux. This effect is called *electromagnetic induction*. This principle can be applied to generate commercial quantities of electricity.

A practical approach to electromagnetic induction is shown, in simplified form, in Figure 6-22. A magnet is moved through an electrical coil. Thus, its flux field cuts across all of the wires in the coil. This movement of the magnetic flux induces a current in the wire. Since the magnetic flux causes a current to be induced, it follows that the more turns there are in a coil, the greater the voltage that will be induced.

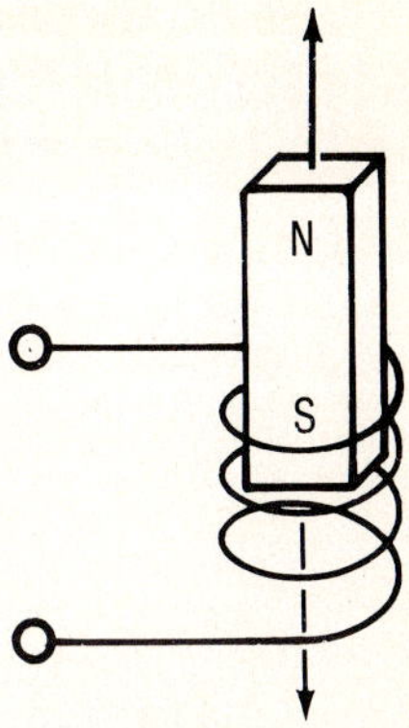

Figure 6-22. Magnet moved through electrical coil induces a current in the coil.

Figure 6-22 introduces a principle only. In practice, a different design approach is taken in the use of electromagnetism to generate electric current. Electric generators are usually built in a round, or cylindrical, pattern. Magnets spin rapidly within wire coils to generate electricity.

Faraday's Law

Michael Faraday (1791–1867), the British physicist and chemist, invented the first electric generator and formulated the laws of magnetic induction. Faraday's law describes the relationship of magnetism and electromagnetic induction. The principles are as follows:

- The more magnetic lines that cut across a conductor, the higher the voltage that is induced.
- The more turns there are in a coil, the higher the voltage will be.
- The faster the movement of the cutting action with magnetic lines, the higher the voltage will be.

Discussion Questions/Problems

1. What property of a magnet induces the flow of current in a wire?
2. What has to be done to a magnet to cause it to induce current in a wire?
3. How does the number of flux lines in a magnet affect the voltage which can be developed across a coil?
4. How does the number of turns in a coil affect the voltage which can be developed across a coil?
5. How does the speed with which a magnet is moved past a coil of wire affect the voltage developed across the coil?

6.8 UNITS OF MAGNETISM

Three systems of units are used for the measurement of magnetism. These are:

- The English units system in terms of inches, pounds, and ampere turns
- The centimetre-gram-second (Cgs) system
- The metre-kilogram-second-ampere (Mksa) system which uses *Systeme International d'Unites* (SI) units.

The relationship between these three systems is shown in Figure 6-23. The system most commonly used in today's electronics industry is the Mksa system using (SI) units.

		SYSTEM OF UNITS		
TERM	SYMBOL	ENGLISH	CGS	MKSA (SI)
magnetic flux	ϕ	line of force	maxwell (Mx)	weber (Wb)
magnetomotive force	mmf	ampere-turn	gilbert (Gb)	ampere-turn
magetizing force, or magnetic field intensity	H	ampere-turn/inch	oersted (Oe) or gilbert/cm	ampere-turn/metre
magnetic flux density	B	line/sq. in.	gauss (G) or maxwell/sq. cm	tesla (T) or weber/sq. metre
permeability	μ	(none)	gauss/oersted	henry/metre
retentivity	(none)		(no name)	
reluctance	$\mathfrak{R}$	(none)	gilbert/maxwell	ampere-turn/weber
permeance	P		(no name)	

Figure 6-23. Units of magnetism in English, Cgs, and Mksa systems.

Magnetic Flux (ϕ)

Units of measure for magnetic flux (ϕ) are discussed below.

Line of force. The line of force is the unit of magnetic flux in the English system.

Maxwell (Mx). The maxwell (Mx) is the unit of magnetic flux in the Cgs system. It is named after James Clerk Maxwell (1831–1879), a Scottish mathematical physicist.

The principle behind this measure is that the stronger the magnet, the more field lines it will create. A maxwell is a unit of magnetic flux equal to one magnetic line of force (magnetic field line). Thus, a magnet with 30 Mx has as much magnetic flux as a magnet with 30 magnetic lines of force.

Weber (Wb). The weber (Wb) is a unit of magnetic flux in the Mksa system. It is named after Wilhelm Weber (1804–1890), a German physicist. The weber (Wb) is used to measure large magnetic flux fields. To compare the weber (Mksa) to the maxwell (Cgs), 1 weber is equal to 1×10^8 Mx. This is such a large amount of magnetic flux that there is rarely need to use a full Wb in describing magnetic fields. Therefore, webers are often stated in terms of the microweber (μWb). A microweber is equal to 10^{-6} Wb. Another way of stating the relationship is that a microweber is equal to 100 Mx.

Magnetomotive Force (mmf)

Units of measure for magnetomotive force (mmf) are discussed below.

Ampere-turn. The ampere-turn is the unit of magnetomotive force in both the English system and the Mksa system. Ampere-turns can be calculated by multiplying the number of turns of wire in a coil by the current (in amperes) flowing through the coil.

Gilbert. The gilbert is the unit of magnetomotive force in the Cgs system. It is named for William Gilbert (1540–1603), an English physicist. The gilbert is the magnetomotive force that would establish 1 maxwell of magnetic flux in a magnetic circuit having a reluctance of 1 unit.

Magnetic Field Intensity (*H*)

The units of measure for magnetic field intensity or magnetizing force (H) are discussed below.

Ampere-turn/inch. The ampere-turn/inch is the English system unit for magnetizing force. It is the magnetic field intensity produced at a distance of 1 inch from a coil of 1 turn through which a current of 1 ampere flows.

Oersted. The oersted is the Cgs system unit for magnetizing force. It is named for the Danish scientist, Hans C. Oersted. The oersted is equal to $1{,}000/4\pi$ ampere-turns per metre. The oersted is the same as a gilbert/cm, and either term is correct.

Ampere-turn/metre. The ampere-turn/metre is the Mksa system unit for magnetizing force. It is the magnetic field intensity (which is the same as magnetizing force) produced at a distance of 1 metre from a coil of 1 turn through which a current of 1 ampere flows. For example, a current of 5 amperes through a 10-turn coil would produce a magnetic field intensity 50 times as great as a current of 1 ampere through 1 turn.

Magnetic Flux Density (*B*)

The units of measure for magnetic flux density are described below.

Line/inch². The line/inch² is the English system unit for magnetic flux density. One line/inch² is equal to 6.45 maxwell/cm² or 6.45 gauss (see below).

Gauss. The gauss is the unit of flux density in the Cgs system. It is named for Karl F. Gauss (1777–1855), a German mathematician. As explained earlier in this chapter, *flux density* is a term describing the concentration of magnetic field lines, in terms of flux lines in a given cross-section area. The closer together the magnetic field lines are, the greater the flux density. One gauss is equal to the flux density of one magnetic field line per square centimetre of area. Stating it another way, 1 gauss is equal to 1 Mx per square centimetre of area.

Flux density is determined by the following formula:

$$B = \frac{\phi}{A}$$

B is the symbol for flux density. The symbol for the area through which magnetic flux lines flow is *A*. The flux lines are in a direction parallel to the direction of the flux field. Area is measured perpendicular to the direction of the flux field. Using the Cgs system as an example, area is measured in terms of centimetres squared (cm²). Magnetic field is measured in terms of maxwells (Mx). The answer is stated in gauss (or the equivalent Mx/cm²):

$$B = \frac{\phi}{\mathrm{A}} = \frac{200\ \mathrm{Mx}}{2\ \mathrm{cm}^2} = 10{,}000\ \text{gauss} = 10{,}000\ \mathrm{Mx/cm}^2$$

Flux density measurements vary with the point at which measurements are taken. A measurement taken close to a pole of a magnet will be higher than one taken at a distance from the pole. Putting it another way, flux density increases at points approaching a magnetic pole.

Tesla. The tesla is the unit of magnetic flux density in the Mksa system. The tesla is named for Nikolai Tesla (1857–1943), an American born in Siberia. One of the great inventors of his time, he made many contributions to the use of AC. He invented the tesla coil. One tesla (Mksa) is equal to 1 weber per square metre.

Permeability, Retentivity, Reluctance, and Permeance

Units of measure for permeability, retentivity, reluctance, and permeance are discussed below.

Permeability. As discussed in the beginning of this chapter, the symbol used for permeability is μ. There is no unit of permeability in the English system. In the Cgs system, the unit of permeability is the gauss/oersted. In the Mksa system, it is the henry/metre.

Permeability is calculated by the following formula:

$$\text{Permeability } \mu = \frac{\text{Flux density}}{\text{Field intensity}}$$

As an example, given a flux density of 10,000 and a field intensity of 200, permeability is:

$$\begin{aligned}\text{Permeability } \mu &= \frac{B}{H}\\ &= \frac{10{,}000}{200}\\ &= 50 \text{ henry/metre}\end{aligned}$$

Retentivity. Retentivity, the ability to retain magnetism, is discussed earlier in this chapter under section 6.1. It has no symbol. There are no special names or terms for units of retentivity.

Reluctance. Reluctance, the opposition to establishing a magnetic field in a magnetic circuit, is also discussed in section 6.1. No name is assigned to the unit of reluctance in the

English system. In the Cgs system, the unit of reluctance is the gilbert/maxwell. In the Mksa system, the unit of reluctance is the ampere-turn/weber.

Permeance. As explained in section 6.1, permeance is the opposite of magnetic reluctance. The relationship is similar to that between siemens and resistance. Permeance measures the ease with which a material permits magnetic field lines to pass through or permeate. For example, steel has an extremely high permeance, since it facilitates the passage of magnetic field lines.

Expressed mathematically, permeance is the reciprocal of reluctance:

$$\text{Permeance} = \frac{1}{\mathcal{R}\ \text{(reluctance)}}$$

If permeance and magnetomotive force of a magnetic field are known, these values can be used to determine the magnetic flux. Units of permeance have not been given names.

Discussion Questions/Problems

1. What does a maxwell measure?
2. How many maxwells make up a weber?
3. What is the principle behind the maxwell?
4. What is a gauss?
5. What does a gilbert measure?
6. What are the units of measure for reluctance?
7. What is permeance?

ALTERNATING CURRENT AND VOLTAGE

7.1 ALTERNATING CURRENT (AC)

Alternating current, abbreviated AC, is the type of electricity most widely used for commercial purposes. In the United States and Canada, virtually all commercial, industrial, educational, and residential electrical outlets use alternating current.

The principles of electricity covered thus far in this book deal with a different type of current. These previous descriptions apply to direct current, or DC. There are differences between AC and DC. The most basic difference is the way the two types of currents flow through conductors.

AC Flow Patterns

Direct current, as illustrated in Figure 7-1 and discussed earlier in this book, flows through electrical circuits in a single direction. Electrons move from a negative pole (source) to a positive pole (sink). In the process, current is delivered to loads. The load shown in Figure 7-1 is the symbol for an electric lamp. The characteristic to remember about the diagram in Figure 7-1 is that direct current flows in one direction only. The flow is constant and even. The characteristics of DC are:

- Single direction of flow
- Steady flow of current
- Relatively steady voltage value.

Figure 7-2 shows the same type of circuit—except that the source is alternating current. As shown, the direction of current flow reverses. AC flows first in one direction, then in the

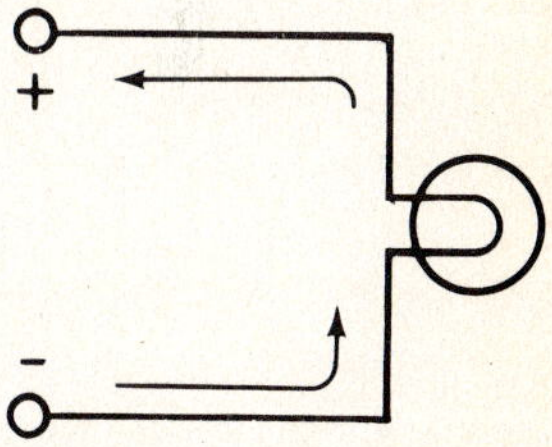

Figure 7-1. Circuit with direct current.

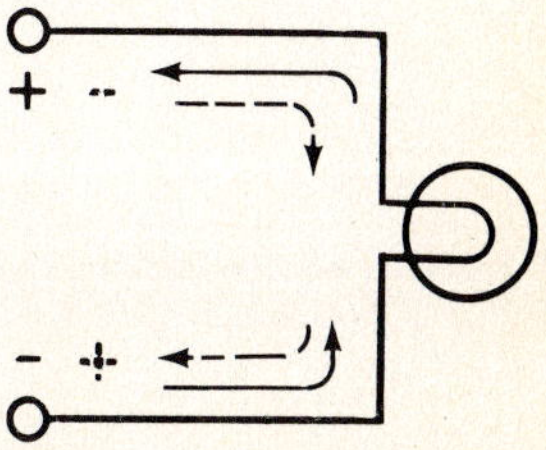

Figure 7-2. Circuit with alternating current.

other; that is, the flow alternates in direction. The time intervals between flow directions may vary. But, for commercially generated power, the rate of alternation will be predictable and constant. For most commercial electrical power in the United States and Canada, the rate of alternation is 60 times per second. The nature and forms of current alternation are discussed later in this chapter.

Another characteristic of AC is that there are changes in voltage polarity to go with the changes in direction of current flow. Since electrons always flow from negative to positive, and since the flow direction changes, the negative source and positive sink must be continually reversing in AC circuits.

Another difference between DC and AC lies in voltage values. With DC, voltage value is relatively constant. With AC, voltage values change continuously. For example, in a 110 V AC circuit, voltage values vary from 0 to 110 V constantly, with alternations 60 times per second in a system delivering commercial power.

AC Versatility

Electricity is easier and more economical to deliver and use in AC form than in DC. As one example, with AC, voltage can readily be increased or decreased or otherwise adapted to the needs of the load or power system. This level of flexibility in voltage is just not possible with DC.

In particular, it is much easier to increase, or step up, AC voltages than DC voltages. Similarly, it is easier to decrease, or step down, AC voltage.

Because of its inherent versatility, AC is used to power sound systems, television sets, and many other devices familiar to everyday life.

To demonstrate AC versatility, consider the processing that takes place in the normal, routine delivery of alternating current from generators to consumers. The steps in this process are shown in simplified form in Figure 7-3.

AC is generated at power plants operated from two basic types of power sources—water or steam. Water power harnesses the weight of falling water to turn generators. Steam comes from heating water, which turns to steam, expands, and turns generators. Sources for heat in steam-operated plants can vary, including coal, oil, natural gas, and nuclear energy. Other, experimental, heat sources include the heat of the sun and the heat of the earth through geothermal wells. Wind is used experimentally to power generators directly.

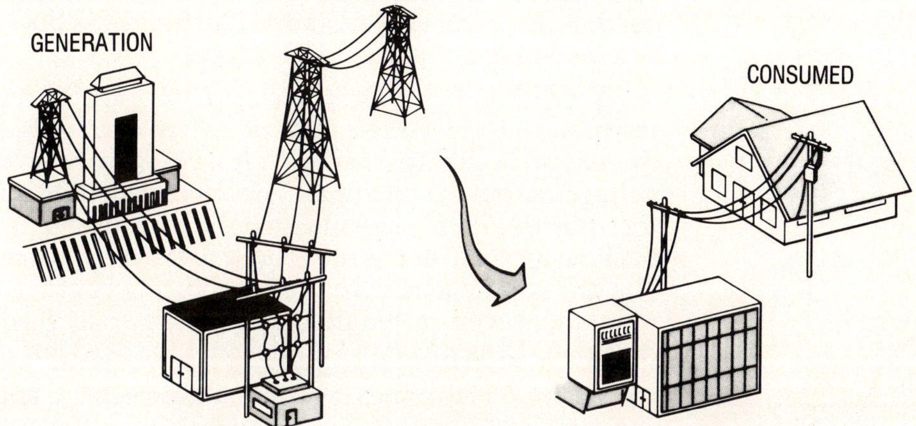

Figure 7-3. Distribution of electric power.

Typically, alternating current leaves a commercial generator with current levels of several thousand amperes and a voltage of 69,000 V. (In some cases, voltage is even greater.) Current is transmitted over power lines. If losses due to line resistance become too great, the voltage can be stepped up along the way by transformers built into the system. At delivery points, the voltage is stepped down by transformers to bring it to the level needed by industrial, commercial, residential, or other consumers.

Power generation, transmission, and tranformation are discussed in greater depth later in this chapter. At this point, the simplified presentation based on Figure 7-3 illustrates the point that AC is widely used and is a standard for commercial power largely because of its inherent versatility.

Discussion Questions/Problems

1. What is alternating current?
2. What happens to voltage polarity of AC?
3. What happens to voltage values of AC?
4. What are some of the ways in which AC is more versatile than DC?

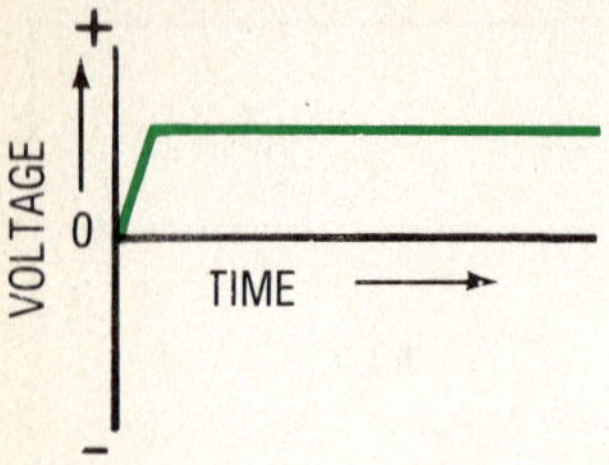

Figure 7-4. DC waveform.

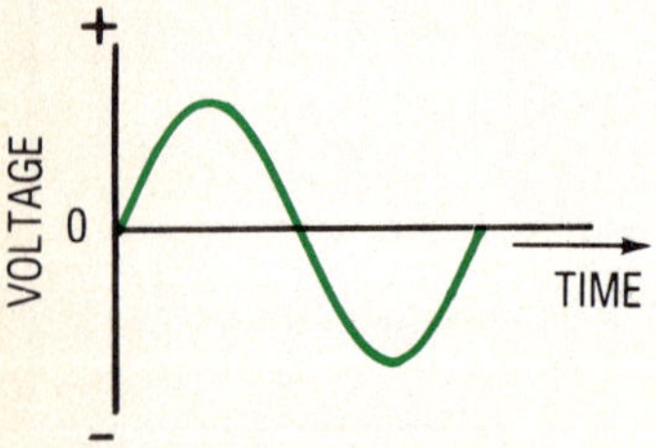

Figure 7-5. AC waveform.

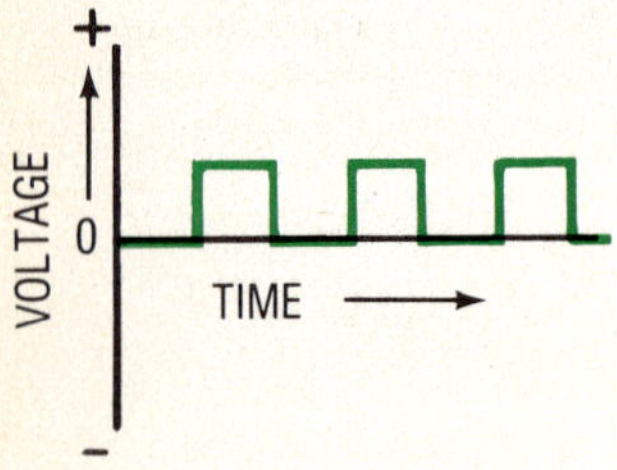

Figure 7-6. Waveform for pulsating (on-off) DC.

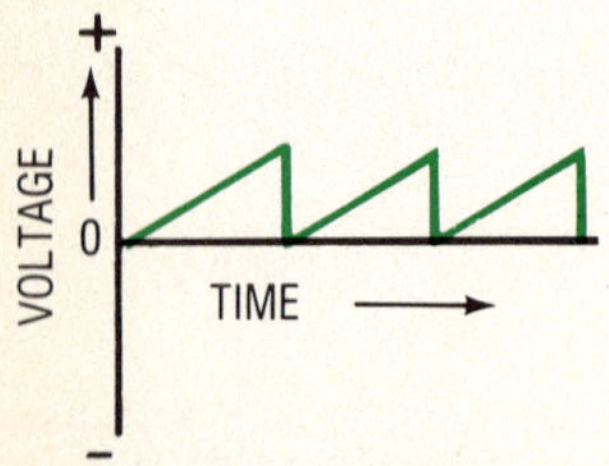

Figure 7-7. Waveform for DC applied gradually and then repeated.

7.2 WAVEFORMS

For the purposes of discussing and understanding current and voltage flow patterns, diagrams called waveforms are used. A *waveform* is a graph that illustrates the flow pattern of a given type of current or voltage.

For example, the DC waveform in Figure 7-4 illustrates the basic nature of direct current. When current is applied, there is an immediate surge, or flow of electrons, to its full voltage output, or potential. The waveform in Figure 7-4 is for ''positive'' DC, meaning that the measurement is at the positive terminal and is referenced to the negative terminal. The same approach can be used in showing negative DC current. The waveform would simply be drawn in the opposite direction, beneath, instead of above, the zero line.

Figure 7-5 illustrates the contrast between DC and AC by showing an AC waveform. In contrast with the essentially constant voltage waveform of the DC flow pattern, AC is depicted by a gradual rising and falling curve. The curve keeps moving in this way, reversing from positive to negative flow continuously.

The waveform in Figure 7-5 is known as a *sine wave*. The sine wave is the most common waveform for AC. Figure 7-5 is a graphic display of the current you receive in your home, office, or school.

Compare the gradual curve in Figure 7-5 with the waveform in Figure 7-6. This illustration shows the flow pattern of pulsating (on-off) DC. Where AC flow is represented by gradual curves, pulsating DC is shown in a series of up and down lines. This indicates that pulsating DC flow is instantaneously on or off under normal conditions.

However, under controlled conditions, it is possible to apply DC gradually. When this is done, the waveform takes the shape shown in Figure 7-7.

Keep in mind that DC waveforms can be positive or negative. Usually, a DC waveform is either positive or negative whereas AC alternates continuously from positive to negative and back.

Formation of Sine Waves

The output of a rotating AC generator is a particular wave shape called a sine wave. It assumes this special shape because of the geometry of the rotating machinery. This generating equipment is usually called an *alternator* because it creates alternating current. (The term *generator* is usually reserved for DC.) The sine wave is the flow pattern created in a single rotation of an AC alternator. To illustrate, a simplified diagram of an AC alternator is shown in Figure 7-8. The drawings, A through E, show a coil of wire that is rotated through the lines of force between magnetic poles. During rotation, the magnets cause a current to flow through the wire of the coils. Electricity is generated as the moving wires cut across the magnetic lines of force.

The polarity and height of the sine waves generated correspond to the position and relationship of the coil to the magnetic field through which it passes. In the drawings in Figure 7-8, a coil is positioned in a uniform magnetic field. When the coil is positioned as shown in Figure 7-8A, it is not moving up past lines of force or down past lines of force. Rather, the coil is moving parallel to the lines of force. Since the wires of the coil are moving parallel to the lines of force, no current is generated.

When the coil is positioned as in Figure 7-8B, it cuts through the maximum number of lines of force. The maximum current is generated when the coil is in this position. When the coil is positioned as shown in Figure 7-8C, the wires are once again parallel to the magnetic lines of force. Again, no current is generated.

As rotation continues and the coil is positioned as shown in Figure 7-8D, the coil again cuts through a maximum number of lines of force. At this point, the current will be at maximum, and flow in a direction opposite from that in Figure 7-8B. The pattern of current flow can be represented as shown in Figure 7-8E.

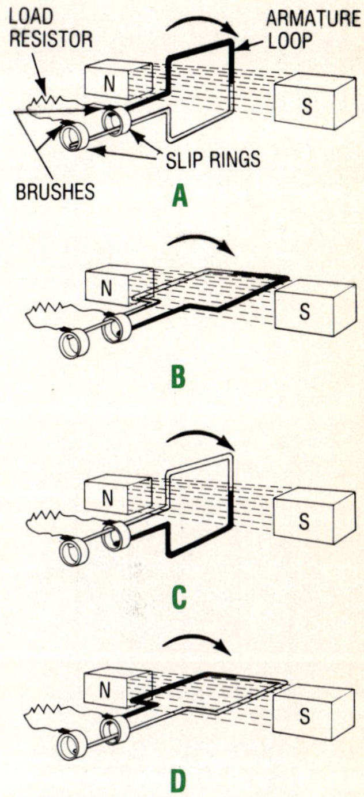

Figure 7-8. Simplified illustration of how AC is generated.

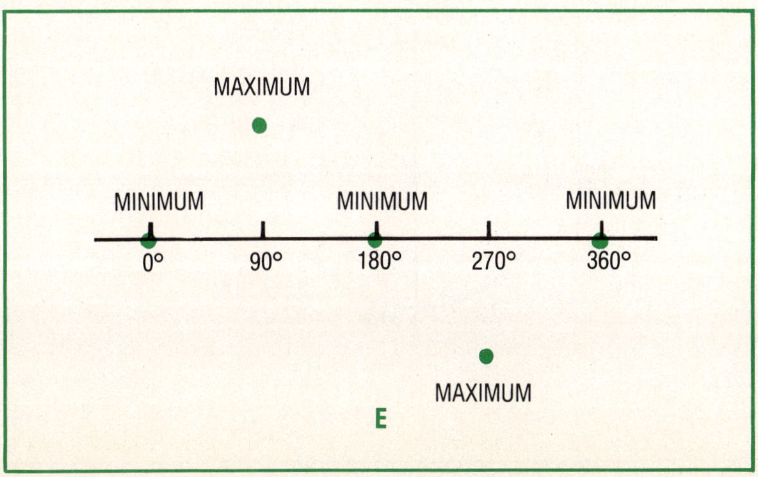

Figure 7-8E. Pattern of AC current flow generated in Figure 7-8.

Note that AC output voltage is proportional to the sine of the angle of rotation. Thus:

- When the coil is positioned as in Figure 7-8A and the angle is zero degrees, the voltage is a minimum (zero) and the sine of zero degrees is zero.
- When the coil is positioned as in Figure 7-8B and the angle is 90 degrees, the voltage is a maximum and the sine of 90 degrees is 1 (the maximum value).
- When the coil is positioned as in Figure 7-8C and the angle is 180 degrees, the voltage is a minimum (zero) and the sine of 180 degrees is zero.
- When the coil is at the position shown in Figure 7-8D, the voltage is a maximum and the sine of 270 degrees is −1 (the maximum value in the negative direction).

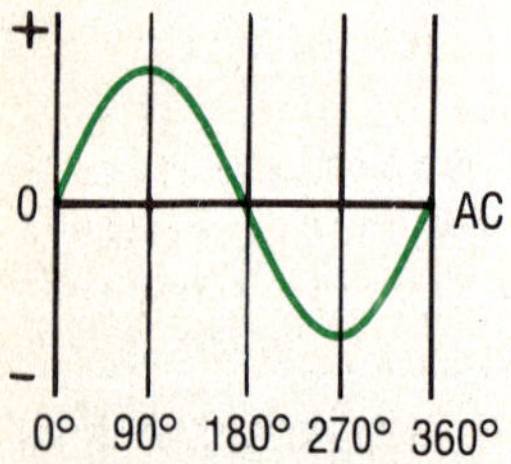

Figure 7-9. Sine wave produced as coil is rotated.

From this data, it is possible to derive the voltage that would be generated when the coil is positioned halfway between the A and B positions in Figure 7-8. The voltage produced would be proportional to the sine of 45 degrees. Plotting the positions for all positions in a coil rotation produces a sine wave like the one in Figure 7-9.

Plotting the voltage that has been shown to be proportional to the sine of the angle of rotation produces the result shown in Figure 7-10. This curve is a plot of the *voltage vector.*

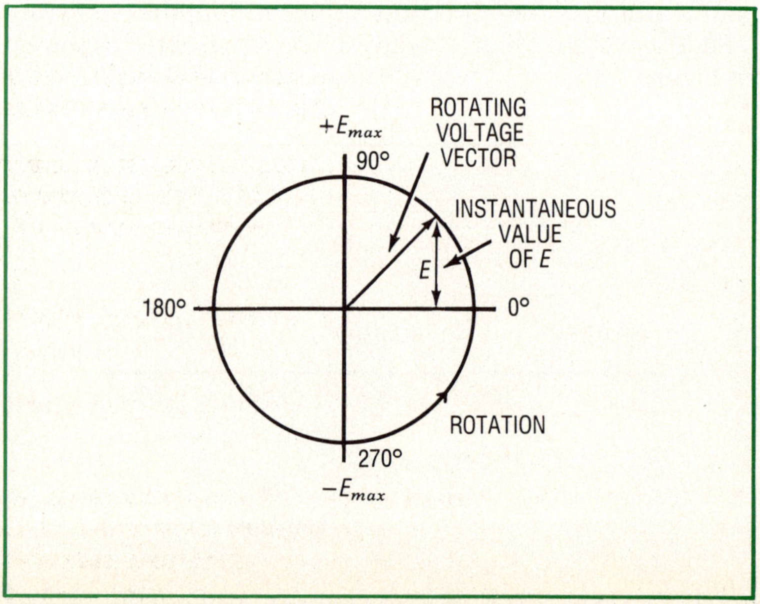

Figure 7-10. Plot of voltage vector.

The AC Cycle

A full rotation of an AC alternator is called a cycle. A cycle is represented by a full sine wave indicating one alternation of current as the alternator turns through a full 360-degree circle. A sine wave of an AC cycle is shown in Figure 7-11. Notice there are two voltage peaks—a positive peak at 90 degrees and a negative peak at 270 degrees.

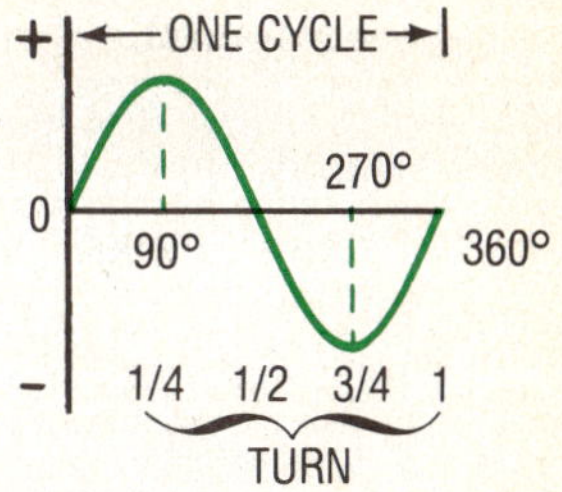

Figure 7-11. Sine wave of AC cycle.

The speed at which an alternator coil rotates is also significant. Devices that use AC current must be compatible with the alternation rate at which current is supplied. One of the basic descriptions that must be given for AC current is its alternation rate. Alternations are measured in numbers of rotations per second, or *cycles per second (cps).* The name given to the rate is *hertz,* abbreviated Hz. A full, 360-degree cycle in one second is known as 1 hertz. The standard for AC generation in the United States and Canada is 60 Hz, which is 60 full 360-degree cycles in one second. Throughout these countries, you can plug an appliance into any wall outlet and feel certain that 60 Hz current will be delivered. In some other countries, particularly Europe, Japan, and parts of Mexico, the standard AC cycle rate is 50 Hz. Individual industries may also have different standards. For example, aircraft and aerospace equipment is usually designed to a standard alternation rate of 400 Hz.

Discussion Questions/Problems

1. How do the voltage characteristics of AC and DC differ?
2. What is a sine wave?
3. What is represented by a sine wave?
4. What is an AC cycle?
5. What does the Hertz measure?
6. What is the standard cycle rate for commercially generated electricity in the United States?

7.3 AC MEASURES

In working with AC, many instruments are used to measure current, voltage, or frequency. Many of these use the pattern of the standard sine wave in some way to produce their readings.

Sine waves are measured on the basis of portions of 360-degree cycles. Angles may be measured in units of:

- Degrees
- Radians.

Angular measures are based upon the relationships of intersecting lines emanating from the center of a circle representing an AC cycle. Given a circle representing a full rotation of an AC generator, angles are measured from the center point. The number of degrees between the lines representing these angles are the units of measure.

A radian is a specific portion of a circle. In mathematical terms, a *radian* is an angle that encloses an arc equal to its radius. The *radius* is the distance from the exact center of the circle to the circumference. A radian can be thought of as a pie-shaped portion of a circle for which the edge is the same length as the side. This is illustrated in Figure 7-12.

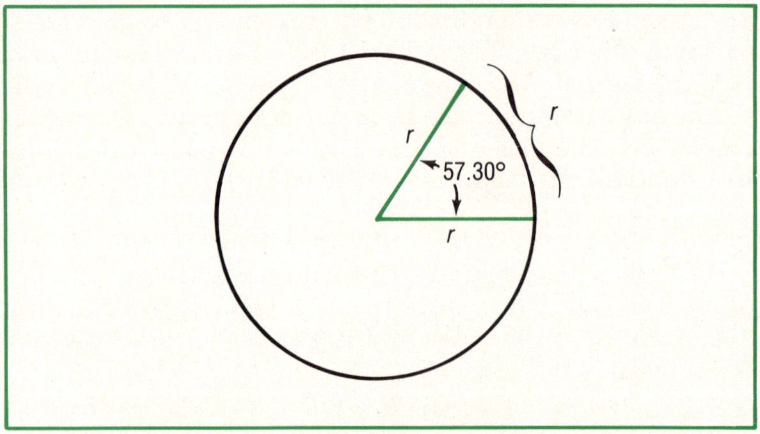

Figure 7-12. Sector of circle equal to one radian.

To determine how many radians there are in a full circle, begin with the formula for the circumference (C) of a circle:

$$C = 2\pi r$$

Since a radian is an angle that marks off a distance of r on the circumference, then 2π radians will mark off a distance of $2\pi r$. Thus, there are 2π radians in a circle. Since a circle encompasses 360 degrees and can be divided into 2π radians, it follows that:

$$360^\circ = 2\pi \text{ radians} \quad \text{Or: } 1 \text{ radian} = 57.30 \text{ degrees}$$

Trigonometry Review

It has been shown that the AC from a rotating alternator can be represented by a sine wave. Given a sine wave representing AC, the instantaneous voltage for any portion of the cycle can be figured through use of the trigonometric function, sine.

To review the terms involved: A *triangle* is a shape formed by three straight lines. A *right triangle* is one which has one right, or 90-degree, angle. In a right triangle, the side opposite the 90-degree angle is called the *hypotenuse*. Figure 7-13 shows a right triangle, where *A* is the right angle and, therefore, *a* is the hypotenuse.

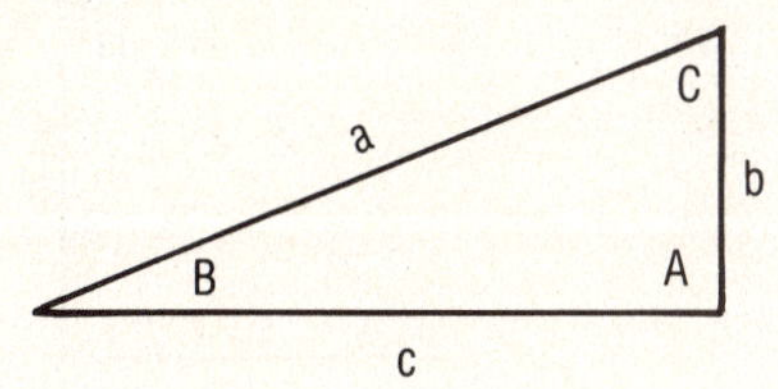

Figure 7-13. Right triangle.

Consider angle *B*. The sine of *B* is the ratio of *b* to the hypotenuse, or *b/a*. The cosine of *B* is the ratio of *c* to the hypotenuse, or *c/a*. When the angle *B* is very small, *b* will be very small and the ratio of *b/a* will be small, and so the sine of *B* will be small. The sine of *B* can be zero. When the angle *B* is large, *b* will be large and the ratio *b/a* will be large. As *B* increases, *b* gets larger until *b* is as large as *a*. So *b/a* can be as great as 1. Thus, the sine of *B*, which is *b/a*, can vary between 0 and 1. The values of sines for angles of varying degrees are shown in sine tables. A sine table is contained in Appendix D.

The formula for determining instantaneous voltage for a sine wave is as follows:

$$v = E_m \sin \theta$$

Theta (θ) = to voltage angle.
Sin stands for sinusoidal or sine.
E_m is the maximum voltage.
v is the instantaneous voltage volume at that angle.

The following is an example showing the determination of instantaneous voltage for a 45-degree angle with a maximum voltage value of 50 V.

$$\text{instantaneous voltage} = v$$
$$= E_m \sin 45^\circ = 50 \times 0.707$$
$$(\sin \text{ of } 45^\circ = 0.707)$$
$$v = 35.35 \text{ V}$$

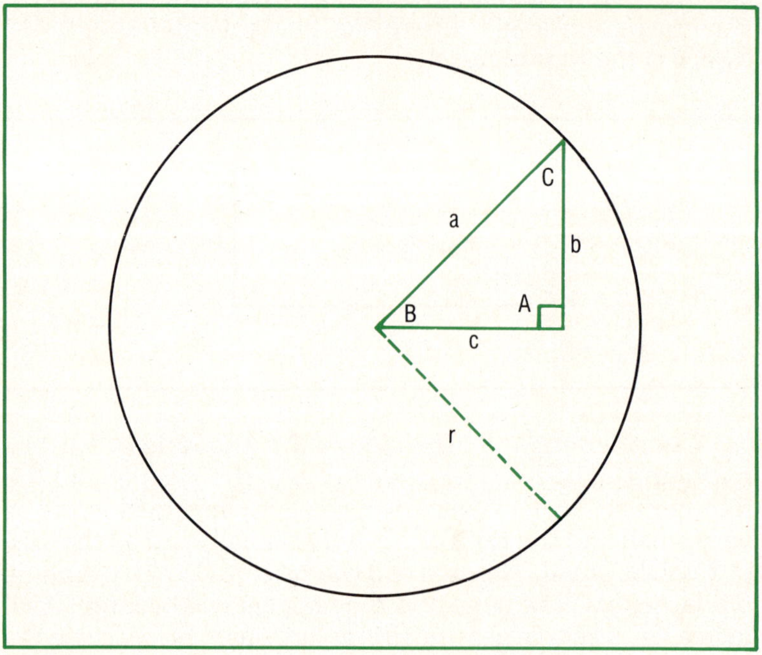

Figure 7-14. Relationship of instantaneous to maximum value.

Figure 7-14 shows the relationship of instantaneous value to maximum value. Another AC value that frequently must be calculated is *instantaneous current*. This is the amount of current (in amperes) that is flowing in response to a given instantaneous voltage. Diagrams illustrating AC circuits use a symbol for an AC generator, or source, as shown in Figure 7-15. A circuit using this symbol is shown in Figure 7-16. The formula for determining instantaneous current in this type of circuit is:

Figure 7-15. Symbol for AC generator or source.

$$i = \text{the instantaneous current}$$
$$i = \frac{v}{R}$$

An example showing how the value of instantaneous current is derived is given below:

$$v = 100 \text{ V}$$
$$R = 10\ \Omega$$
$$i = \frac{v}{R} = \frac{100 \text{ V}}{10\ \Omega} = 10 \text{ A}$$

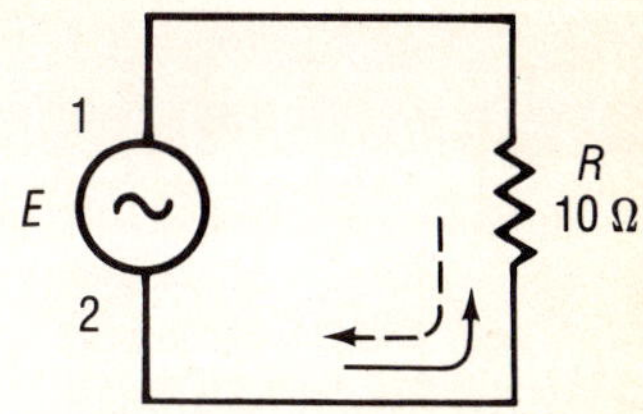

Figure 7-16. Circuit with AC source.

The height of a given sine wave at any point is called *amplitude.* The maximum amplitude is the peak voltage. See Figure 7-17.

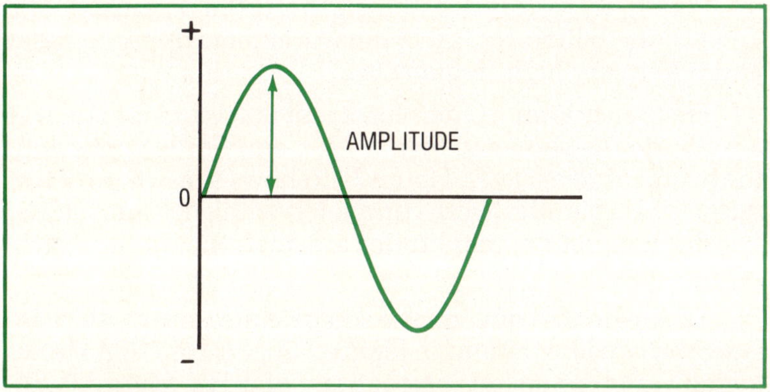

Figure 7-17. Amplitude of sine wave.

In considering sine waves, note that the same waveform applies to both current and voltage. This is illustrated in Figure 7-18.

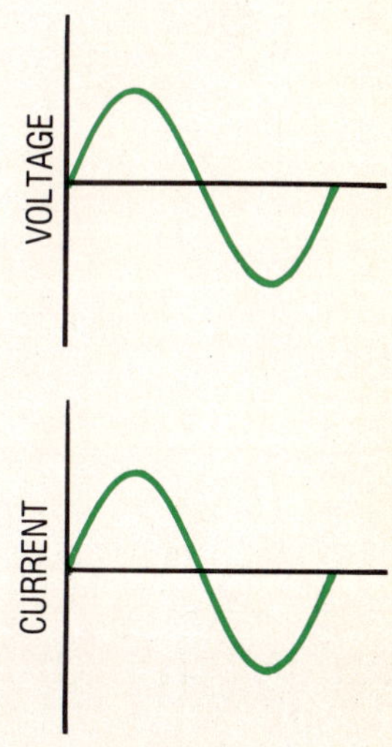

Figure 7-18. Waveforms for current and voltage are the same.

Discussion Questions/Problems

1. What is a radian?
2. What is instantaneous voltage?
3. What is the hypotenuse of a triangle?
4. What ratio is represented by the value of sine?
5. What is the sine of 75 degrees?
6. What is instantaneous current?
7. If instantaneous voltage is 50 V and resistance is 5 Ω, what is the instantaneous current?

7.4 FREQUENCY

The number of AC cycles completed during a specific time period is known as the *frequency*. The term usually used in describing AC frequency is *cycles per second*, or *hertz*. For convenience, separate measures and symbols are used in describing AC at high frequencies. The kilohertz (kHz) designates 1,000 Hz. In general, kHz measurements are used to describe ranges that include AM radio, loran guidance frequencies, and short-wave radio ranges.

Another unit, the megahertz (MHz) is used to designate millions of hertz. MHz measurements are used to describe FM radio, television, and UHF broadcasting.

For higher ranges still, the gigahertz (GHz) is used. The GHz is one billion Hz. GHz measurements apply to radar and other very high frequencies.

Frequency is an AC measurement that is independent of amplitude. As shown in Figure 7-19, amplitude varies with the height of sine waves. Figure 7-19 shows sine waves of two different amplitudes, both completed within the same time. For this illustration, amplitudes are different but frequency is the same.

To illustrate frequency as a separate measure of AC, consider the two illustrations in Figure 7-20. Figure 7-20A shows a sine wave for AC at a frequency of 1 Hz. Figure 7-20B represents AC at a frequency of 6 Hz.

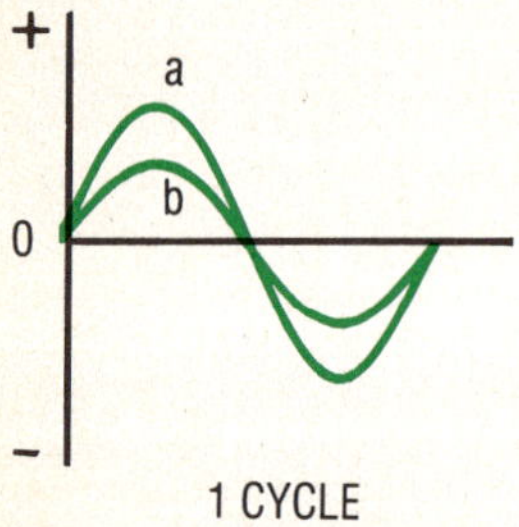

Figure 7-19. Sine waves of two different amplitudes, both with same frequency.

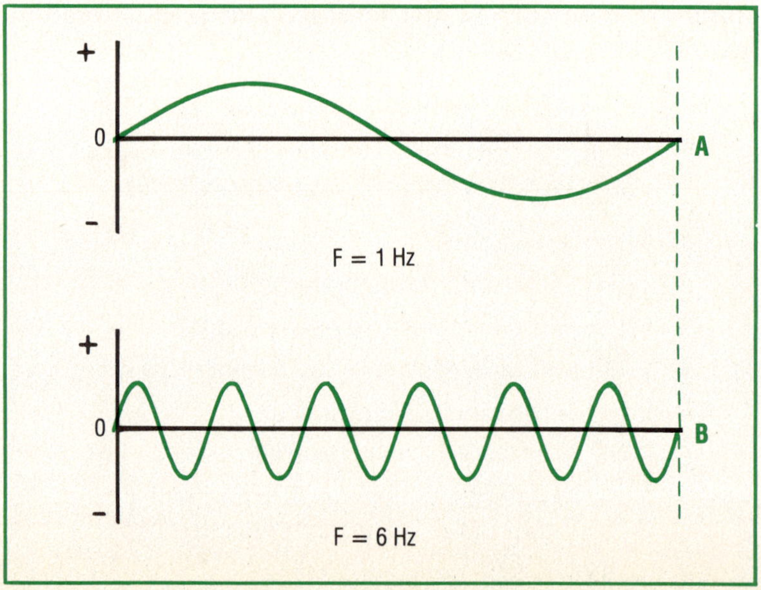

Figure 7-20. Sine waves with different frequencies.

At different frequencies, waveforms have different characteristics. The total range of AC frequencies runs from 0 Hz up through 10^{22} Hz. This full range is known as the electromagnetic spectrum. A chart showing the electromagnetic spectrum is presented in Figure 7-21.

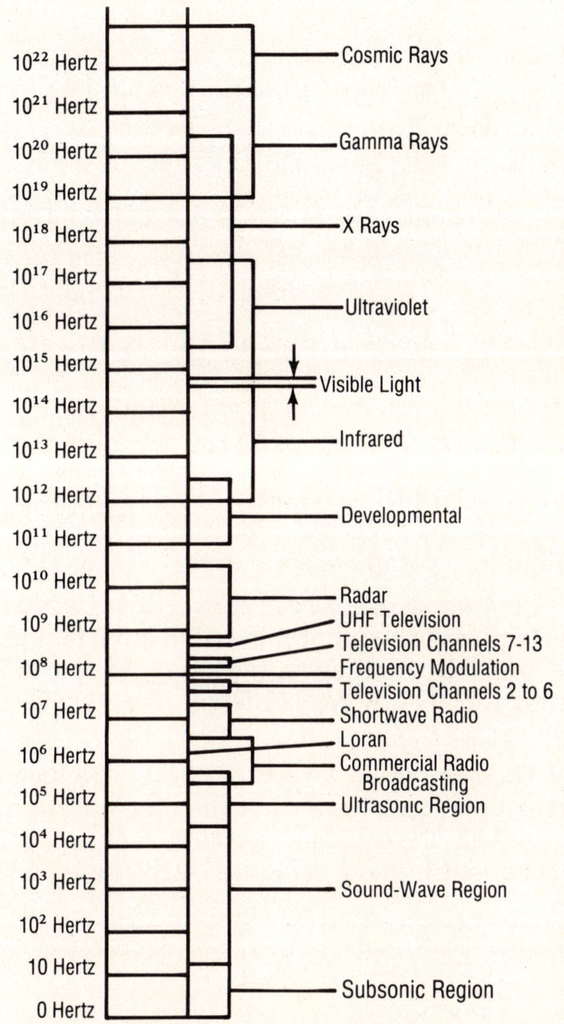

Figure 7-21. Electromagnetic spectrum.

Included in the electromagnetic spectrum are sound waves, radio waves, radar, light, x-rays, gamma rays, and cosmic rays. At higher frequencies, the electromagnetic spectrum includes nuclear radiation.

The different segments within the electromagnetic spectrum have different, important uses, highlighted in Figure 7-21. These include the range of human hearing, a full range of radio and television broadcasting, and radar for communication and navigation. As the spectrum approaches its higher ranges, it also includes visible light and, then, radiation waves used for medical diagnosis and treatment.

Discussion Questions/Problems

1. How does AC frequency compare with AC amplitude?
2. What is the common time period used for frequency measure?
3. What is the meaning of the frequency designation 60 Hz?

7.5 PERIOD

The *period* is the time required for one complete AC cycle. The cycle includes a regular, repeating series of events, as represented by a complete sine wave.

Period as a measure of AC relates directly to frequency: The higher the frequency, the shorter the period. Conversely, the lower the frequency, the longer the period will be. The period for 50-cycle AC is 1/50 sec. For 60-cycle, the period is 1/60 sec.

The symbol for an AC period is T, which is the capital Greek letter *tau*. T indicates the time duration per cycle. In Figure 7-22, T is equal to 4 seconds, as shown in Figure 7-22A. For the 3-cycle wave in Figure 7-22B, T is equal to 3 seconds.

Figure 7-22. Waves of different periods.

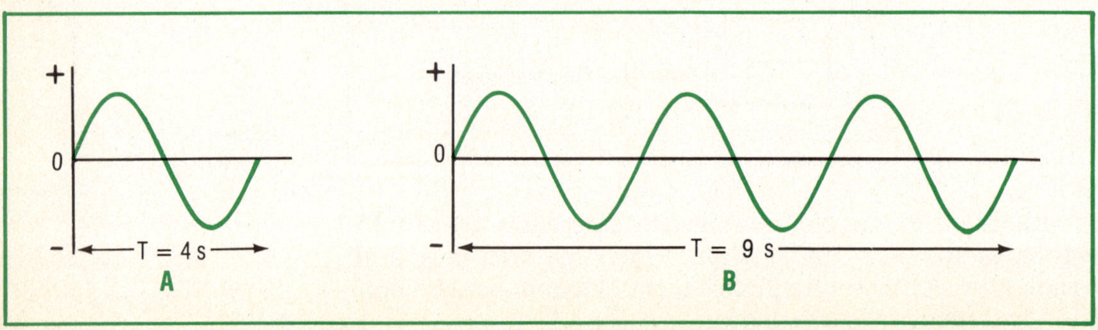

Discussion Questions/Problems

1. How does the period for AC current relate to frequency?
2. How does the period of AC current relate to amplitude?
3. How does the period of AC current relate to the AC cycle?

7.6 WAVELENGTH

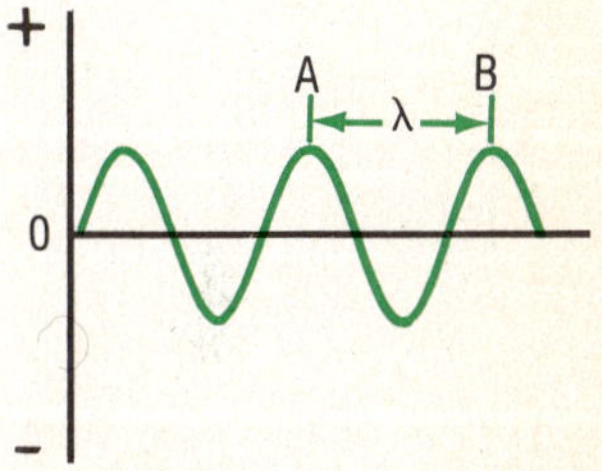

Figure 7-23. Wavelength of an AC cycle.

The length of an AC cycle as represented by a sine wave is called the *wavelength*. Wavelengths are illustrated, as shown in Figure 7-23, as the distance (length) between the peaks of consecutive sine waves. Actually, the wavelengths can be measured between any corresponding points on consecutive sine waves—not just the peaks.

The symbol for wavelength is the Greek letter *lambda* (λ).

The formula for determining wavelength is based on the velocity and frequency of the wave. Velocity is the speed of travel for a signal in a given medium. For example, the velocity of light in a vacuum is 186,000 miles per second. The same velocity applies to radio signals traveling through air. The metric equivalent of this velocity is 3×10^{10} cm per second. Metric measurements are usually used in all calculations involving wavelengths. The symbol for velocity is the lowercase Greek *upsilon* (υ).

The symbol for frequency is f. The formula for determining wavelength when velocity and frequency are known is as follows:

$$\lambda = \frac{\upsilon}{f}$$

The example below shows how this formula is applied. The frequency used is 30 MHz. Since this is a radio frequency, the velocity is the speed of light. The result, as shown, gives the length of the wave in cm.

$$\lambda = \frac{\upsilon}{f} = \frac{3 \times 10^{10}\ \text{cm/s}}{30 \times 10^{6}} = \frac{3 \times 10^{4}}{30} = 1{,}000\ \text{cm}$$

When frequency and wavelength are known and velocity is unknown, use the following formula:

$$v = f\lambda$$

An example of how this formula is applied is shown below. In this example, the frequency is 30 MHz and the wavelength is 1,000 cm. The equation shows a solution for velocity in centimetres per second.

$$v = f\lambda = 30 \times 10^6 \times 1{,}000 \text{ cm} = 3 \times 10^{10} \text{ cm/s}$$

When velocity and wavelength are known but frequency is unknown, the formula is:

$$f = \frac{v}{\lambda}$$

The equation below applies this formula. In this example, velocity is the speed of light and wavelength is 1,000 cm. The solution is the frequency of the signal.

$$\begin{aligned} f &= \frac{v}{\lambda} \\ &= \frac{3 \times 10^{10} \text{ cm/s}}{1{,}000 \text{ cm}} \\ &= 30{,}000{,}000 \text{ Hz} \\ f &= 30 \text{ MHz} \end{aligned}$$

Discussion Questions/Problems

1. What does wavelength measure?
2. How are wavelength, frequency, and velocity related?
3. If frequency is 300 MHz and velocity is the speed of light, what is the wavelength of the signal?

7.7 AC VALUES

Suppose you were told that a circuit has 100 V of AC. This can mean any of four different voltages, depending upon how values are determined. There are four basic ways for establishing the value of an AC voltage:

- Peak
- Peak-to-peak
- Root-mean-square (also called *effective value*)
- Average.

The methods of measuring these voltage values for AC are illustrated in the sine wave diagram in Figure 7-24. Methods of developing individual values are discussed below.

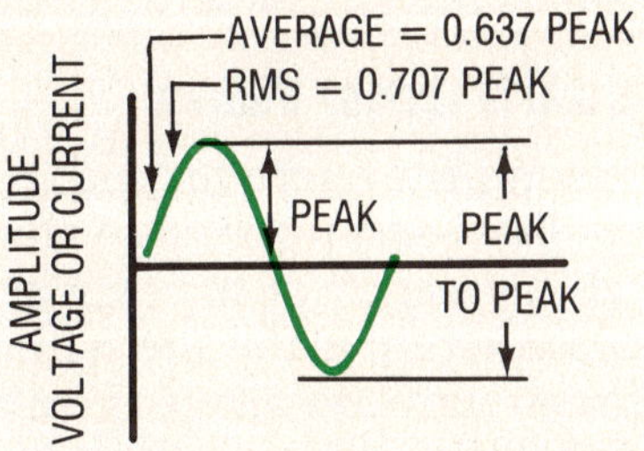

Figure 7-24. Several methods of measuring AC voltage.

Peak Value

The *peak* value is the maximum voltage when sine B = 1 (the top of the sine wave), as shown in Figure 7-25.

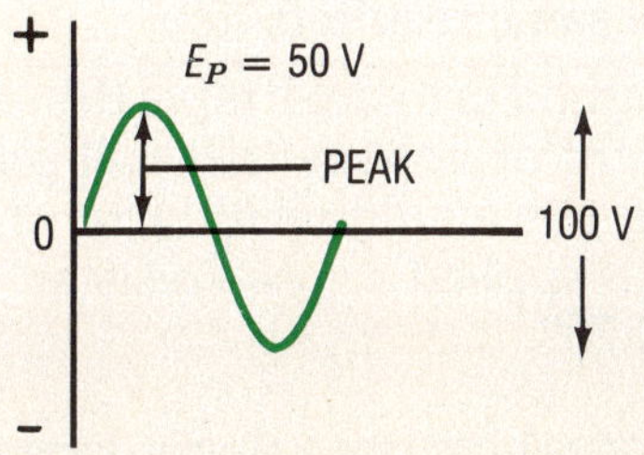

Figure 7-25. The peak value.

Peak-to-Peak Value

The *peak-to-peak* value is the voltage potential from positive peak to negative peak of the sine wave. The peak-to-peak value is always double the peak value. The peak value for the same wave would be 50 V, as shown in Figure 7-26.

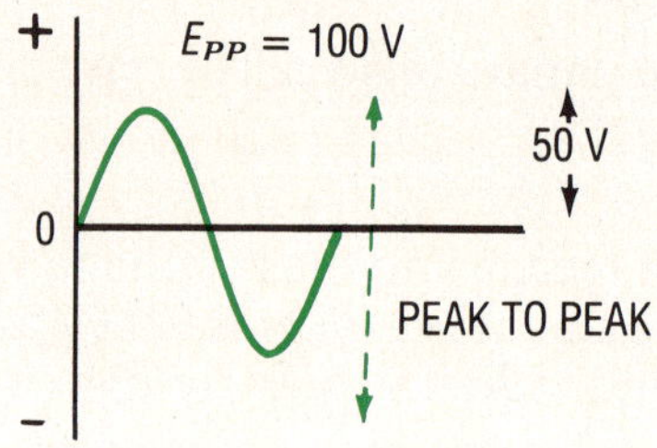

Figure 7-26. The peak-to-peak value.

Root-Mean-Square (rms) Value

The value that presents the effective capacity of electricity to do work is the *root-mean-square*, or *rms*, value. In effect, rms is a mathematical "average" of the ever-changing instantaneous AC values inherent in the alternating current cycle. The rms value for any given voltage or current is equal to 0.707 times the peak value. This mathematical value of rms is derived through three different types of calculations:

- One method is to take the sine for each angle within the sine wave, then square this value. All the squares are added together and then divided by the number of values that have been added. The answer is 0.707.
- Another method is to develop the square root of the average of the squares of values for each instantaneous point within one complete cycle. This calculation delivers the same result: 0.707.
- Another method is to determine the value of the alternating current or voltage that will produce the same heating effect as would result when an equal value of DC voltage is applied across a resistor. The result is still the same: 0.707.

Figure 7-27 illustrates how rms value is determined for a wave with a peak-to-peak value of 100 V. This diagram illustrates a principle that is important in specifying components for electronic circuits. Because most AC voltages are given in rms values, some allowance must be made to handle peak voltages. For example, household electricity is probably rated at between 110 and 120 V. This reading is an rms value. Therefore, the light bulbs and other electrical devices used in homes are designed to withstand peak voltages of up to 200 V.

The allowance for peak voltages is needed because, as illustrated in Figure 7-27, the peak value is greater than the rms value. Actually, it is 1.44 times the average value.

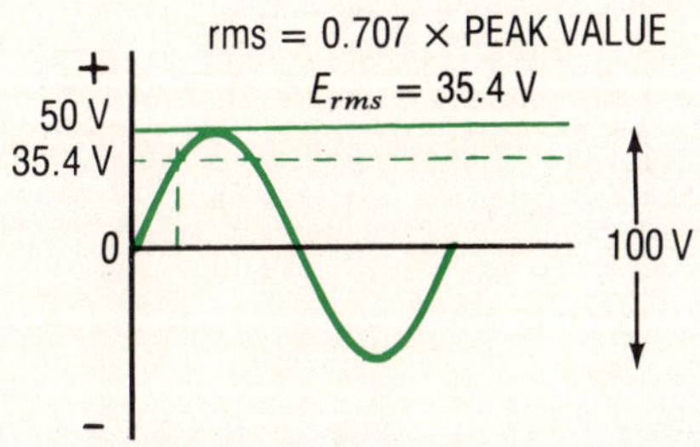

Figure 7-27. The rms value.

In designing or applying safety values for electrical circuits, it is frequently necessary to compute peak value or peak-to-peak value when only the rms value is known. The formula for determining peak value when the rms value is given is:

$$\text{peak value} = \frac{1}{0.707} \times \text{rms} = 1.414 \times \text{rms}$$

The formula for determining peak-to-peak value when rms is known is:

$$\begin{aligned}\text{peak-to-peak value} &= 2 \times \text{peak value} \\ &= 2\,(1.414) \times \text{rms} \\ &= 2.828 \times \text{rms}\end{aligned}$$

Average Value

The arithmetic average of one complete alternation is known as the *average* value. Mathematically, the average value is

0.637 percent of the peak value. To determine the average value, multiply 0.637 by the peak value.

Figure 7-28 is a diagram showing determination of average value. The computation for average value of a circuit with a peak-to-peak value of 100 V is included in Figure 7-28.

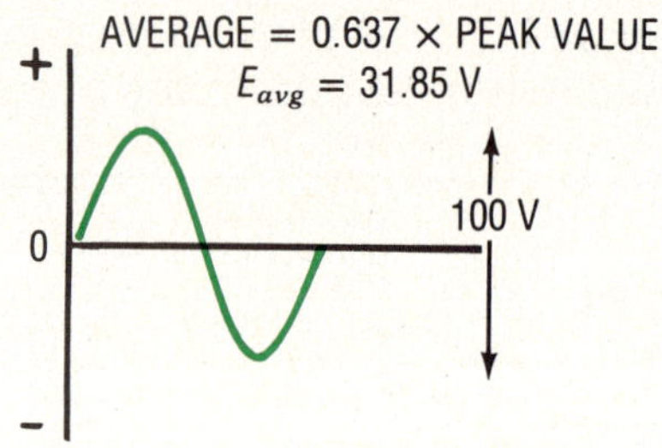

Figure 7-28. The average value.

Discussion Questions/Problems

1. What is peak value?
2. What is peak-to-peak value?
3. What percentage of peak value is the root-mean-square value?
4. What percentage of peak value is the average value?
5. If a circuit has an rms value of 110 V, what peak value must be handled by its components?
6. Why is rms value particularly significant?

7.8 OHM'S LAW FOR AC

All of the formulas you have learned for application of Ohm's law up to this point have been for DC. All of these rules and equations also apply to AC, with an exception. There will be slight differences in the application of the equations for Ohm's law in AC. The formula to use depends upon what measure of AC you are talking about. There are separate equations for each way to measure AC current.

For instantaneous current, voltage, or resistance, use the following equations:

$$i = \frac{v}{R} \qquad v = iR \qquad R = \frac{v}{i}$$

For peak values of AC, use these equations:

$$I_p = \frac{E_p}{R} \qquad E_p = I_pR \qquad R = \frac{E_p}{I_p}$$

For peak-to-peak AC values, use the following:

$$I_{pp} = \frac{E_{pp}}{R} \qquad E_{pp} = I_{pp}R \qquad R = \frac{E_{pp}}{I_{pp}}$$

For the root mean square (rms) values of AC, use these equations:

$$I_{rms} = \frac{E_{rms}}{R} \qquad E_{rms} = I_{rms}R \qquad R = \frac{E_{rms}}{I_{rms}}$$

For the average values of AC, these are the equations to use:

$$I_{avg} = \frac{E_{avg}}{R} \qquad E_{avg} = I_{avg}R \qquad R = \frac{E_{avg}}{I_{avg}}$$

Notice that, in all of the above sets of equations, the basic formulas are the same. The idea is to be careful not to mix the types of voltage or current. Do not mix peak current with average voltage, for example. To illustrate application of Ohm's law to AC, the following example shows development of the value of rms voltage for the circuit in Figure 7-29. Note that resistance in this circuit is 10 Ω. Current is 5 A. The solution for voltage in this circuit is:

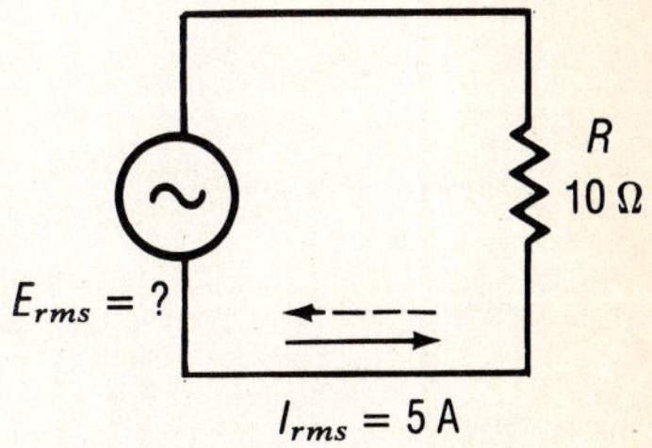

Figure 7-29. Circuit used to show how rms voltage is calculated.

$$E_{rms} = I_{rms}R = 5\ \text{A} \times 10\ \Omega = 50\ \text{V}_{rms}$$

Discussion Questions/Problems

1. What is the instantaneous voltage value in a circuit with instantaneous current of 2 A and resistance of 2K Ω?
2. What is the peak-to-peak current in a circuit with a peak voltage of 50 V and a resistance of 10K Ω?
3. What is the average voltage in a circuit with a peak-to-peak current of 2 A and a resistance of 1.5K Ω?

7.9 AC NETWORK THEOREMS

Network theorems, as introduced in Chapter 3, provide shortcuts to analyze circuits. Network theorems make it possible to simplify complex circuits so that electrical properties and values can be determined with relative ease. The theorems presented in Chapter 3 deal with DC current. The same techniques can be used to simplify and solve AC circuits, as long as you always remember this rule: *In solving AC circuits, it is necessary to determine what type of AC current you are dealing with.*

To illustrate, two of the basic network analysis tools are discussed below.

Kirchhoff's Law of Current

Kirchhoff's law of current holds that the amount of current leaving any point, or node, of the circuit is equal to the current that entered that node.

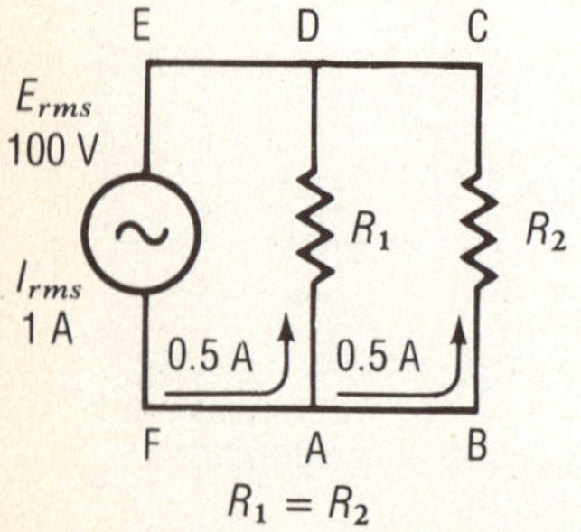

Figure 7-30. Circuit used to illustrate Kirchhoff's law of current for AC.

In Figure 7-30, rms values are used to illustrate this principle. Remember, when you deal with AC, most of your examples will be given in rms values. In Figure 7-30, it is given that the resistances of the two circuit branches are equal. Thus, the current divides equally between the branches. If the circuit has total current of 1 A, each branch will have 0.5 A of current.

Remember that the direction of current flow in the circuit of Figure 7-30 continually reverses. The current starts at point F with a value of 1 A. The current then divides at point A into two equal parts. These divided currents then come together again at point D. On the first portion of the cycle, the current flows from point F to point E. Then, when the current reverses, a 1 A flow initiates at point E. On the reverse cycle, the current divides at point D and is recombined at point A.

Norton's Theorem

Norton's theorem is used to analyze a network in terms of current. The concept calls for reducing a network into a simple parallel circuit. This is illustrated in Figure 7-31. Figure 7-31A shows a circuit with three resistors, two in parallel. Using the process explained in Chapter 3, this circuit can be simplified so that total resistance is represented in a single resistor, as shown in Figure 7-31B. For mathematical purposes, Figure 7-31B shows a single, simple parallel circuit, with resistance expressed in terms of the Norton equivalent. Current can then be computed using the Norton theorem.

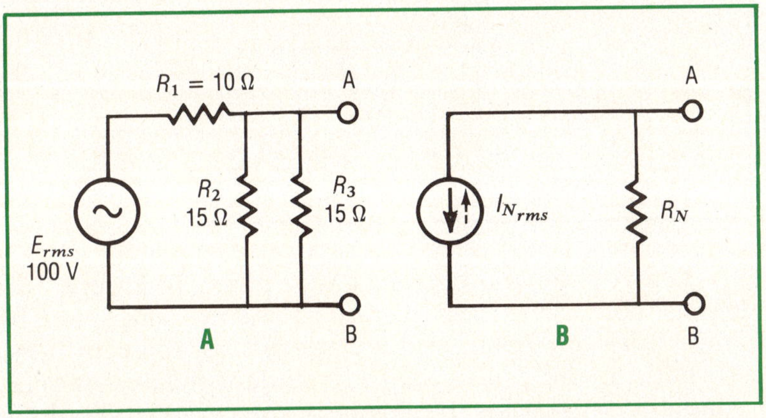

Figure 7-31. Circuit simplified so AC current can be computed using Norton's theorem.

Note that Figure 7-31 identifies a circuit as having rms values. In working with AC, you have to know what is being measured.

Discussion Questions/Problems

1. Using the circuit in Figure 7-31A, show the Thevenin equivalent of the circuit.
2. Using the circuit in Figure 7-31A, trace the current flow through use of Kirchhoff's law.
3. Using Kirchhoff's law for voltage, trace the voltages around Figure 7-31A, determining the average value for the voltage across each resistor.

7.10 PHASE RELATIONSHIPS

Figure 7-32. Resistive circuit.

The term *phase* describes the time relationship of the current and voltage waveforms for an AC signal. Up to this point, you have seen single sine waves representing AC. In reality, there may be different sine waves representing voltage and current.

For example, consider the circuit in Figure 7-32. This is a resistive circuit. The rule for resistive circuits is: *In resistive circuits, current and voltage are always in phase.* Thus, the sine waves for this circuit would look like the ones in Figure 7-33.

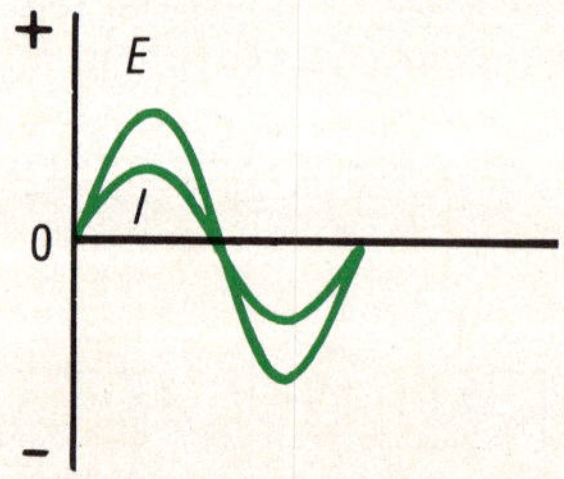

Figure 7-33. Voltage and current for resistive circuit.

The voltage and current are in phase in Figure 7-33 because both waves cross the zero point at the same spot. The amplitudes of current and voltage are not relevant here. Amplitude and phase are independent measurements. Amplitude measures the height of the curve while phase refers entirely to the time correspondence (or congruence) of the waves for voltage and current.

If the waveforms in Figure 7-33 can be said to be *in phase,* then those in Figure 7-34 are clearly out of phase. Remember, waves are out of phase when waveforms representing them

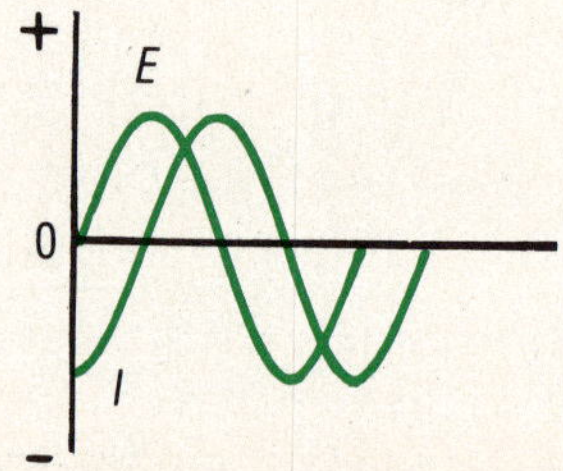

Figure 7-34. Out-of-phase waveforms.

cross the zero point at different times. In Figure 7-34, current is said to be "lagging voltage." This is because the waveform for current crosses the zero point after, or behind, that for voltage.

In other situations, voltage lags current. In any case, the amount, or degree, to which current lags or leads voltage is an important factor in the design of AC circuits. This degree to which voltage and current are out of phase is referred to as the *phase angle*.

Voltage and current may be out of phase in any amount from a fraction of a degree up to 360 degrees. A 180-degree phase angle occurs when voltage and current cross the same zero points in opposite directions. A 180-degree phase angle is illustrated in Figure 7-35.

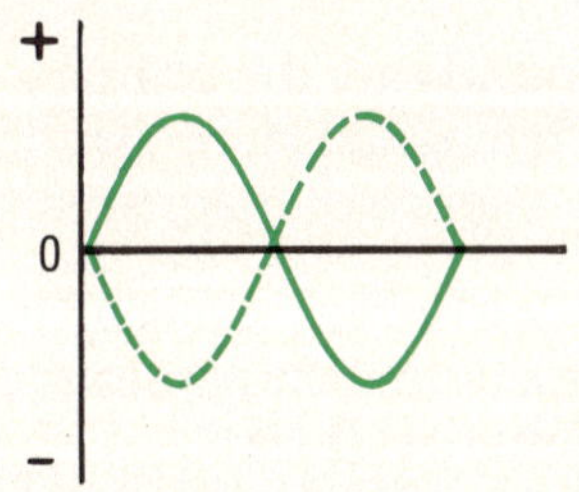

Figure 7-35. Waveforms 180 degrees out of phase.

As indicated, phase angle is an important consideration in determining the behavior of AC circuits. Phase angles and their characteristics are discussed further in the chapters that follow.

Discussion Questions/Problems

1. What is phase in AC current?
2. What is meant when a current is described as being in phase?
3. What is meant when a current is described as being out of phase?
4. What is the phase angle of AC current?

REVIEW

CHAPTERS 6 AND 7

SUMMARY

1. Ferrous compounds are the primary substances, or materials, in which magnetism occurs. Iron-based materials facilitate the presence of magnetism. (6.1)

2. Magnets occur naturally. The earth contains materials which are magnetic in their natural state. The best known of these is magnetite. (6.1)

3. Different types of ferrous materials have different properties for acquiring magnetism. The ability to conduct magnetic lines of force is called the *permeability* of a material. (6.1)

4. The permeability of materials determines their value and potential use in magnetic applications. (6.2)

5. Magnetic flux is the group of magnetic field lines that surround a magnet. (6.3)

6. Reluctance ($\mathfrak{R}$)is the opposition to magnetic flux. (6.3)

7. An electromagnet is a type of temporary magnet. The magnetism transferred electrically to ferrous materials, either temporarily or permanently, comes from a property known as *electromagnetism*. (6.4)

8. The performance of an electromagnet depends upon a number of variables. These variables are the electrical and physical characteristics of an electromagnet's construction. (6.5)

9. The Hall effect is the small electrical potential that is created in a wire when a stationary magnetic field is applied perpendicularly to the wire. (6.6)

10. If a magnet is moved across a wire, the movement of the flux field generates electricity. A current is created in the wire by electromagnetic induction. (6.7)

11. Three systems of units are used for the measurement of magnetism:
    1. English units in terms of inches, pounds, and ampere turns
    2. Centimetre-gram-second (Cgs) system
    3. Metre-kilogram-second-ampere (Mksa) system using Systeme International d'Unites (SI) units.

    The system most commonly used in today's electronics industry is the Mksa system using SI units. (6.8)

12. Alternating current, abbreviated AC, is the type of electricity most widely used for commerical purposes. (7.1)

13. Electricity is easier and more economical to deliver and use in AC form than in DC. As one example, AC can be readily stepped up or down in voltage. (7.1)

14. A waveform is a graph that illustrates the flow pattern versus time of a given type of current or voltage. (7.2)

15. The output of a rotating AC generator is a particular wave shape called a *sine wave*. (7.2)

16. A full rotation of an AC alternator is called a *cycle*. (7.2)

17. Sine waves are measured on the basis of portions of 360-degree cycles. Angles may be measured in units of degrees or radians. (7.3)

18. The number of AC cycles completed during a specific time period is known as the *frequency*. (7.4)

19. The period is the time required for one complete AC cycle. (7.5)

20. The length of an AC cycle, as represented by a sine wave, is called the *wavelength*. (7.6)

21. There are four basic ways for establishing the magnitude of an AC voltage:
    1. peak
    2. peak-to-peak
    3. root-mean-square (also called *effective value*)
    4. average. (7.7)

22. Ohm's law also applies to AC. (7.8)
23. Network theorems, as introduced in Chapter 3, provide shortcuts to analyze both DC and AC circuits. (7.9)
24. The term *phase* describes the time relationship of the current and voltage waveforms for an AC signal. (7.10)

PROBLEMS

1. If a material has a permeability of 40, what is its permeability classification? (6.2)
2. What is the permeability of a material having a B of 1,200 and an H of 20? (6.2 and 6.3)
3. What is the permeability of a material having a B of 2,000 and an H of 50? (6.2 and 6.3)
4. Find the magnetic flux for a magnetomotive force (mmf) of 1,800 and a reluctance ($\mathcal{R}$) of 9. (6.3)
5. What is the magnetic flux density (B) for 50,000 Mx through an area of 2 cm^2? (6.3 and 6.8)
6. What is the magnetizing force (H) produced by a current of 20 A at a distance of 5 cm? (6.3 and 6.8)
7. If a coil of 400 turns has 1 A flowing through it, what is NI? (6.5)
8. What is the magnetizing force (H), given 400 NI and r of 0.5? (6.5)
9. What is the magnetomotive force (mmf) of an iron-core coil having 400 NI? (6.5)
10. What permeance corresponds to a reluctance of 5? (6.8)
11. How many times per second does AC change polarity in the United States? (7.1)
12. What does a 60 Hz waveform look like? (7.2)
13. What is the instantaneous voltage value given a maximum voltage of 100 V and an angle of 40 degrees? (7.3)
14. What is the instantaneous current value given an instantaneous voltage value of 64.28 V and a resistance of 12 Ω? (7.3)
15. What is the wavelength if the velocity is 3×10^{10} cm/s and the frequency is 90 MHz? (7.6)
16. What is the frequency given a velocity of 6×10^{10} cm/s and a wavelength of 500 cm? (7.6)

17. If an rms value of 100 V is given, what is the peak-to-peak value? (7.7)
18. If an rms value of 100 V is given, what is the peak value? (7.7)
19. Given an I_{pp} of 4 A and an R of 20 Ω, what is the E_{pp}? (7.8)
20. Given an E_{rms} of 90 V and an R of 30 Ω, what is the I_{rms}? (7.8)
21. Given an I_{rms} of 5 A and an R of 50 Ω, what is the E_{pp}? (7.8)
22. Given an I_{avg} of 4 A and an E_{avg} of 12 V, what is the R? (7.8)

(Answers to the odd-numbered problems can be found at the back of the book.)

INDUCTANCE AND INDUCTORS

8.1 THE INDUCTOR

Inductance is the property of a circuit that enables it to oppose any change in current. An inductor is a device that opposes any change in current in an AC circuit. An inductor is also a device which can store a magnetic field.

You already know the basic principle behind inductance: *As current flows through a conductor, magnetic field lines build up around the conductor.* These field lines are concentric to the conductor. This means the field lines and the conductor have a common center or axis. As the magnetic field lines build up, they create opposition to the flow of current. The magnetic force turns back against the conductor from which it is generated, attempting to obstruct the flow of current. This opposition to changes in current is known as *counter electromotive force*, or *counter emf*.

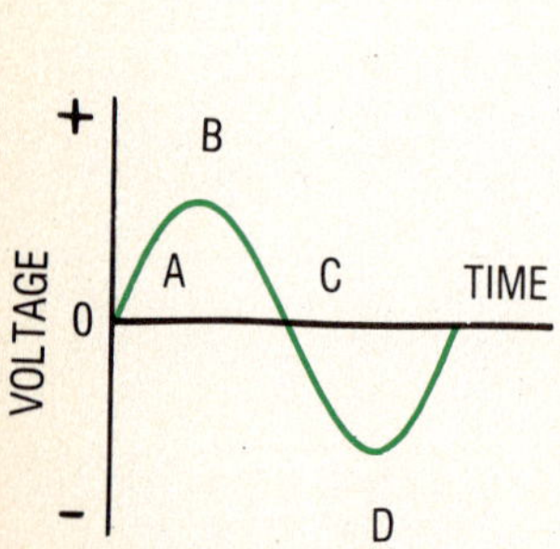

Figure 8-1. Sine wave showing counter emf.

The amount of counter emf is in proportion to the rate of AC alternation. The faster the rate of alternation, the greater the counter emf will be. This principle is illustrated in Figure 8-1, which shows the nature of counter emf. As AC flows through an inductor, it sets up a magnetic field that generates a counter electromotive force. During the sine wave pattern of AC flow, counter emf is greatest when the rate of change is at its highest. At point A in Figure 8-1, counter emf is higher than it is at point B, where the rate of change is relatively low. Similarly, counter emf is stronger at point C than at point D. At point B, where for a brief moment there is neither an increase nor a decrease in voltage, the rate of voltage change is zero. At this point, counter emf stops building. Instead, the magnetic field begins to collapse at the time the direction of current flow changes. As the magnetic field collapses, it generates current in the conductor in a direction that promotes continued current flow. As a result, the inductor works to prevent an increase in current in the interval between points A and B and a decrease in the interval between points B and C. Notice that this has the effect of opposing any change in current—exactly what the opening sentence of this chapter states.

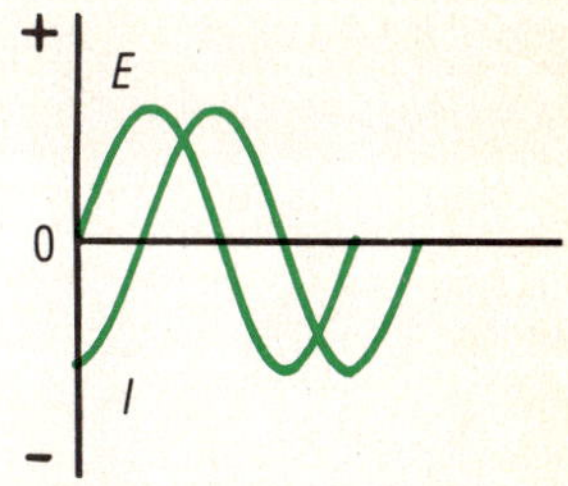

Figure 8-2. Inductance causes current to lag voltage and creates counter emf.

In the case of an AC sine wave, the effect is for current to lag voltage due to the inductance present in the circuit. In a pure inductive AC circuit, current lags voltage by 90 degrees, as shown in Figure 8-2.

In Figure 8-2, inductance also creates a counter emf as the current is rising on the upward path of the sine wave. This counter emf serves to maintain the relationship in which current lags voltage by 90 degrees.

All conductors have some inductance. The total amount of inductance depends upon the shape or formation of the conductor itself. Straight wire has small amounts of inductance. Coils of wire have much more.

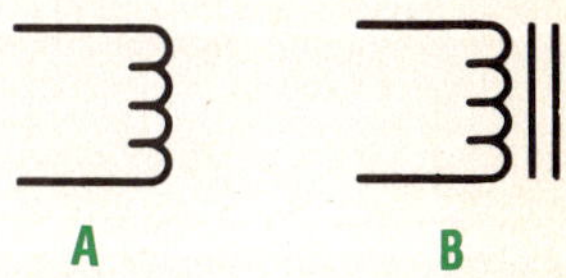

Figure 8-3. Symbols for inductors.

Symbols for inductance are shown in Figure 8-3. Figure 8-3A is for an air core inductor and Figure 8-3B, for an iron core inductor. Figure 8-4 shows an air core and an iron core inductor.

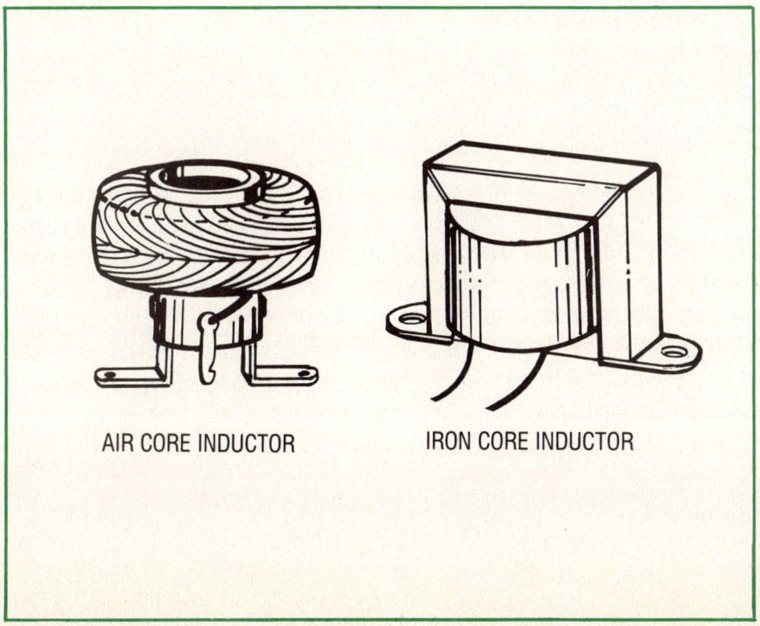

Figure 8-4. Inductors.

Discussion Questions/Problems

1. What is inductance?
2. At what points in a sine wave does the phenomenon of inductance have maximum effect?
3. How does the counter emf effect of inductance differ on the downward path of a sine wave?
4. What is the relationship of current to voltage in a typical AC inductive circuit?
5. How does coiling wire affect the inductance of the wire?

8.2 MEASURING INDUCTANCE

The basic unit of measurement for inductance is the henry. The abbreviation is H. This unit is named for Joseph H. Henry (1797-1878), an American physicist who made many important discoveries in electromagnetism.

Since inductance is a property of a coil, the standard henry is defined in terms of a standard coil. A coil with one henry of inductance will produce a counter emf of 1 volt when a one-cycle-per-second AC current of 1 ampere is passed through the coil.

The formula for determining inductance is as follows:

$$L = \frac{e_L}{\Delta i / \Delta \tau}$$

In this formula, L is the inductance of the coil in henrys. e_L is the voltage across the coil. $\Delta i/\Delta \tau$ is the change in current (Δi) for a given time interval ($\Delta \tau$). It has already been shown that the counter emf is proportional to the rate of AC alternation. Thus, the corresponding current (which lags the voltage by 90 degrees) will also be proportional to the rate of AC alternation. Figure 8-5 gives an example of how inductance is computed.

The inductance of this circuit is determined as follows:

$$\Delta i = 6\text{ A} - 4\text{ A} = 2\text{ A} \qquad \Delta \tau = 1\text{ s}$$

$$L = \frac{e_L}{\Delta i / \Delta \tau} = \frac{10\text{ V}}{2\text{ A/s}} = 5\text{ H}$$

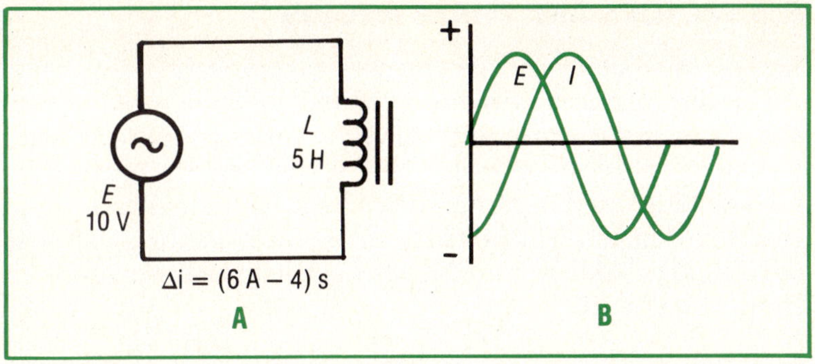

Figure 8-5. Inductive circuit.

Remember, in an inductive circuit like the one illustrated in Figure 8-5A, current lags voltage by 90 degrees as shown in Figure 8-5B. This is an important difference between inductive and resistive circuits. In a resistive circuit, such as in Figure 8-6A, the sine waves for voltage and current are exactly in phase, as shown in Figure 8-6B.

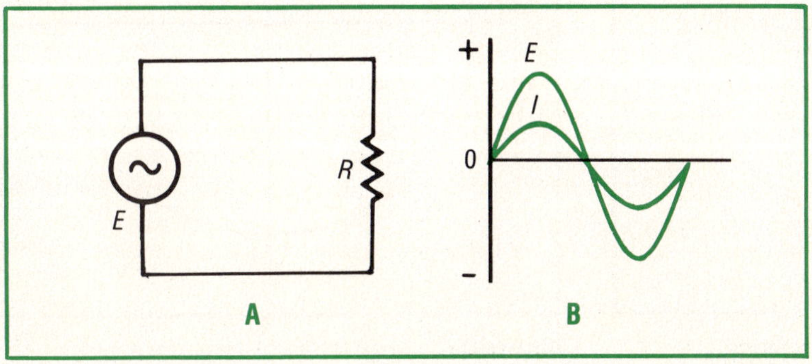

Figure 8-6. Resistive circuit.

Discussion Questions/Problems

1. What is a henry and what does it represent?
2. How does the rate of change, or alternation, affect the voltage across an inductor?
3. How does an inductive circuit compare with a resistive circuit?
4. If a circuit has a voltage of 5 V and a Δi of 2 A at a $\Delta \tau$ of 1 second, what is the inductance of the coil in henrys?

8.3 VECTORS AND PHASE ANGLES

Remember that a vector can be used to represent a sine wave. A rotating vector is sometimes called a *phasor*. The properties of inductance are shown in vector form in Figure 8-7 with a pair of rotating vectors. One of the vectors represents voltage; the other, current. The current vector in this diagram is at zero degrees. The voltage vector is at 90 degrees. Because current lags voltage by 90 degrees, this must represent an inductive circuit.

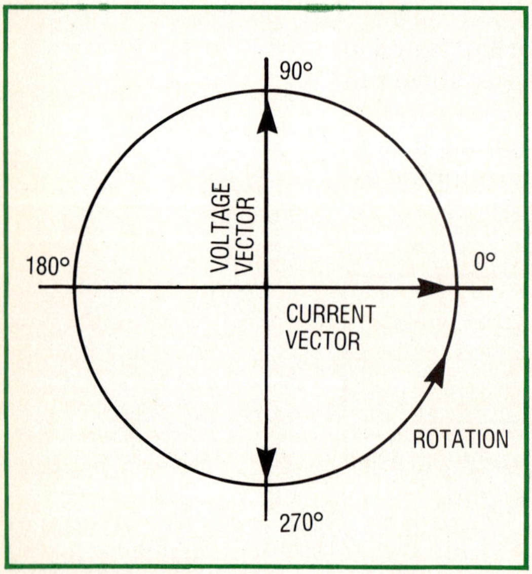

Figure 8-7. Properties of inductance.

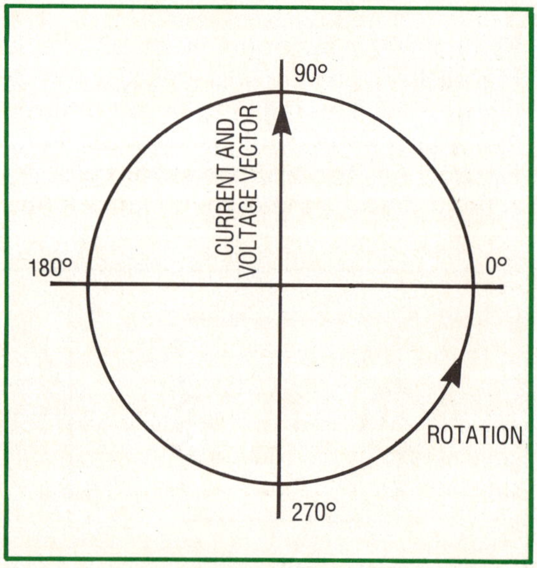

Figure 8-8. Properties of resistance.

Figure 8-8 shows a vector representation for a resistive circuit at the point the voltage vector is at 90 degrees. In a resistive circuit, current and voltage are in phase throughout the waveform.

The vector drawings in Figures 8-7 and 8-8 provide an excellent basis for comparing the properties of resistance and inductance. Both resistance and inductance oppose the flow of electricity, but in different ways. Resistance consumes power and creates heat without affecting the properties or patterns of AC. Inductance, on the other hand, does not dissipate energy in the way that resistance does. However, inductance causes a phase angle between voltage and current.

As current is increased in a circuit, electrical energy in an inductive circuit is transformed to a magnetic field. As current is decreased, the magnetic field collapses and the current flow is maintained. Energy, therefore, is not consumed; it is merely changed from current to magnetic field, and then back to current.

The inductance of a coil depends on its physical design. This is illustrated in Figure 8-9. In Figure 8-9A, there are comparatively few turns in an air core coil, and inductance is low. In Figure 8-9B, inductance increases with two dimensions of the coil. There are more turns, adding inductance. Also, the coil is longer. The longer the coil, the greater the inductance.

Length and number of turns must be considered together. For example, given the same number of turns but different lengths in two coils, the coil that is longer in length will have lower inductance. This is because the shorter coil concentrates the magnetic fields, thus increasing inductance. Therefore, concentrating turns adds to inductance; spreading out the turns in a coil decreases inductance.

Figure 8-9C shows a small iron core inductor. This has relatively little inductance because of the comparatively few turns in the short length of the coil. By comparison, the iron core inductor in Figure 8-9D has a greater inductance because it has more turns in its longer length.

Figure 8-9 illustrates the two factors that determine the inductance of coils. First, inductance increases as the number of turns in the coil increases. Second, the greater the area occupied by each turn of a coil, the greater the inductance will be.

There is also a third physical factor that affects inductance of a coil. This applies to iron (and other ferromagnetic) core inductors only. The greater the permeability of the iron core, the higher the inductance will be.

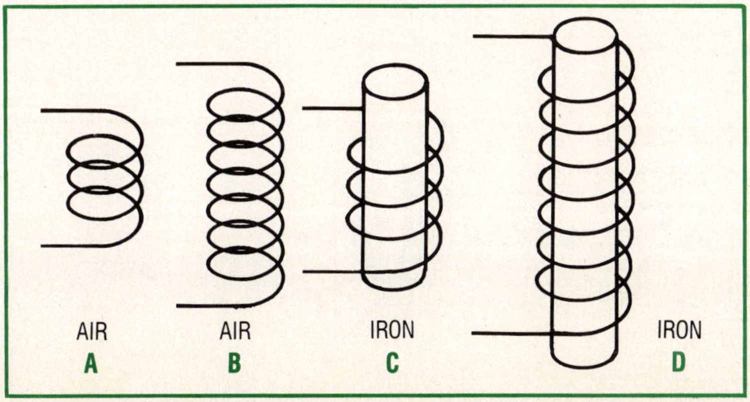

Figure 8-9. Physical design factors determine inductance of coil.

Discussion Questions/Problems

1. What is a rotating vector called?
2. What is the significance of the arrow showing direction of rotation in a vector diagram?
3. If the vector diagram for a normal inductive circuit showed the voltage vector at 120 degrees, what would be the position of the current vector?
4. What determines inductance?
5. What are the major differences in the properties of resistance and inductance?
6. How does the number of turns in a coil affect inductance?
7. How does permeability of an iron core affect the inductance of a coil?

8.4 LENZ'S LAWS

The basic law of electromagnetic induction was discovered and stated by H. F. E. Lenz in 1833. Lenz's law of electromagnetic induction states that a current which is caused by a change in a magnetic flux will be directed so as to oppose the act which caused the current.

This is illustrated in Figure 8-10, which shows a bar magnet moving toward a coil of wire. As the N pole of the bar

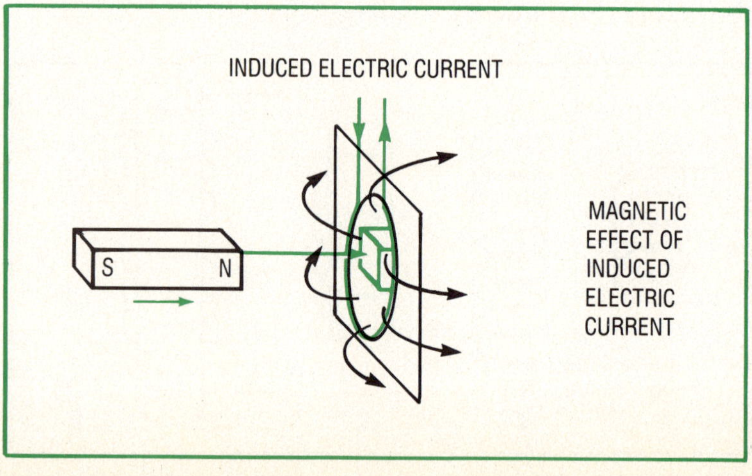

Figure 8-10. Simplified illustration of Lenz's law of electromagnetic induction.

magnet moves toward the coil, the magnet's lines of flux will intersect the coil, inducing a current. According to Lenz's law, the current induced in the coil will oppose the motion of the magnet. In other words, the current will cause the coil to have its N pole on the side toward the bar magnet, opposing the action of the bar magnet.

Lenz's law means that in an AC circuit, inductance opposes changes in current. This effect is illustrated in Figure 8-11A, which shows an AC source and an inductor. Figure 8-11B shows that, as the amplitude of positive current increases, inductive opposition also increases. As the amplitude of positive current decreases, the inductive opposition decreases. As the sine wave drops below the zero point, there is again inductive opposition.

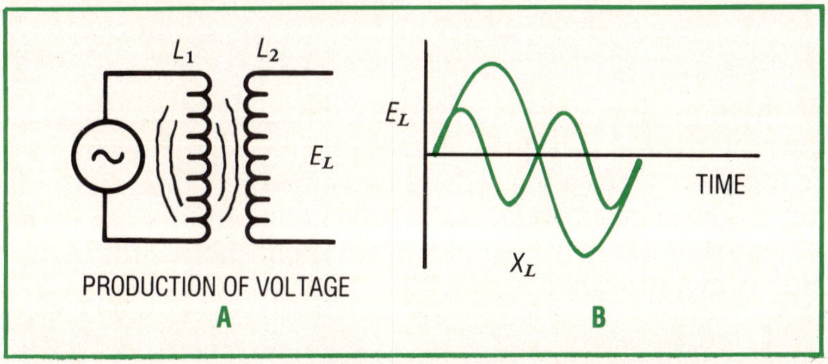

Figure 8-11. Effect of inductance opposing changes in current in AC circuit.

Because of the inductive opposition to changing current, a coil to which AC is applied generates a voltage. The formula that governs the voltage generated across a coil is:

$$E_L = L\frac{\Delta i}{\Delta \tau}$$

In this equation, L is the inductance of the coil, in henrys. Δi is the change in current through an alternation. $\Delta \tau$ is the time for a single alternation.

Now consider an example of the application of this formula. Suppose $L = 5$ H; $\Delta i = 2$ A; and $\Delta \tau = 1$ second.

$$E_L = L\frac{\Delta i}{\Delta \tau} = 5 \times \frac{2}{1} = 10 \text{ V}$$

In this situation, a coil with 2 A of applied AC and a period of 1 second will develop 10 V AC across it.

Mutual Inductance

Magnetic flux establishes a magnetic field that occupies space. Magnetism penetrates the air around an electromagnetic conductor, such as a coil. When two conductors are placed close to each other, the magnetic field of one can affect the other. For example, if a coil with an applied AC current were placed close to a coil with no applied current, a voltage would be created in the second coil. This phenomenon is known as *mutual induction*.

The principle of mutual induction is illustrated in Figure 8-11. This shows a coil with AC applied alongside a coil that is not carrying any current. A voltage is induced in the second coil. The formula for the induced voltage is:

$$e_L = \frac{N\Delta\phi}{\Delta\tau}$$

In this equation, e is electromotive force. N is the number of turns in the coil. ϕ is flux in the coil L_2, measured in Wb. (Wb is the abbreviation for webers explained in Chapter 6.) τ is time, given in seconds. $\Delta\phi$ is the change in flux. $\Delta\tau$ is the change in time. An example of the application of this equation is given below:

$$\Delta\phi = 0.05 \text{ Wb} \qquad \Delta\tau = 1 \text{ s} \qquad N = 20 \text{ turns}$$

$$e_L = \frac{20 \times 0.05}{1} = 1 \text{ V}$$

In this example, $\Delta\phi = 0.05$ Wb, $\Delta\tau = 1$ second, and $N = 20$ turns. All of these figures apply to L_2. Given these factors, $e = 1$ V. Even though there is no voltage directly applied to L_2, a voltage will appear across it because of the current flowing through L_1, and the mutual induction between the two coils.

Coefficient of Coupling ($\varkappa$)

To apply the principle of mutual inductance, the relationship between inductive coils must be known. This relationship is the coefficient of coupling. The symbol for the coefficient of coupling is $\varkappa$. The formula for determining $\varkappa$ is:

$$\varkappa = \frac{\text{flux linking coil 1 and coil 2}}{\text{total flux produced by coil 1}}$$

To illustrate this principle, suppose half the flux from coil 1 is transferred into coil 2. If this happens, $\varkappa = 0.5$. This is a ratio. There is no unit of measure involved. An example of an equation determining the coefficient of coupling is given below. In this equation, it is assumed that the flux linking the two coils is equal to 10 μWb and that the total flux of coil 1 is 20 μWb.

$$\varkappa = \frac{10\ \mu\text{Wb}}{20\ \mu\text{Wb}} = 0.5$$

Once the coefficient of coupling is known, it becomes possible to apply a mathematical formula to determine the mutual inductance of two coils. The formula is:

$$L_m = \varkappa\sqrt{L_1 L_2}$$

As an example, assume that $\varkappa = 0.5$, $L_1 = 15\ \mu\text{H}$, and $L_2 = 25\ \mu\text{H}$. The solution is as follows:

$$L_m = 0.5\sqrt{15\ \mu\text{H} \times 25\ \mu\text{H}} = 9.68\ \mu\text{H}$$

Sometimes the terms *loose* and *tight* are used to describe the coupling. Tightly coupled coils are those with a high coefficient of coupling. When the coefficient of coupling is low, the coils are said to be loosely coupled.

Discussion Questions/Problems

1. Under Lenz's law, what is the effect of an inductance on AC flowing through the inductor?
2. What is mutual inductance?
3. How does the number of turns in a coil affect the voltage induced in a second coil?
4. If you have coils with inductances of 10 mH and 30 mH, and the coefficient of coupling is 0.4, what is the mutual inductance?

8.5 INDUCTOR CHARACTERISTICS

Inductors are never perfect. There is always some resistance in a coil that results from the wire itself. The lower this resistance, the better the inductor will perform. Therefore, inductors are rated according to quality. The symbol used for the quality factor is Q.

The value of Q for an inductive circuit is determined by applying the following formula:

$$Q = \frac{X_L}{R}$$

Assume that an inductive circuit has an inductive reactance of 1,000 Ω and a resistance of 25 Ω. The quality factor is determined as follows:

$$Q = \frac{X_L}{R} = \frac{1{,}000\ \Omega}{25\ \Omega} = 40$$

The higher the quality of a coil, the better will be its performance in most applications. As a rule of thumb, quality coils have a Q of 20 or better.

Discussion Questions/Problems

1. How is the quality factor of an inductor determined?
2. What is the symbol for the quality factor of an inductor?
3. If an inductor has inductive reactance of 500 Ω and a resistance of 100 Ω, what is the quality factor?
4. Based on your answer to the previous question, would the inductor for which you found the Q be of high or low quality?

8.6 INDUCTORS IN SERIES AND IN PARALLEL

When several coils are connected in series and there is no mutual inductance among the coils, total inductance is determined by adding the inductances of the individual coils. This is similar to the method used to determine the resistance of several resistors in series. The principle for determining total inductance in a series of coils is as follows:

$$L_T = L_1 + L_2 + L_3 + \ldots + L_N$$

Below is an example of how this formula is used to determine total inductance of two coils in series, one of 2 H and one of 3 H:

$$L_T = L_1 + L_2$$
$$L_T = 2\text{ H} + 3\text{ H} = 5\text{ H}$$

When several coils are connected in parallel, total inductance is equal to the reciprocal of the sum of the reciprocals of all of the individual inductances. The formula is:

$$L_T = \frac{1}{\frac{1}{L_1} + \frac{1}{L_2} + \frac{1}{L_3} + \cdots + \frac{1}{L_N}}$$

This is similar to the method used to determine the resistance of several resistors in parallel. An example of the application of the formula for determining total inductance of coils connected in parallel is given below:

$$L_1 = 2\text{ H} \quad L_2 = 3\text{ H}$$
$$L_T = \frac{1}{\frac{1}{L_1} + \frac{1}{L_2}} = \frac{1}{\frac{1}{2} + \frac{1}{3}} = 1.2\text{ H}$$

Series-Aiding Inductance

In practice, it becomes impossible to eliminate all mutual inductance in a transformer. There is almost always some mutual inductance present. Of course, the closer the coils are placed to each other, the greater the mutual inductance will be.

When mutual inductance occurs between coils with turns in the same direction, as shown in Figure 8-12A, a series-aiding effect takes place. That is, there is an increase in out-

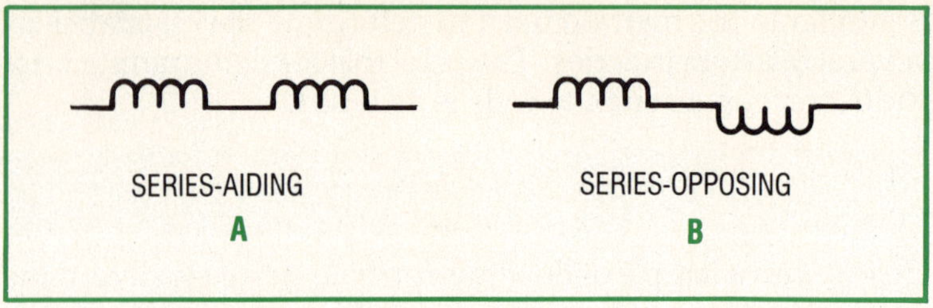

Figure 8-12. Series-aiding and series-opposing inductance.

put due to the effect of mutual inductance. This increase is expressed mathematically as follows:

$$L_T = L_1 + L_2 + 2L_m$$

The formula for determining mutual inductance is given earlier in this chapter. For the purposes of this discussion, mutual inductance is a given. The following is an example of how to determine total inductance for two series-aiding coils:

$$L_1 = 2\text{ mH} \quad L_2 = 3\text{ mH} \quad L_m = 2\text{ mH}$$
$$L_T = L_1 + L_2 + 2L_m$$
$$L_T = 2\text{ mH} + 3\text{ mH} + 2\,(2\text{ mH}) = 9\text{ mH}$$

Series-Opposing Inductance

When coils of a transformer are wound in opposite directions, as in Figure 8-12B, a series-opposing effect exists. The formula for series-opposing inductance subtracts, rather than adds, the term for mutual inductance. This formula is:

$$L_T = L_1 + L_2 - 2L_m$$

When two coils are connected in series so that their mutual inductance opposes rather than aids, the result is significantly different. Using the same values as in the series-aiding example above, the equation below shows how total inductance is determined:

$$L_T = L_1 + L_2 - 2L_m$$
$$L_T = 2\text{ mH} + 3\text{ mH} - 2\,(2\text{ mH}) = 1\text{ mH}$$

Note that, using the same values, a series-aiding circuit has an inductance of 9 mH whereas a series-opposing circuit has an inductance of only 1 mH.

Discussion Questions/Problems

1. What is the effect of connecting coils in series within a single circuit?
2. What is the effect of connecting coils in parallel within a single circuit?
3. What happens when mutually coupled series coils in a circuit have turns that are wound in the same direction?
4. What happens when mutually coupled series coils in the same circuit have turns that are wound in opposing directions?
5. If a series-opposing circuit has coils with inductances of 3 mH and 4 mH, as well as a mutual inductance of 1 mH, what is total inductance for the circuit?

8.7 INDUCTIVE REACTANCE

An inductor displays a special property when alternating current is applied: An inductor opposes the flow of AC through the inductor. This property of inductors is called *inductive reactance.*

The amount of inductive reactance is proportional to the amount of inductance: the greater the inductance, the greater the inductive reactance.

Interestingly, direct current meets no opposition in passing through an inductor. Inductive reactance has no effect on DC.

The frequency of the AC itself affects the amount of inductive reactance: the higher the frequency, the greater the inductive reactance.

The symbol for inductive reactance is X_L. The formula for inductive reactance is:

$$X_L = 2\pi f L \qquad \pi = 3.14$$

Inductive reactance is measured in ohms (Ω). An example of how inductive reactance is calculated is given below. In reviewing these equations, remember that inductive reactance increases as the AC frequency increases.

$$f = 1{,}000 \text{ Hz} \quad L = 0.15 \text{ H}$$

$$X_L = 2\pi f L$$

$$X_L = 6.28\ (1{,}000)\ (0.15) = 942\ \Omega$$

The formula for inductance (measured in henrys) is as follows:

$$L = \frac{X_L}{2\pi f}$$

To illustrate, the inductance of a coil with an inductive reactance of 1,500 Ω, at a frequency of 5,000 Hz, is:

$$L = \frac{X_L}{2\pi f} = \frac{1{,}500}{6.28\ (5{,}000)} = 0.048 \text{ H}$$

To find frequency in a circuit where the inductive reactance and the inductance are both known, use the following formula:

$$f = \frac{X_L}{2\pi L}$$

To illustrate, the frequency which would cause an inductive reactance of 3,000 Ω in an inductance of 0.30 H is:

$$f = \frac{X_L}{2\pi L} = \frac{3{,}000}{6.28\ (0.30)} = 1{,}592.36 \text{ Hz}$$

Ohm's law applies to inductive reactance just as it applies to resistance. The relationships among current, voltage, and inductive reactance are:

$$I = \frac{E}{X_L} \qquad E = IX_L \qquad X_L = \frac{E}{I}$$

The following is an example of how Ohm's law is applied to determine current in a circuit with inductive reactance:

$$E = 120\text{ V} \quad X_L = 1{,}000\ \Omega$$
$$I = \frac{E}{X_L} = \frac{120}{1{,}000} = 0.12\text{ A}$$

Discussion Questions/Problems

1. What does inductive reactance do?
2. What is the relationship between inductance and inductive reactance?
3. What is the relationship between AC frequency and inductive reactance?
4. What effect does inductive reactance have upon direct current?

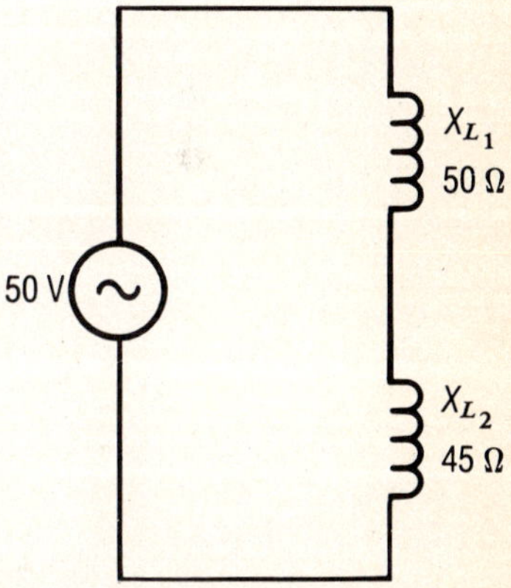

Figure 8-13. Circuit with inductors in series.

8.8 CONFIGURING INDUCTIVE REACTANCES

When inductors are connected in series, their inductive reactance is additive, or cumulative. A circuit with inductors in series is shown in Figure 8-13. The formula that applies to such circuits is:

$$X_{L_T} = X_{L_1} + X_{L_2} + X_{L_3} + \ldots + X_{L_N}$$

Below is an example of how this formula is used to determine inductive reactance for the circuit in Figure 8-13:

$$X_{L_T} = X_{L_1} + X_{L_2} = 50\ \Omega + 45\ \Omega = 95\ \Omega$$

A circuit with inductors in parallel is shown in Figure 8-14. Although given previously, the formula for determining the inductive reactance of inductors in parallel is repeated here:

$$X_{L_T} = \cfrac{1}{\cfrac{1}{X_{L_1}} + \cfrac{1}{X_{L_2}} + \cfrac{1}{X_{L_3}} + \cdots + \cfrac{1}{X_{L_N}}}$$

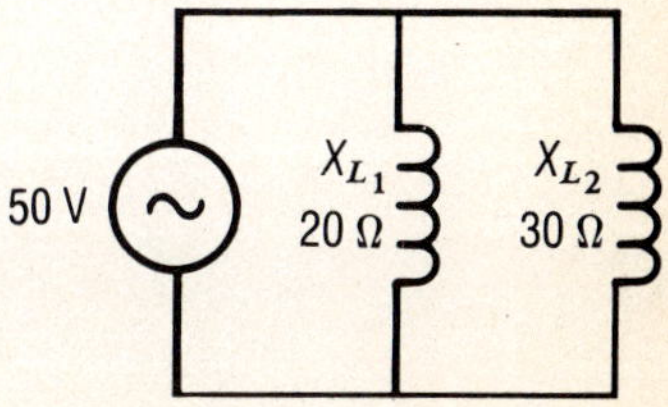

Figure 8-14. Circuit with inductors in parallel.

Shown below are the computations to determine inductive reactance for the circuit in Figure 8-14:

$$X_{L_T} = \frac{1}{\frac{1}{X_{L_1}} + \frac{1}{X_{L_2}}}$$

$$= \frac{1}{\frac{1}{20\ \Omega} + \frac{1}{30\ \Omega}}$$

$$= \frac{1}{\frac{3}{60} + \frac{2}{60}}$$

$$= \frac{1}{\frac{5}{60}}$$

$$X_{L_T} = 12\ \Omega$$

Chapter 2 presents a special formula for applying Ohm's law to determine the resistance of two resistors connected in parallel. The same basic formula can be used to determine the inductive reactance of two inductors connected in parallel. The formula, applied to determine inductive reactance for the circuit in Figure 8-14, is as follows:

$$X_{L_T} = \frac{X_{L_1} \times X_{L_2}}{X_{L_1} + X_{L_2}} = \frac{20\ \Omega \times 30\ \Omega}{20\ \Omega + 30\ \Omega} = \frac{600}{50} = 12\ \Omega$$

Ohm's law and the equations above can be used to determine current in the circuit shown in Figure 8-14.

$$I = \frac{E}{X_{L_T}} = \frac{50\ V}{12\ \Omega} = 4.17\ A$$

Applying Inductive Reactance

Inductive reactance can be a useful property because it is frequency sensitive. Because of this property, inductive reactance can be used as a means of blocking, or screening, unwanted frequencies. This characteristic is valuable in many aspects of electronics. Examples are the designs of electronic filters and of radio and television antennas.

Electronic filters are special circuits that allow certain frequencies to pass while blocking others. Examples of filters commonly used include low-pass units. These units pass only those currents with frequencies lower than a certain (design) frequency. There are also high-pass filters, which pass only those frequencies above the design frequency. Other filters are called band-pass units. These permit only a specified band of frequencies to pass.

Figure 8-15 illustrates a low-pass filter. This diagram shows a coil connected in series to the load. Since higher frequencies increase inductive reactance, this circuit has the effect of allowing low frequencies to pass while blocking higher frequencies.

A high-pass filter design is shown in Figure 8-16. This circuit has a resistor in series with the load, which is a coil. This circuit has the opposite effect of the one in Figure 8-15. The circuit in Figure 8-16 permits high frequencies to pass because they are not shunted to ground by the coil.

Band-pass filters require electronic components not yet covered. They are described in a later chapter.

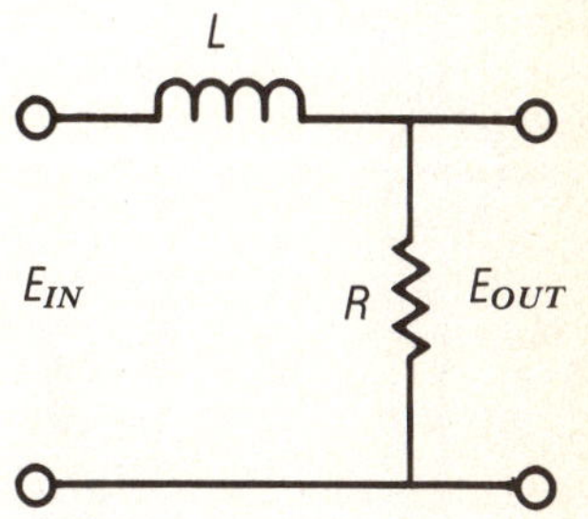

Figure 8-15. Circuit with low-pass filter.

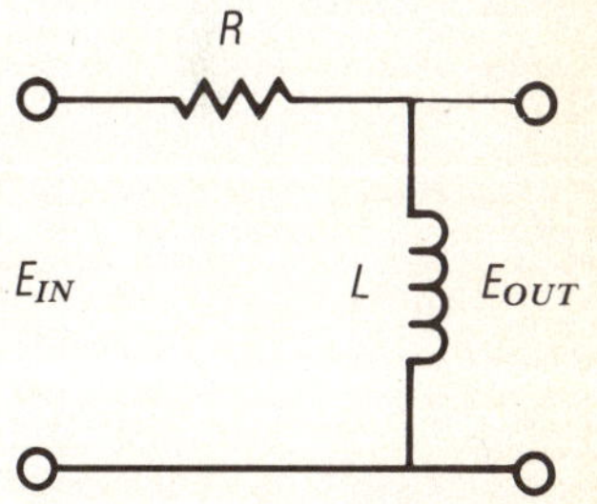

Figure 8-16. Circuit with high-pass filter.

Discussion Questions/Problems

1. How is equivalent inductive reactance determined when inductors are connected in series?
2. How is equivalent inductive reactance determined when inductors are connected in parallel?
3. A circuit has inductors of 75 Ω and 25 Ω connected in parallel. What is the equivalent inductive reactance for this circuit?
4. A circuit has inductors of 35 Ω and 45 Ω connected in series. What is the equivalent inductive reactance?
5. What is the inductive reactance of a coil with an applied frequency of 15 kHz and an *L* of 2 mH?

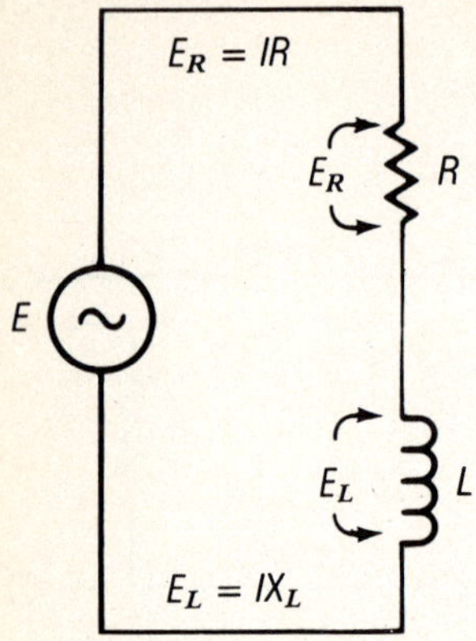

Figure 8-17. *RL* series circuit.

8.9 *RL* CIRCUITS

An *RL* circuit is one that includes both resistance and inductance. There must be at least one resistor and one inductor in an *RL* circuit, though there can be more of each. A common *RL* circuit uses a resistor and an inductor in series, as shown in Figure 8-17. In an *RL* series circuit, current lags voltage by some phase angle between 0 and 90 degrees. The phase angle depends upon the values of resistance and inductance.

Pythagorean Theorem

To determine phase angles in *RL* circuits, a formula known as the Pythagorean theorem is used. This theorem gets its name from Pythagoras, an ancient Greek philosopher, teacher, and mathematician. As originally devised, the Pythagorean theorem was an aid to geographic measurements and was often used in the geometric solutions to navigation problems. Today, the theorem is used in algebra and trigonometry. The theorem is a means of determining the length of one side of a right triangle when the length of the other two sides are known.

In electronics, the Pythagorean theorem is a tool for determining phase angles when working with AC. The basic formula for the Pythagorean theorem is:

$$c^2 = a^2 + b^2$$
$$c = \sqrt{a^2 + b^2}$$

The principle of the Pythagorean theorem is that *the square of the hypotenuse of a right triangle is equal to the sum of the squares of the other two sides*. Figure 8-18 illustrates a right triangle, identifying the sides as applied in the formula above.

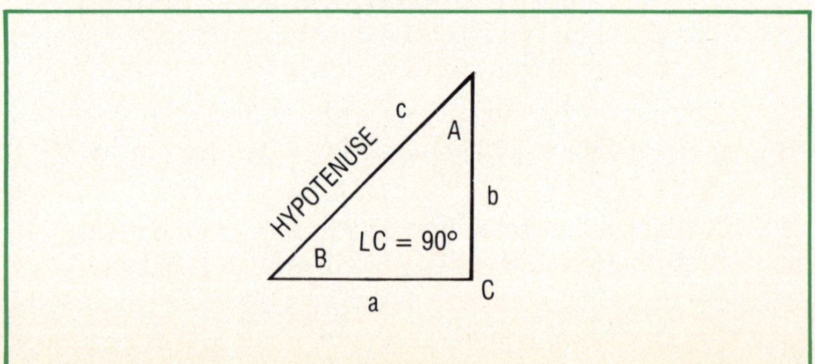

Figure 8-18. Right triangle.

To illustrate, consider the series *RL* circuit in Figure 8-17. Since this is a series circuit, the same current flows through both *R* and *L*. In accordance with Ohm's law, the voltage across *R* will be *IR*, and the voltage across *L* will be IX_L. Since current through the inductor always lags the current through the resistor by 90 degrees, it is possible to construct the drawing in figure 8-19 (in which *E* is the supply voltage).

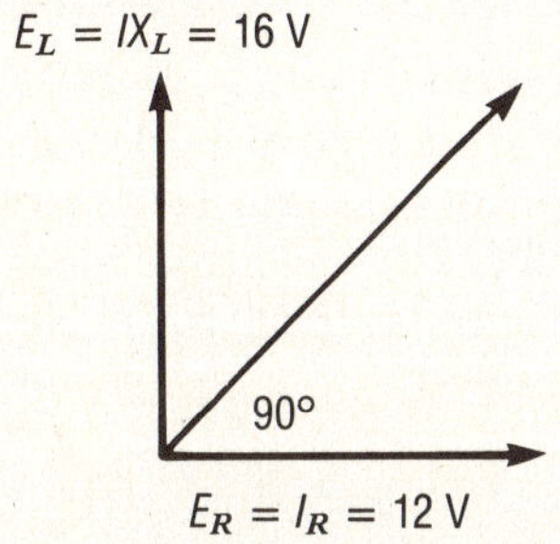

Figure 8-19. Phase angle for *RL* series circuit of Figure 8-17.

Using the Pythagorean Theorem, *E* is given by:

$$E = \sqrt{(E_R)^2 + (E_L)^2}$$

$$E = \sqrt{(12)^2 + (16)^2} = \sqrt{144 + 256} = \sqrt{400} = 20 \text{ V}$$

Similarly, in cases where *E* and E_R are known, or *E* and E_L are known, the formulas used to determine the voltage across the resistor and the inductor in Figure 8-19 are shown below:

$$E_R = \sqrt{(E)^2 - (E_L)^2} \qquad E_L = \sqrt{(E)^2 - (E_R)^2}$$

Discussion Questions/Problems

1. What is an *RL* circuit?
2. In an *RL* circuit, does voltage lag or lead current?
3. What phase angles should you expect to find in an *RL* circuit?
4. State the Pythagorean theorem.
5. Refer to Figure 8-17. In this circuit, if $E_R = 4$ V and $E_L = 3$ V, what is the total voltage of the circuit?

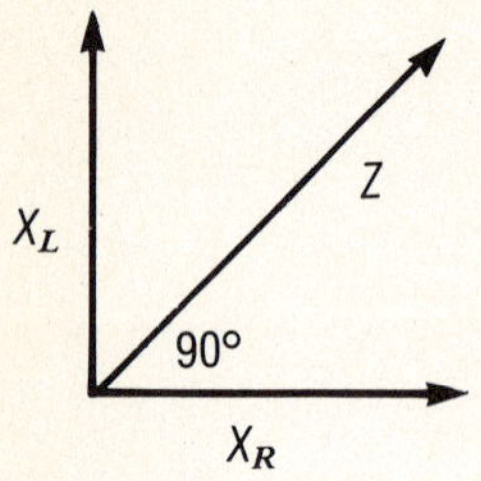

Figure 8-20. Phase angle for *RL* series circuit used to determine impedance.

8.10 IMPEDANCE

The total opposition to current flow in an *RL* circuit is called *impedance*, and includes a component due to resistance and a component due to inductive reactance. The symbol for impedance is Z. The formula for determining impedance is:

$$Z = \frac{E}{I}$$

Impedance is represented by the symbol Z in the diagram in Figure 8-20. Notice that Figure 8-20, for impedance, is the same as Figure 8-19, for voltage. To determine the values of impedance, resistance, and inductive reactance in a series circuit, the following formulas (each a different form of the same formula) would be used:

$$Z = \sqrt{(R)^2 + (X_L)^2}$$
$$R = \sqrt{(Z)^2 - (X_L)^2}$$
$$X_L = \sqrt{(Z)^2 - (R)^2}$$

Suppose the values in Figure 8-20 were 3 Ω of resistance and 4 Ω of inductive reactance. To find Z for this series circuit, use:

$$\begin{aligned} Z &= \sqrt{(R)^2 + (X_L)^2} \\ &= \sqrt{(3)^2 + (4)^2} \\ &= \sqrt{9 + 16} \\ &= \sqrt{25} \\ Z &= 5\ \Omega \end{aligned}$$

Suppose Figure 8-20 shows an impedance of 20 Ω and an inductive reactance of 12 Ω. Resistance is determined as follows:

$$\begin{aligned} R &= \sqrt{(Z)^2 - (X_L)^2} \\ &= \sqrt{(20)^2 - (12)^2} \\ &= \sqrt{400 - 144} \\ &= \sqrt{256} \\ R &= 16\ \Omega \end{aligned}$$

Notice that both Figure 8-19 and Figure 8-20 show right triangles. It is known that there is a 90 degree angle (right angle) because the current in an inductor lags the current in a resistor by 90 degrees. Of course, the triangle includes two other angles, and the specific value of these angles has a particular importance. Only one angle must be referred to, because the other angle will simply be 90 degrees minus this angle. This angle is called the phase angle. The symbol for the phase angle of an *RL* circuit is θ, which is the lowercase Greek letter *theta*. The formula to determine the phase angle in a series *RL* circuit is:

$$\theta = \text{arc tan}\left(\frac{X_L}{R}\right)$$

Note that this equation uses the arc tangent of the θ angle. First, it is necessary to use a calculator or an arctan table to find the value of the arc tangent. For the example in Figure 8-20, with an inductive reactance of 4 Ω and a resistance of 3 Ω, the phase angle is determined as follows:

$$\theta = \text{arc tan}\left(\frac{X_L}{R}\right) = \text{arc tan}\,\frac{4}{3} = \text{arc tan}\ 1.33 = 53^\circ$$

This calculation establishes that current lags voltage by 53 degrees in a series impedance circuit with the given values.

Impedance in Parallel Circuits

Figure 8-21 shows a parallel *RL* circuit, in which *R* and X_L are the oppositions of the resistive and inductive elements. The

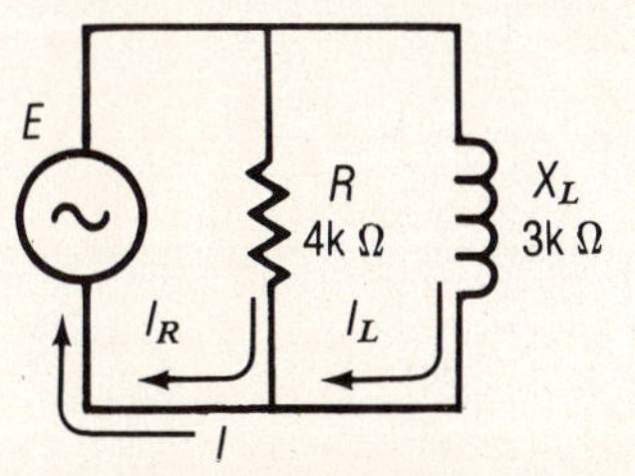

Figure 8-21. Parallel *RL* circuit.

voltage across *R* equals the voltage across *L* and, in turn, equals *E*. Thus:

$$I_R X_R = I_L X_L = E$$

And since $X_R = R$:

$$I_R R = I_L X_L = E$$

Also, $E = IZ$ where Z is the equivalent impedance of R and L in parallel. Therefore, note also that:

$$I^2 = (I_L)^2 + (I_R)^2$$

Then:

$$\begin{aligned}
Z &= \frac{E}{I} \\
&= \frac{I_R X_R}{I} \\
&= \frac{I_R R}{\sqrt{(I_L)^2 + (I_R)^2}} \\
&= \frac{\dfrac{I_L X_L}{R} \times R}{\sqrt{(I_L)^2 + (I_R)^2}} \\
&= \frac{RX_L \times \dfrac{I_L}{R}}{\sqrt{\dfrac{(I_R)^2 R^2}{(X_L)^2} + \dfrac{(X_L)^2}{(X_L)^2} \times (I_R)^2}} \\
&= \frac{RX_L \times \dfrac{I_L}{R}}{\dfrac{I_R}{X}\sqrt{R^2 + (X_L)^2}} \\
Z &= \frac{RX_L}{\sqrt{R^2 + (X_L)^2}}
\end{aligned}$$

An example of a parallel impedance RL circuit is shown in Figure 8-21. For the values in this diagram, impedance is calculated as follows:

$$Z = \frac{RX_L}{\sqrt{R^2 + (X_L)^2}}$$

$$= \frac{(4K)(3K)}{\sqrt{(4K)^2 + (3K)^2}}$$

$$= \frac{12K}{\sqrt{25K}}$$

$$= \frac{12K}{5{,}000}$$

$$Z = 2{,}400\ \Omega$$

The formula for determining the phase angle of a parallel impedance *RL* circuit is:

$$\theta = \text{arc tan}\left(\frac{R}{X_L}\right)$$

The phase angle for the circuit and values in Figure 8-21 are determined as follows:

$$\theta = \text{arc tan}\left(\frac{R}{X_L}\right) = \frac{4K}{3K} = 1.33 = 53^{\circ}$$

Discussion Questions/Problems

1. What is impedance?
2. How is impedance determined in a series *RL* circuit?
3. What is the phase angle of a series *RL* circuit with an inductive reactance of 8 Ω and a resistance of 6 Ω?
4. How is impedance determined in a parallel *RL* circuit?
5. How is the phase angle of a parallel impedance *RL* circuit determined?

8.11 TYPES OF INDUCTORS AND TRANSFORMERS

A *transformer* is a device which uses the principle of mutual inductance to transfer electrical energy from a source to a destination. A transformer has at least two coils. Each transformer has one *primary* coil, which is connected to the energy source. There is also a *secondary* coil, to which energy can be transferred. There may be multiple secondary coils within a transformer. This depends on the application and design.

Figure 8-22 shows schematics of two basic transformers. In Figure 8-22A, the two coils are simply close to each other, with only air in between. This is called an air core transformer. In Figure 8-22B, there is an iron core between the coils. This core becomes a temporary magnet, assisting the inductive process.

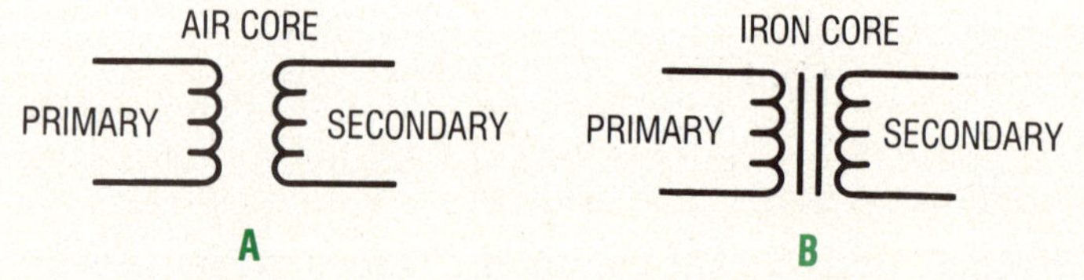

Figure 8-22. Schematics of transformers.

Transformers built for commercial electric power applications are usually constructed with iron cores. Most of these cores are built up of laminations. A laminated core, in simple terms, has a number of thin sheets of iron material bonded together, as shown in Figure 8-23A. Generally, laminated iron cores work better in transformers than solid iron cores. The reason is that laminated cores have smaller eddy current losses, and also less magnetic reluctance.

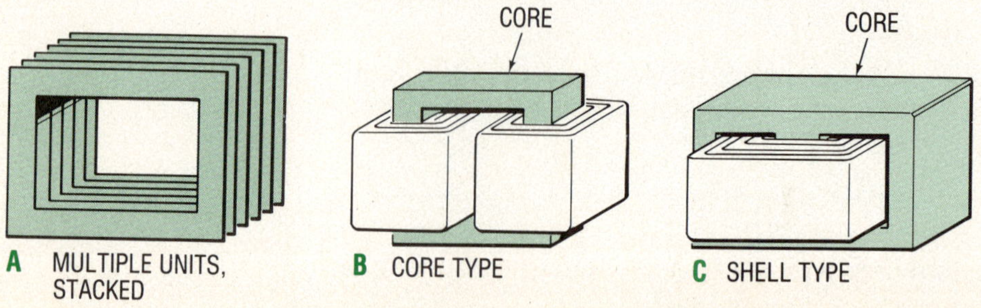

Figure 8-23. Laminated iron core construction of transformers.

The type of laminated core shown in Figure 8-23B is an open type. Transformers using this type of core have two coils wrapped around the outside portions of the core.

Another configuration of iron core used commercially is the shell type, shown in Figure 8-23C. This type of core has a center member. The coils can then be wrapped around the center member, between the outside structures. Surrounding the coil with core material in this way increases magnetic induction in either a primary or secondary coil.

For additional precision in inductance devices (both transformers and inductors), toroidal cores are often used. Figure 8-24A shows a basic circular, or toroidal core, shape. Figure 8-24B shows a coil added to the core. Toroidal structures often result in greater electrical precision, with less precision necessary in the manufacturing process. Another core configuration is shown in Figure 8-25. The basic core structure is illustrated in Figure 8-25A which shows a central iron member, called a *slug*, inserted into a ceramic holder. In Figure 8-25B, the ceramic form has been wrapped with a coil. The slug may be screwed into or out of the coil. This is a way

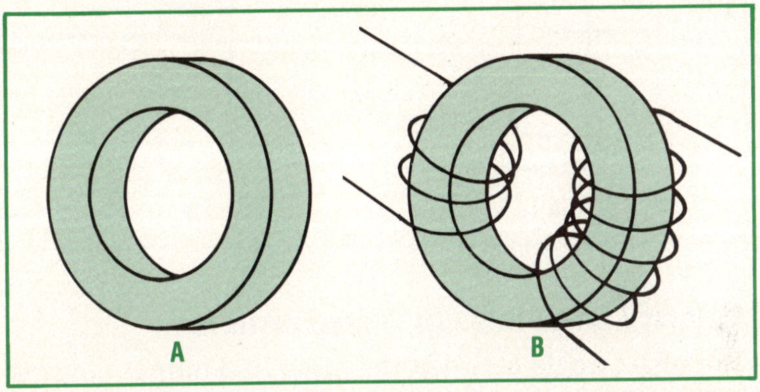

Figure 8-24. Toroidal structures used for transformers and inductors.

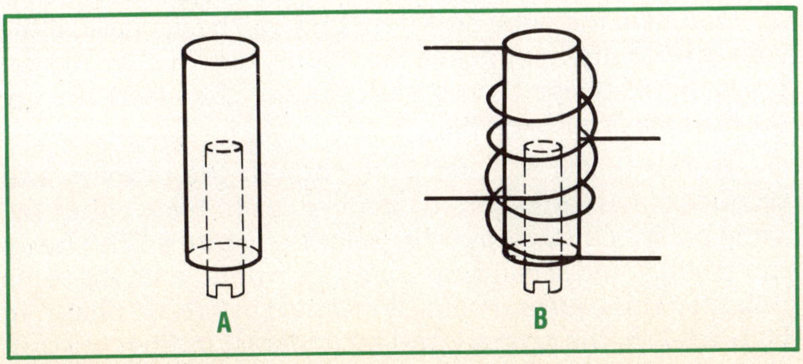

Figure 8-25. Iron slug inserted into ceramic holder used as transformer core.

to vary the inductance of a coil. This type of transformer is used largely in communication devices, including television sets and radios.

As a general principle, the ''tighter'' a transformer coil is wound, the greater its inductance will be. ''Tight'' means that the turns of the coil are very close together. Conversely, the looser the windings, the lower the inductance. Figure 8-26 shows loose and tight coils. In Figure 8-26B, the tightly wound coils are superimposed on each other.

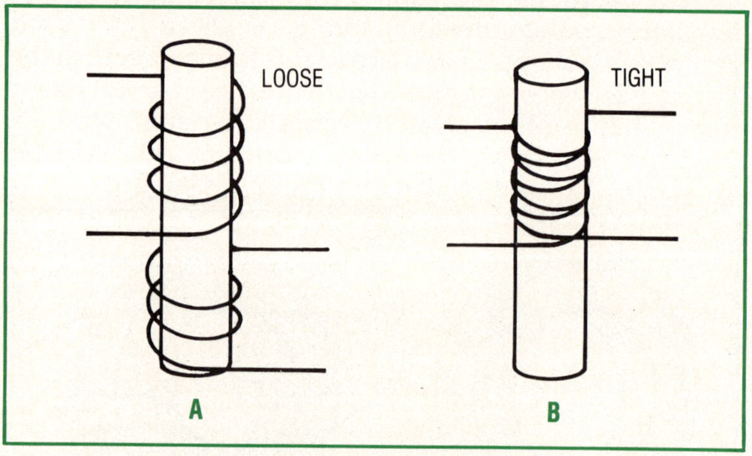

Figure 8-26. Loosely and tightly wound coils.

Polarity and Phase in Transformers

The phase angle between the input and output of transformers is determined by the direction of the coil windings. If the primary and secondary coils are wound in the same direction, the secondary coil will have the same polarity as the primary. If the coils are wound in opposite directions, the polarity of the secondary coil will be opposite to that of the primary coil.

The same relationship holds for phase angles. If the coils are wound in the same direction, they will be in phase. If the coils are wound in opposite directions, they will be 180 degrees out of phase. These performance characteristics are diagrammed in Figure 8-27.

Notice that it is possible to connect the primary and secondary windings into a circuit so as to have either the same polarity or the opposite polarity. Figure 8-27A shows the primary and secondary windings wound with the same polarity. This causes the input and output to be in phase. In Figure 8-27B, the primary and secondary windings are con-

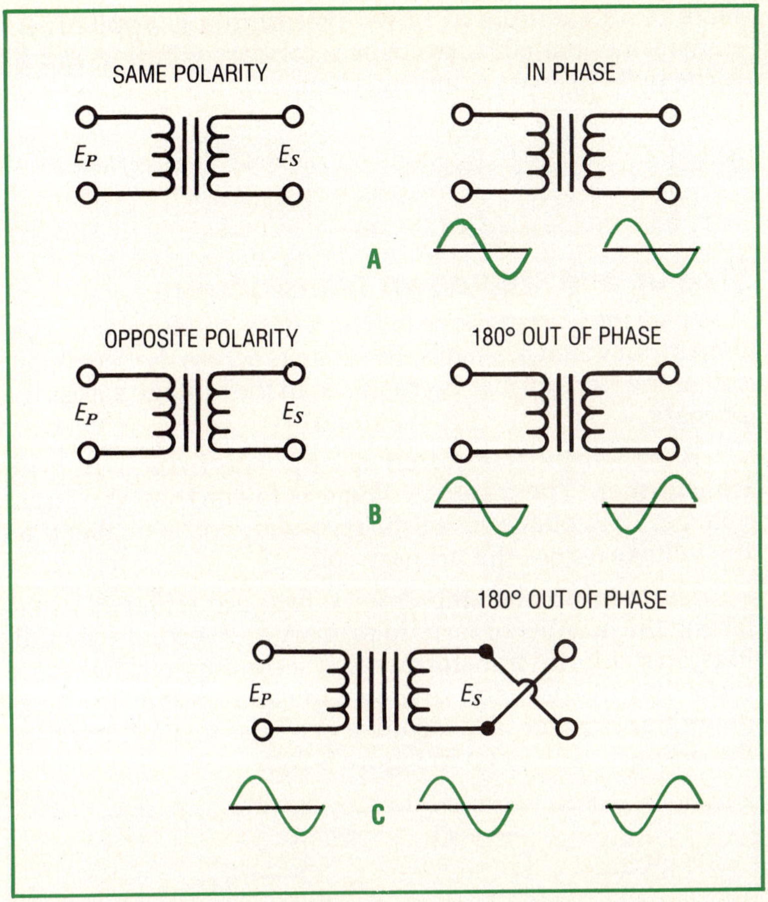

Figure 8-27. Performance characteristics of coil windings.

nected with opposite polarities. This causes the input and output to be out of phase 180 degrees. Figure 8-27C shows the primary and secondary windings connected with the same polarity. However, the connection to the secondary winding has been crisscrossed, reversing the output. The primary coil is, therefore, out of phase with the output.

Turns Ratio

The turns ratio of the coils used in transformers determines the output voltage. In designing a transformer it is necessary to determine the ratio between the number of turns in the primary and the secondary coils. The formula used in determining turns ratio is:

$$\text{Turns Ratio} = \frac{N_s}{N_p}$$

Here is an example of how this formula is applied. If a transformer has 600 turns in its secondary winding and 100 turns in its primary winding, then:

$$\text{Turns Ratio} = \frac{N_s}{N_p} = \frac{600}{100} = 6$$

Step-up and Step-down Transformers

A *step-up transformer* is one in which output voltage is greater than input voltage. That is, the voltage across the secondary coil of the transformer is greater than the voltage across the primary.

A *step-down transformer* is opposite from a step-up transformer. The output voltage is lower than the input voltage. The voltage across the secondary coil is smaller than the voltage across the primary coil.

The step-up and step-down effects are achieved by selecting the number of turns in primary and secondary windings. In a step-up transformer, the secondary coil has more turns than the primary, as shown in Figure 8-28A. In a step-down transformer, the secondary coil has fewer turns than the primary, as shown in Figure 8-28B.

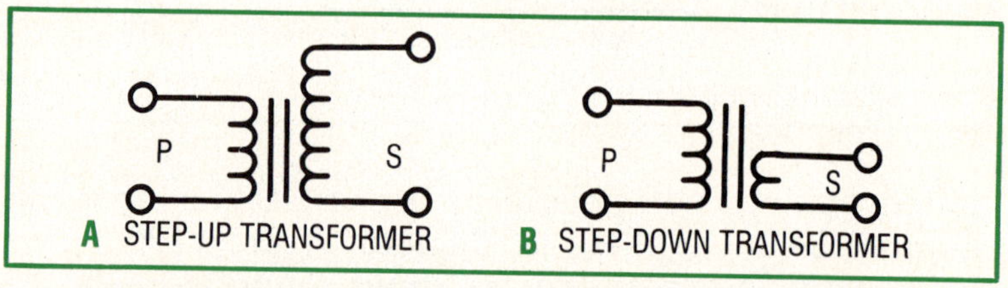

Figure 8-28. Step-up and step-down effects of coil windings.

The voltages in the primary and secondary coils will vary in direct proportion to the turns ratio. The relationship is indicated by the following equation:

$$\frac{E_s}{E_p} = \frac{N_s}{N_p}$$

Thus, if the voltage across one of the coils in a transformer is known, along with either the turns ratio or the number of turns in both the primary and secondary windings, it is possible to determine the voltage in the other coil. For example, assume that the voltage across a primary coil of a step-up transformer is 120 V. Also assume that there is a turns ratio of 6. The voltage across the secondary coil is determined as follows:

$$E_s = \frac{N_s}{N_p} E_p = 6 \times 120 \text{ V} = 720 \text{ V}$$

A similar relationship holds true for current in primary and secondary coils. However, to keep the same power in both coils, current must be in inverse relationship. The relationship is:

$$\frac{I_p}{I_s} = \frac{N_s}{N_p}$$

To illustrate how this equation is used to determine current, assume that a primary coil has a current of 1 A. The step-up transformer has a turns ratio of 6. The current in the secondary coil is determined as follows:

$$I_s = \frac{N_p}{N_s} I_p = 6 \times 1 = 6 \text{ A}$$

Where power is concerned, the relationship between primary and secondary coils is different. As discussed above, the rule for power is that the secondary wattage is equal to the primary wattage. Since power is computed by multiplying current and voltage, even though there are inverse relationships for current and voltage, the multiplications always come out the same. This principle for power transfer by a transformer is demonstrated mathematically in the equations below:

$$P_s = P_p$$

$$P_p = E_p I_p \qquad\qquad P_s = E_s I_s$$

$$E_s = \frac{N_s}{N_p} E_p \qquad\qquad I_s = \frac{N_p}{N_s} I_p$$

$$P_s = \frac{N_s}{N_p} E_p \times \frac{N_p}{N_s} I_p$$

$$P_s = E_p I_p = P_p$$

Efficiency of Transformers

Transformers do have losses, so the secondary winding does not receive all of the power that goes into the primary winding. Still, the transformer is one of the most efficient electrical devices. The power output is proportionately high as compared with power input. Efficiencies of 80 to 90 percent are common.

The power losses that do exist result from magnetic hysteresis losses, from resistance in the wire coils, and from eddy currents.

In many applications it is important to be able to determine efficiency of a transformer. In a transformer, efficiency is determined as follows:

$$\text{Efficiency} = \frac{P_{out}}{P_{in}} \times 100\%$$

As this equation indicates, a perfect transformer would be one in which power output was exactly equal to power input. Since this never happens, the efficiency falls off in direct proportion to the loss of power in output.

Autotransformers and Variable Inductances

An autotransformer is a transformer that has a single coil and several additional connections, so that it behaves somewhat as two separate coils. Figure 8-29A shows an autotransformer in which the section of the coil from A to B acts as the primary, and the section from A to C acts as the secondary.

A–B–C is wound as one continuous coil, with a *tap* at B. The section AB carries both the primary current and the secondary current. As with a regular transformer, the secondary voltage is proportional to the ratio of turns in AC to turns in AB. In an autotransformer the primary and secondary voltages are not isolated from each other as they would be in a transformer. This is because they share a common winding section, AB.

Autotransformers can be designed and built so the output voltage is adjustable. As in Figure 8-29B, this is done by making the *tap* at B adjustable to one of several positions, B_1–B_4. In some types of autotransformers, the adjustment is by means of a sliding contact on the coils, producing an output voltage that is adjustable over a fairly wide range. In a device of this type, there is always a minimum number of

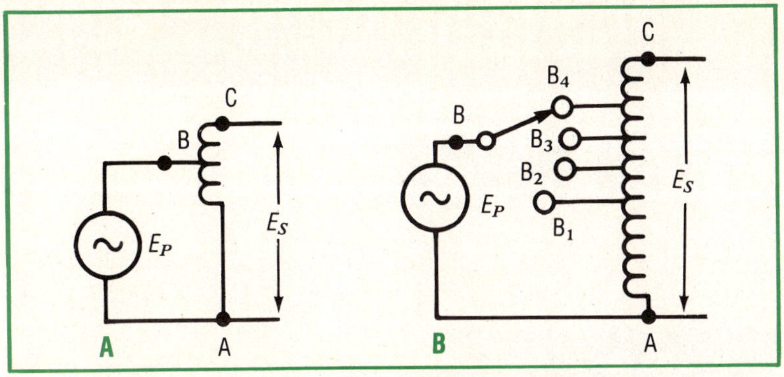

Figure 8-29. Autotransformer.

turns (B_1), and the sliding contact cannot be moved below this point. This assures that there is always at least a minimum number of turns in the primary part of the winding.

Discussion Questions/Problems

1. What is the relationship between primary and secondary windings in a transformer?
2. Why are the cores of most commercial transformers laminated?
3. How does the direction of turns in coils affect the way a transformer is used?
4. How does the turns ratio relate to voltage, current, and power in transformers?
5. If a given transformer has 1,000 turns in its secondary coil and 500 turns in the primary coil, what is the turns ratio?
6. Given a turns ratio of 5 and a primary voltage of 120 V, what is the voltage of the secondary coil?
7. What determines the output current of a secondary coil in a transformer?

CAPACITANCE AND CAPACITORS

9.1 THE CAPACITOR

Capacitance is the property of a circuit that allows electrons to be stored as an electric charge. A capacitor is a device that opposes any change in voltage in an AC circuit. In addition, a capacitor is a device that can store an electric charge. Note the complementary properties of capacitors and inductors:

- A capacitor opposes any change in *voltage* in an AC circuit. An inductor opposes any change in *current* in an AC circuit.
- A capacitor stores *electric charge*. An inductor stores a *magnetic field*.

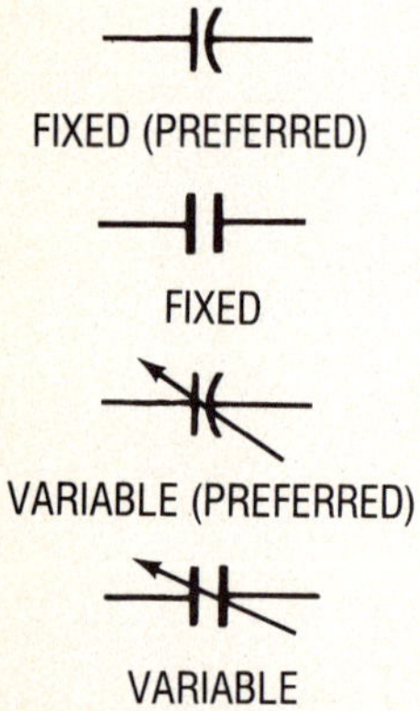

Figure 9-1. Symbols for fixed and variable capacitors.

A simple capacitor consists of two plates of conducting material, separated by an insulator. When the plates cannot be moved, they form a *fixed capacitor*. When the distance between the plates can be adjusted, the device is known as a *variable capacitor* (sometimes called a *condenser*). Figure 9-1 shows the symbols used to represent fixed and variable capacitors.

The insulation between the plates of a capacitor may be air or some other type of insulation and is known as a *dielectric*. A dielectric is a nonmetallic insulating material. The dielectric between the plates of a capacitor permits the plates to accumulate electric charge.

Figure 9-2 illustrates the functioning of a simple capacitor. When a DC source is applied to the plates of a capacitor, positive charges accumulate on one plate and negative charges accumulate on the other plate. Thus, a difference in electrical

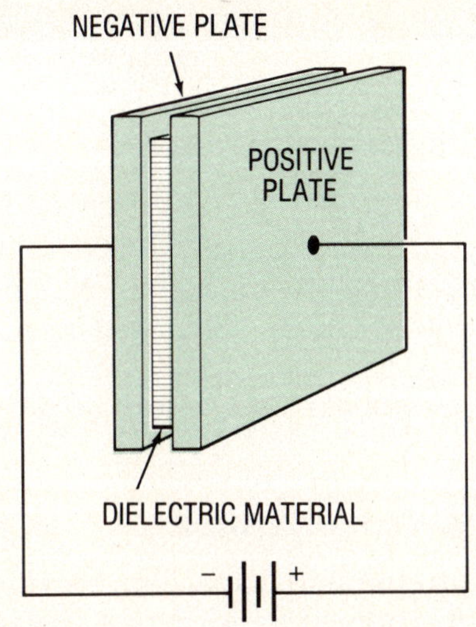

Figure 9-2. Design of simple capacitor.

potential builds up across the capacitor. When the potential difference equals the source of voltage, the capacitor stops charging. The plates of the capacitor will then retain their charge even if the electrical source is removed. These charges will not be retained indefinitely by the capacitor. Eventually the charges are dissipated. A capacitor does, however, retain its electrical potential for some time.

Capacity

The ability of a capacitor to accept charge depends upon its capacitance and upon the voltage applied to build the charge. The basic formula for measuring the charge on a capacitor is:

$$Q = CE$$

In this equation, the letter *Q* indicates charge, which is measured in coulombs. The letter *C* is the symbol for capacitance, which is measured in farads. (The farad is discussed below.) *E* is the symbol for voltage.

The equation for capacity can be restructured to find the values of voltage or capacitance. These other forms of the basic equation are:

$$E = \frac{Q}{C} \qquad C = \frac{Q}{E}$$

Discussion Questions/Problems

1. What is capacitance?
2. How does a capacitor work?
3. What determines the full charge of capacitor plates?
4. What would be the charge in coulombs on a capacitor that has 10 V applied and a capacitance of 100 μF?

9.2 MEASURING CAPACITANCE

The unit of measurement for capacitance is the farad, abbreviated F. The unit is named for Michael Faraday. A capacitor has a capacitance of 1 farad when a potential difference of 1 volt will charge it with 1 coulomb of electricity. For electronic applications, the farad is too large for practical use. Therefore, capacitance measurements are usually given in smaller units, as shown in the table in Figure 9-3.

FARAD EQUIVALENCIES			
MICROFARAD (μF)	=	one millionth of a farad	$= 1 \times 10^{-6}$ F
PICOFARAD* (pF)	=	one-millionth of a microfarad	$= 1 \times 10^{-12}$ F
NANOFARAD (nF)	=	one-thousanth of a microfarad	$= 1 \times 10^{-9}$ F

**Also called* MICROMICROFARAD ($\mu\mu$ F)

CAPACITANCE VALUES		
DIELECTRIC	CAPACITANCE	BREAKDOWN VOLTAGE
AIR	10–400 pF	400
CERAMIC	0.001–1 μF	500–20,000
ELECTROLYTIC	0.01–5,000 μF	5–500
MICA	5–5,000 pF	500–20,000
PAPER	0.001–1 μF	200–1,600

Figure 9-3. Farad equivalencies and capacitance values.

Determining Charge, Capacity, and Voltage

Examples of the use of the capacitance formula are given in the paragraphs that follow.

Here is an example of the formula to determine charge:

$$C = 5\ \mu F \qquad E = 100\ V$$

$$Q = CE = 5 \times 10^{-6} \times 100$$

$$Q = 500 \times 10^{-6}\ \text{Coulombs}$$

Next, is an example showing how capacitance is calculated:

$$Q = 40 \times 10^{-6}\ C \qquad E = 25\ V$$

$$C = \frac{Q}{E} = \frac{40 \times 10^{-6}}{25} = 1.6 \times 10^{-6}\ \text{F or } 1.6\ \mu F$$

Voltage calculations are as follows:

$$Q = 50 \times 10^{-6}\ C \qquad C = 50 \times 10^{-6}\ F$$

$$E = \frac{Q}{C} = \frac{50 \times 10^{-6}}{50 \times 10^{-6}} = 1\ V$$

The capacitance of a capacitor is determined by the area of the plates, the thickness of the dielectric, and the ability of the dielectric to hold an electric charge.

Generally, the larger the plate area, the greater the capacitance will be. Conversely, the thinner the dielectric, the greater the capacitance.

Another factor that affects the properties of capacitors is the material used as a dielectric. Two indicators are used to describe dielectric materials. One is called the *dielectric constant*, represented by K_ϵ. The dielectric constant is a ratio that measures insulation properties in relation to a vacuum. The dielectric constant of a capacitor consisting of two conducting plates in a vacuum is assigned a value of 1.0. Other dielectric materials are then rated comparatively.

The dielectric constant (also known as *relative permittivity*) has no units of dimension. It is a relative measure that compares the absolute permittivity of a given material (ϵ) to the absolute permittivity of a vacuum, (ϵ_0).

$$\epsilon_v = \frac{\epsilon}{\epsilon_0}$$

The absolute permittivity of any material is:

$$\epsilon = \epsilon_v \epsilon_0 \text{ where}$$
$$\epsilon_0 = 8.85 \times 10^{-12} \text{ F/m}$$

Permittivity is the ability of a dielectric to hold a static charge. It is comparable to permeability, the ability of a ferrous material to hold or conduct a magnetic field.

The other indicator of dielectric performance is known as *dielectric strength*. This is a rating of dielectric materials according to their ability to withstand voltage. Ratings are in terms of 1 mil (thousandth of an inch) of thickness. Dielectric strength is measured in terms of volts per mil, i.e., the voltage required to ''spark through'' 1 mil of insulation.

The dielectric constants and strengths for a number of materials are given in Figure 9-4.

MATERIAL	DIELECTRIC CONSTANT (K_ϵ)	DIELECTRIC STRENGTH (Volts per Mil))
VACUUM	1.0	
AIR	1.0	31
POLYSTYRENE	2.4	300–710
RUBBER	2–4	470
OIL	2–5	375
PAPER	2–6	1,250
BAKELITE (PAPER BASE)	3.8	250–585
MICA	3–8	600–1,500
PORCELAIN	6.0	200–400
GLASS	8.0	760
CERAMICS	80–1,200	600–1,200

Figure 9-4. Dielectric constants and strengths.

Dielectric materials for use in capacitors are selected for specific performance characteristics and applications. The best material to do any given job is then chosen. The characteristics evaluated are both physical and electrical. They include capacity, voltage rating, operating temperatures, frequency, and polarity.

Capacitance of a Capacitor

The capacitance of any given capacitor is determined by applying the following formula:

$$C = \frac{K_\epsilon \times A}{d} (8.85 \times 10^{-12}\ \text{F/m})$$

In the above equation, A is the area of one of the capacitor's plates, measured in square metres. The symbol d is the distance between the plates, measured in metres. K_ϵ represents the dielectric constant. The constant in parentheses represents the absolute permittivity of the dielectric used in the capacitor. Note that increasing the area of the plates increases capacitance and that increasing the distance between plates decreases capacitance. An example of the application of the formula for determining the capacity of a capacitor is given below. These calculations are based on a capacitor with a paper dielectric having a dielectric constant of 4, an area of 3 m^2, and a spacing, or distance between plates, of 0.03 m.

$$\begin{aligned} C &= \frac{K_\epsilon \times A}{d} (8.85 \times 10^{-12}\ \text{F/m}) \\ &= \frac{4 \times 3}{0.03} \times (8.85 \times 10^{-12}\ \text{F/m}) \\ &= 400 \times 8.85 \times 10^{-12} \\ &= 3{,}540 \times 10^{-12} \\ C &= 3{,}540\ \text{pF} \end{aligned}$$

Discussion Questions/Problems

1. How does plate size affect capacitance?
2. How does the thickness of the dielectric affect capacitance?
3. What is the dielectric constant of a material?
4. What is dielectric strength?
5. What is the capacitance of a capacitor with a polystyrene dielectric, an area of 0.5 m and a distance of 1 mm?

9.3 PHASE ANGLES

Phase angles for capacitors follow rules that are complementary to those for inductors. In an inductor, voltage leads current. However, in a capacitor, voltage lags current by 90 degrees. For a capacitor, as shown in Figure 9-5, current

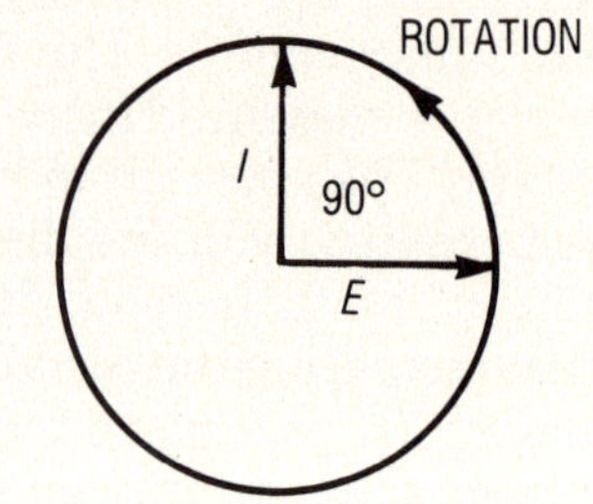

Figure 9-5. Rotation vector for capacitor.

and voltage vectors are 90 degrees out of phase, and voltage lags current (current leads voltage) by 90 degrees. Figure 9-6 shows phase relationships for current and voltage for a purely capacitive circuit.

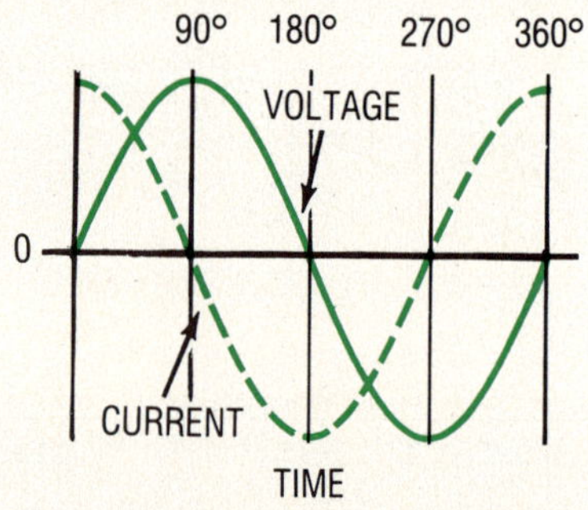

Figure 9-6. Phase relationships for current and voltage in capacitive circuit.

This phase angle information about capacitive circuits means that, when a capacitor is fully charged, it has its highest voltage and lowest (zero) current levels. The current is zero because the capacitor is fully charged and cannot accept any more charge. Conversely, a peak level of current means a low level of voltage and implies that the capacitor is just beginning to charge. In capacitors, current always leads voltage.

The relationship of the magnitudes of current and voltage in capacitive circuits can be described as follows:

$$i_c = C \frac{\Delta v}{\Delta \tau}$$

In this equation, i_c represents instantaneous current in the circuit, with C being capacitance, measured in farads. Δv is the change in voltage. $\Delta \tau$ is the change in time.

For an example of the application of this formula, assume a capacitance of 10 pF, a voltage difference of 50 V, and a time difference of 1 μsec. The solution is as follows:

$$i_c = 10 \times 10^{-12} \times \frac{50}{1 \times 10^{-6}} = 0.05 \text{ mA}$$

Discussion Questions/Problems

1. What are the phase angle relationships of current and voltage for a capacitor?
2. How do phase angles of current and voltage for capacitors compare with phase angles of current and voltage for inductors?
3. When a capacitor is fully charged, what is the relationship of current to voltage?
4. When a capacitor is just beginning to charge, what is the relationship of current to voltage?

9.4 CAPACITOR CHARGE STATES

A capacitor can be in any of these states:

- Uncharged
- Charging
- Charged
- Discharging.

The states of a capacitor are illustrated in Figure 9-7. In Figure 9-7A, the capacitance is uncharged. The switch is open and no current flows in the circuit.

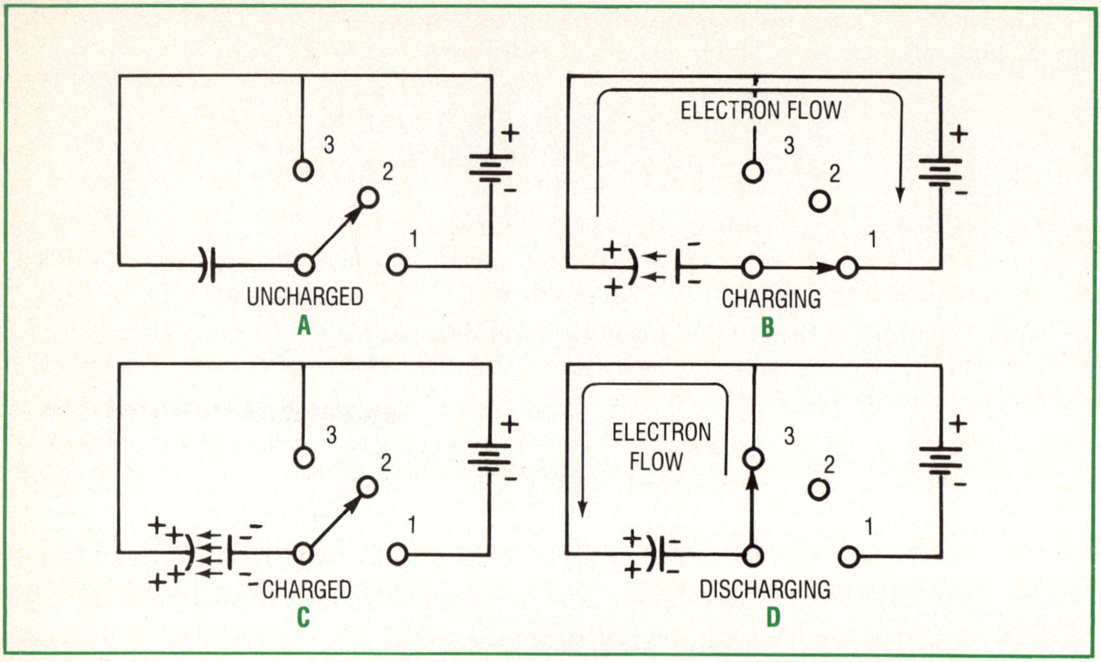

Figure 9-7. States of a capacitor.

In Figure 9-7B, the switch is moved to apply a DC source. Electrons build a potential difference between the plates of the capacitor.

When the capacitor is fully charged, the maximum potential difference has been accumulated. At this point the circuit can be opened, as is the case in Figure 9-7C, and the capacitor will retain its charge.

Figure 9-7D shows the circuit switched into the discharging mode. The capacitor is short-circuited. Electrons flow from the negative plate to the positive plate, thus neutralizing the charge on the capacitor.

Discussion Questions/Problems

1. What are the four states in which a capacitor can be?
2. When has the maximum potential difference been accumulated for a capacitor?
3. What happens when a capacitance circuit is switched into the discharge mode?

9.5 CAPACITOR CHARACTERISTICS

Several characteristics of a capacitor affect its performance. These include:

- Temperature coefficient
- Leakage
- Power factor
- Behavior with AC and DC.

Temperature Coefficient

Performance of a capacitor varies with ambient (surrounding) temperature. Capacitance can respond to increases in temperature in a positive or a negative manner. A positive temperature coefficient means that capacitance increases when temperature increases. A negative coefficient means that capacitance decreases when temperature increases.

Temperature coefficients are given in parts per million (ppm) per degree of temperature on the Celsius scale. In effect, there is an increase of 1×10^{-6} μF in capacitance per degree Celsius of temperature increase, for a positive temperature coefficient of 1 ppm.

In stating a temperature coefficient, a reference point is used. This reference point, which is applied to many electronic components, is 25 degrees C. As an example, a positive temperature coefficient of 100 ppm for a 1 μF capacitor would cause an increase of 100 pF for every degree of increase in temperature above 25 degrees C.

Leakage

No dielectric material is perfect, so no dielectric can cause a capacitor to retain its charge indefinitely. There is always a small amount of leakage. For this reason, the charge on a capacitor will eventually dissipate after a certain amount of time.

Power Factor

In a capacitor, the power factor is a measure of the amount of charge lost over a period of time. The power factor is used to determine the quality factor (Q) of a capacitor. Q is the reciprocal of the power factor. Thus, if a power factor for a given capacitor is stated as 0.02, the value of Q for this capacitor is 50. The higher the Q value, the better the capacitor will be.

Since capacitance losses reflect the quality of the dielectric, the power factor of a capacitor is a measure of the insulating capabilities of the dielectric.

Behavior with AC and DC

Direct current is totally blocked by a capacitor. This is because the insulation of the dielectric material prevents DC from flowing in the circuit. A capacitor acts exactly like an insulator. In effect, a capacitor is a break in a DC circuit.

By contrast, AC appears to flow through a capacitor in a circuit although the flow of current changes with each alternation. This effect is demonstrated in Figure 9-8. Figure 9-8A shows that, during the positive half-cycle, electrons flow from one plate of a capacitor to the other plate. When the direction of current flow reverses, so do the charges of the capacitor plates, as shown in Figure 9-8B.

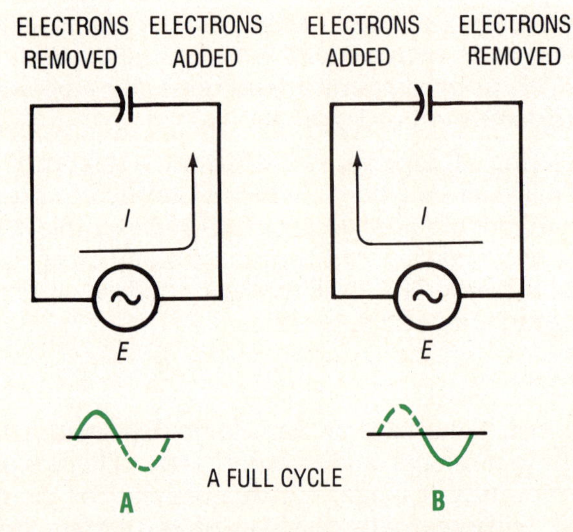

Figure 9-8. Effect of capacitor on AC.

This effect is further illustrated in Figure 9-9, which shows a capacitive circuit to which an AC source has been applied. Each plate of the capacitor receives the electrons from one alternation within the AC cycle. Since AC continues to flow, it appears to pass through the circuit. This apparent current is called displacement current. In reality, each alternation is stopped at the capacitor's dielectric. Electrons build up on first one plate, then the other, discharging on the off

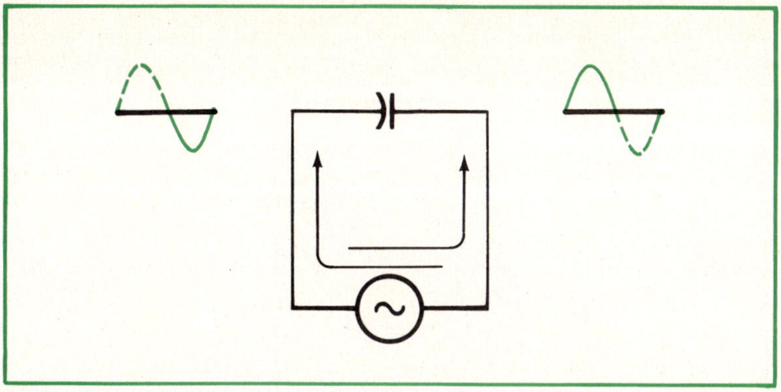

Figure 9-9. Capacitive circuit with AC source.

alternation. Although the displacement current is simply bouncing back and forth between the plates, the effect is the same as if the current were actually flowing through the capacitor.

Discussion Questions/Problems

1. What is the effect of a positive or negative temperature coefficient on the characteristics of a capacitor?
2. What effect would a 10 degree increase in temperature cause on a capacitor with a negative temperature coefficient of 250 ppm and a nominal value of 1 μF?
3. Why does a capacitor lose its charge when the charging source is removed?
4. How is the quality of a capacitor determined?
5. What is the difference in the way a capacitor behaves with DC and with AC?

9.6 CAPACITORS IN SERIES AND IN PARALLEL

When capacitors are connected in series, their value is determined by the formula shown below:

$$C_T = \frac{1}{\frac{1}{C_1} + \frac{1}{C_2} + \frac{1}{C_3} + \cdots + \frac{1}{C_N}}$$

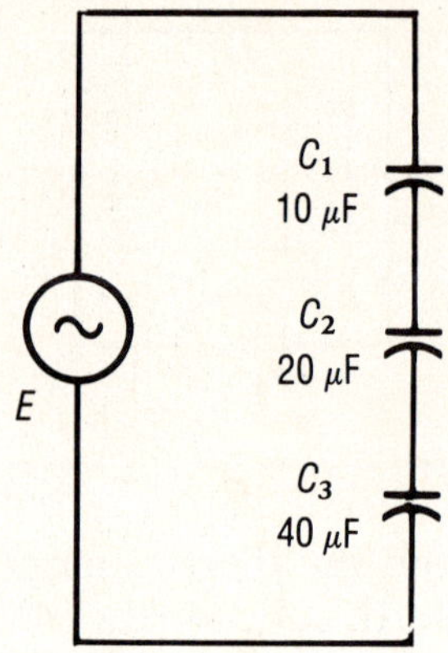

Figure 9-10. Series capacitive circuit with AC source.

To demonstrate the effect of capacitors connected in series, see Figure 9-10. Assuming the values given for the three capacitors in this figure, total capacitance is determined as follows:

$$C_T = \frac{1}{\frac{1}{C_1} + \frac{1}{C_2} + \frac{1}{C_3}}$$

$$= \frac{1}{\frac{1}{10} + \frac{1}{20} + \frac{1}{40}}$$

$$= \frac{1}{\frac{4}{40} + \frac{2}{40} + \frac{1}{40}}$$

$$= \frac{1}{\frac{7}{40}}$$

$$C_T = 5.71\ \mu F$$

The resultant capacitance of several capacitors connected in series is always less than the smallest capacitance. When two capacitors of the same value are connected in series, a shortcut formula may be used:

$$C_T = \frac{C_1 \times C_2}{C_1 + C_2}$$

Notice the similarity between the formula for two capacitors in series and the formula for two resistors in parallel. A series

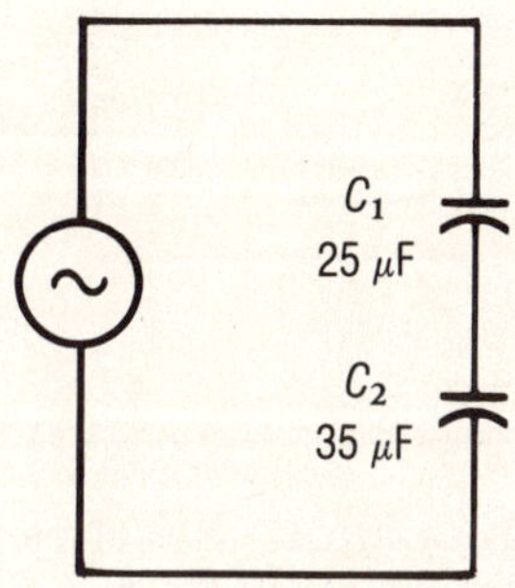

Figure 9-11. Two capacitors connected in series with AC source.

circuit with two capacitors is shown in Figure 9-11. The calculation for total capacitance in this circuit is as follows:

$$C_T = \frac{C_1 \times C_2}{C_1 + C_2}$$

$$= \frac{25 \times 35}{25 + 35}$$

$$= \frac{875}{60}$$

$$C_T = 14.58\ \mu F$$

When non-equal capacitors are connected in series, voltage does not divide equally across the units. A circuit diagram illustrating how voltage divides is shown in Figure 9-12. The voltage divides inversely in proportion to the capacitance values of the capacitors in a series circuit.

As seen in Figure 9-12, all capacitors in a series circuit must be part of the same current path. In any series circuit,

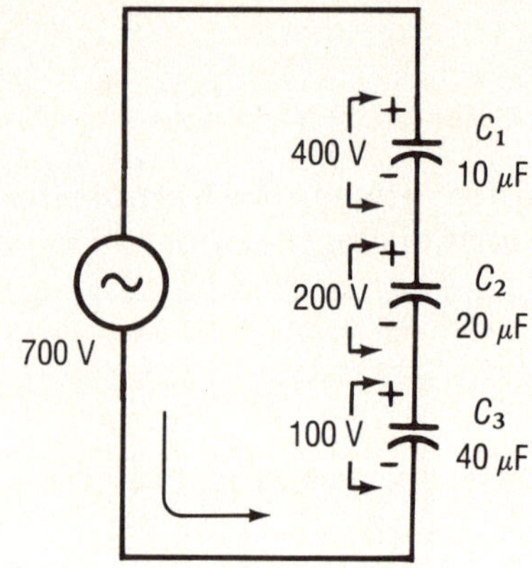

Figure 9-12. Non-equal capacitors connected in series with AC source.

current is the same throughout the series path. It follows that all capacitors contained within that series circuit must also have the same charge. Since $E = Q/C$, the unit with the smallest capacitance has the highest potential across it. Voltage divides in inverse proportion to the capacitance values of capacitors connected in series.

Capacitors in Parallel

When capacitors are connected in parallel, total capacitance for the circuit is equal to the sum of the individual capacitances. Total capacitance can be determined by adding all capacitances:

$$C_T = C_1 + C_2 + C_3 + \ldots + C_N$$

A circuit in which capacitors are connected in parallel is shown in Figure 9-13. The calculation for the total capacitance of this circuit is:

$$C_T = C_1 + C_2 + C_3$$

$$= 10 + 20 + 40$$

$$C_T = 70\ \mu F$$

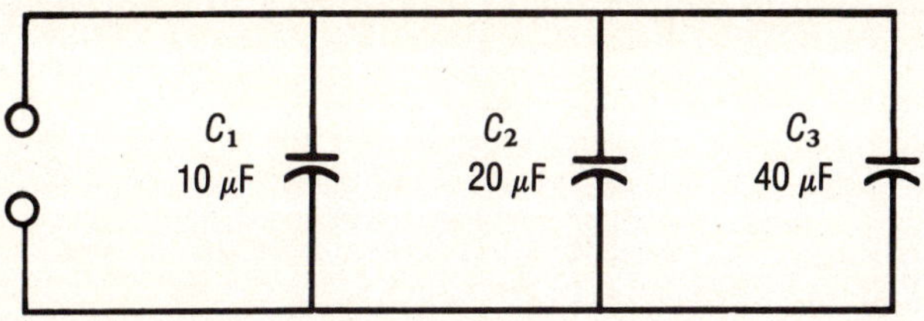

Figure 9-13. Circuit with three capacitors connected in parallel.

Discussion Questions/Problems

1. What circuit conditions must be present to charge a capacitor?
2. What circuit conditions must be present to discharge a capacitor?
3. Three capacitors are connected in series. Each has a value of 60 μF. What is the equivalent capacitance of the circuit?
4. Assume the same capacitors as in Question 3 except that they are connected in parallel. What is the equivalent capacitance?

9.7 CAPACITIVE REACTANCE

A capacitor displays a special property when alternating current is applied: A capacitor opposes the flow of AC through the capacitor. This property of capacitors is called *capacitive reactance*.

The amount of capacitive reactance is inversely proportional to the capacitance of the capacitor. The greater the capacitance, the less the capacitive reactance.

The frequency of the AC itself affects the amount of capacitive reactance: The higher the frequency, the less the capacitive reactance.

The symbol for capacitive reactance is X_C. Capacitive reactance is measured in ohms. The formula for capacitive reactance is:

$$X_C = \frac{1}{2\pi fC}$$

In the example below, capacitive reactance is determined for a circuit with a frequency of 1,000 Hz and a capacitance of 5 μF.

$$X_C = \frac{1}{6.28 \times 1{,}000 \times 5 \times 10^{-6}}$$

$$= \frac{1 \times 10^6}{6.28 \times 1{,}000 \times 5}$$

$$= \frac{1 \times 10^6}{31{,}400}$$

$$X_C = 31.85\ \Omega$$

To find the capacitance of a circuit when capacitive reactance and frequency are known, use the following formula:

$$C = \frac{1}{2\pi f X_C}$$

The example below shows how this formula is used to determine capacitance for a circuit with a frequency of 5,000 Hz and a capacitive reactance of 10 Ω.

$$C = \frac{1}{6.28 \times 5{,}000 \times 10}$$

$$= \frac{1}{314{,}000}$$

$$C = 3.185\ \mu\text{F}$$

To determine frequency when the other characteristics of a capacitive reactance circuit are known, use this formula:

$$f = \frac{1}{2\pi C X_C}$$

To determine frequency in a circuit with a capacitance of 100 μF and a capacitive reactance of 20 Ω, apply the formula as follows:

$$f = \frac{1}{6.28 \times 100 \times 10^{-6} \times 20}$$

$$= \frac{1 \times 10^6}{6.28 \times 100 \times 20}$$

$$= \frac{1 \times 10^6}{12{,}560}$$

$$f = 79.62 \text{ Hz}$$

Capacitive Reactances in Series Circuits

When capacitors are connected in series, as shown in Figure 9-14, capacitive reactance is cumulative. To find capacitive reactance for the entire circuit, add the reactances of all of the capacitors. The formula is:

$$X_{C_T} = X_{C_1} + X_{C_2} + X_{C_3} + \ldots + X_{C_N}$$

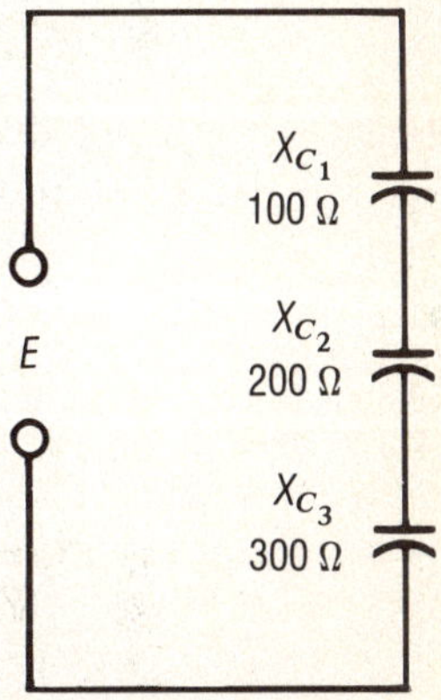

Figure 9-14. Circuit with three capacitors connected in series.

Thus, the total capacitive reactance for the circuit in Figure 9-14 is determined as follows:

$$X_{C_T} = 100 + 200 + 300$$

$$X_{C_T} = 600\ \Omega$$

Note that capacitive reactance for capacitors in series is determined in the opposite manner from the way capacitance is determined. To calculate the reciprocal of total capacitance, you add reciprocals of the individual capacitances. To calculate total capacitive reactance, you add the individual reactances. Capacitive reactances are treated like resistances in a resistance circuit.

Capacitive Reactances in Parallel Circuits

When capacitors are connected in parallel, their equivalent capacitive reactance is determined through the reciprocal formula. A typical circuit in which capacitors are connected in

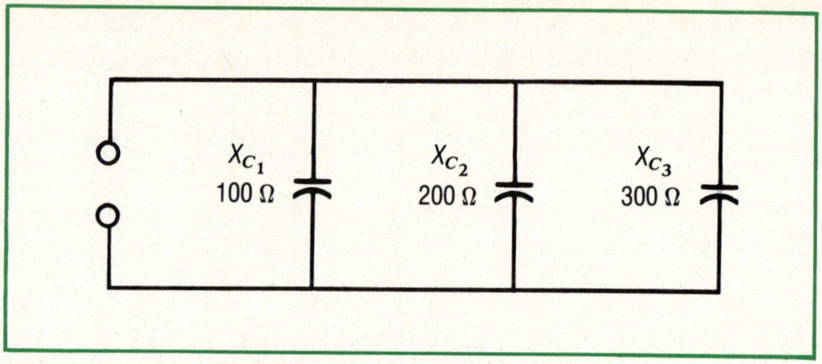

Figure 9-15. Circuit with three capacitors connected in parallel.

parallel is shown in Figure 9-15. The formula for determining total capacitance in this circuit is:

$$X_{C_T} = \frac{1}{\frac{1}{X_{C_1}} + \frac{1}{X_{C_2}} + \frac{1}{X_{C_3}} + \cdots + \frac{1}{X_{C_N}}}$$

The calculations below demonstrate how equivalent capacitive reactance is determined for the circuit with capacitors connected in parallel, as shown in Figure 9-15.

$$X_{C_T} = \frac{1}{\frac{1}{100} + \frac{1}{200} + \frac{1}{300}}$$

$$= \frac{1}{\frac{6}{600} + \frac{3}{600} + \frac{2}{600}}$$

$$= \frac{1}{\frac{11}{60}}$$

$$X_{C_T} = 54.55\ \Omega$$

When a circuit has only two capacitors in parallel, the total capacitive reactance may be calculated with the following "short-cut" formula:

$$X_{C_T} = \frac{X_{C_1} \times X_{C_2}}{X_{C_1} + X_{C_2}}$$

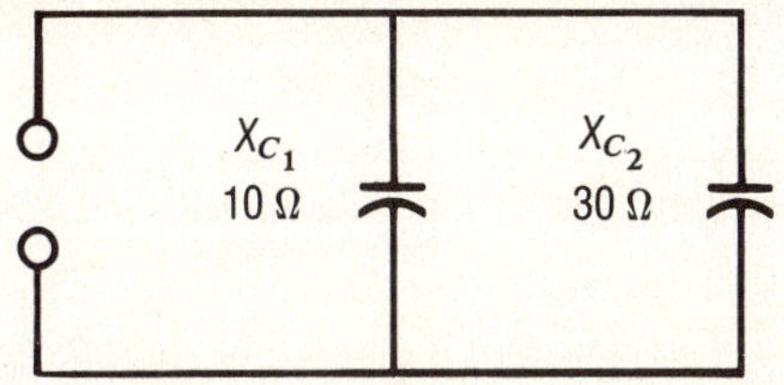

Figure 9-16. Circuit with two capacitors connected in parallel.

The diagram for a circuit with two capacitors connected in parallel is shown in Figure 9-16. Computation of total capacitive reactance for this circuit is as follows:

$$X_{C_T} = \frac{10 \times 30}{10 + 30}$$

$$= \frac{300}{40}$$

$$X_{C_T} = 7.50\ \Omega$$

Notice the similarities and differences between the equations for resistances in series and parallel and capacitors in series and parallel. Also note the similarities and differences between the equations for inductors in series and parallel and capacitors in series and parallel.

Ohm's Law for Capacitive Reactance

The three forms of the formula used for applying Ohm's law to determine capacitive reactance are as follows:

$$E = IX_C \qquad I = \frac{E}{X_C} \qquad X_C = \frac{E}{I}$$

To illustrate how Ohm's law is used in capacitive reactance circuits, the equation below computes current for a circuit with 100 V of voltage and 10 Ω of capacitive reactance.

$$I = \frac{E}{X_C} = \frac{100}{10} = 10\ \text{A}$$

Discussion Questions/Problems

1. What happens to capacitive reactance in a circuit as frequency increases?
2. When capacitance of a circuit is reduced, what happens to capacitive reactance?
3. How is capacitive reactance determined in a circuit in which capacitors are linked in series?
4. How is capacitive reactance of a circuit determined when capacitors are connected in parallel?

9.8 *RC* CIRCUITS

An *RC* circuit is one that has both resistance and capacitance. Circuits of this type are used frequently in electronic applications. *RC* circuits have several unique properties which are discussed below.

Phase Angle

In *RC* circuits, current always leads voltage and, as is the case in *RL* circuits, phase angles are not always 90 degrees. Rather, current leads voltage by some value between 0 and 90 degrees.

Series *RC* Circuits

When a circuit has a capacitor and a resistor in series, as shown in Figure 9-17A, the formula for computing total voltage is as follows:

$$E_T = \sqrt{(E_R)^2 + (E_C)^2}$$

Notice that E_T is the vector sum of E_R and E_C, as shown in Figure 9-17B. The calculations below apply this formula to determine total voltage for the circuit in Figure 9-17A.

$$E_T = \sqrt{(20)^2 + (15)^2}$$

$$= \sqrt{400 + 225}$$

$$= \sqrt{625}$$

$$E_T = 25\ \Omega$$

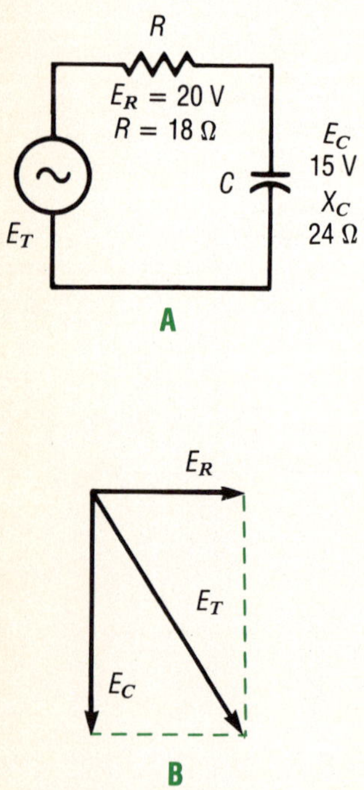

Figure 9-17. Circuit with a capacitor and a resistor in series.

Impedance in Series *RC* Circuits

Impedances for series *RC* circuits are computed with the following formula:

$$Z = \sqrt{R^2 + (X_C)^2}$$

The value of impedance in the circuit in Figure 9-17 is computed as follows:

$$\begin{aligned} Z &= \sqrt{(18)^2 + (24)^2} \\ &= \sqrt{324 + 576} \\ &= \sqrt{900} \\ Z &= 30\ \Omega \end{aligned}$$

Phase Angles in Series *RC* Circuits

In series *RC* circuits, phase angles are determined by the following formula:

$$\theta = \text{arc tan}\left(\frac{X_C}{R}\right)$$

The relationship represented by this equation is illustrated in the diagram in Figure 9-18. The phase angle for the circuit represented by this diagram is calculated as:

$$\theta = \text{arc tan}\ \frac{1{,}000}{1{,}000} = \text{arc tan}\ 1 = 45^\circ$$

For the circuit represented by Figure 9-18, voltage lags current by 45 degrees.

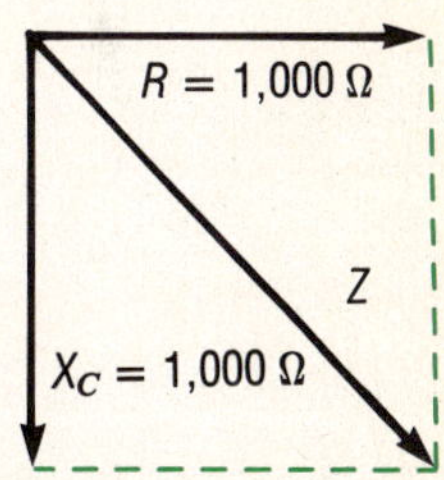

Figure 9-18. Phasor relationship of voltage and current for *RC* circuits.

Impedance in Parallel *RC* Circuits

When resistors and capacitors are connected in parallel within the same circuit, as shown in Figure 9-19, impedance is determined by applying the following formula:

$$Z = \frac{RX_C}{\sqrt{R^2 + (X_C)^2}}$$

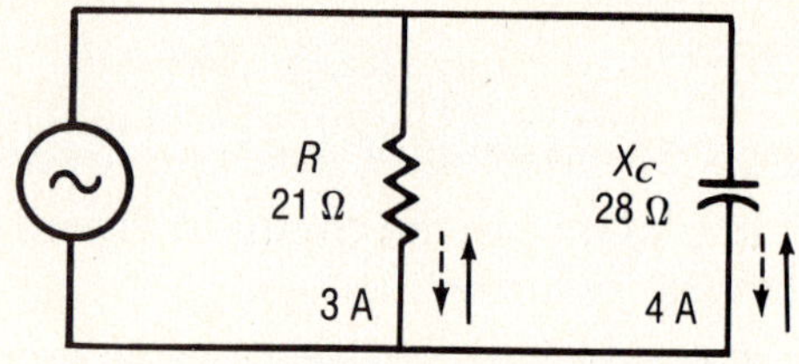

Figure 9-19. Circuit with capacitor and resistor in parallel.

Notice the similarity between this formula and the one in Chapter 8 for calculating impedance of a circuit with inductors and resistors in parallel. Impedance for the circuit in Figure 9-19 is computed as follows:

$$Z = \frac{21 \times 28}{\sqrt{(21)^2 + (28)^2}}$$

$$= \frac{588}{\sqrt{441 + 784}}$$

$$= \frac{588}{\sqrt{1{,}225}}$$

$$= \frac{588}{35}$$

$$Z = 16.8\ \Omega$$

Current in Parallel *RC* Circuits

To determine current in a parallel *RC* circuit, use the following formula.

$$I_T = \sqrt{(I_R)^2 + (I_C)^2}$$

Total current for the circuit illustrated in Figure 9-19 is computed as shown below:

$$I_T = \sqrt{(3)^2 + (4)^2} = \sqrt{9 + 16} = \sqrt{25} = 5\ \text{A}$$

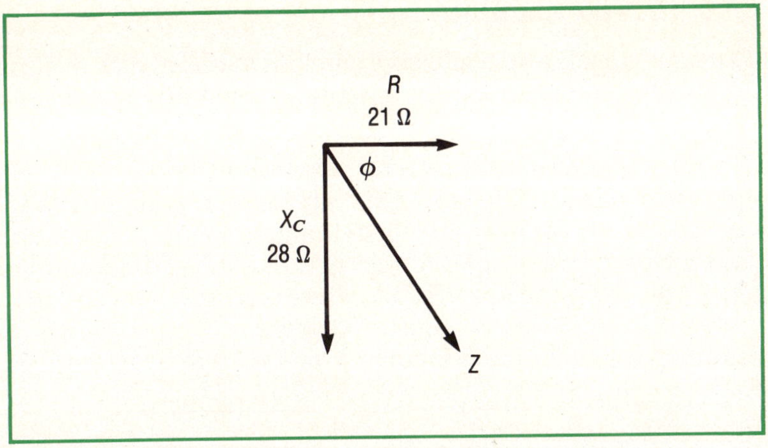

Figure 9-20. Phase angle diagram for circuit of Figure 9-19.

Figure 9-20 diagrams phase angle values for the circuit illustrated in Figure 9-19. To compute phase angles for parallel *RC* circuits, use the following formula:

$$\theta = \text{arc tan}\left(\frac{X_C}{R}\right)$$

Computation of the phase angle for the diagram in Figure 9-20 is shown below:

$$\theta = \text{arc tan}\left(\frac{-28}{21}\right) = \text{arc tan } -1.33 = -53^{\circ}$$

Discussion Questions/Problems

1. What is the phase angle relationship between *R* and *C* in an *RC* circuit?
2. What is the relationship between phase angles for current and voltage in an *RC* circuit?
3. If a series circuit has an E_R of 8 V and an E_C of 6 V, what is the resultant voltage?
4. What is the impedance of a series *RC* circuit if the *R* is 32 Ω and X_C is 24 Ω?
5. In a parallel *RC* circuit, what is the impedance if *R* is 18 Ω and X_C is 24 Ω?

9.9 VECTOR ALGEBRA

Vector algebra is a specialized branch of mathematics dealing with phase angles. Vector algebra is also known as *phasor algebra*.

One application for vector algebra techniques is to express the relationship between voltages and currents that are not in phase. For example, vector algebra would be used to determine phase angles for circuits like those diagrammed in Figure 9-21. In Figure 9-21A, there is a capacitive circuit in which current leads voltage by 90 degrees. Figure 9-21B is an inductive circuit in which current lags voltage by 90 degrees.

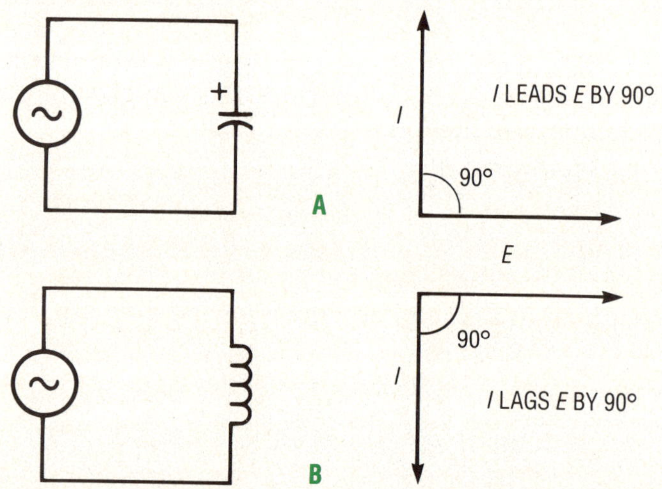

Figure 9-21. Capacitive and inductive circuits.

Figure 9-21 illustrates situations in which vector algebra can be applied to show relationships between two quantities for circuits that can be described in terms of phase angles. Vector algebra can also be applied to problems involving more than two quantities.

In applying vector algebra, two phase quantities are combined to give a phase sum that is different from the algebraic sum. Results are derived through the use of trigonometry. To illustrate, consider the Pythagorean theorem, described and illustrated in Chapter 8. Figure 9-22 illustrates the basic relationships covered in the Pythagorean theorem. The equation for the Pythagorean theorem is repeated here for reference:

$$c^2 = a^2 + b^2 \qquad c = \sqrt{a^2 + b^2}$$

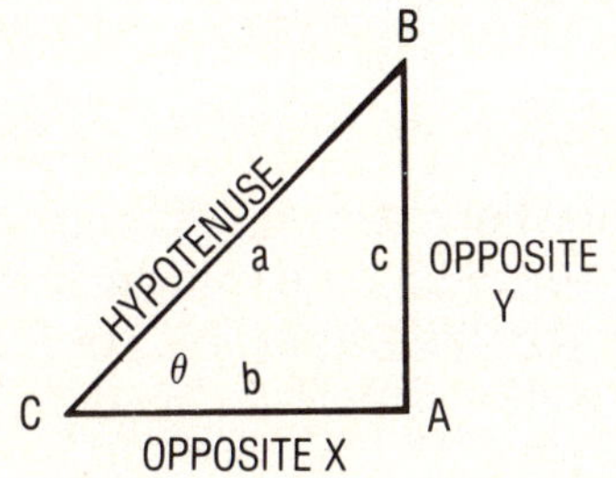

Figure 9-22. Right triangle showing basic relationships covered in Pythagorean theorem.

Discussion Questions/Problems

1. To what types of circuits can vector algebra be applied?
2. How do the Pythagorean theorem and vector algebra techniques relate to each other?

9.10 IMPEDANCE AND POWER

Building upon the Pythagorean theorem makes it possible to explore power concepts within *RC* circuits. These are the principles that apply:

- In a pure resistive circuit, power is dissipated in the form of heat.
- In a pure capacitive circuit, power is not dissipated, but is returned to the power source.
- Within an *RC* circuit, both effects take place. Some power is dissipated as heat. Some power is returned to the source.

The total power dissipated by a resistor can be described by the following equation:

$$P_R = I^2R$$

The total reactive power in a capacitive reactive circuit is determined as follows:

$$P_{X_C} = I^2 X_C$$

The relationship between X_C, R, and Z would be as diagrammed in Figure 9-23. The same relationships can also be shown for other circuit properties. For example, Figure 9-24 diagrams the relationships for I^2R, I^2X_C, and I^2Z.

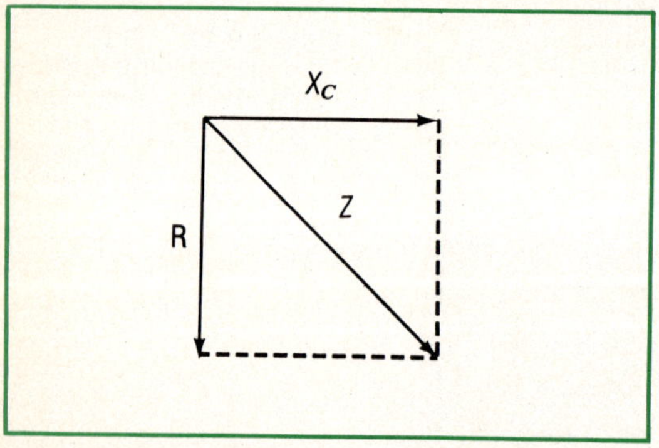

Figure 9-23. Phasor relationship for reactive power in *RC* circuit.

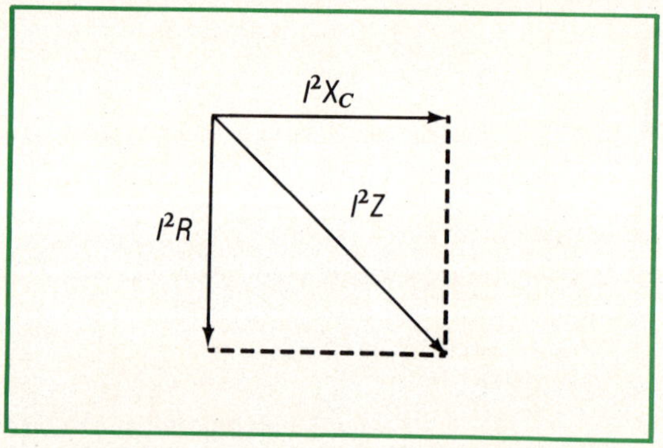

Figure 9-24. Phasor relationships for *RC* circuit properties.

Combining the vectors illustrated in Figure 9-24 develops a triangle that shows the apparent power usage for the circuit. This triangle is illustrated in Figure 9-25. The formula for apparent power in any AC circuit, whether resistive, capacitive, inductive, or any combination of these, is:

$$P_a = I^2 Z$$

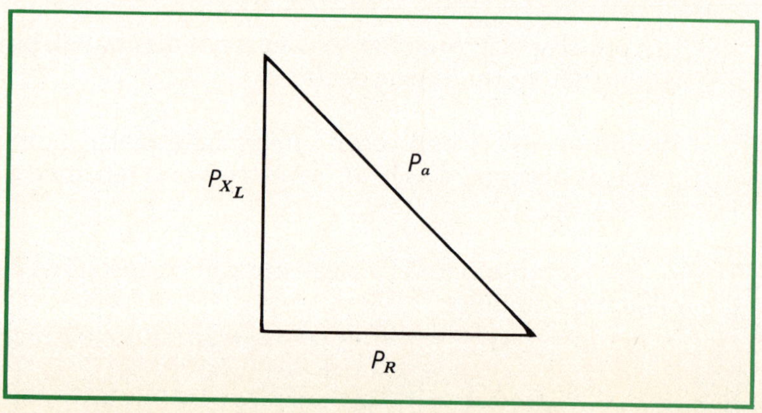

Figure 9-25. Triangle showing apparent power usage in AC circuit.

An example of the application for this equation in a circuit with 2 A of current and 5 Ω of impedance is:

$$P_a = I^2Z = (2)^2 \times 5 = 4 \times 5 = 20 \text{ W}$$

Discussion Questions/Problems

1. What happens to power in a pure resistive circuit?
2. What happens to power in a pure capacitive circuit?
3. What happens to power within an *RC* circuit?
4. If a circuit has 4 A of current and 10 Ω of resistance, what is the apparent power?
5. If a circuit has 4 A of current and 10 Ω of impedance, what is the apparent power?

9.11 TYPES OF CAPACITORS

Capacitors are classified according to their dielectrics. The major types of capacitors are described and illustrated below.

Air Capacitors

Capacitors that use air as their dielectric are usually variable. Typical applications include tuning circuits for radios, television sets, and other communications equipment. Several air capacitors are illustrated in Figure 9-26.

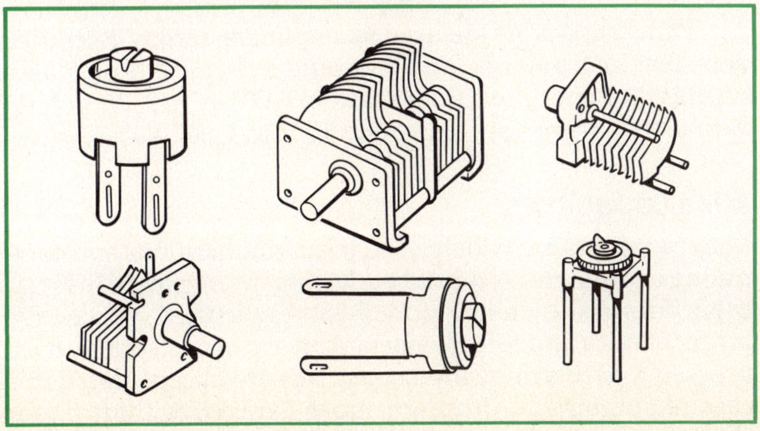

Figure 9-26. Air capacitors.

Paper Capacitors

Paper capacitors are the most widely used in the field of electronics. The construction of a typical paper capacitor, along with two kinds of paper capacitor packages, is shown in Figure 9-27. These capacitors are formed by wrapping successive layers of wax paper and foil into a cylinder. Wire leads are soldered to the ends of the foil wrappings. Leads can either be extended straight out from the capacitor or bent.

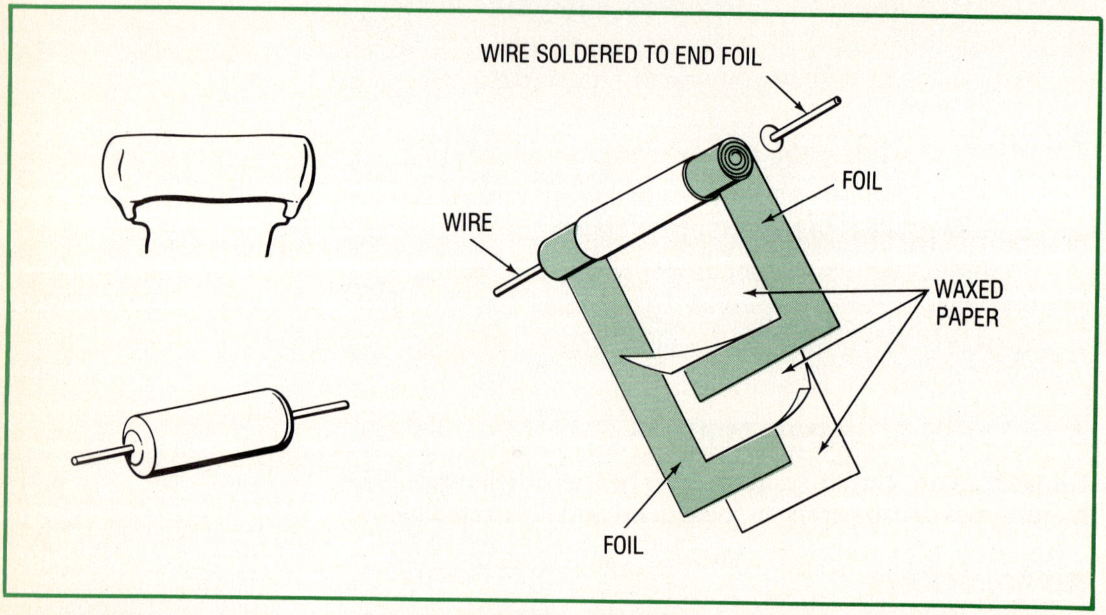

Figure 9-27. Paper capacitors and their construction.

In packaging, a black band is usually drawn at one end of paper capacitors. This black marking indicates the lead that is connected to the outside foil wrapping. The other lead is connected to the inside foil wrapping.

Paper capacitors are used in amplifiers, radios, television sets, test instruments, and computers. Typical capacitance ratings for paper capacitors are from 0.001 μF to 2 μF. Voltage ratings run from approximately 50 V to 1,600 V.

Mica Capacitors

Mica capacitors are widely used in communications and computing equipment. A typical packaging arrangement for mica capacitors is shown in Figure 9-28. As pictured, successive layers of mica and foil are stacked on top of each other, then pressed together. Finished capacitors are encapsulated in a ceramic material. Although more expensive than paper

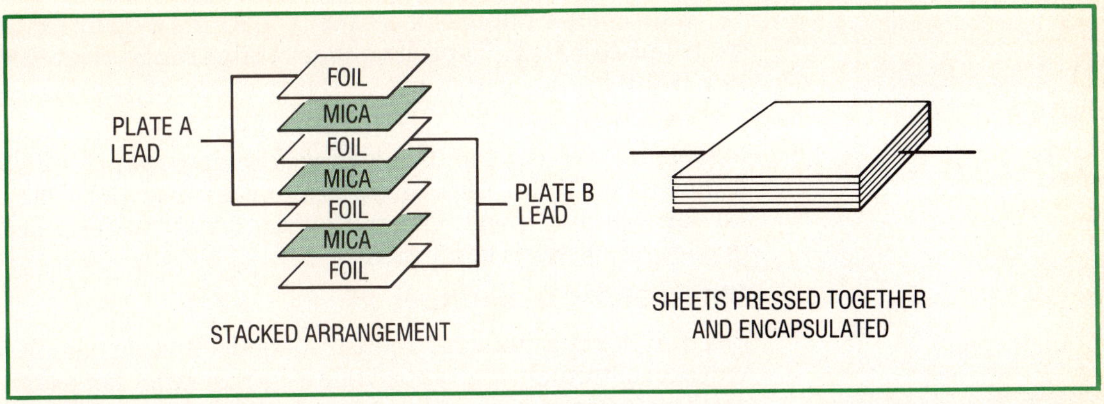

Figure 9-28. Construction and packaging arrangement for mica capacitor.

capacitors, mica units can perform in wider temperature ranges. Mica capacitors also tend to have greater stability and reliability. Capacitance ranges from 3 pF to 47,000 pF. Voltage ratings range from 100 V to 1,000 V.

Ceramic Capacitors

Ceramic capacitors use a form of clay as the dielectric. Clay is pressed into molds, then fired. Ceramic capacitors use a variety of shapes, including the round and rectangular forms shown in Figure 9-29. In addition, some ceramic capacitors have tubular shapes. The dielectric materials are layered with foil or other plate materials and packaged in molded containers, usually made of plastic.

Figure 9-29. Ceramic capacitors and their construction.

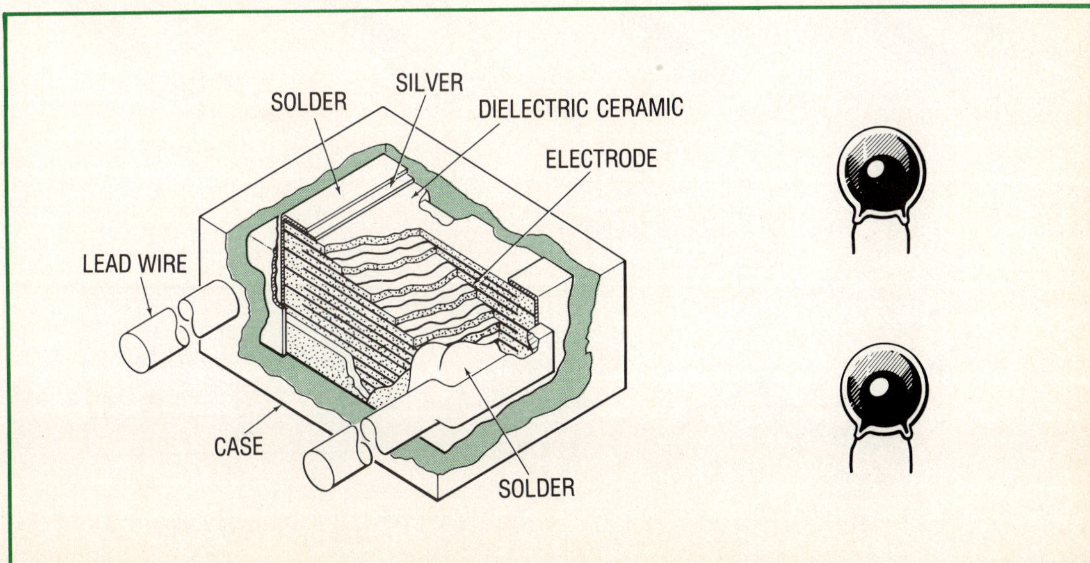

Capacitance ratings for ceramic capacitors usually range from 1 pF to 900 pF. Typical voltage ratings range from 50 V to 2,000 V.

An advantage of ceramic capacitors is that they can tolerate high voltages and wide ranges of temperature. Quality and reliability are not as good as for mica capacitors. Costs, however, are far lower. Ceramic capacitors are widely used in communications equipment.

Electrolytic Capacitors

Electrolytic capacitors use a construction technique that provides an extremely high capacitance-to-size ratio. For example, an electrolytic capacitor might be 10 percent as large as a paper capacitor with the same rating. This is due largely to the use of extremely thin dielectric materials. Construction methods, along with examples of several electrolytic capacitors, are shown in Figure 9-30. Note that an oxide-coated film is used as the dielectric, and that the construction is not symmetric. The electrolyte is in contact with the negative plate, whereas the oxide dielectric is in contact with the positive plate. When used in a circuit, it is important to connect an electrolytic capacitor in the right polarity. The connections of an electrolytic capacitor are marked + and −.

Figure 9-30. Electrolytic capacitors and their construction.

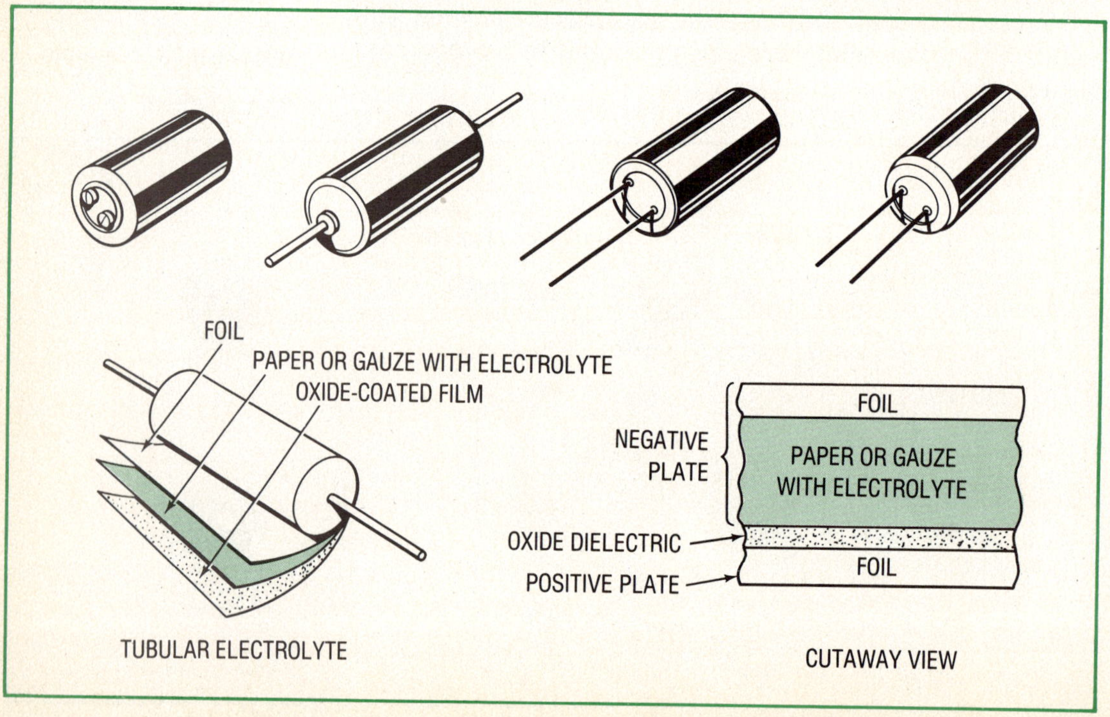

Electrolytic capacitors are offered in a wide range of capacitance values. For example, a standard catalog shows electrolytic capacitors with capacitances of 1 μF and also with 270,000 μF. Voltage ratings typically vary from 3 V to 450 V. In use, electrolytic capacitors are generally installed in low-frequency devices in which they serve also as filters. Examples include amplifiers, power supplies, and computers.

Electrolytic capacitors have a limitation not present in most other types. They have limited shelf lives. The electrolyte tends to deteriorate, much as is the case with batteries. Because of this, electrolytic capacitors are sometimes dated and should not be used beyond the expiration date. Shelf lives of electrolytic capacitors tend to be about 10 years. In actual use, the functional life of an electrolytic capacitor is extended greatly. Very little deterioration occurs as long as they are in active circuits.

Paper-Oil Capacitors

Paper-oil capacitors use oil-soaked (impregnated) paper as their dielectric. The advantage of this type of dielectric is that it has an extremely high voltage rating.

Paper-oil capacitors are packaged in canister-type units like those shown in Figure 9-31. Capacitance values range from 0.5 μF to 125 μF. Voltage ratings vary from 200 V to 5,000 V. These capacitors are often used in high-voltage applications. Examples include motor starters, air conditioners, and business machines.

Figure 9-31. Paper-oil capacitors.

Tantalum Capacitors

Tantalum capacitors are similar to electrolytic capacitors in that they are polarized. Differences are primarily in the type of material used. Electrolytic capacitors use an aluminum foil. The foil in tantalum capacitors is made from the metal tantalum. Tantalum capacitors are shown in Figure 9-32.

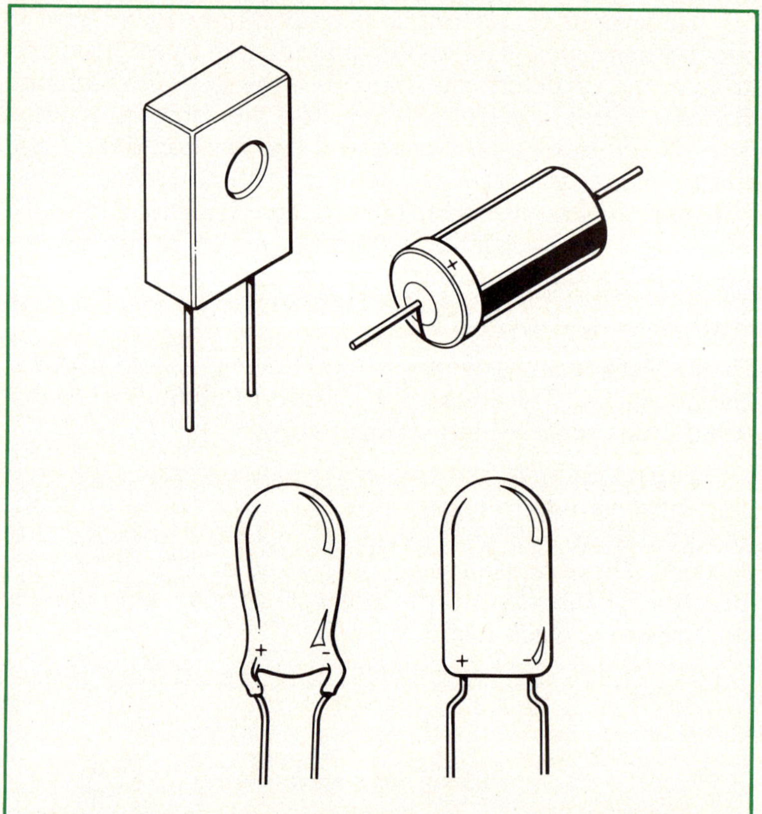

Figure 9-32. Tantalum capacitors.

Two types of tantalum capacitors are produced. One is the solid, or slug, type. These are small units with molded tantalum slugs that function as plates. The other uses etched foil. Construction techniques, except for the difference in foil, are similar to those of electrolytic capacitors.

Capacitance values range from 0.0047 μF to 560 μF. Voltage ratings are from 6 V to 150 V.

As a general rule, tantalum capacitors are used in transistor-type circuits. This is because tantalum capacitors are extremely small and, therefore, can fit into miniature devices. Tantalum capacitors are much smaller than electrolytic capacitors with similar values. In addition, tantalum capacitors have advantages of longer shelf life, lower probability of leakage, and other durability features.

Polystyrene Capacitors

As a dielectric, polystyrene has many advantages. It is easy to handle in fabrication. The basic material is a resin that can be cured during production. The finished dielectric is insensitive to humidity and extremely rugged. Finished capacitors, such as the one shown in Figure 9-33, have high performance in comparison with their size, weight, and price.

Figure 9-33. Polystyrene capacitor.

In manufacture, polystyrene is fused to an aluminum film that is rolled into a capacitor body. Connecting wires are fitted into end caps of the capacitors.

Capacitance values range from 5 pF to 10,000 pF. Voltage ratings are from 200 V to 600 V.

Applications include situations in which light weight and high resistance to humidity are required. Examples include aircraft and marine devices.

Polycarbonate Film Wrapped Capacitors

Construction of polycarbonate film wrapped capacitors is similar to that used for polystyrene capacitors. The differences are primarily in the characteristics of the materials involved. Polycarbonate units are slightly more costly and can be built to larger capacitance values.

Capacitance values range from 0.01 μF to 2 μF. Voltage ratings run from 50 V to 200 V.

Applications are virtually the same as those for paper capacitors. Polycarbonate capacitors are used in radios, television sets, and other communication devices for which designers or users require larger capacitance than is available with paper units. A polycarbonate film wrapped capacitor is shown in Figure 9-34.

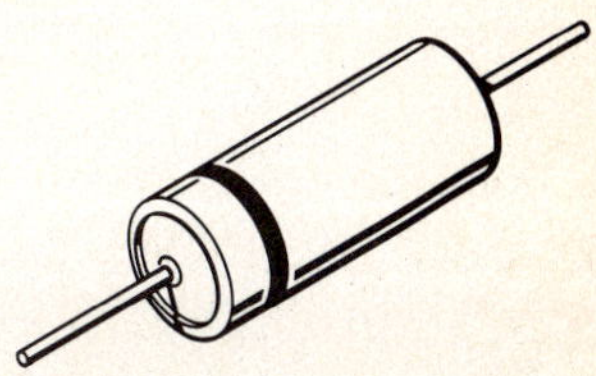

Figure 9-34. Polycarbonate film wrapped capacitor.

Mylar® Capacitors

Mylar® film (a product of the DuPont Company) is used as a dielectric in place of paper or other films for these capacitors. The finished capacitors have high stability and reliability. They can also be wrapped in a variety of shapes, such as the oval profile shown in Figure 9-35, to fit into small packages of components.

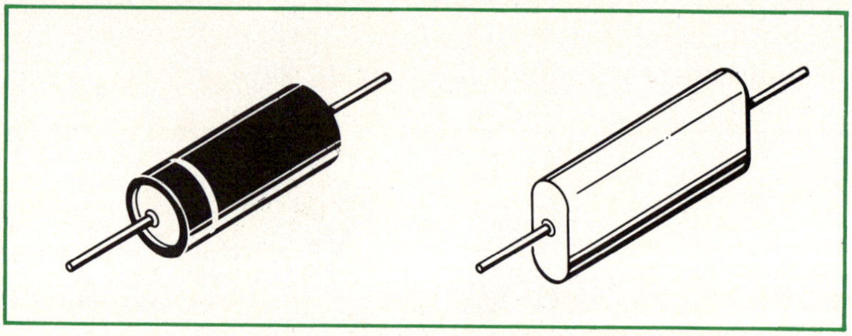

Figure 9-35. Mylar® capacitors.

Capacitance values available range from 0.001 μF to 1 μF. Voltage ratings range from 50 V to 600 V.

Mylar® capacitors are often used in miniaturized parts or systems.

Other special films, including Teflon®, metalized paper, metalized acetate film, and others are also used as dielectrics in a variety of capacitors.

Discussion Questions/Problems

1. What are the main advantages of paper capacitors?
2. What are the main advantages of mica capacitors?
3. Why are electrolytics used in capacitors?
4. How do tantalum and electrolytic capacitors compare in cost and performance?

CHAPTERS 8 AND 9

SUMMARY

1. An inductor is a device that opposes any change in current in an AC circuit. This opposition to changes in current is known as *counter electromotive force* or *counter emf*. (8.1)

2. The basic unit of measurement for inductance is the henry. The abbreviation is H. (8.2)

3. A vector can be used to represent a sine wave. A rotating vector is sometimes called a *phasor*. (8.3)

4. The basic law of electromagnetic inductance was discovered by H. F. E. Lenz in 1833. Lenz's law of electromagnetic induction states that a current which is caused by a change in a magnetic flux will be directed so as to oppose the act which caused the current. (8.4)

5. Inductors are never perfect. There is always some resistance in a coil that results from the wire itself. (8.5)

6. Inductors can be connected in series and in parallel. (8.6)

7. An inductor displays a special property when alternating current is applied. An inductor opposes the flow of AC through the inductor. This property of inductors is called *inductive reactance*. (8.7)

8. An *RL* circuit is one that includes both resistance and inductance. (8.9)

9. The total opposition to flow in an *RL* circuit is called *impedance* and includes a component due to resistance and a component due to inductive reactance. The symbol for impedance is Z. (8.10)
10. A transformer is a device that uses the principle of mutual inductance to transfer electrical energy from a source to a destination. (8.11)
11. A capacitor is a device that opposes any change in voltage in an AC circuit. Also, a capacitor can store an electrical charge. (9.1)
12. The unit of measurement for capacitance is the farad, abbreviated F. (9.2)
13. Phase angles for capacitors follow rules that are complementary to those for inductors. (9.3)
14. Several characteristics of a capacitor affect its performance. These include temperature coefficient, leakage, power factor, and behavior with AC and DC. (9.4 and 9.5)
15. Capacitors can be connected in series and in parallel. (9.6)
16. A capacitor displays a special property when alternating current is applied: A capacitor opposes the flow of AC through the capacitor. This property is called *capacitive reactance*. (9.7)
17. An *RC* circuit is one that has both resistance and capacitance. (9.8)
18. Vector algebra is a specialized branch of mathematics dealing with phase angles. Vector algebra is also known as *phasor algebra*. (9.9)
19. In a pure capacitive circuit, power is not dissipated but is returned to the power source. In an *RC* circuit some power is dissipated as heat in the resistance and some is returned to the source. (9.10)
20. Capacitors are classified according to their dielectrics. Paper capacitors are the most widely used. (9.11)

PROBLEMS

1. What is the value, in henrys, of a coil having a voltage across it of 24 V and for which there is a change in current of 2 A for an interval of 1 second? (8.2)
2. What is the voltage across a coil of 6 H when the current through it varies by 1 A in an interval of 2 seconds? (8.4)

3. What is the coefficient of coupling ($\varkappa$) between two coils having 60 μWb produced by coil 1 and 30 μWb of flux linking? (8.4)
4. What is the mutual inductance between two coils, one having a value of 5 μH and the other a value of 3 μH, where $\varkappa$ is 0.3? (8.4)
5. What is the Q of a coil given an X_L of 110 Ω and an R of 5 Ω? (8.5)
6. What is the total inductance of two coils connected in series having values of 5 μH and 4 μH and a mutual inductance of 1 μH in a series-aiding circuit? (8.6)
7. What is the total inductance of two coils connected in series having values of 5 μH and 4 μH and a mutual inductance of 1 μH when connected to be series-opposing? (8.6)
8. What is the inductive reactance of a coil having a frequency of 5 kHz and an L of 3 mH? (8.7)
9. What is the inductive reactance of a coil having a frequency of 20 kHz and an L of 6 μH? (8.7)
10. If a circuit has an X_L of 5 Ω and a current of 0.5 A, what is the voltage? (8.8)
11. Using Figure R4-1, find the total voltage, given that the voltage across the coil is 32 V and the voltage across the resistor is 24 V. (8.9)

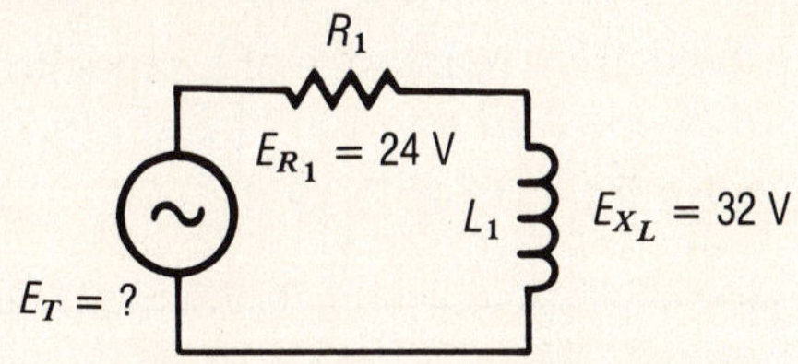

Figure R4-1. Series *RC* circuit for Problem 11.

12. Using Figure R4-2, find Z. (8.10)

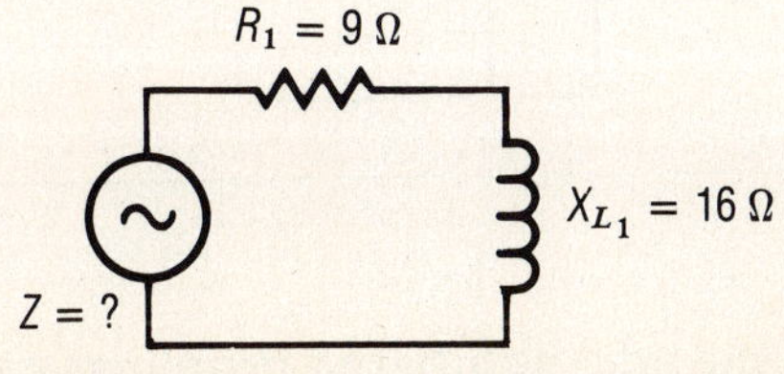

Figure R4-2. Series *RC* circuit for Problem 12.

13. What is the secondary voltage of a transformer given that the E_p equals 100 V and the N_s/N_p is 5? (8.11)
14. What is the secondary current of a transformer given that I_p equals 2 A and N_p/N_s is 6? (8.11)
15. What is the voltage across a capacitor that has a charge of 12×10^{-6} and a capacitance of 20 μF? (9.1)
16. The power factor for a given capacitor is 0.03. What is the value of Q? (9.5)
17. Two 10 μF capacitors are connected in series. What is the total capacitance? (9.6)
18. Three capacitors of 5 μF, 10 μF, and 15 μF are connected in series. What is the total capacitance? (9.6)
19. Figure R4-3 shows three capacitors connected in parallel. What is the total capacitance? (9.6)

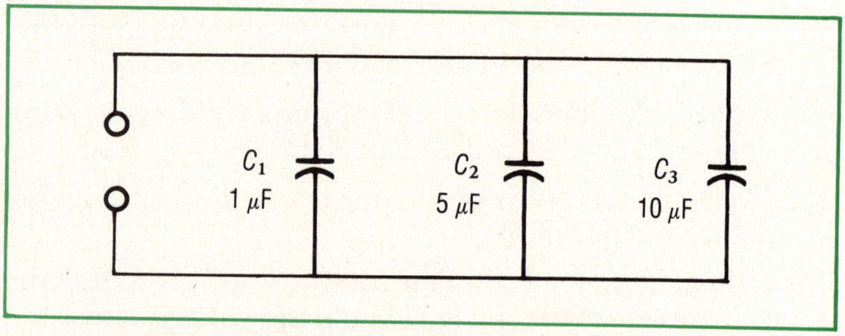

Figure R4-3. Circuit with three capacitors connected in parallel for Problem 19.

20. Find the capacitive reactance of a capacitor having a frequency of 1,000 Hz and a value of 10 μF. (9.7)
21. Find the capacitive reactance of a capacitor having a frequency of 60 Hz and a value of 2 μF. (9.7)
22. Using Figure R4-4, find Z. (9.8)

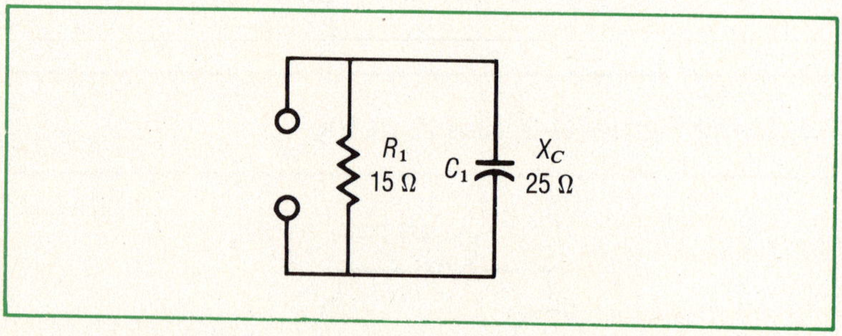

Figure R4-4. Parallel *RC* circuit for Problem 22.

(Answers to the odd-numbered problems can be found at the back of the book.)

10 ALTERNATING CURRENT CIRCUITS

10.1 INDUCTANCE, CAPACITANCE, AND RESISTANCE

This chapter deals with advanced concepts and methods using AC circuits. Specifically, you will be dealing with the properties of inductance, capacitance, and resistance as they apply to AC circuits. In preparation, the summary that follows reviews the characteristics of these electrical properties.

Characteristics of Inductance

Inductance is an electrical property inherent to coils and is important to their use in circuits. An inductance:

- Opposes any change in current
- Can store electricity in the form of an electromagnetic field around the inductor
- Limits the flow of AC in a circuit
- Allows DC to pass readily
- Causes the current through the inductor to lag voltage across the inductor by a phase angle difference of 90 degrees.

Characteristics of Capacitance

Capacitance is an electrical property inherent to metal plates separated by an insulator. A capacitor:

- Opposes any change in voltage
- Can store energy in the form of charge on the capacitor plates
- Limits the flow of AC in a circuit

- Blocks the flow of DC
- Causes the current through the capacitor to lead the voltage across the capacitor by a phase angle difference of 90 degrees.

Characteristics of Resistance

Resistance is an electrical property of all current-carrying materials. All current-carrying materials resist the flow of electrons to a certain degree. Designers choose the amount of resistance in a circuit by selecting resistors, devices that incorporate current-limiting materials. A resistor:

- Limits the flow of current
- Produces no electromagnetic or electrostatic fields
- Passes both AC and DC readily, within the limits of the resistance
- Causes no phase angle difference in response to alternating current.

Discussion Questions/Problems

1. In which type of circuit does current lead voltage?
2. In which type of circuit does current lag voltage?
3. Which type of circuit has no phase angle difference?
4. Which type of circuit creates a magnetic field?
5. Which type of circuit creates an electrostatic field?

10.2 SERIES CIRCUITS WITH INDUCTANCE, CAPACITANCE, AND RESISTANCE

Chapters 8 and 9 establish that, in an inductive circuit, current lags voltage by 90 degrees and, in a capacitive circuit, current leads voltage by 90 degrees. Thus, an inductive reactance causes current to lag voltage by 90 degrees and a capacitive reactance causes current to lead voltage by 90 degrees. It is, therefore, apparent that these two properties have a *complementary* relationship.

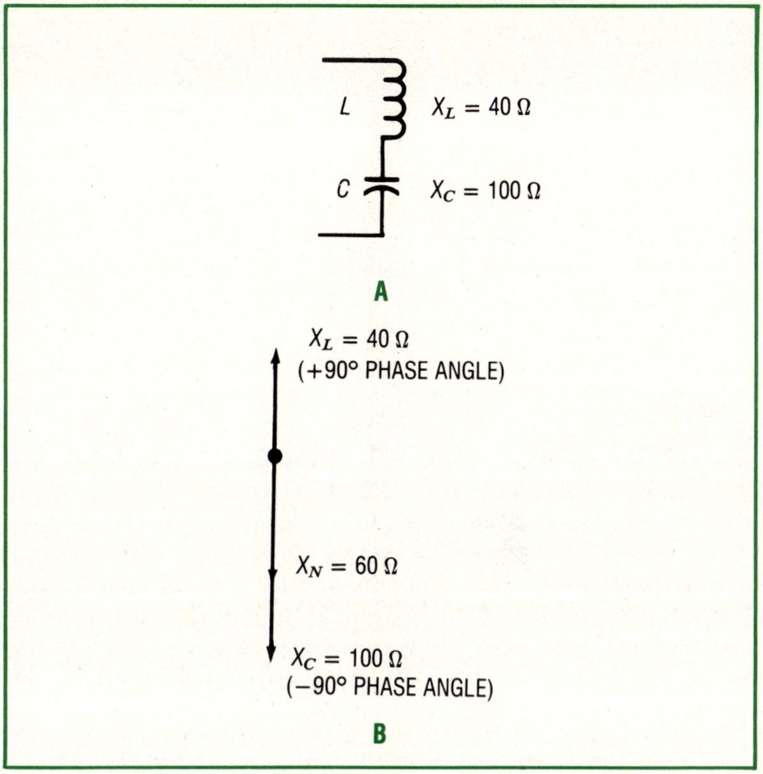

Figure 10-1. Series circuit schematic and phasors.

Figure 10-1A shows a circuit with an inductance that has an impedance of X_L and a capacitance that has an impedance of X_C. The *net reactive component* is the difference of the two. In this case the net reactive component X_N is capacitive because the capacitive reactance is greater than the inductive reactance. This is shown in Figure 10-1B.

Remember: To find inductive reactance, the following formula is used:

$$X_L = 2\pi f L$$

To find capacitive reactance, this is the formula:

$$X_C = \frac{1}{2\pi f C}$$

These two equations are useful when the frequencies at which the inductance and capacitance operate are known.

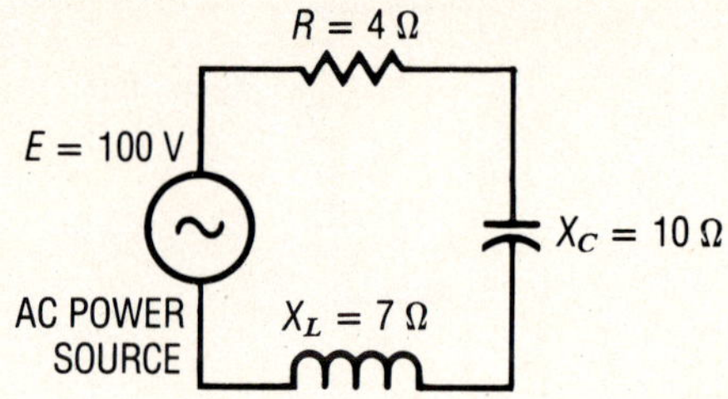

Figure 10-2. Series *RC* circuit with AC source.

Figure 10-2 shows a circuit with inductance, capacitance, and resistance. Since a circuit with only capacitance and inductance has a phase angle of either +90 degrees or −90 degrees, a circuit which has resistance will also have a phase angle somewhere between +90 degrees and −90 degrees. To find impedance, a formula is needed which depends on the *net* reactance in the circuit. If inductive reactance is greater than capacitive reactance, the following formula is used:

$$\text{if } X_L > X_C \text{ then } Z = \sqrt{R^2 + (X_L - X_C)^2}$$

If capacitive reactance is greater than inductive reactance, this formula is used:

$$\text{if } X_C > X_L \text{ then } Z = \sqrt{R^2 + (X_C - X_L)^2}$$

The example below determines impedance for the circuit diagrammed in Figure 10-2.

$$\begin{aligned} Z &= \sqrt{R^2 + (X_C - X_L)^2} \\ &= \sqrt{4^2 + (10 - 7)^2} \\ &= \sqrt{16 + 9} \\ &= \sqrt{25} \\ Z &= 5\ \Omega \end{aligned}$$

If voltage for a circuit is known, Ohm's law can be used to find current. The formula used is as follows:

$$I = \frac{E}{Z}$$

Given a voltage of 100 V for the circuit in Figure 10-2, current is determined as follows:

$$I = \frac{E}{Z} = \frac{100}{5} = 20 \text{ A}$$

Ohm's law can also be used to determine the voltage drop across each component of the circuit in Figure 10-2. The voltage drops across the resistor, the coil, and the capacitor are determined by:

$$E_R = IR = 20 \times 4 = 80 \text{ V}$$

$$E_{X_C} = IX_C = 20 \times 10 = 200 \text{ V}$$

$$EX_L = IX_L = 20 \times 70 = 140 \text{ V}$$

The Power Factor in Series AC Circuits

The power factor for an AC circuit is the ratio of real power to apparent power. To find the power factor for a series AC circuit, divide its resistance (representing real power) by its impedance (representing apparent power). The formula is as follows:

$$PF = \frac{R}{Z}$$

The power factor for the circuit in Figure 10-2 is determined as follows:

$$PF = \frac{4}{5} = 0.80$$

Given the power factor, the phase angle for a series AC circuit is determined through use of the following formula, applied to find the phase angle for the circuit in Figure 10-2:

$$\theta = \text{arc tan } 0.80 = 38.66^\circ$$

A diagram of the phase relationships helps in understanding the power factor. Remember that the net reactive component of the voltage across the inductor and capacitor is the difference of the two voltages. In this example, it is:

$$E_{X_C} - E_{X_L} = 200 \text{ V} - 140 \text{ V} = 60 \text{ V}$$

Figure 10-3 shows the phase relationships of the voltages in this circuit.

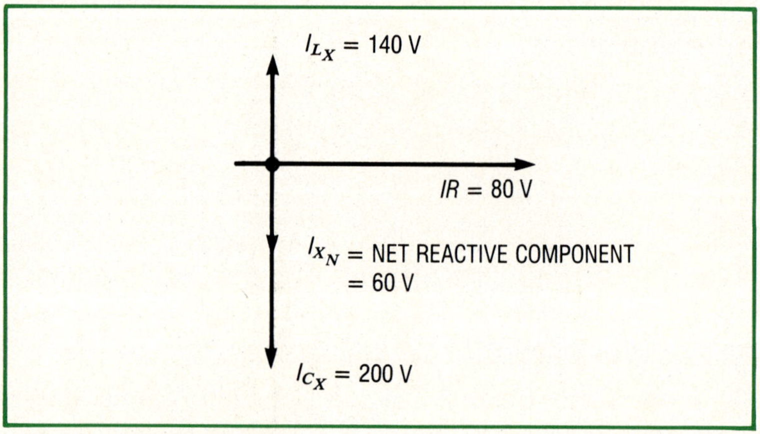

Figure 10-3. Phase relationships for circuit of Figure 10-2.

The resultant phase angle can be found from the resistive and net reactive components of voltage, as shown in Figure 10-4.

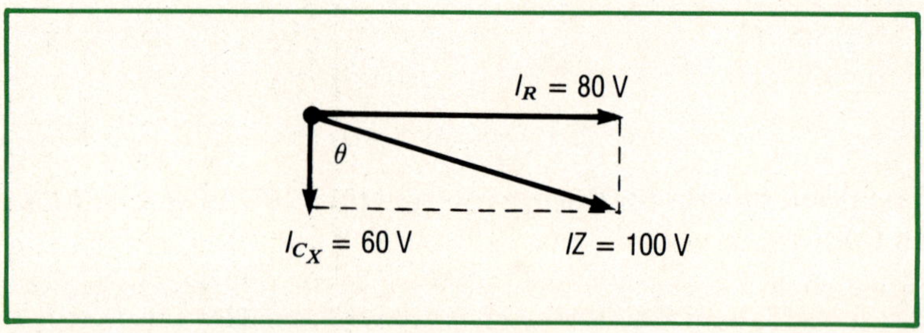

Figure 10-4. Phase angle diagram for circuit of Figure 10-2.

The phase angle is the angle whose tangent is the voltage across the resistor (representing real power) divided by the voltage across the total circuit (representing apparent power). Thus, the phase angle is:

$$\theta = \text{arc tan } \frac{I_R}{I_Z} = \frac{80}{100} = \frac{4}{5} = 38.66^{\circ}$$

Discussion Questions/Problems

1. How is impedance determined for a series AC circuit?
2. In a series AC circuit with a resistance of 3 Ω, a capacitive reactance of 7 Ω, and an inductive reactance of 3 Ω, what is the impedance of the circuit?
3. What would be the power factor for the same circuit?
4. What would be the phase angle for the same circuit?

10.3 PARALLEL CIRCUITS WITH INDUCTANCE, CAPACITANCE, AND RESISTANCE

To determine the impedance of an AC parallel circuit, use Ohm's law just as it was used for the series AC circuit. The formula is:

$$Z = \frac{E}{I_T}$$

Notice that this formula requires the value for total current as one of the factors. Thus, Ohm's law is not useful if the total current is not known.

The total impedance can be determined from the impedances of the individual components. Since the components are in parallel, you cannot simply add and subtract their impedances, as you could for series components. Instead, you must work with the reciprocals of the impedances. You used a similar approach in an earlier chapter when you worked with conductance. Conductance is the reciprocal of resistance. The symbol for conductance is G. Conductance is measured in siemens (S).

In determining properties of parallel AC circuits, you will use two other reciprocals, susceptance and admittance. *Susceptance* is the reciprocal of reactance. The symbol for susceptance is B. Susceptance is measured in siemens. The formulas for determining susceptance when either capacitive reactance or inductive reactance is known are as follows:

$$B_C = \frac{1}{X_C} \text{ or } B_L = \frac{1}{X_L}$$

Admittance is the reciprocal of impedance. The symbol is Y. Admittance is measured in siemens. The formula for determining admittance is:

$$Y = \frac{1}{Z}$$

Figure 10-5 shows how resistance (R), susceptance (B), and admittance (Y) relate to each other in a parallel circuit.

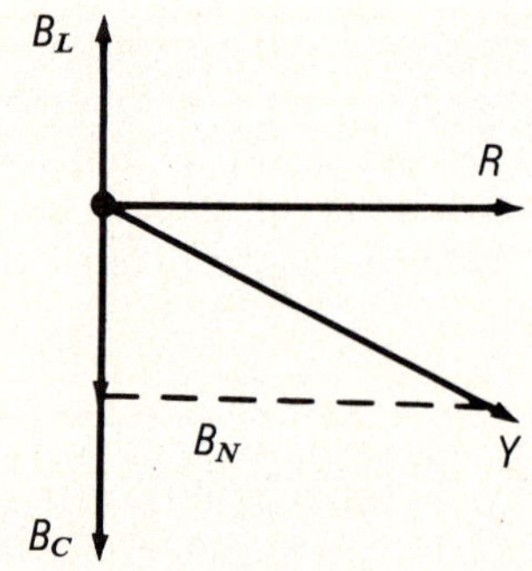

Figure 10-5. Relationships of resistance, susceptance, and admittance in parallel circuit.

Parallel Circuit—Capacitors

An AC circuit with three capacitors in parallel is shown in Figure 10-6. The formula for determining susceptance in a circuit of this type is:

$$B_T = \frac{1}{X_{C_1}} + \frac{1}{X_{C_2}} + \frac{1}{X_{C_3}} + \cdots + \frac{1}{X_{C_N}}$$

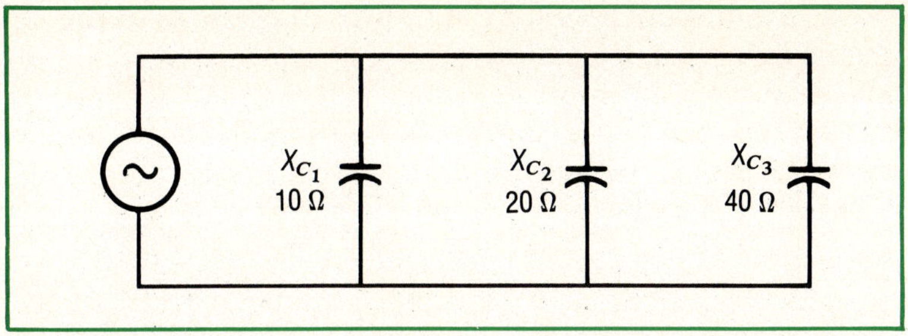

Figure 10-6. AC circuit with three capacitors in parallel.

The computations for finding the susceptance of the circuit in Figure 10-6 are:

$$B_T = \frac{1}{10} + \frac{1}{20} + \frac{1}{40} = \frac{4}{40} + \frac{2}{40} + \frac{1}{40} = \frac{7}{40} \text{ S or } 0.175 \text{ S}$$

Parallel Circuit—Capacitors and Inductors

Figure 10-7, incorporating both capacitance and inductance, contains two capacitors and two coils, all in parallel. To determine susceptance of this circuit, determine capacitive susceptance and inductive susceptance separately, then add the two

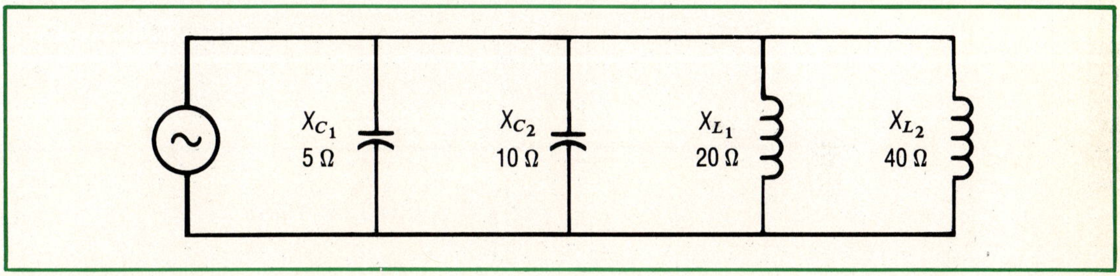

Figure 10-7. Parallel AC circuit with two capacitors and two coils.

together. The computations for determining capacitive susceptance are:

$$B_{T_C} = \frac{1}{X_{C_1}} + \frac{1}{X_{C_2}}$$

$$= \frac{1}{5} + \frac{1}{10}$$

$$= \frac{2}{10} + \frac{1}{10}$$

$$B_{T_C} = \frac{3}{10} \text{ S or } 0.30 \text{ S}$$

Next, the susceptance for the inductive portions of the circuit is determined as follows:

$$B_{T_L} = \frac{1}{X_{L_1}} + \frac{1}{X_{L_2}}$$

$$= \frac{1}{20} + \frac{1}{40}$$

$$= \frac{2}{40} + \frac{1}{40}$$

$$B_{T_L} = \frac{3}{40} \text{ S or } 0.075 \text{ S}$$

The net susceptance for the circuit is the difference between the larger and the smaller of total capacitive and inductive susceptances. The computations for determining net susceptance for the circuit in Figure 10-7 are as follows:

$$B_N = B_{T_L} - B_{T_C}$$

$$B_T = \frac{3}{10} - \frac{3}{40} = \frac{12}{40} - \frac{3}{40} = \frac{9}{40} \text{ S or } 0.225 \text{ S}$$

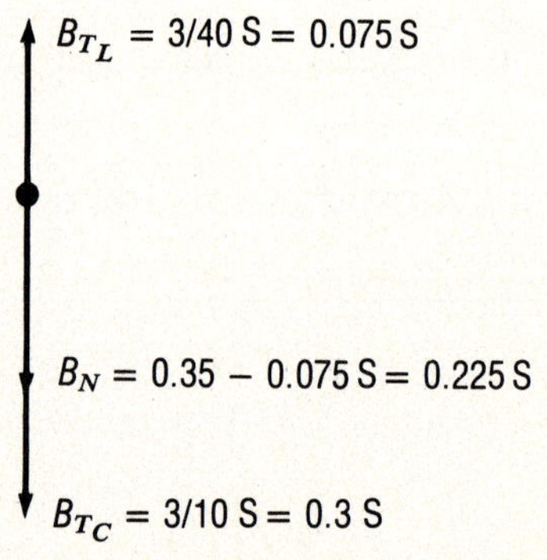

Figure 10-8. Phasor diagram to use to determine net susceptance of circuit of Figure 10-7.

Figure 10-8 shows how the capacitive and inductive susceptances are added graphically to get net susceptance. Notice that the susceptances are added mathematically as well as graphically. Also, you must remember to take the sign of each component into account.

Because capacitive susceptance in Figure 10-7 is greater than inductive susceptance, the circuit has a net capacitive susceptance. To determine the net reactance of this circuit, take the reciprocal of the susceptance. This is done as follows:

$$X_N = \frac{40}{9} \text{ or } 4.45\ \Omega$$

This is the net reactance of the circuit of Figure 10-7. Now consider whether it is an inductive or capacitive reactance.

Notice that $B_C = 1/X_C$ and $B_L = 1/X_L$. If $B_C > B_L$, as is the case, then $X_C > X_L$.

The inductive reactance is greater than the capacitive reactance. The circuit is inductive. Notice the general rule which results from the reciprocal relationship of susceptance and reactance: A capacitive susceptance corresponds to an inductive reactance and vice-versa.

Parallel Circuit—Capacitors, Inductors, and Resistors

A circuit that includes a capacitor, a coil, and a resistor all in parallel is shown in Figure 10-9. To determine impedance in

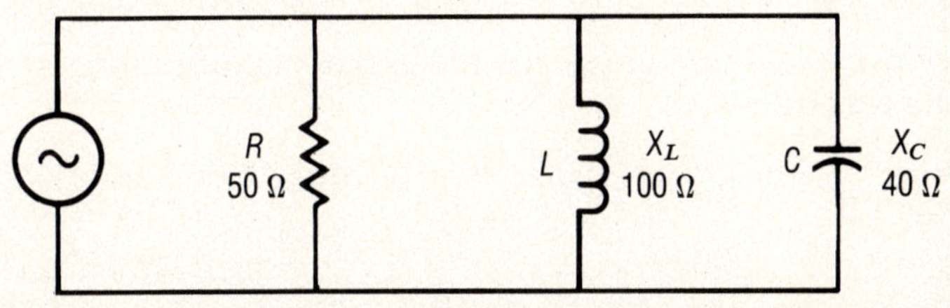

Figure 10-9. Circuit with capacitor, coil, and resistor all in parallel.

this circuit, you begin by calculating the susceptance for each of the reactive components and the conductance for the resistive components. This is demonstrated through a series of equations, with the givens being as follows:

$$R = 50\ \Omega \qquad G = \frac{1}{50}\ \text{S} = 0.02\ \text{S}$$

$$X_L = 100\ \Omega \qquad B_L = \frac{1}{100}\ \text{S} = 0.01\ \text{S}$$

$$X_C = 40\ \Omega \qquad B_C = \frac{1}{40}\ \text{S} = 0.025\ \text{S}$$

The capacitive and inductive susceptances are added mathematically to find the net susceptance.

$$B_N = B_L - B_C = 0.01 \text{ S} - 0.025 \text{ S} = 0.015 \text{ S}$$

Then the admittance is calculated from the net susceptance and the conductance. The reactive component (susceptance) and the resistive component (conductance) are used to calculate the *complex* resultant (admittance).

$$Y = \sqrt{G^2 + B^2}$$

The calculations used to determine admittance for the circuit of Figure 10-9 are:

$$\begin{aligned} Y &= \sqrt{0.02^2 + 0.015^2} \\ &= \sqrt{0.0004 + 0.000225} \\ &= \sqrt{0.000625} \\ Y &= 0.025 \text{ S} \end{aligned}$$

For the total impedance, the reciprocal of admittance is used. The formula is:

$$Z = \frac{1}{Y}$$

The calculation for determining total impedance for the circuit in Figure 10-9 is as follows:

$$Z = \frac{1}{0.025 \text{ S}} = 40 \ \Omega$$

The process of finding the impedance for the circuit in Figure 10-9 can be illustrated by the following diagrams. Figure 10-10A shows how the susceptances B_L and B_C combine into a net susceptance B_N. Figure 10-10B shows how the net susceptance B_N is combined with the conductance G to result in admittance Y. The final step is to take the reciprocal of Y, because $Z = 1/Y$.

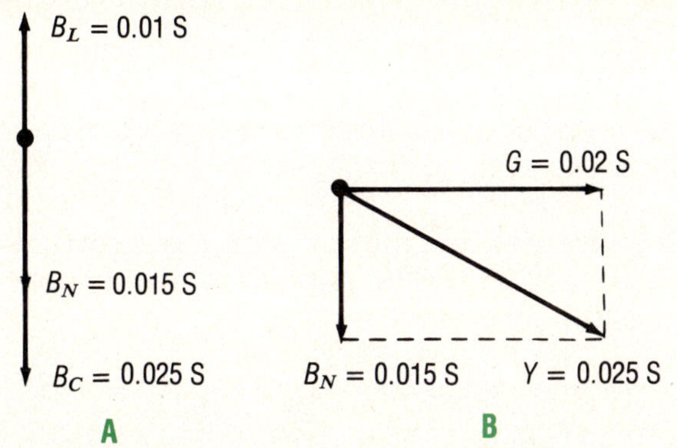

Figure 10-10. Phasor and phase angle diagrams for circuit of Figure 10-9.

Discussion Questions/Problems

1. What is susceptance?
2. Describe admittance and explain its role in determining impedance for a parallel AC circuit.

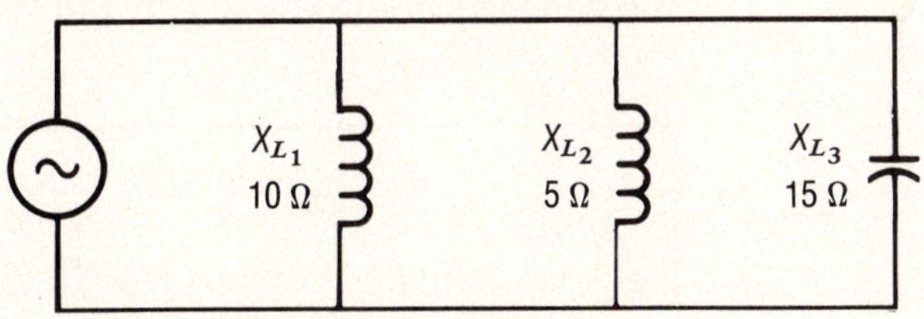

Figure 10-11. Circuit schematic for Problems 3, 4, and 5.

3. Determine the susceptance for the circuit in Figure 10-11.
4. Determine the admittance for the circuit in Figure 10-11.
5. Determine the total impedance for the circuit in Figure 10-11.

10.4 POWER AND THE POWER FACTOR

The *power* loss of any given circuit is a measure of the rate at which energy is dissipated through resistance in the circuit. The formula for determining power is as follows:

$$P = I^2R$$

To illustrate, assume that a circuit has a current of 2 A and a resistance of 25 Ω. The power loss for this circuit is computed as follows:

$$P = 2^2 \times 25 = 100 \text{ W}$$

Another measure of power in an AC circuit is *reactive power*. Reactive power is a hypothetical quantity. It is the power that would be delivered to the reactive components of the circuit. Reactive components do not absorb power, however. On the positive half-cycle a capacitor would be charged, drawing power from the source. On the negative half-cycle the capacitor would be discharged, returning power to the source. The result is that there is no net power loss. The same is true for an inductor, which would draw power on the positive half-cycle and return power on the negative half-cycle, again for no net power usage.

There is also apparent power, which is the vector sum of the real power and the reactive power. This is illustrated by Figure 10-12.

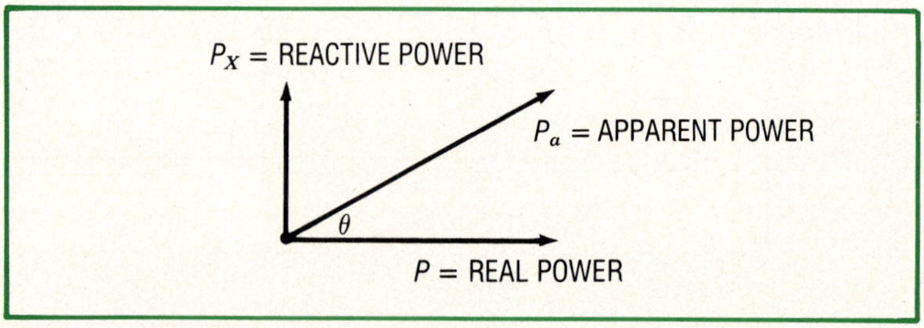

Figure 10-12. Vector diagram for apparent power.

Note that apparent power applies to AC circuits only. In DC circuits, apparent power and real power are the same. Apparent power is derived simply by multiplying voltage by current for the circuit. The formula is:

$$P_a = EI$$

Both the power and apparent power figures are needed to determine the *power factor* in an AC circuit. The power factor is a measure of the real power actually delivered to the circuit. It is a mathematical factor used in circuit design. The power factor is an expression of the quality of a circuit. The power factor is determined by dividing real power by apparent power. Mathematically, the formula for determining the power factor is as follows:

$$PF = \cos\theta = \frac{P}{P_a}$$

Another way to think of power factor can be derived from looking at Figure 10-12 in a different way. Notice that the real power P is the component of the apparent power which is along the horizontal axis.

$$P = P_a \cos\theta$$

Since Cos θ is the factor by which apparent power must be multiplied to get real power, Cos θ is known as the power factor. Also, since Cos θ can vary between zero and 1, the power factor can vary between zero and 1.

Different equivalents for the power factor formula are used in special types of circuits. For example, in a series-reactive circuit, the power factor is determined as follows:

$$PF = \frac{R}{Z}$$

To illustrate, the power factor for a circuit with a resistance of 50 Ω and an impedance of 100 Ω would be computed as follows:

$$PF = \frac{50}{100} = \frac{1}{2} = 0.5$$

In an AC circuit with parallel reactances, the formula for determining the power factor is to divide conductance by admittance. This mathematical formula is shown below:

$$PF = \frac{G}{Y}$$

As an example, in a circuit with a conductance of 0.03 S and an admittance of 0.06 S, the power factor is determined as follows:

$$PF = \frac{0.03}{0.06} = \frac{1}{2} = 0.5$$

A general rule for AC series reactive circuits is that the power factor is the ratio of the resistive component of voltage to the source voltage. A method for determining the power factor of an AC circuit with parallel reactances is to divide the current through the resistances by total current through the circuits. The formula is as follows:

$$PF = \frac{I_R}{I_T}$$

Discussion Questions/Problems

1. What is true, or real, power and how is it measured in AC circuits?
2. What is apparent power and how is it measured in AC circuits?
3. What is the power factor of an AC circuit?
4. If a parallel AC circuit has a susceptance of 0.04 S and an admittance of 0.08 S, what is its power factor?

10.5 COMPLEX NUMBERS AND THE *J*-OPERATOR

In computations involving AC circuits an additional variable must often be considered. This variable is the phase angle, a factor that can affect the performance of a circuit significantly. When the phase angle must be considered in circuit computations, circuit analysis becomes more complex. The greater the number of reactive parts in a circuit, the greater the complexity of the computations can be.

Mathematical techniques can be used to simplify computation of these circuits. One technique is to use a mathematical tool called a *j-operator*. The *j*-operator is the imaginary component of complex numbers in mathematical formulas. In mathematics, the letter i is generally used as an abbreviation to indicate an imaginary number. However, in electronic calculations, the letter i could be confusing because it usually represents current. Therefore, in working with imaginary numbers associated with electronic circuits, the letter j is substituted.

To understand the concept of imaginary numbers, begin with a basis for determining real numbers. In mathematics, any number multiplied by itself (its square) will give a positive number. This applies whether the real number is positive or negative, as shown below:

$$(+2) \times (+2) = 2^2 \text{ or } 4$$
$$(-2) \times (-2) = 2^2 \text{ or } 4$$

Imaginary numbers are a useful concept when you are faced with an equation that would normally defy solution. Consider the equation below:

$$x^2 + 3 = 0$$
$$x^2 = -3$$
$$x = \sqrt{-3}$$

In the above equation, there is no normal solution because there is no real number with a square root equal to -3. Therefore, the square root of a negative number is not a real number. In working with such numbers mathematically, they are identified as imaginary numbers.

To solve equations in which real numbers are presented under a radical sign, you can always use the technique of *factoring*. Examples are given below:

$$\sqrt{64} = \sqrt{16} \times \sqrt{4} = 4 \times 2 = 8$$
$$\sqrt{8} = \sqrt{4} \times \sqrt{2} = 2\sqrt{2}$$

The same technique can be used with imaginary numbers:

$$\sqrt{-16} = \sqrt{16} \times \sqrt{-1} = 4\sqrt{-1}$$
$$\sqrt{-8} = \sqrt{4} \times \sqrt{2} \times \sqrt{-1} = 2\sqrt{2}\,\sqrt{-1}$$

In the two examples of factoring imaginary numbers, the quantity represented by the square root of -1 can be replaced by the letter j. The following values can then be applied:

$$\text{if } j = \sqrt{-1} \text{ then } j^2 = -1$$
$$\text{and } j^3 = j^2 \times j = -1 \times j = -1$$
$$\text{and } j^4 = j^2 \times j^2 = +1$$

Since you can replace the square root of -1 with j, it is also true that:

$$\sqrt{-16} = 4j$$
$$\sqrt{-\ 5} = j\sqrt{5}$$
$$\sqrt{-\ 8} = 2j\sqrt{2}$$

A major use of the j-operator in dealing with AC circuits is in computations involving phase angles. The factor j can be shown to correspond to a phase angle of 90 degrees. As shown in Figure 10-13, exponents of j can be used to indicate the values of 180°, 270°, and 0° or 360°.

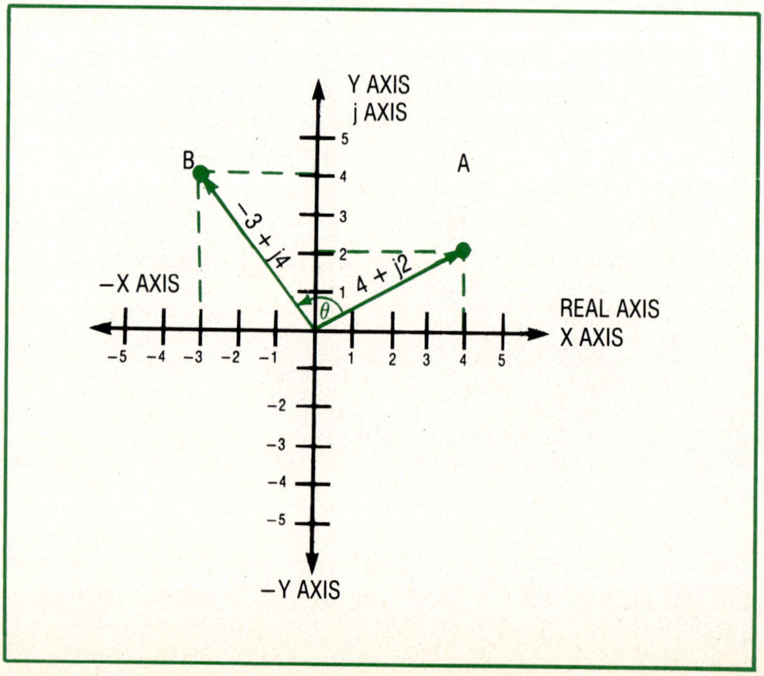

Figure 10-13. Complex plane diagram for the *j*-operator.

Complex Numbers

To determine some of the properties of AC circuits, it is necessary to combine values for real and imaginary numbers. The result is complex numbers using *j*-operators and real numbers in combinations like those shown below:

$$4 + j2$$
$$3 - j3$$
$$5a + jb$$

Both the real and imaginary parts of complex numbers can be either positive or negative. Consider the example $4 + j2$. This means that a point is at four units on the x, or real, axis of the phase angle coordinates and at two units (out of phase) on the y axis, or, as it is sometimes called, the j axis.

Mathematically, the terms of complex numbers are added as phasors, or vectors.

Discussion Questions/Problems

1. What is a complex number in terms of the properties of AC circuits?
2. What is a *j*-operator and why is the letter *j* used?
3. What are the negative or positive properties of complex numbers?
4. How would you write the square root of -9 as a *j*-operator?

10.6 THE COORDINATE SYSTEM

Mathematical values of the properties of AC circuits can be shown graphically as well as through complex numbers. For this purpose, a graphic representation known as a *complex plane* is used. A complex plane diagram is shown in Figure 10-13.

Note that, in Figure 10-13, there are two axis lines that form the basis for the complex plane. These are the x and the y axes. Each of the main axes has a negative and a positive direction, with a zero point (called the *origin*) where the lines intersect. Each of the axis lines can be divided into segments with numeric values. In Figure 10-13, the values 1 through 5 are used for both negative and positive directions.

The x axis can be thought of as the real axis. It is the equivalent of the real portion of a complex number. Real numbers are always presented by their positions on the x axis.

Similarly, the y axis represents the imaginary numbers, and is the equivalent of the imaginary portion of a complex number. For this reason, the y axis can also be thought of as the j axis. Figure 10-13 shows two values for complex numbers. For example, consider the vector, or phasor, drawn to point A. This vector represents the complex number $4 + j2$. The position of the vector is derived from the coordinates, as shown by dotted lines on the diagram in Figure 10-13. The vector to point B is similarly derived. Following this method, any complex number can be shown as a point on the coordinate system of a complex plane.

Representations on a complex plane give both the magnitude and direction of a vector. Consider, for example, the vector line from 0 to point B. The magnitude is the distance from 0 to B. The vector magnitude is determined through use of the Pythagorean theorem, as shown below:

$$\overrightarrow{(OB)^2} = -3^2 + 4^2$$

$$= 9 + 16$$

$$\overrightarrow{(OB)^2} = 25$$

$$\overrightarrow{OB} = 5$$

The vector direction is determined by trigonometry. The vector is at an angle where:

$$\text{arc tan } \theta = \frac{4 \text{ (distance on } j \text{ axis)}}{-3 \text{ (distance on } x \text{ axis)}}$$

$$= \frac{-4}{3}$$

$$= -1.33\ \theta$$

$$\text{arc tan } \theta = 143^{\circ}$$

The complex plane can be used to represent the value of impedances, complex voltages, and complex currents.

Discussion Questions/Problems

1. On which coordinate of a complex plane is the *j*-operator represented?
2. On which coordinate of a complex plane are real numbers represented?
3. How is the magnitude of a vector shown on a complex plane?
4. How is the direction of a vector determined after the vector has been represented on a complex plane?

10.7 ADDITION, SUBTRACTION, MULTIPLICATION, AND DIVISION USING THE *J*-OPERATOR

To add complex numbers with *j*-operators, add the real parts and the imaginary parts separately, as shown below:

$$(5 + j3) + (4 - j2) = (5 + 4) + (j3 - j2) = 9 + j$$

To subtract complex numbers, subtract the real and imaginary parts separately to determine the difference. An example is shown below:

$$(4 + j2) - (3 + j4) = (4 - 3) + (j2 - j4) = 1 - j2$$

To multiply complex numbers, a cross-multiplication step is necessary. Each of the values of each complex number is multiplied by each of the values of the other complex number. This process is illustrated below:

$$\begin{aligned}(7 + j2) \times (2 - j3) &= 7(2 - j3) + j2(2 - j3) \\ &= 14 - j21 + j4 - j^2 6 \\ &= 14 - j17 + 6 \\ &= 20 - j17\end{aligned}$$

To divide complex numbers, convert the denominator to a real number. Do this by using a multiplication factor that makes it possible to factor out the *j*-operator of the denominator. Remember, if both numerator and denominator are multiplied by the same factor, the factor is unaffected. An example of division of complex numbers is shown below:

$$\frac{8 + j2}{2 + j3} = \frac{8 + j2}{2 + j3} \times \left(\frac{2 - j3}{2 - j3}\right) = \frac{16 - j24 + j2 - j^2 6}{4 - j^2 9}$$

$$= \frac{16 - j22 + 6}{4 + 9}$$

$$= \frac{22 - j22}{13}$$

$$= 1.69 - j1.69$$

Discussion Questions/Problems

Solve the complex number equations given below:

1. $(6 + j2) + (5 + j)$
2. $(5 - j3) - (1 + j2)$
3. $(2 + j6) \times (3 - j)$
4. $\dfrac{(4 + j2)}{(2 + j)}$

10.8 RECTANGULAR NOTATION AND POLAR NOTATION

The complex plane is assumed to be rectangular. That is, the axes of the complex plane would, if all of the value lines were extended, become a graph, as shown in Figure 10-14. When the values of complex numbers are presented in this manner, the format is known as *rectangular notation*. This format can also be described as a *rectangular coordinate system*.

Another method for presenting the values of complex numbers as encountered in analysis of AC circuits is through

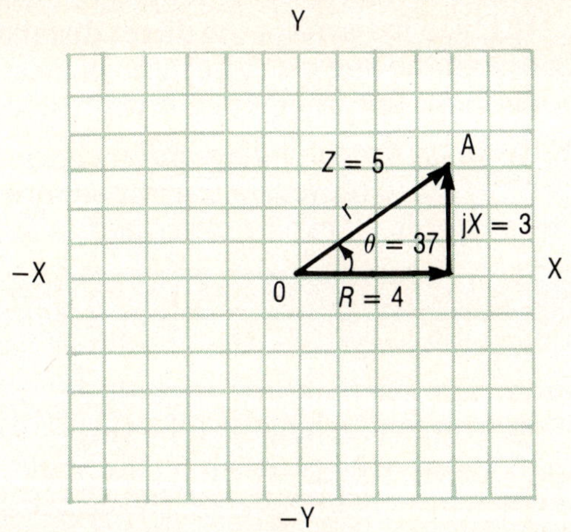

$$Z = \sqrt{R^2 + N^2}$$
$$\theta = \arctan (X/R)$$

Figure 10-14. Rectangular notation for magnitude and angle of a complex number.

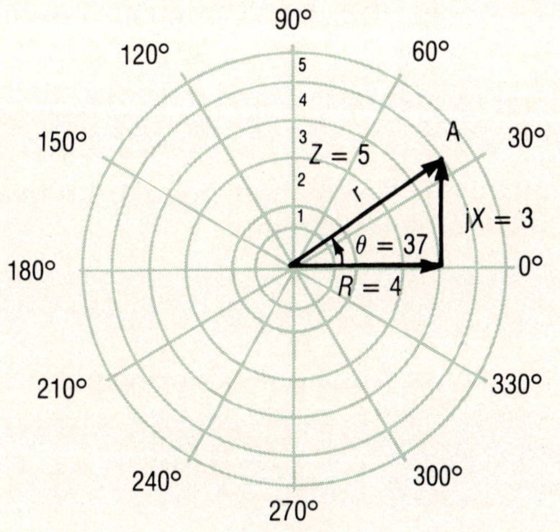

$$R = Z \cos \theta$$
$$X = Z \sin \theta$$

Figure 10-15. Polar notation for values of complex numbers.

use of a format called *polar notation*. The polar notation method is illustrated in Figure 10-15.

Rectangular and polar notation differ in how a point is described on the graph. Rectangular notation is based on measuring up or down and right or left from a central origin. Dimensions are linear along the axis. Polar notation, is based on angular measurement and the length of a vector. Measurements are given in terms of the direct distance from the central origin, and the angle of direction as measured from the right horizontal. The angle here is represented as θ.

Look at the polar notation diagram in Figure 10-15. The vector 0A can be said to represent the impedance present in a circuit. The vector itself can be described as follows:

$$r \angle \theta$$

In these symbolic presentations, r would be the absolute value of the number, or the length of the vector. θ represents the angle ($\angle$). Note that emphasis is on the angle. When the vector is used to represent impedance, the vector is described as $Z \angle \theta$.

The notation does not mean that Z is multiplied by θ. Instead, it is a short-cut way to say ''a vector of Z ohms at an angle of θ degrees.''

The vector 0A represents the vector sum of the values of the real axis and the j axis. In rectangular coordinates, this value would be:

$$x + jy$$

The basis for this representation stems from two sources. From the Pythagorean theorem:

$$r^2 = x^2 + y^2$$

And from some basic trigonometric principles:

$$\tan \theta = \frac{y}{x} \quad x = r \cos \theta \quad y = r \sin \theta$$

Thus, the values in Figure 10-15 can be presented and derived mathematically as follows:

$$4 + j3$$

$$r = \sqrt{16 + 9} = 5$$

$$\tan \theta = \frac{3}{4} \quad \theta = 37^\circ$$

$$4 + j3 = 5 \angle 37^\circ$$

This example shows that it is easy to switch to polar notation from rectangular notation. They are simply two different ways to say the same thing, although there are cases where one or the other will be more useful.

With polar notation it is easy to identify phase angles. Because of this, polar notation is often a more convenient form for working with complex numbers in AC electrical problems. Multiplication is simple with polar notation. To multiply complex numbers in polar notation, all you have to do is add the angles and multiply the vector magnitudes:

$$6 \angle 32^\circ \times 3 \angle 24^\circ = 18 \angle 56^\circ$$

To divide complex numbers given in polar notation, simply subtract the angles and divide the vector magnitude:

$$\frac{12 \angle 50^\circ}{6 \angle 25^\circ} = 2 \angle 25^\circ$$

Though multiplication and division are more convenient when numbers are in polar notation, polar notation is not convenient for addition or subtraction. To add or subtract functions given in polar notation, the values must first be converted to rectangular notation format.

As an illustration, consider the circuit in Figure 10-16. This shows a circuit with both inductive reactance and resistance. The impedance of this circuit can be represented in polar notation as shown in Figure 10-17. The x axis presents

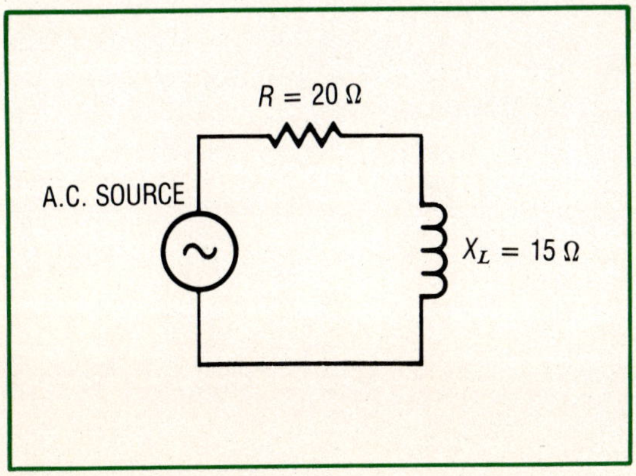

Figure 10-16. Series AC circuit with inductive reactance and resistance.

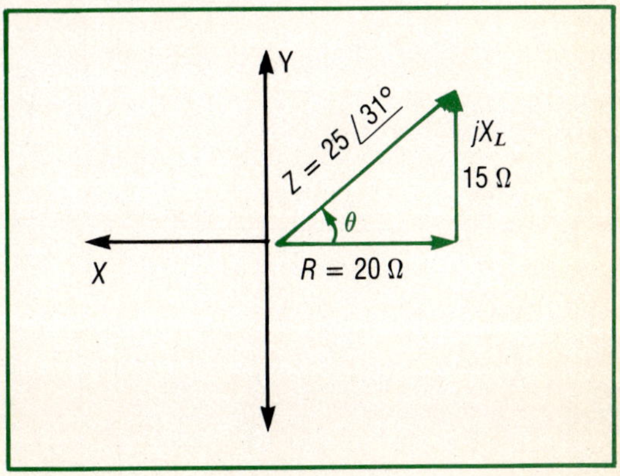

Figure 10-17. Polar notation for impedance of circuit of Figure 10-16.

the real value for R. The impedance can be shown in rectangular notation by converting the polar values to their equivalent rectangular values:

$$R = Z \cos \theta$$

Similarly, the value of the y, or j, axis can be related to the polar values as:

$$X_L = Z \sin \theta$$

Applying these principles, the value of impedance for the circuit represented in Figure 10-16 can be derived as follows:

$$Z \angle 0^\circ = 25 \angle 37^\circ$$

sin for 37° is 0.6

cos for 37° is 0.8

$$R = Z \cos \theta = 25 \times 0.8 = 20$$

$$X_L = Z \sin \theta = 25 \times 0.6 = 15$$

$$\therefore 25 \angle 37^\circ = 20 + j15$$

Discussion Questions/Problems

1. Under what circumstances is it best to use rectangular notation?
2. Under what circumstances is it best to use polar notation?
3. What is the meaning of the following term as applied to polar notation?

$$Z \angle \theta$$

TIME CONSTANTS, RESONANCE, AND FILTERS

11.1 THE TIME CONSTANT

A capacitor will store an electric charge. It can store this charge for an extended period of time if the dielectric is of high quality. If the terminals of a charged capacitor are connected to each other, the charge will neutralize rapidly. If a charge lasts a very long time with an open circuit and dissipates in a very short time with a short circuit, there is an intermediate time for discharge if an intermediate resistance (neither zero nor extremely large) is placed across the capacitor terminals. The time required for the charge to neutralize is an exact function of the capacitance and resistance. This time is called the *time constant* and is an important feature of many circuits where timing determines the function and performance of the circuits. In resistor-capacitor circuits it is called an *RC time constant*.

Another form of time constant circuit is the *RL time constant,* in which an inductor and resistor are used. When a current is flowing in an inductor there is a magnetic field built up around the inductor. If the current is interrupted, the magnetic field collapses very quickly. There must be an intermediate condition between maintaining the magnetic field and allowing it to collapse rapidly. This intermediate condition occurs when the magnetic field is allowed to collapse at a controlled rate. Again, a resistor can determine the rate at which the magnetic field collapses.

Thus, the *time constant* is a measure of the time required to charge (or discharge) a capacitor, or to build (or collapse) a magnetic field around a coil. The time constant is the specific amount of time required to attain 63.2 percent of a desired charge (capacitor) or a desired magnetic field (inductor).

Although the figure of 63.2 percent may seem unusual, it is mathematically correct and is the result of the exponential behavior of *RC* and *RL* circuits. If the charge on a capacitor is measured as it charges through a resistor, it will be seen that the charge is not linear with time. That is, if it takes one minute for the capacitor to charge fully, it does *not* take one-quarter minute for one-quarter charge, nor one-half minute for one-half charge. Instead, the charge will increase exponentially over the one-minute interval.

An exponent is a power of a number—for example, the square of a number (exponent = 2) or some other power. An exponential function is one in which a particular item—for example, the charge on a capacitor—changes as the power of the number changes. Figure 11-1A shows exponential functions in which the exponent is a positive number. (Notice the curve for the case when the exponent is 1. This is a straight line, and is called a *linear* curve. It is a special case of the more general exponential curve.) Figure 11-1B shows exponential functions when the exponent is a negative number. These are the exponential curves for the charging and discharging of capacitors. The same curves apply to the build-up and decay of the magnetic fields around a coil.

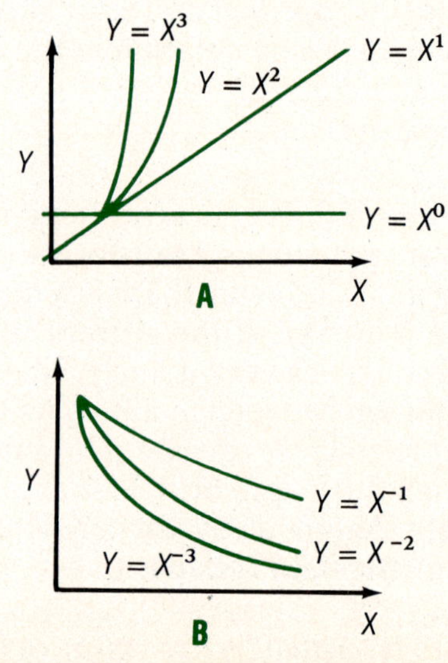

Figure 11-1. Curve diagrams for (A) positive and (B) negative exponent numbers.

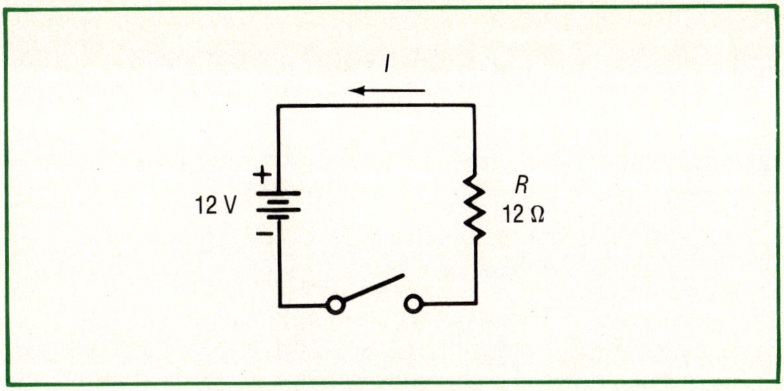

Figure 11-2. Resistive circuit.

Time constants apply to capacitive or inductive circuits only. In a purely resistive circuit, there is no time factor because, for all intents and purposes, current flows in all parts of the circuit instantaneously. A resistive circuit is shown in Figure 11-2. This circuit has a 12 V source and a resistor of 12 Ω. As you know, the current in this circuit will be 1 A because, according to Ohm's law, current is equal to voltage divided by resistance. Thus, a curve of current versus time for this circuit, shown in Figure 11-3, would consist of a straight line. This indicates that the current in the circuit is present immediately and remains at the same value as long as the switch is closed.

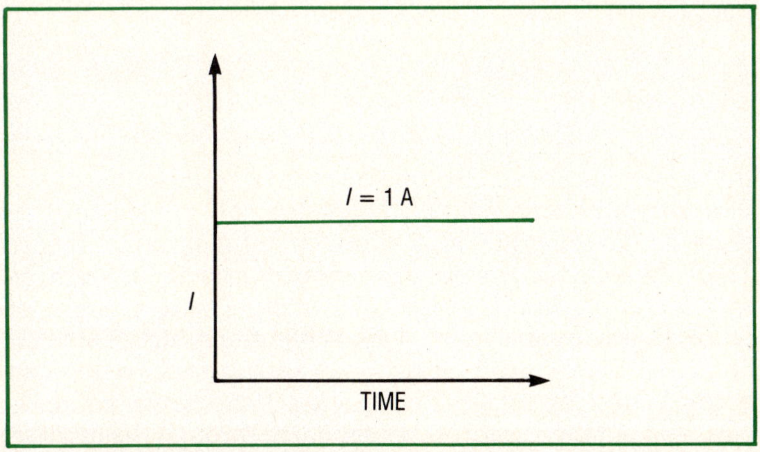

Figure 11-3. Time constant for resistive circuit.

Time constants can be quite different in *RC* and *RL* circuits. Depending upon the values of resistance, capacitance, and/or inductance, time constants can vary widely. The sections that follow in this chapter discuss *RC* and *RL* time constants.

Discussion Questions/Problems

1. What is the significance of the percentage 63.2 in definition of a time constant?
2. What is an exponential quantity?
3. What is the difference between a linear and an exponential curve?
4. What are the differences between purely resistive circuits and circuits with capacitance and/or inductance as well as resistance, as concerns the time constant?

11.2 THE *RC* TIME CONSTANT

A capacitor charges or discharges through the resistance in an *RC* circuit. This means the charge placed on a capacitor moves through the circuit's resistance. Similarly, when the voltage applied to a capacitor is removed, the charge is dissipated through the circuit's resistance—as well as through normal resistances inherent in conductors. The rate of charge or discharge is determined by the time constant of the individual *RC* circuit. The time constant can be computed by multiplying the resistance by capacitance. The formula is as follows:

$$T = RC$$

In the formula above, the time constant is represented by the Greek capital letter *tau* (T). The value of T is measured in seconds. To illustrate, look at the circuit diagram in Figure 11-4A. This circuit contains a three-position switch. When the switch is at position 1, the circuit contains a battery, a resistor of 15K Ω, and a capacitor of 1 μF. When the switch is at position 2, the circuit is completely open and no electron flow can occur. When the switch is at position 3, the power source is removed but the capacitor can discharge through the resistor. The formula for computing a time constant can be applied to this circuit as follows:

$$T = RC = (15 \times 10^3) \times (1 \times 10^{-6}) = 0.015 \text{ seconds}$$

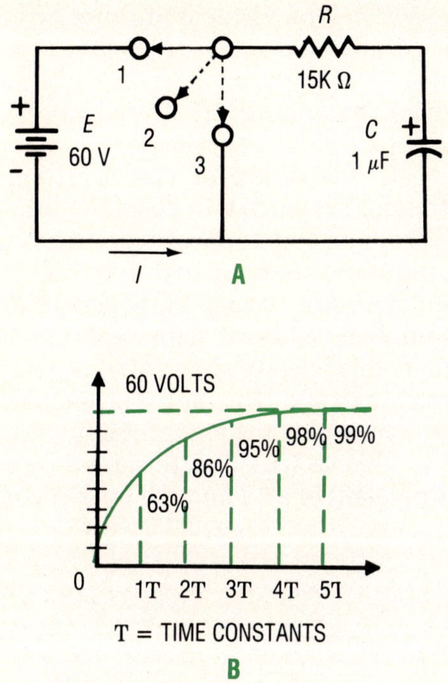

Figure 11-4. Schematic for *RC* circuit with three-position switch and diagram of charging-cycle time constant.

The Capacitance Charging Cycle

In discussing the cycle by which a capacitor is charged it is important to recall the basic relationship between charge, capacitance, and voltage. This relationship, $Q = CE$, is discussed in Chapter 9.

The charge (Q) on a capacitor is a function of its capacitance (C) and of the voltage (E) applied to the capacitor. In the discussions that follow, it is important to distinguish between charge and voltage. In addition, ''charging'' a capacitor does not simply mean putting a charge on a capacitor. It also means that the capacitor has a certain potential difference (voltage) between the plates.

During any one time constant, the charge built up across a capacitor changes by 63.2 percent. This means that a totally uncharged capacitor will charge to a voltage of 63.2 percent of the eventual steady state voltage in the first time constant. In the second time constant, the capacitor will add 63.2 percent of the difference between the voltage at the end of the first cycle and the eventual steady state voltage. With each succeeding time constant, 63.2 percent of the remaining steady state voltage is added. At this rate, the charging process could proceed over a virtually infinite number of time constants, with a smaller charge being added each time. For practical purposes, however, a capacitor reaches a full charge during five time constants, as illustrated below.

Time constant 1. Consider the circuit in Figure 11-4A. Assume that the circuit switch is in position 2 and that the circuit has a 60 V voltage source. The capacitor is uncharged. The earlier computation established that the time constant for this particular circuit, under these conditions, is 0.015 seconds. In 15 milliseconds the capacitor voltage builds to 63.2 percent of its final steady state voltage. The 15 milliseconds represents one time constant. For a 60 V circuit, the voltage charge achieved at the end of the first time constant is equal to 38 V. This voltage is determined by multiplying the decimal equivalent of 63.2 percent by 60 V, as illustrated below.

63.2% of 60 V

$$0.632 \times 60\text{ V} = 38\text{ V}$$

Time constant 2. During the second time constant, the capacitor charges another 63.2 percent of the remaining charge capacity. Thus, in the next 15 milliseconds, the capacitor builds voltage equivalent to 86 percent of a full charge. This computation is shown below:

(A) $60\text{ V} - 38\text{ V} = 22\text{ V}$

(B) 63.2% of 22 V

(C) $0.632 \times 22\text{ V} = 14\text{ V}$

(D) $38\text{ V} + 14\text{ V} = 52\text{ V}$

(E) $\frac{52}{60} = 0.86$ or 86%

From the first line (A) of the computation above it is possible to determine the amount of charge remaining to be added to the capacitor. In terms of voltage, this is determined by subtracting the amount of voltage due to the first time constant from the eventual steady state voltage. The result is 22 V. Then, as line B shows, 63.2 percent is to be taken from the remaining 22 V. Line C shows how the decimal equivalent of 63.2 percent is multiplied by the remaining 22 V. The result is 14 V. On line D, the voltage for the second time constant is added to the voltage for the first time constant. The result is 52 V. This means that, at the end of the second time constant, the capacitor voltage has reached 52 V.

The percentage of cumulative charge is then computed on line E. Since a full charge is 60 V and there are now 52 volts on the capacitor, it is now 86 percent charged.

Time constant 3. During the third time constant the capacitor charges yet another 63.2 percent of the remaining voltage, as follows:

(A) $60\text{ V} - 52\text{ V} = 8\text{ V}$

(B) 63.2% of 8 V

(C) $0.632 \times 8\text{ V} = 5\text{ V}$

(D) $52\text{ V} + 5\text{ V} = 57\text{ V}$

(E) $\frac{57}{60} = 0.95$ or 95%

Line A in the computations above shows the determination of the amount of the unfilled charge for the capacitor: 8 V. Line B indicates that the computation for the third time constant is 63.2 percent of 8 V. On line C, 8 V is multiplied by the decimal equivalent of 63.2 percent. The increase in potential during this time constant is 5 V.

On line D, the total charge is determined by adding 5 V to the previous charge of 52 V. The present level of charge is 57 V. Line E shows that the capacitor is 95 percent charged after three time constants.

Time constant 4. The computations for the fourth time-constant period of 15 milliseconds, shown below, indicate that an additional 1.89 V is added. At the conclusion of the fourth time constant, the capacitor is charged to 98 percent of its capacity.

(A) $60\text{ V} - 57\text{ V} = 3\text{ V}$

(B) 63.2% of 3 V

(C) $0.632 \times 3\text{ V} = 1.89\text{ V}$

(D) $57\text{ V} + 1.89\text{ V} = 58.89\text{ V}$

(E) $\frac{58.89}{60} = 0.98$ or 98%

Time constant 5. During the fifth time constant, as shown in the computations below, an additional 1.11 V of charge is added. At this point, the charge level is 99 percent.

(A) $60\text{ V} - 58.89\text{ V} = 1.11\text{ V}$

(B) 63.2% of 1.11 V

(C) $0.632 \times 1.11\text{ V} = 0.7\text{ V}$

(D) $58.89\text{ V} + 0.7\text{ V} = 59.59\text{ V}$

(E) $\dfrac{59.59}{60} = 0.99$ or 99%

A 99 percent charge is considered to be a full charge. It is considered that a capacitor receives and builds a full charge during five time constants. The value 5 T is often called the *transient* time.

Figure 11-4B shows the curve just derived for the charge of the capacitor (C) through resistor (R).

The Capacitance Discharging Cycle

When a capacitor discharges through a resistance, it loses 63.2 percent of its charge during each time-constant interval. This means that after each time-constant interval, the remaining voltage is 36.8 percent of the voltage at the beginning of the period. As with charging, the discharging could, theoretically, go on indefinitely, with increasingly small amounts of discharge taking place over each time interval. However, for practical purposes, it is assumed that a capacitor is completely discharged after five time-constant periods. For an illustration of the discharging cycle, refer to Figure 11-5A. This shows a circuit with the same basic configuration as the one used to illustrate the charging cycle. There is a three-position switch in a circuit with a power source, a resistor,

Figure 11-5. Schematic for *RC* circuit and diagram of time constant for discharging cycle.

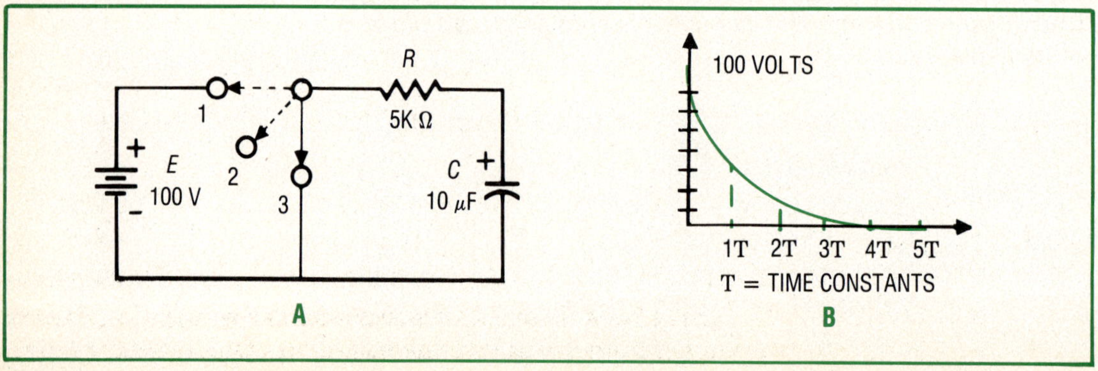

and a capacitor. However, the values of the components are changed. Voltage on this circuit is 100 V. Resistance is 5K Ω, and capacitance is 10 μF. Assuming that the capacitor starts with a full charge of 100 V, the time constant for discharging is determined as follows:

$$T = RC = (5 \times 10^3) \times (10 \times 10^{-6}) = 0.05 \text{ s}$$

As indicated, the time constant for the circuit in Figure 11-5A is 0.05 seconds. This means that the charge remaining on the capacitor will be 36.8 percent of the prior charge for each time interval of 0.05 seconds after the voltage source is removed.

Time constant 1. The charge level following the first time-constant interval, as computed below, is 36.8 V.

$$36.8\% \text{ of } 100 \text{ V}$$

$$0.368 \times 100 \text{ V} = 36.8 \text{ V}$$

Time constant 2. During the second time constant, the charge on the capacitor is reduced to 36.8 percent of 36.8 V. The remaining charge, as computed below, is 13.54 V. This is approximately 14 percent of the original charge.

$$\text{(A)} \quad 36.8\% \text{ of } 36.8 \text{ V}$$

$$\text{(B)} \quad 0.368 \times 36.8 \text{ V} = 13.54 \text{ V}$$

$$\text{(C)} \quad \frac{13.54}{100} = 0.14 \text{ or } 14\%$$

Time constant 3. During the third time constant, the capacitor discharges to the level of 36.8 percent of 13.54 V. The remaining charge after this time constant is approximately 5 percent. The computations are shown below:

$$\text{(A)} \quad 36.8\% \text{ of } 13.54 \text{ V}$$

$$\text{(B)} \quad 0.368 \times 13.54 \text{ V} = 4.98 \text{ V}$$

$$\text{(C)} \quad \frac{4.98}{100} = 0.05 \text{ or } 5\%$$

Time constant 4. During the fourth time constant of the discharging cycle, the charge is reduced to a total of 1.83 V. The computations are shown below:

(A) 36.8% of 4.98 V

(B) $0.368 \times 4.98\ \text{V} = 1.83\ \text{V}$

(C) $\frac{1.83}{100} = 0.02$ or 2%

Time constant 5. After the fifth time constant, the capacitor is discharged fully—for practical purposes. The capacitor retains only approximately 1 percent of the original charge. This is accepted as being equivalent to a fully discharged state. The computations for the fifth time constant are shown below:

(A) 36.8% of 1.83 V

(B) $0.368 \times 1.83\ \text{V} = 0.67\ \text{V}$

(C) $\frac{0.67}{100} = 0.01$ or 1%

Figure 11-5B shows the curve for the discharge of the capacitor (C) through the resistor (R).

Discussion Questions/Problems

1. What is the time constant for a circuit with a resistance of 100K Ω and a capacitance of 50 μF?
2. Using the same circuit properties as in Question 1 and a source voltage of 50 V, give the charging voltage status for each of the first five time-constant intervals.
3. Using the values in Figure 11-4A, compute the discharge levels for each of the five time-constant intervals for this circuit.

11.3 EXPONENTIAL FORM

An exponential form is a curve that shows exponential relationships in the properties of current or voltage. Figure 11-4B is an exponential curve showing the buildup of voltage across a capacitor during charging. Figure 11-5B is an exponential curve showing the discharging of a capacitor.

It is important to learn to examine and derive meaning from exponential curves. This is the most common way of explaining the behavior of components in *RC* and *RL* circuits. There are very few linear curves in electronics. These are usually limited to DC circuits with resistance only. The capacitive and inductive components you encounter and use in *RL* and *RC* circuits will have characteristics which include exponential curves.

Mathematically, exponential curves for voltage and current are derived by using two general equations. The first of these is:

$$v = E_F + (E_i - E_F)\, e^{-\frac{t}{T}}$$

This equation develops the value of instantaneous voltage. A similar formula is used to calculate the value of instantaneous current:

$$i = I_F + (I_i - I_F)\, e^{-\frac{t}{T}}$$

Either of the equations—for instantaneous voltage or instantaneous current—can be used in developing data for exponential curves which represent either charging or discharging. In the equations above, E_F and I_F are the final values for voltage and current. *E* and *I* are the initial values for voltage and current. The letter *t* represents the value of time. T is equal to *RC* in the case of an *RC* circuit, or *L/R* in the case of an *RL* circuit. The letter *e* is the base of natural logarithms and has a constant value of 2.71828. It has a base 10 equivalent of 0.314. The base 10 equivalent of a natural logarithm is known as a *common logarithm*.

A *logarithm* of a number is the exponent to which a base is raised to equal that number. Logarithms are applied when an exponent cannot be represented by a whole number. For example, suppose you have a problem in which it is necessary to raise the whole number 2 to a value of 12. Two raised to the third power is 8. The logarithm of 8 to the base 2 is 3. Similarly, 2 raised to the fourth power is 16, and the logarithm of 16 to the base 2 is 4. So, there must be some number (N) between 3 and 4 such that 2 raised to the N power will be 12. The logarithm of 12 to the base 2 must be a number (N) between 3 and 4. Therefore, you cannot use a whole number exponent. Instead, you would calculate or use a table to find the exponent (logarithm) that will derive the value of 12 from a base number of 2.

There are two principal systems of logarithms:

- The common system using base 10
- The natural system using base *e*, in which $e = 2.71828$.

Natural logarithms are also called *napierian logarithms*. The name comes from the developer of the system, John Napier (1550–1617), a Scottish mathematician.

Only one set of tables is necessary since it is possible to convert a logarithm in one base (b) to a logarithm in any other base (a) by a simple formula:

$$\text{Log}_{a^y} = \text{Log}_{b^y} \times \text{Log}_{a^b}$$

Thus, for this example,

$$\begin{aligned} \text{Log}_{2^{12}} &= \text{Log}_{10^{12}} \times \text{Log}_{2^{10}} \\ &= 1.0792 \times 3.3219 \\ \text{Log}_{2^{12}} &= 3.5850 \end{aligned}$$

Formulas for Charging a Capacitor

The formula used to calculate the charge on a capacitor as a function of time, when being charged, is adapted from the one given for instantaneous voltage. In this formula, it is assumed that the initial, or starting, voltage is equal to 0. A capacitor is considered for practical purposes to be discharged when it retains anything less than 1 percent of its initial charge. Given an initial voltage of 0, the formula used to determine charging data is derived in the equations listed below. Line A repeats the basic formula. The remaining two lines show the derivation for an initial voltage of 0.

$$\text{(A)} \quad v = E_F + (E_i - E_F)\, e^{-\frac{t}{T}}$$

$$\text{(B)} \quad v = E_F + (0 - E_F)\, e^{-\frac{t}{RC}}$$

$$\text{(C)} \quad v = E_F\,(1 - e^{-\frac{t}{RC}})$$

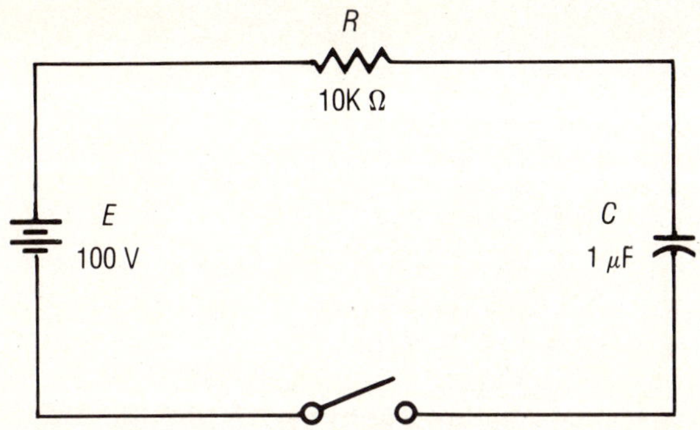

Figure 11-6. *RC* circuit to use to determine time constant for charging cycle.

To demonstrate the use of the formula for a charging capacitor, consider the circuit diagrammed in Figure 11-6. This shows a circuit with a voltage source of 100 V, a resistor of 10K Ω, and a capacitor of 1 μF. The first step is to determine the time constant, as shown below:

$$\mathrm{T} = RC$$
$$= (10 \times 10^{3}) \times (1 \times 10^{-6})$$
$$\mathrm{T} = 0.01 \text{ s}$$

Now, to illustrate use of the exponential form equation, assume you have been asked to determine the voltage across the capacitor at 0.015 seconds. In solving equations like the one shown below, it is best to use either a calculator or log tables. For this problem, tables of natural logs would be used. The calculations are as follows:

$$\text{(A)} \quad E_C = E_F\,(1 - e^{-\frac{t}{RC}})$$
$$\text{(B)} \quad = 100 \text{ V}\,(1 - e^{-\frac{0.015\text{ s}}{0.01\text{ s}}})$$
$$\text{(C)} \quad = 100 \text{ V}\,(1 - e^{-1.5})$$
$$\text{(D)} \quad = 100 \text{ V}\,(1 - 0.223)$$
$$\text{(E)} \quad E_C = 77.7 \text{ V}$$

The curve derived from this equation is illustrated below in Figure 11-7.

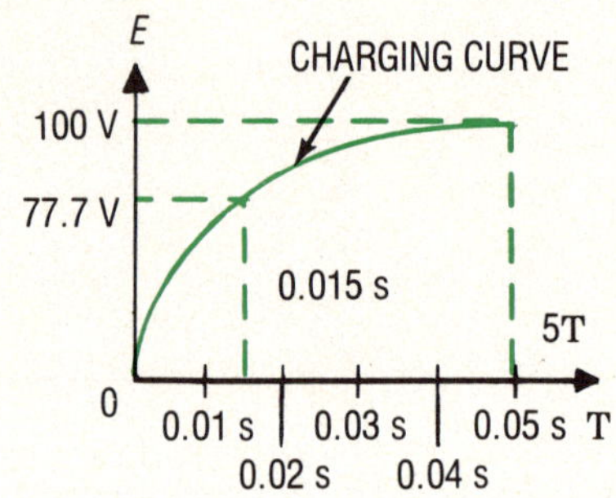

Figure 11-7. Time constant curve for charging cycle of circuit of Figure 11-6.

Formulas for Discharging

To illustrate use of the same formula to develop data for an exponential curve showing discharging, consider the circuit

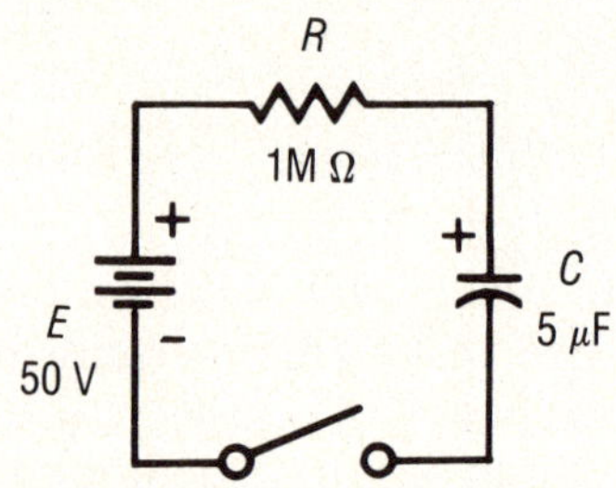

Figure 11-8. *RC* circuit for determining time constant for discharging cycle.

in Figure 11-8. Assume you have been asked to solve the circuit for a time of 4 seconds, i.e., to determine the voltage remaining across the capacitor after 4 seconds. You begin by determining the value of T, as follows:

$$
\begin{aligned}
T &= RC \\
&= (1 \times 10^{6}) \times (5 \times 10^{-6}) \\
T &= 5 \text{ s}
\end{aligned}
$$

With the time constant determined (5 s), you can then apply the formula to determine the voltage level after an elapsed time of 4 seconds. These calculations are as follows:

$$E_C = E_i e^{-\frac{T}{RC}}$$

$$= 50e^{-\frac{4\text{ s}}{5\text{ s}}}$$

$$= 50e^{-0.8}$$

$$= 50(0.45)$$

$$E_C = 22.5\text{ V}$$

The exponential curve plotting this discharge is shown in Figure 11-9.

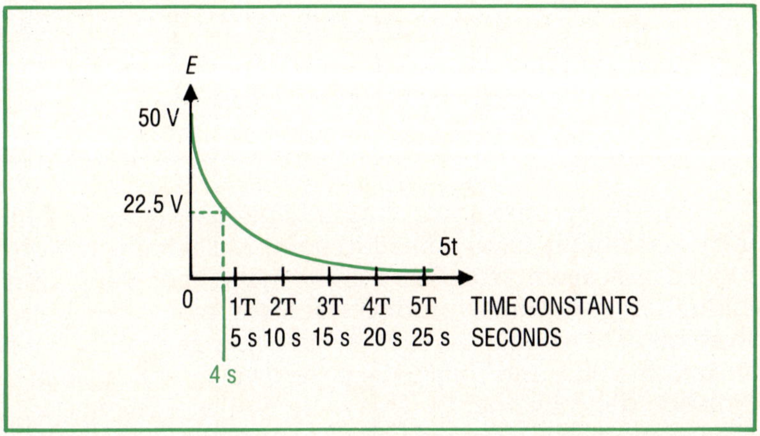

Figure 11-9. Time constant curve for discharging cycle of circuit of Figure 11-8.

Discussion Questions/Problems

1. What is meant by exponential relationships?
2. For the circuit in Figure 11-6, determine the voltage level at 0.02 seconds.
3. For the circuit in Figure 11-8, determine the remaining voltage level at 3 seconds.

11.4 *RC* CIRCUIT WAVE SHAPES

Figure 11-10A shows a circuit with just a three-position switch, a capacitor, and a source of DC voltage.

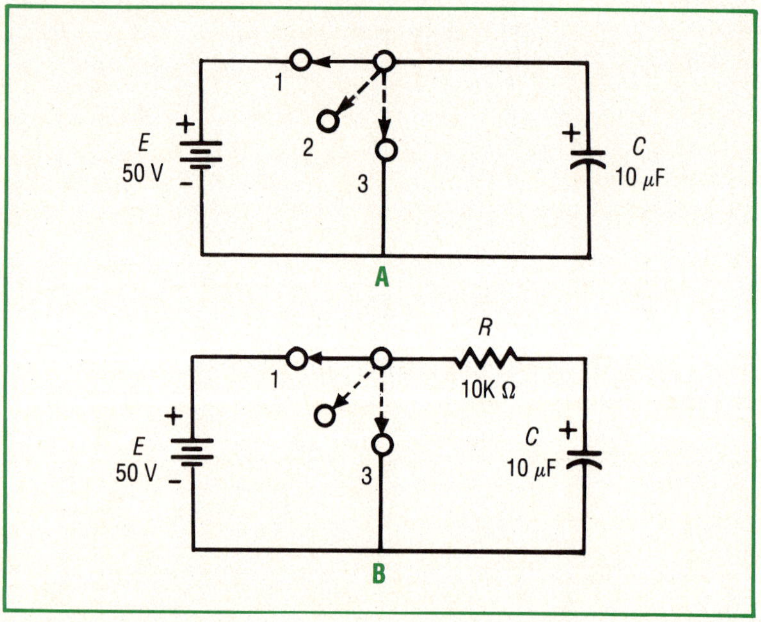

Figure 11-10. Schematics of (A) circuit with a capacitor, DC voltage source, and three-position switch and (B) with resistor added.

Now, suppose the switch of the circuit in Figure 11-10A is moved rapidly from point 1 to point 3. Then it is rapidly moved back again, from 3 to 1. If the movement were rapid enough, the waveform produced across *C* would be square in shape. The voltage across *C* would change rapidly from zero to 50 volts, and then back to zero volts, with plateaus between the movements. This is known as a *square wave*. Now consider the circuit of Figure 11-10B. When the switch is moved rapidly from 1 to 3, and then from 3 to 1, the voltage across *C* will be of the form shown in Figure 11-11. This type

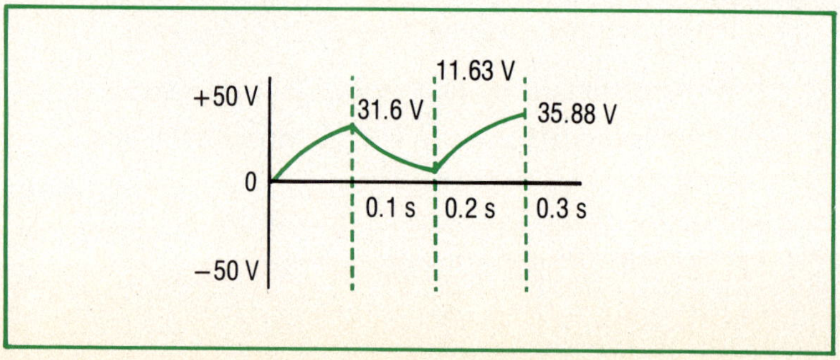

Figure 11-11. Waveform for voltage for *RC* circuit of Figure 11-10.

of waveform, because it is other than a perfect sine wave, is sometimes called a *non-sinusodal wave*. Notice the exponential rise and fall of the voltage across the capacitor (C). The exponential curve will be steeper if *R* or *C* is reduced. The exponential curve will be more gradual if *R* or *C* is increased.

Square waves can be produced by a special device, called a *square wave generator*. In Figure 11-12, the symbol for a square wave generator is drawn in the middle of the diagram. The diagram shows an *RC* circuit driven by a square wave generator. When a circuit device of this type is used to drive an *RC* series circuit, waveforms like the one diagrammed in Figure 11-11 will be generated across the capacitor (C).

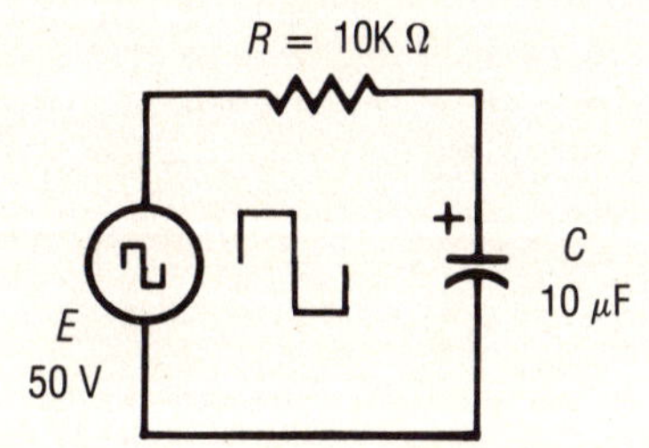

Figure 11-12. *RC* circuit driven by square wave generator.

The waveform in Figure 11-11 assumes that voltage is applied, then disconnected, at each time constant. During the first time constant, the charge on the capacitor builds to 63.2 percent of the 50 V source. This means that, after the first time constant, the charge on the capacitor is equal to 31.6 V. When the current flow is interrupted for the next time constant, the capacitor discharges at the rate of 36.8 percent of the established charge of 31.6 V. Thus, after the second time constant, the charge on the capacitor is at 11.63 V. For the third time constant, the 50 V source is reapplied. The charge over the third time constant is explained, along with the first two, in the calculations below:

ON — CHARGE	63.2% of 50 V
	0.632 × 50 V = 31.6 V
OFF — DISCHARGE	36.8% of 31.6 V = 11.63 V
ON — CHARGE	50 V − 11.63 V = 38.37 V
	63.2% of 38.37 V = 24.25 V
	24.25 V + 11.63 V = 35.88 V

After three full time constants, the charge on the capacitor is equal to 35.88 V.

The charge and discharge pattern shown in Figure 11-11 corresponds with the calculations above. The illustration in Figure 11-11, as well as the computations above, represent a circuit with a time constant of 0.1 s. The current is on for 0.1 s, then off for 0.1 s. The time constant for Figure 11-11 is calculated below:

$$T = RC = 10K\ \Omega \times 10\ \mu F = 0.1\ s$$

Current Shapes

Remember that current through the capacitor leads voltage across the capacitor by 90 degrees in an *RC* circuit. Thus, the waveform for current in the circuit diagrammed in Figure 11-10 would be significantly different from the waveform for voltage. Figure 11-13 shows a waveform for current in this circuit. To derive the values shown in this figure, separate calculations are needed for the charging and discharging cycles. For the charging cycle, the formula used is:

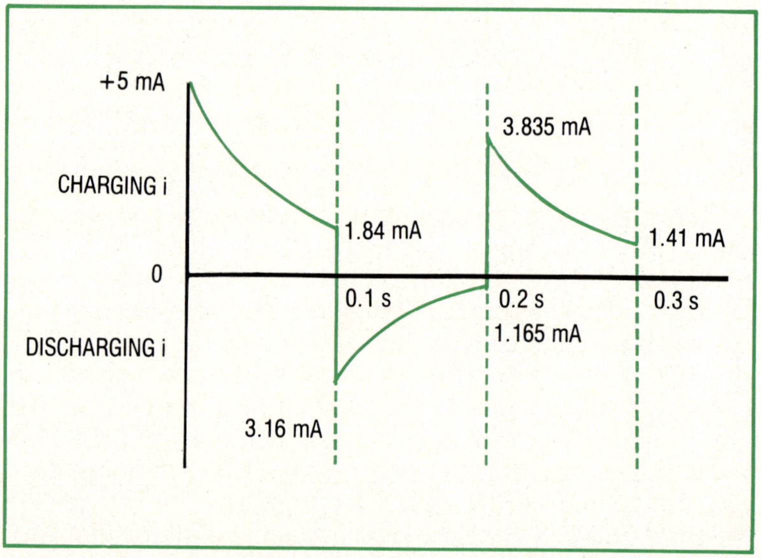

Figure 11-13. Waveform for current of circuit of Figure 11-10.

$$i = \frac{(E_T - E_C)}{R}$$

For the discharging cycle, this formula is used:

$$i = \frac{E_C}{R}$$

For both equations, E_C is the voltage across the capacitor, and E_T is the total battery voltage. Note an important difference between the equations for the charging and discharging cycles. For the charging cycle, the charging current equals the total battery voltage minus the voltage across the resistor, all divided by the resistance (R). This is because voltage is applied during the charging cycle and must therefore be included in the computations. In the discharging cycle, current equals only the capacitor voltage divided by the resistance (R). The reason for this is that during the discharge cycle there is no applied voltage on the circuit and the only voltage is the capacitor voltage. Also notice that the current flow on discharge is, predictably, opposite to the direction of flow on the charge part of the cycle.

The computations supporting the waveform in Figure 11-13 are given below:

$$36.8\% \text{ of } 5 \text{ mA} = 1.84 \text{ mA}$$
$$5 \text{ mA} - 1.84 \text{ mA} = 3.16 \text{ mA}$$
$$36.8\% \text{ of } 3.16 \text{ mA} = 1.165 \text{ mA}$$
$$5 \text{ mA} - 1.165 \text{ mA} = 3.835 \text{ mA}$$
$$36.8\% \text{ of } 3.835 \text{ mA} = 1.41 \text{ mA}$$

Discussion Questions/Problems

1. What is a square wave and how is it formed?
2. What is the difference between a square wave and a typical pulse wave?
3. What is the difference between a pulse wave and a sinusoidal wave?
4. What is the time constant for a circuit that contains a 100K Ω resistor in series with a 100 μF capacitor?

11.5 *RC* INTEGRATION AND DIFFERENTIATION

The voltage across a capacitor in a series *RC* circuit at any given moment is called the *integrated voltage*. The name comes from the fact that voltage charges are accumulated across the capacitor over a period of time.

The voltage across a resistor in a series *RC* circuit is called a *differentiated voltage*. This name indicates that voltage across a resistor represents the rate of change of the voltage input to the *RC* circuit.

Integration and differentiation are properties that must be considered in connection with outputs from *RC* circuits. *RC* circuits can integrate and differentiate voltages in a direct analog to the mathematical integration and differentiation of calculus.

The *RC* Integrator

The *RC* integrator is a particular circuit which reacts in a specific way to pulse voltages. To illustrate, refer to the circuit diagram in Figure 11-14. This shows an *RC* circuit in

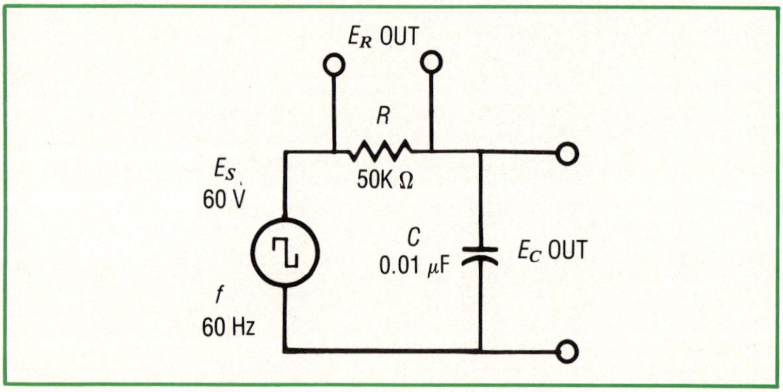

Figure 11-14. *RC* circuit diagram for integrated output.

which voltage output can be measured across the capacitor. In this diagram, the integrated output is marked E_C. Properties of integrated output include:

- The capacitor will charge and discharge according to the time constant of the circuit.
- The capacitor charges when the source voltage is applied.
- The capacitor discharges when the source voltage is removed. (The capacitor will discharge through an external resistance in the circuit, or through its own dielectric.)

Note that the circuit in Figure 11-14 includes a square wave generator. This type of generator functions something like a switch which can instantaneously apply or disconnect current from a battery into a circuit.

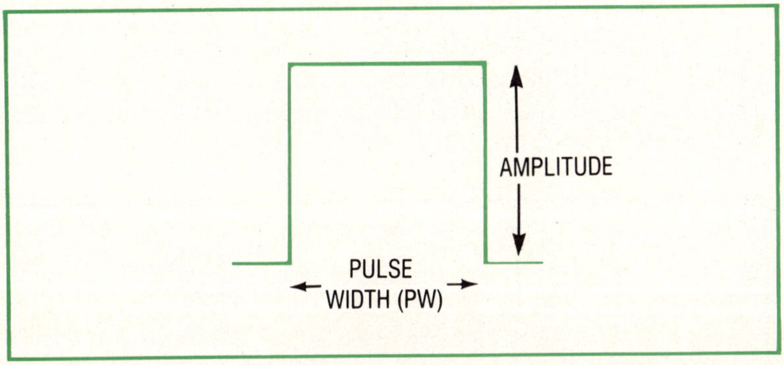

Figure 11-15. Waveform created by square wave generator.

A characteristic waveform created by a square wave generator is shown in Figure 11-15. This illustration identifies the main terms used in describing square waves. The height of the wave is called the *amplitude*. Amplitude is a measure of the amount of voltage applied. The other dimension used in describing a square wave is the *pulse width*. The pulse width is obviously a measurement of the time duration of the square wave. Thus, the lower the frequency of applied current, the wider the pulse width will be. A slower frequency means that current is applied longer between alternations. Therefore, the square waveform for lower frequencies will have wider pulse widths. Conversely, of course, narrower pulse widths indicate a shorter time duration or faster alternation of the current.

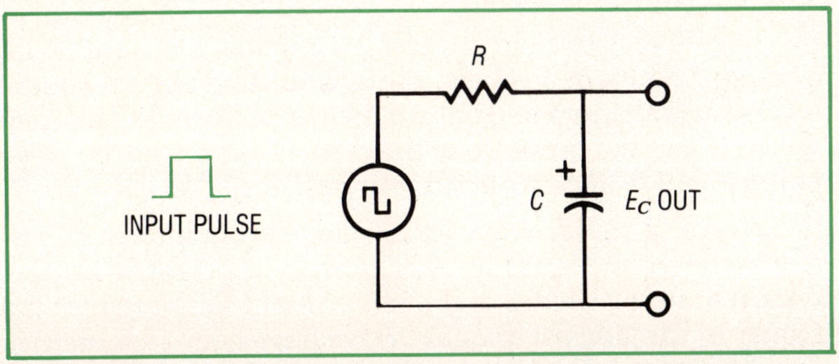

Figure 11-16. *RC* circuit with square wave generator input.

As shown in Figure 11-16, a square wave is applied to the input of an *RC* circuit. Square waves can be used to observe the characteristic behavior of certain types of circuits. When square waves are applied to these circuits they modify the square waves in special ways. By observing the output of these circuits, it is possible to learn about their characteristic behavior. Remember, the square wave generator acts like a simple on/off switch. The capacitor begins to charge the moment current is applied. The capacitor begins to discharge when the square wave goes to zero. The output wave (Figure 11-17) looks entirely different from the input wave. This diagram shows the characteristic exponential increase and decrease in the voltage across the capacitor.

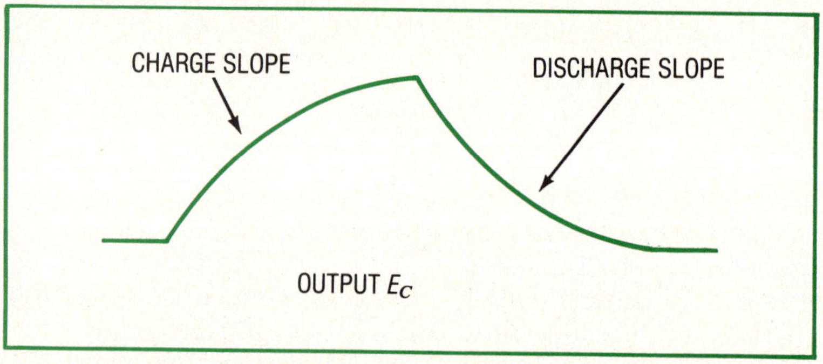

Figure 11-17. Effects of capacitance.

Comparing the input and output curves clearly shows the effects of capacitance. The input pulse is represented by a square wave. This means that the applied voltage reaches its peak instantaneously and remains at this level throughout the duration of the pulse. Because the capacitor charges and discharges through the resistor R, the effect is to cause the characteristic exponential slope during the charge and discharge.

The *RC* Differentiator

In a purely resistive circuit, there is no time constant. With a resistor alone in a circuit, there would be perfect square waves on both input and output. However, in an *RC* differentiator circuit, the presence of the capacitor changes the wave pattern which appears across the resistor.

A typical square wave for a voltage input into an *RC* circuit is shown in Figure 11-18. The wave is perfectly square, with two vertical sides and a horizontal top. Note that the names of the sides of a square wave have been identified in this illustration. The initial, or front, vertical line is called the *leading edge* of the square wave. The final, or rear, vertical line is called the *trailing edge*. These—like amplitude and pulse width—are standard terms used in describing square waves.

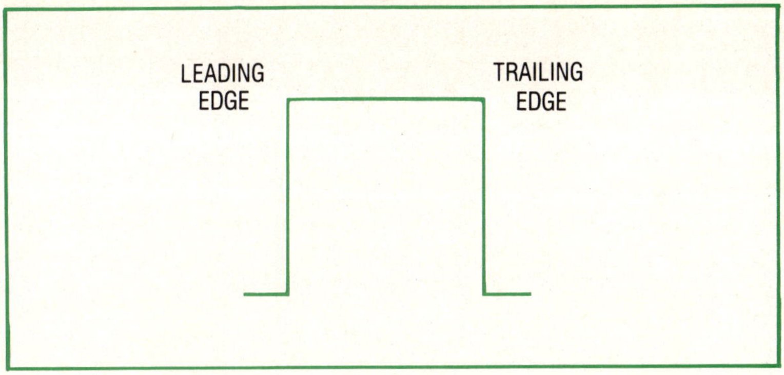

Figure 11-18. Square wave generated for voltage input.

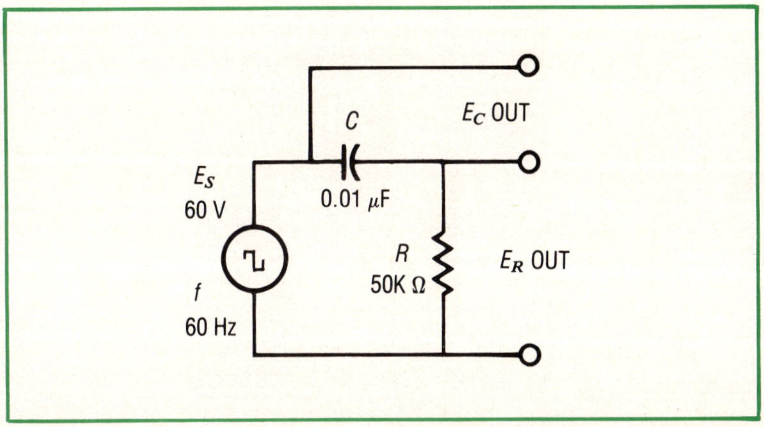

Figure 11-19. *RC* circuit driven by a square wave generator, showing outputs across both capacitor and resistor.

Figure 11-19 shows a circuit with outputs across both the capacitor and resistor of an *RC* circuit with a short time constant that is driven by a square wave generator. The outputs across the capacitor and the resistor, as well as the input voltage and the current in the circuit, are compared in Figure 11-20. Figure 11-20A shows a normal square wave as input. Figure 11-20B shows the integrated output across the capacitor and the charging and discharging cycles typical for a capacitor. Figure 11-20C shows the curve for the current through the *RC* circuit. Compare the waveforms in Figure 11-20B and 11-20C. These demonstrate that the voltage lags the current in an *RC* circuit. When the capacitive voltage is increasing, the current of the circuit is decreasing within the same time period. Figure 11-20D shows the voltage across the resistor—the differentiated voltage. This corresponds closely with the pattern of the current output for the circuit. This is to be expected because the voltage across the resistor will equal the current through it multiplied by the resistance.

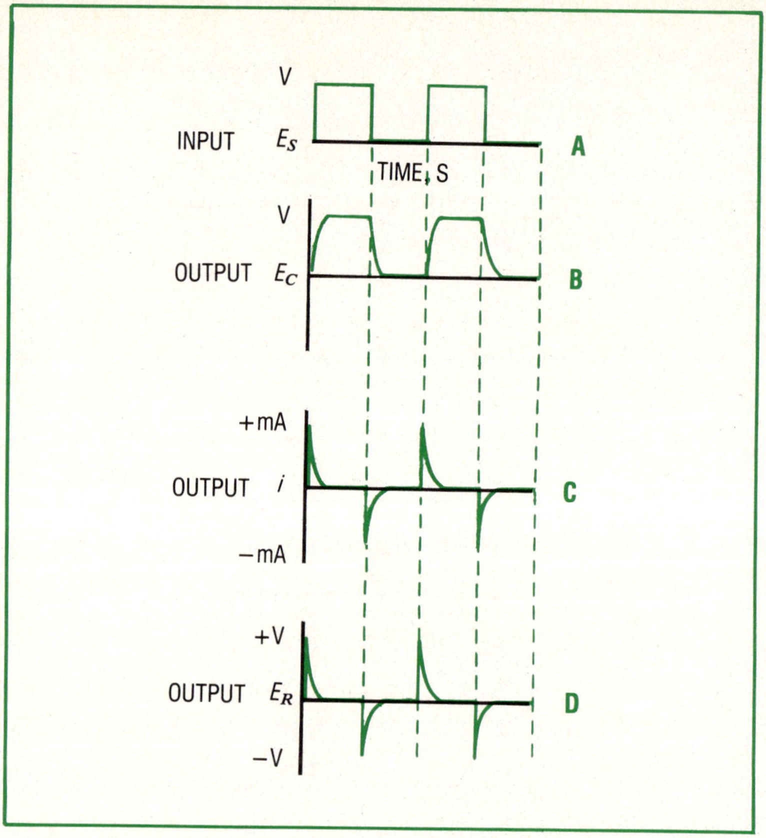

Figure 11-20. Waveforms of charging and discharging cycles for current and voltage of *RC* circuit of Figure 11-19.

Note the wave patterns in Figure 11-20 as they apply to the charging and discharging cycles of the capacitor. At the beginning of the charging cycle, voltage across the resistor (E_R) is equal to the full source voltage (E_S). On completion of the discharging cycle, the differentiated voltage is essentially zero.

Discussion Questions/Problems

1. What is the meaning of voltage integration?
2. What is the meaning of voltage differentiation?
3. Why are the waveforms across resistors in *RC* circuits different from the waveforms for purely resistive circuits?

11.6 THE *L/R* TIME CONSTANT

Inductors, you will remember, oppose changes in the current flowing through them. When a coil is included in a circuit, current flow cannot be changed instantaneously. Instead, if current has been flowing through a coil and the current is interrupted, the inductor causes a time interval to elapse as the current decreases and the magnetic field collapses. This time interval is known as the time constant for an *RL* circuit.

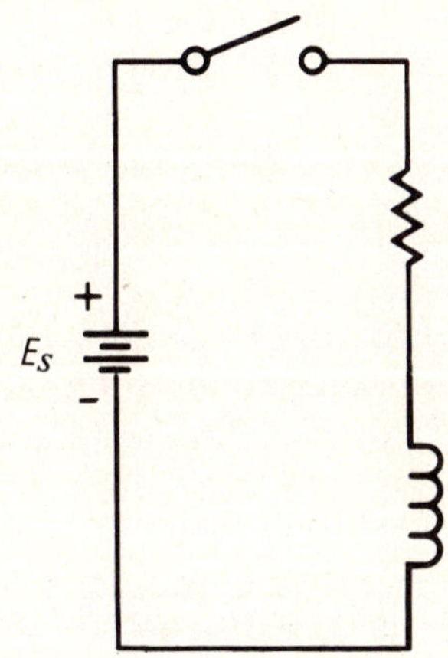

Figure 11-21. Schematic for typical *L/R* circuit.

A typical *RL* circuit is diagrammed in Figure 11-21. This circuit is also referred to as an *L/R* circuit. You will understand the reason for this other designation from the formula shown below. This formula determines the inductive time constant for an *RL*, or *L/R*, circuit.

$$T = \frac{L}{R}$$

To illustrate how the time constant for an *L/R* circuit is determined, consider the circuit in Figure 11-22. This diagram provides the values needed to calculate the time constant, as shown below:

$$T = \frac{L}{R} = \frac{15 \text{ mH}}{10\text{K } \Omega} = \frac{15 \times 10^{-3}}{10 \times 10^{3}} = 1.5 \times 10^{-6} = 1.5\ \mu\text{s}$$

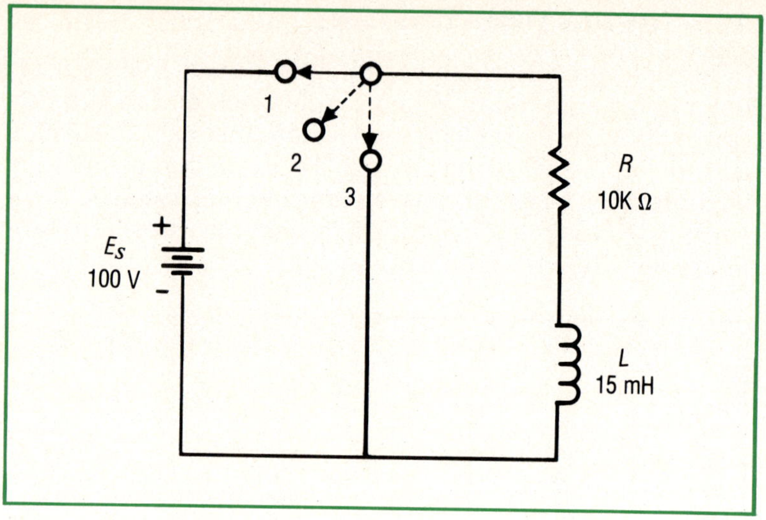

Figure 11-22. *L/R* circuit to use to determine inductive time constant.

Time constants for inductors have both similarities and differences as compared with those for capacitors. A major similarity is that there are five time constants in the cycle for reaching full charge or discharge status. A difference lies in the fact that in a charging inductor it is current which builds up, not voltage, as is the case with a capacitor. Another difference is that it is customary to speak of the growth and decay (rather than the charge and discharge) of an *RL* time constant.

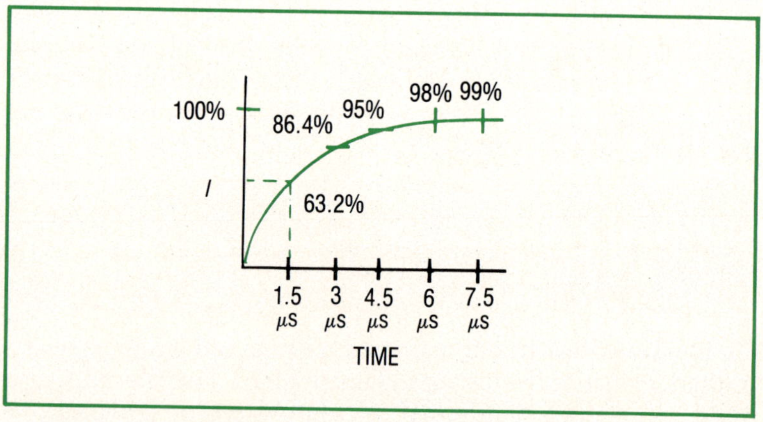

Figure 11-23. Curve of current buildup for time constants of *L/R* circuit of Figure 11-22.

Inductance Current Growth

Figure 11-23 is a curve showing current buildup for the five time constants of the inductor/resistor in the circuit diagrammed in Figure 11-22. The current buildup shown in Figure 11-23 assumes zero current in the inductor at the outset. At the outset also, it is assumed that the switch is open, in position 2, as shown in Figure 11-22. The first step in determining current values for the circuit in Figure 11-22 is to determine the eventual steady state current. The steady state current is the current that will exist in the circuit after the short-term transient effects due to the inductance are over. Note that the values given are for voltage, resistance, and inductance. To determine total current, Ohm's law is used:

$$I_T = \frac{E_S}{R}$$

The calculation to determine steady state current in this circuit is given below:

$$I_T = \frac{E_S}{R}$$

$$= \frac{100 \text{ V}}{10\text{K V}}$$

$$I_T = 0.01 \text{ A}$$

For an inductive/resistive circuit, it is assumed that full voltage is applied instantaneously. In reality there is an extremely small delay as a result of the self-inductance of the wire that makes up the coil and circuit. However, this delay is too insignificant to be a factor in calculating time constants.

Time constant 1. When the switch of the circuit in Figure 11-22 is moved to position 1, full voltage (100 V) is applied to the circuit. The amount of current will depend upon the resistance and the time constant of the circuit. The time constant for this circuit has been calculated. It is 1.5 μs. During the first time constant, current will reach 63.2 percent of its eventual steady state value. The computations are as follows:

$$i = 63.2\% \text{ of } 0.01 \text{ A}$$

$$= 0.632 \times 0.01 \text{ A} = 0.00632 \text{ A} = 6.32 \text{ mA}$$

Time constant 2. During the second time constant, there will be an additional current buildup of 63.2 percent of the remaining portion needed to reach the eventual steady state current. At the conclusion of the second time constant, then, the current buildup will be 86 percent of the steady state current. The computations for this buildup are as follows:

$$0.01\ \text{A} - 6.32\ \text{mA} = 3.68\ \text{mA}$$
$$0.632 \times 3.68\ \text{mA} = 2.32\ \text{mA}$$
$$6.32\ \text{mA} + 2.32\ \text{mA} = 8.64\ \text{mA}$$

Time constant 3. During the third time constant, with a buildup of an additional 63.2 percent of the remaining portion of steady state current, the current reaches 95 percent. The computations are as follows:

$$0.01\ \text{A} - 8.64\ \text{mA} = 1.36\ \text{mA}$$
$$0.632 \times 1.36\ \text{mA} = 0.86\ \text{mA}$$
$$8.64\ \text{mA} + 0.86\ \text{mA} = 9.5\ \text{mA}$$

Time constant 4. During the fourth time constant, the current reaches 98 percent of steady state value. The calculations are as follows:

$$0.01\ \text{A} - 9.5\ \text{mA} = 0.5\ \text{mA}$$
$$0.632 \times 0.5\ \text{mA} = 0.3\ \text{mA}$$
$$9.5\ \text{mA} + 0.3\ \text{mA} = 9.8\ \text{mA}$$

Time constant 5. During the fifth time constant, the current exceeds 99 percent of steady state value. This is considered a full buildup. The computations for this fifth time constant are as follows:

$$0.01\ \text{A} - 9.8\ \text{mA} = 0.2\ \text{mA}$$
$$0.632 \times 0.2\ \text{mA} = 0.1\ \text{mA}$$
$$9.8\ \text{mA} + 0.1\ \text{mA} = 9.9\ \text{mA} \approx 0.01\ \text{A}$$

By this point it should be apparent that the same numbers appear after each time-constant interval: 63.2 percent after the first time-constant interval, 86.4 percent after the second time-constant interval, 95 percent after the third time-constant interval, 98 percent after the fourth, and 99 percent after the fifth. Although the actual time in seconds (or fractions of a second) for the time constant may vary depending on the circuit, the percentages at the end of each interval are always the same.

Since the value at each interval depends on the preceding interval(s), the values, as shown in Figure 11-23, are:

1. time constant $1 - \frac{1}{e} = 1 - \frac{1}{2.718} = 63.2\%$
2. time constants $1 - \frac{1}{e^2} = 1 - \frac{1}{(2.718)^2} = 86.4\%$
3. time constants $1 - \frac{1}{e^3} = 1 - \frac{1}{(2.718)^3} = 95\%$
4. time constants $1 - \frac{1}{e^4} = 1 - \frac{1}{(2.718)^4} = 98\%$
5. time constants $1 - \frac{1}{e^5} = 1 - \frac{1}{(2.718)^5} = 99\%$

Inductance Current Decay

When current through an *L/R* circuit is interrupted, the current begins to decay. Decay of the current in an inductor corresponds with the discharge of a capacitor. That is, the current that has been built up dissipates.

To illustrate the process of current decay, consider the circuit diagrammed in Figure 11-24.

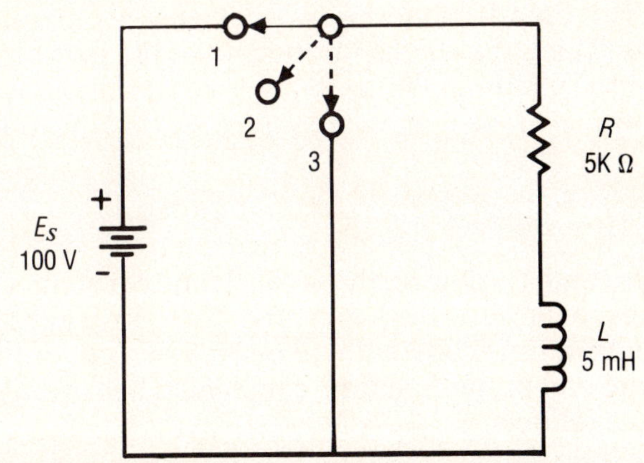

Figure 11-24. *L/R* circuit to use to determine time constant for inductance current decay.

To determine the decay over a series of time constants, you begin by determining the time constant for the circuit.

The time constant for the circuit in Figure 11-24 is computed as follows:

$$T = \frac{L}{R}$$
$$= \frac{5 \text{ mH}}{5\text{K } \Omega}$$
$$= \frac{5 \times 10^{-3}}{5 \times 10^{3}}$$
$$= 1 \times 10^{-6}$$
$$T = 1\ \mu\text{s}$$

It is also necessary to determine the steady state current for a circuit before either buildup or decay can be computed. The computations to determine steady state current for the circuit in Figure 11-24 are as follows:

$$I_T = \frac{E_S}{R} = \frac{100 \text{ V}}{5\text{K } \Omega} = 0.02 \text{ A}$$

Time constant 1. As was the case with capacitor discharge, current through an inductor decays each time constant at a fixed rate of 36.8 percent of the current at the start of the time period. This decay is uniform—as a percentage—for each time constant. Thus, the current at the end of the first time constant for the circuit in Figure 11-24 is computed as follows:

$$i = 36.8\% \text{ of } 0.02 \text{ A}$$
$$0.368 \times 0.02 \text{ A} = 7.36 \text{ mA}$$

Time constant 2. During the second time constant, current continues to decay at the same rate. The current at the end of the second time constant will be 36.8 percent of the current at the end of the first time constant. The current is calculated as follows:

$$0.368 \times 7.36 \text{ mA} = 2.71 \text{ mA}$$

Time constant 3. By now, the current has fallen to 13.6 percent (0.368 × 0.368) of the initial steady state current. For the third time constant, the current decays to 5 percent of the initial steady state current. The computations for the current at the end of this third time constant are:

$$0.368 \times 2.71 \text{ mA} = 1 \text{ mA}$$

Time constant 4. During the fourth time constant, the decay level drops to 2 percent of the initial steady state current buildup. The computations for this current decay cycle are:

$$0.368 \times 1 \text{ mA} = 0.368 \text{ mA}$$

Time constant 5. Following the fifth time constant, decay is considered complete. The remaining current is equal to approximately 1 percent of the initial steady state current. The computations are:

$$0.368 \times 0.368 \text{ mA} = 0.13 \text{ mA} \approx 0 \text{ A}$$

Figure 11-25 shows the current decay over five time constants.

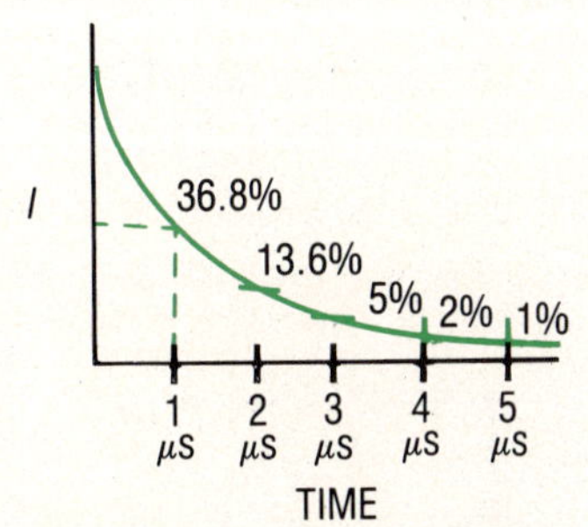

Figure 11-25. Curve of current decay for time constants of *L/R* circuit of Figure 11-24.

Discussion Questions/Problems

1. What is meant by an *L/R* time constant?
2. What is meant by current buildup?
3. What is current decay?
4. If a circuit has a resistor of 15K Ω in series with a 20 mH inductor, what is the time constant?
5. What would be the decay for the first time constant for the circuit in Figure 11-22?

11.7 *L/R* INTEGRATION AND DIFFERENTIATION

An *L/R* circuit can act as an integrator, and integrated current output is measured through the resistor. When an *L/R* circuit is used as a differentiator, the output is measured as voltage across the inductor.

The *L/R* Integrator

Figure 11-26A shows an *L/R* integrator circuit. Notice that a resistor and an inductor are connected in series. This series arrangement, when driven by square wave voltage, shows the characteristics of the inductor as a current integrator. An inductor, of course, prevents sudden changes in current.

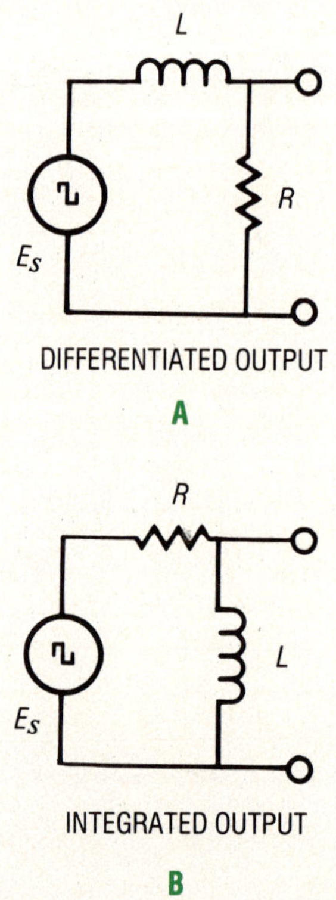

Figure 11-26. Circuit schematics showing inductor as (A) current integrator and (B) differentiator.

Because of this characteristic, the inductor acts as an open for the circuit when the leading, or rising, edge of the square wave is applied. Some time after the signal reaches the plateau of the square wave, the current begins to build up in the inductor. Then, when the pulse drops at the trailing edge of the square wave, an induced voltage is created across the coil to help keep the current flowing. The result is that current flow increases and decreases exponentially. Measured across the resistor, the current flow is seen as a voltage which increases and decreases exponentially. This voltage is shown in Figure 11-27.

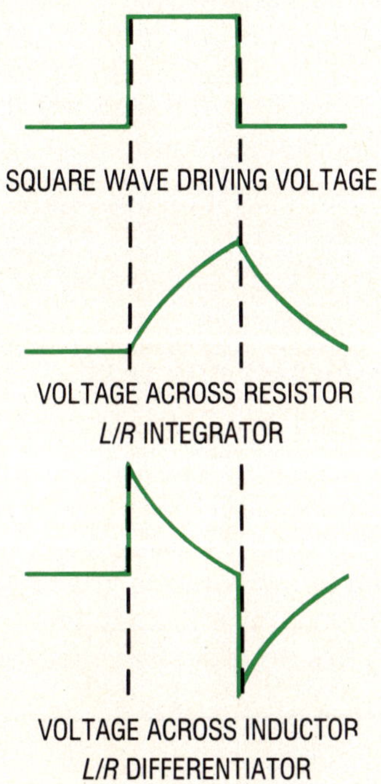

Figure 11-27. Waveforms of voltages for circuits of Figure 11-26.

The *L/R* Differentiator

Figure 11-26B shows the same circuit as Figure 11-26A, but the *R* and the *L* have been interchanged to illustrate the concept of the *L/R* differentiator. In an *L/R* circuit of this type, the inductor is the differentiator. An inductor, of course, prevents sudden changes in current direction. Because of this characteristic, the inductor acts as an open circuit for the leading or rising edge of the square wave. Thus, the voltage across the inductor is at a maximum at the time of the leading edge. The voltage then drops exponentially as the current increases exponentially. This is to be expected because the current through an inductor lags the voltage across it by

90 degrees. That is, the initial voltage across the inductor is maximum. It is not until 90 degrees later that the current through the inductor is maximum. Similarly, the trailing edge of the square wave causes a negative voltage pulse across the inductor. The voltage and current relationships for the *L*/*R* differentiator and the *L*/*R* integrator are shown in Figure 11-27.

Discussion Questions/Problems

1. Describe the differences between *RC* and *L*/*R* circuits.
2. What component functions as an *L*/*R* integrator?
3. What component functions as an *L*/*R* differentiator?

11.8 SERIES RESONANT CIRCUITS

As established in earlier chapters, the reactances of capacitors and inductors vary with frequency. For example, consider the circuit in Figure 11-28. This circuit has an inductor and a capacitor connected in series with an AC source. When the reactances of the two components, the capacitor and the inductor, are equal in magnitude, an interesting phenomenon occurs. Since they are of opposite signs the two reactances

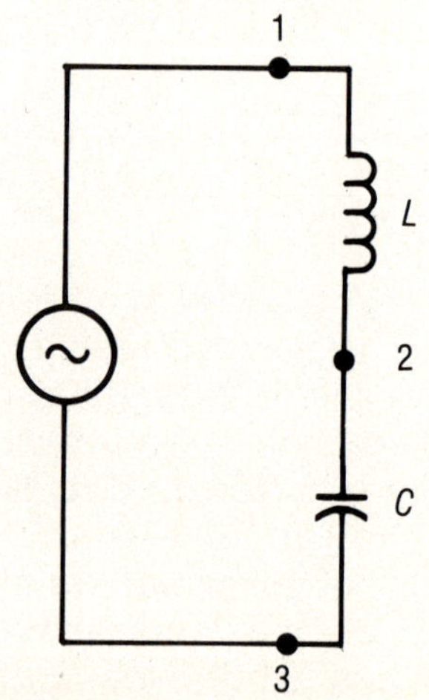

Figure 11-28. AC circuit with inductor and capacitor connected in series.

cancel out, and there is essentially zero impedance along the path 1–2–3. This condition of zero impedance in a series *LC* circuit is called *resonance*.

In most practical applications, there is some resistance as well as the inductance and capacitance. A typical *RLC* series resonanace circuit is diagrammed in Figure 11-29. The resistive, inductive, and capacitive values of the components are also shown. A circuit of this type will be in series resonance when X_L is equal to X_C. The components for this circuit may be selected so that series resonance will occur at a given desired frequency.

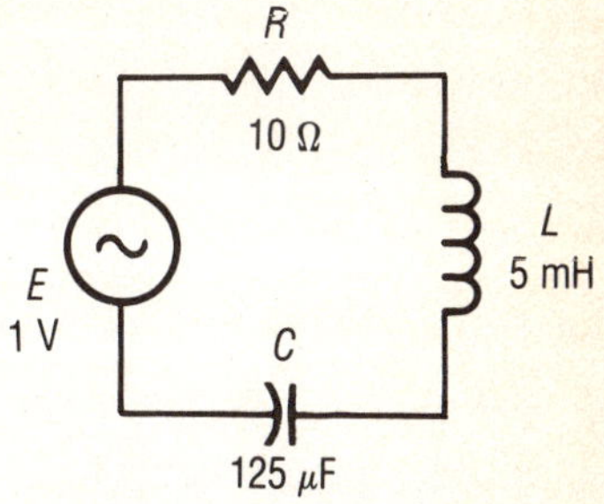

Figure 11-29. Schematic for *RLC* series resonant circuit.

The formula for calculating the resonant frequency of a series circuit is as follows:

$$f_r = \frac{1}{2\pi\sqrt{LC}}$$

In this formula, *L* represents inductance, measured in henrys, and *C* is the capacitance measured in farads. f_r is the series resonant frequency, measured in hertz. The calculation to determine the resonant frequency of the circuit in Figure 11-29 is shown below. Note that the series resistance (*R*) is not used in this calculation.

$$\begin{aligned} f_r &= \frac{1}{2\pi\sqrt{LC}} \\ &= \frac{1}{2\pi\sqrt{(5 \text{ mH})(125 \ \mu\text{f})}} \\ &= \frac{1}{2\pi\sqrt{625 \times 10^{-6}}} \\ &= \frac{10^3}{2\pi \times 25} \\ &= \frac{1{,}000}{50\pi} \\ f_r &= 6.37 \text{ Hz} \end{aligned}$$

A series *RLC* circuit has a characteristic set of phase relationships: Inductive voltage leads the current by 90 degrees. Capacitive voltage lags the current by 90 degrees. Therefore,

X_L and X_C are out of phase with each other by 180 degrees. Current through the resistance would be in phase, at a zero-degree angle. Thus, when X_L is equal to X_C, the circuit would be purely resistive.

At frequencies different from the resonance frequency, the circuit assumes the property of the component with the larger reactive value. For example, when capacitive reactance is greater than inductive reactance, the frequency at which the circuit is driven below is the resonant frequency, and the circuit will be capacitive. If frequencies are above the resonant frequency, the inductive reactance is greater than the capacitive reactance, and the circuit will be inductive. A curve demonstrating these circuit properties is shown in Figure 11-30.

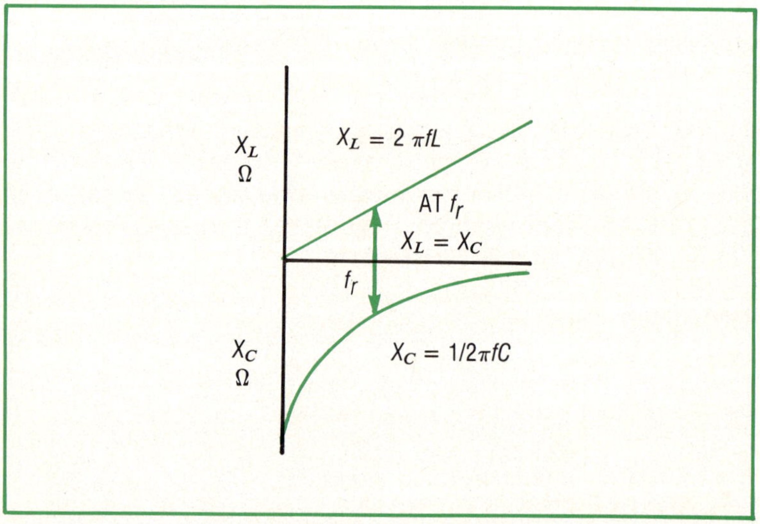

Figure 11-30. Diagram of properties of *RLC* series resonant circuit.

The current through a series *RLC* circuit is greatest when it is in resonance. The current at resonance in a series *RLC* circuit, like the one in Figure 11-31, can be determined through use of the Ohm's law formula. This is because the X_L and X_C cancel each other out, leaving only the *R* effective in the circuit.

$$I = \frac{E_T}{R}$$

Note that this formula is applicable only for a circuit that is in a condition of resonance. For the particular circuit values

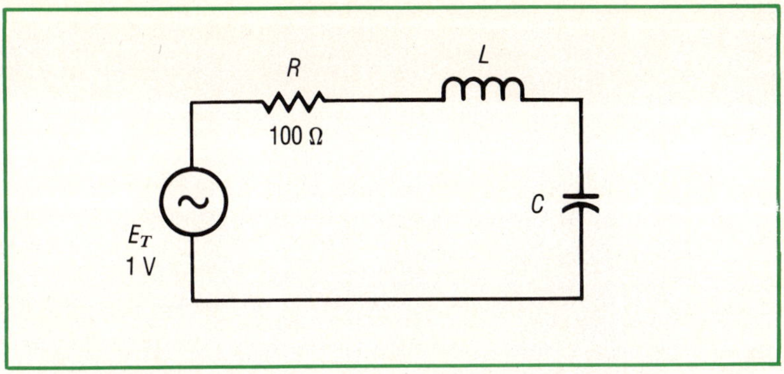

Figure 11-31. Series *RLC* circuit to use to determine current at resonance.

illustrated in Figure 11-31, the current at resonance would be determined as follows:

$$I = \frac{1\ \text{V}}{100\ \Omega} = 0.01\ \text{A}$$

Note again that the current of 0.01 A would exist only at the resonant frequency.

Given the conditions described above, the current in this circuit will be in phase with the applied voltage. The relationships between current and voltage in an *RLC* series circuit depend upon the applied frequency, as follows:

- When current in the circuit is at the resonant frequency of the circuit, the current will be at maximum.
- When the frequency applied to the circuit is above or below the resonant frequency, the current decreases because impedance increases.

These principles are useful in electronic applications because they provide a way to relate frequency to the level of voltage across circuit components. Assume a voltage is applied across a series *LC* circuit which is at the resonant frequency. The current through the circuit will be at a maximum. Also, the voltage across both the inductor and the capacitor will be at maximum. That is, the voltage across each of the components is equal and is at maximum. Also, $X_L = X_C$. But the voltages across the two components will be 180 degrees out of phase with each other. The voltages will cancel. Individually, however, the voltage across either the inductor or the capacitor can be much greater than the source voltage.

Discussion Questions/Problems

1. What conditions must exist to produce resonance in a series *RLC* circuit?
2. What is the phase relationship between inductance and reactance in a series *RLC* circuit?
3. Under what conditions does inductive reactance exceed capacitive reactance in a series *RLC* circuit?
4. Under what conditions does capacitive reactance exceed inductive reactance in a series *RLC* circuit?
5. What impedance condition indicates resonance in a series *RLC* circuit?
6. What is the phase relationship between an inductor and a capacitor in a series *RLC* circuit?

11.9 PARALLEL RESONANT CIRCUITS

Circuits where capacitors, inductors, and resistors are connected in parallel in virtually any combination, have many of the same properties as series resonant circuits. For example, a condition of resonance exists when $X_L = X_C$.

However, when a parallel *LC* circuit is driven at its resonant frequency, the currents are equal and are 180 degrees out of phase with each other, rather than the voltages (as was the case for the series resonant circuit). The result is minimal current flow through the circuit, and the parallel impedance is at its maximum level. Thus, the current circulates back and forth between the capacitor and the inductor. Virtually no current is drawn from the power source when the capacitance and inductance are balanced exactly and there is a low level of electrical friction. In practice, none of the circuit elements —capacitors, inductors, resistors, or wire—is exact or ideally matched. Therefore, there is usually some energy loss.

Figure 11-32 shows a parallel *RLC* circuit. In Figure 11-33 the series resistance (R_S) of the coil has been replaced by an equivalent parallel resistance (R_P) in which $R_P = 1/R_S \times L/C$. This configuration is also known as a *tank* circuit. The name is probably derived from the fact that the circuit acts as a reservoir for electrical energy. When reactances of the induc-

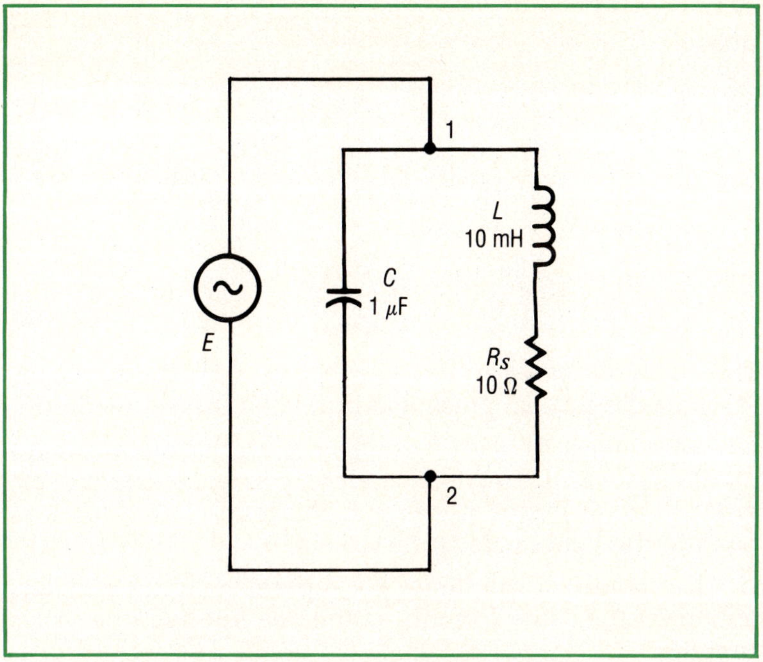

Figure 11-32. Parallel *RLC* circuit.

tor and the capacitor are equal, resonance occurs. As shown in Chapter 8, the magnitude of the impedance (Z) of a reactance (X) in parallel with a resistance (R) is:

$$|Z| = \frac{RX}{\sqrt{X^2 + R^2}}$$

When the reactance is capacitive:

$$|Z| = \frac{R}{\sqrt{1 + \frac{R^2}{(X_C)^2}}}$$

And when the reactance is inductive:

$$|Z| = \frac{R}{\sqrt{1 + \frac{R^2}{(X_L)^2}}}$$

When L and C are in parallel, since they are of opposite phase:

$$|Z| = \frac{R}{\sqrt{1 + R^2(X_C - X_L)^2}}$$

At resonance $X_L = X_C$ and:

$$|Z| = R$$

Thus, a parallel resonant circuit, when tuned to the frequency at which it is driven, presents a purely resistive impedance.

Note the word *tuned*. Tuning and filtering are major applications for tank circuits. Therefore, you will encounter this type of circuit frequently as you continue your study and work in the field of electronics.

The resonant frequency for a parallel circuit is determined by the same formula as the one applied in a series circuit:

$$f_r = \frac{1}{2\pi\sqrt{LC}}$$

Notice that Figure 11-33 showed the R, L, and C all in parallel. Remember that the resistance R_P is the parallel equivalent of the small resistance R_S in the inductor. Since the parallel

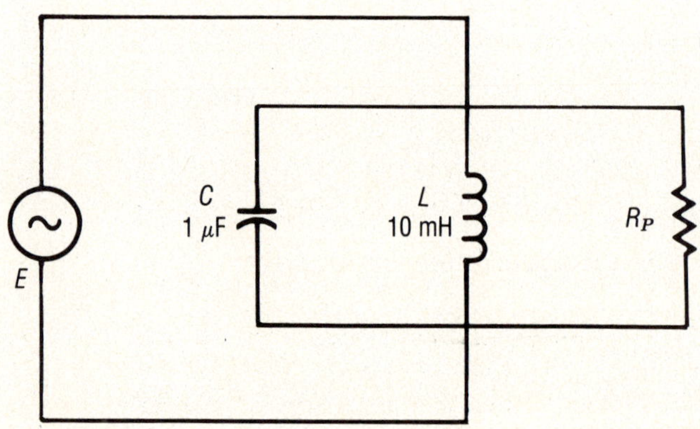

Figure 11-33. Tank circuit.

The peak of the curve is at the frequency where the *LC* circuit is in resonance.

In electronic applications it is important to select a particular portion of the frequency range—a specific bandwidth. The term *bandwidth* has a specific meaning. It is the range of frequencies for which the current is equal to or greater than 70.7 percent of its resonant value. The lower edge of this range of frequencies is shown as f_l in Figure 11-34, and the upper edge as f_u. In effect, all frequencies below f_l and above f_u are considered outside the bandwidth.

To determine bandwidth for any given resonant circuit, simply take the difference of the cutoff frequencies, thus:

$$BW = f_u - f_l$$

To illustrate, suppose a circuit has one cutoff frequency at 20 MHz and another at 10 MHz. Then:

$$BW = 20\text{ MHz} - 10\text{ MHz}$$
$$BW = 10\text{ MHz}$$

To determine current cutoff points for a given bandwidth, simply multiply the current at the resonant frequency by 0.707. The formula is:

$$I_{f_1} = I_{f_2}$$
$$I_{f_1} = 0.707\ I_{MAX}$$

Applying this formula to a circuit that has a current of 1 A at the resonant frequency, the currents at the upper (and lower) edges of the bandwidth are:

$$0.707 \times 1\text{ A} = 0.707\text{ A}$$

Since the impedance of a parallel *RLC* circuit is greatest at its resonant frequency, then the impedance at the upper and lower edges of the bandwidth is given by:

$$Z_T = 0.707\ Z_{MAX}$$

The resonant frequency for a parallel resonant circuit lies between the upper and lower cutoff frequencies, f_u and f_l. When these two bandwidth limits (f_u and f_l) are known, the resonant frequency can be determined through use of the following formulas:

$$f_r = f_l + \frac{BW}{2} \qquad\qquad f_r = f_u - \frac{BW}{2}$$

The resonant frequency is (approximately) halfway between f_u and f_l. For example, in a parallel resonant circuit with band edges of 20 MHz and 10 MHz, the resonant frequency is:

$$\begin{aligned} f_r &= f_l + \frac{BW}{2} \\ &= 10\text{ MHz} + \frac{10\text{ MHz}}{2} \\ &= 10\text{ MHz} + 5\text{ MHz} \\ f_r &= 15\text{ MHz} \end{aligned}$$

In terms of Q, the resonant frequency is:

$$f_o = Q(f_u - f_l)$$

This equation is true for values of Q greater than about 50. (Notice that Q for the example above is about 1.5.) In this equation f_o represents the center of frequency.

The parallel resonant circuit can *select* certain frequencies and reject others. For a resonant circuit, the term *selectivity* refers to the ability to accept a specific, desired signal of a particular frequency and reject other signals of a nearby but different frequency. The principle of selectivity is illustrated in Figure 11-35. As a general rule, the narrower the bandwidth

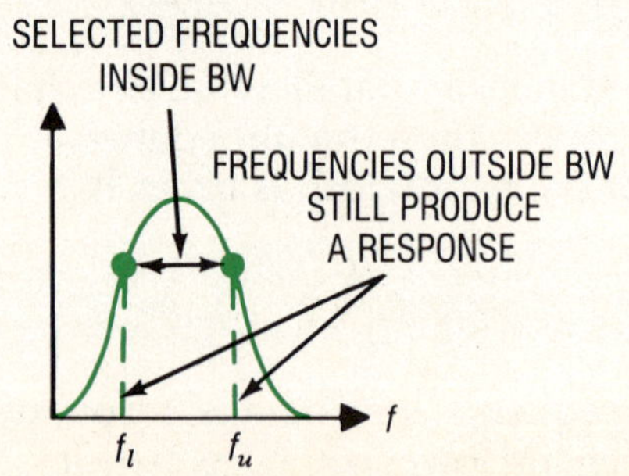

Figure 11-35. Bandwidth for parallel resonant circuits.

of a given circuit, the greater the ability to select among frequencies which are close together.

Discussion Questions/Problems

1. What is meant by the term *bandwidth*?
2. What are cutoff frequencies and how are they determined?
3. If a given circuit has cutoff frequencies of 50 kHz and 25 kHz, what is the bandwidth?
4. Using the same values as in the previous question, what is the resonant frequency for the circuit?
5. What is the meaning of the term *selectivity* as applied to bandwidth?

11.12 THE QUALITY FACTOR IN RESONANT CIRCUITS

Resonant circuits can be rated according to a uniform quality factor (Q). The Q of a resonant circuit is the ratio of its reactive value to its resistive value. Mathematically, this relationship for a parallel resonant circuit, in which r is the *series* resistance of the inductor, is expressed as follows:

$$Q = \frac{X_L}{r}$$

This same formula applies for a series resonant circuit, where r is again the series resistance of the inductor. To illustrate, assume that a given circuit has an X_L of 20 Ω and an r of 5 Ω. The Q would be determined as follows:

$$Q = \frac{20\ \Omega}{5\ \Omega} = 4\ \Omega$$

Since $X_L = X_C$ at resonance, Q may also be determined from the formula:

$$Q = \frac{X_C}{r} = \frac{1}{2\pi fCR}$$

The value of Q for any circuit relates to its selectivity and bandwidth. A high Q factor usually indicates a sharp narrow bandwidth and, for a parallel resonant circuit, a relatively high impedance. These factors make for better selectivity. Selectivity for a parallel resonant circuit varies with Q, as illustrated in Figure 11-36A. Figure 11-36B shows the curves for a series resonant circuit. A high Q means a sharp narrow bandwidth and a relatively low impedance.

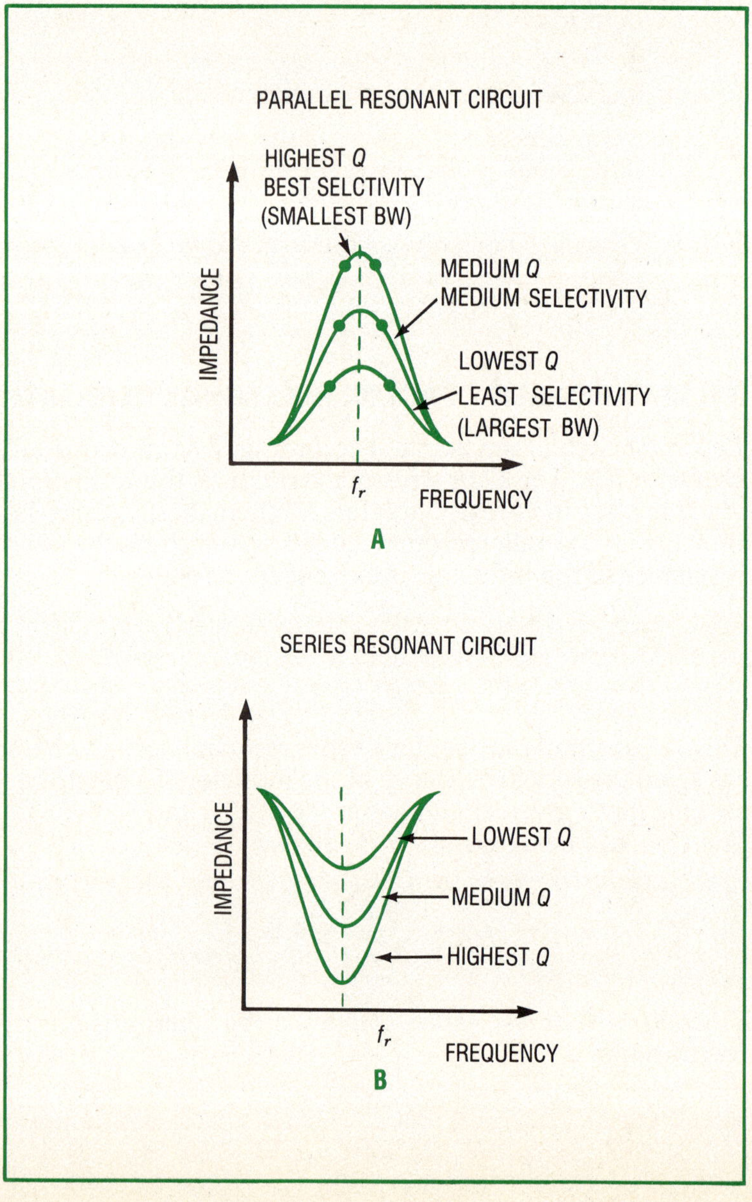

Figure 11-36. Values of Q in (A) parallel resonant circuits and (B) series resonant circuits.

Mathematically, bandwidth for a given resonant circuit can be shown in terms of the quality factor. The formula for this bandwidth calculation is as follows:

$$BW = \frac{f_r}{Q}$$

Assuming that a circuit has a Q of 4 and a center frequency of 10 kHz, bandwidth would be computed as follows:

$$BW = \frac{10 \text{ kHz}}{4} = 2{,}500 \text{ Hz}$$

Discussion Questions/Problems

1. What is the quality factor in a resonant circuit?
2. If a resonant circuit has an X_L of 50 Ω and an R of 10 Ω, what is the Q?
3. What is the bandwidth of a resonant circuit with a center frequency of 50 kHz and a Q of 25?

11.13 FILTERS

In electricity and electronics, the term *filter* describes a specific type of device. A filter is a device which incorporates resistors and either inductors or capacitors or both inductors and capacitors. These, in combination, offer comparatively little opposition to certain frequencies (which may include DC). At the same time, a filter attenuates other frequencies. This means that a filter allows certain frequencies to pass and blocks others. The designer can select the frequencies to pass or attenuate by selecting the *RLC* values and the particular circuit.

There are many types of electrical and electronic filters. Some of the most useful include:

- Low-pass filters
- High-pass filters
- Band-pass filters
- Band-stop filters.

Low-Pass Filters

A low-pass filter, as its name implies, permits the flow of current below a given frequency. Frequencies above the specific frequency are blocked. Low-pass filters are mentioned in earlier chapters in connection with capacitors. Figure 11-37 shows a circuit for a simple low-pass filter. The circuit contains a capacitor and a resistor.

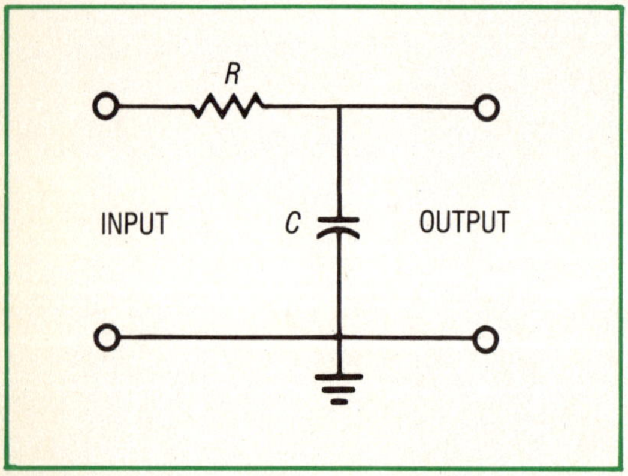

Figure 11-37. Series *RC* circuit for low-pass filter.

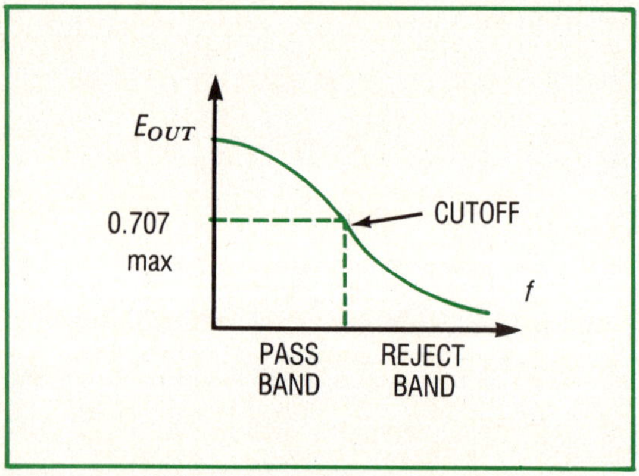

Figure 11-38. Performance curve for low-pass filter circuit.

Figure 11-38 diagrams the performance of this circuit. Note that the cutoff point is defined to be at 70.7 percent of maximum voltage, although higher frequencies will be passed to some extent. The higher frequencies are blocked, or attenuated. The range (and relative amplitude) of the signals that will be passed is enclosed in the curve.

Refer to Figure 11-37. The capacitor in this schematic is called a *bypass capacitor*. For frequencies greatly above the pass level, the capacitor acts as a short circuit, or shunt. In electronics, a *shunt* is any alternate path of lower impedance for a current or signal. In the low-pass circuit of Figure 11-37, high-frequency signals are shunted to ground. Frequencies near the cutoff frequency are only partially shunted to ground. This explains the gradual slope of the curve. Low-frequency signals are transmitted through the circuit.

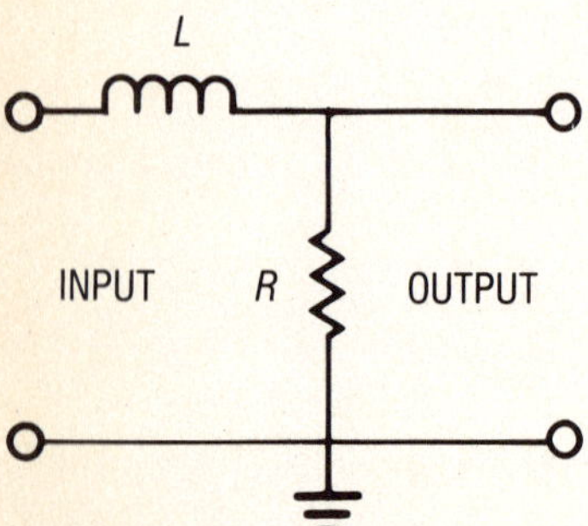

Figure 11-39. Circuit for low-pass filter with inductor.

A low-pass filter using an inductor is illustrated in Figure 11-39. In this type of circuit, the inductor (*L*) acts, with the resistor, as a voltage divider. The inductor would have a high reactance at high frequencies. However, at low frequencies, the inductor would have a low reactance. Low-frequency signals pass through this circuit. Signals above the cutoff frequency encounter increased reactance and cannot pass.

The same principles can be applied in a number of other configurations to form low-pass filters. Three such approaches are illustrated in Figure 11-40. Figure 11-40A is known as an *inverted inductor*, or *inverted L* circuit. In this configuration, an inductor and a capacitor function in series. The inductor serves to improve the filtering capability of the circuit, providing a sharper cutoff point. Figure 11-40B shows a *T-type* low-pass filter circuit. As indicated, this adds a second inductor to the basic inverted *L* circuit, enhancing the filtering effect even further by steepening the skirts of the characteristic filter curve. In the *π-type* circuit, Figure 11-40C, two capacitors are combined with an inductor—again to give steeper skirts at the upper and lower cutoff points.

Figure 11-40. Circuit configurations for low-pass filters.

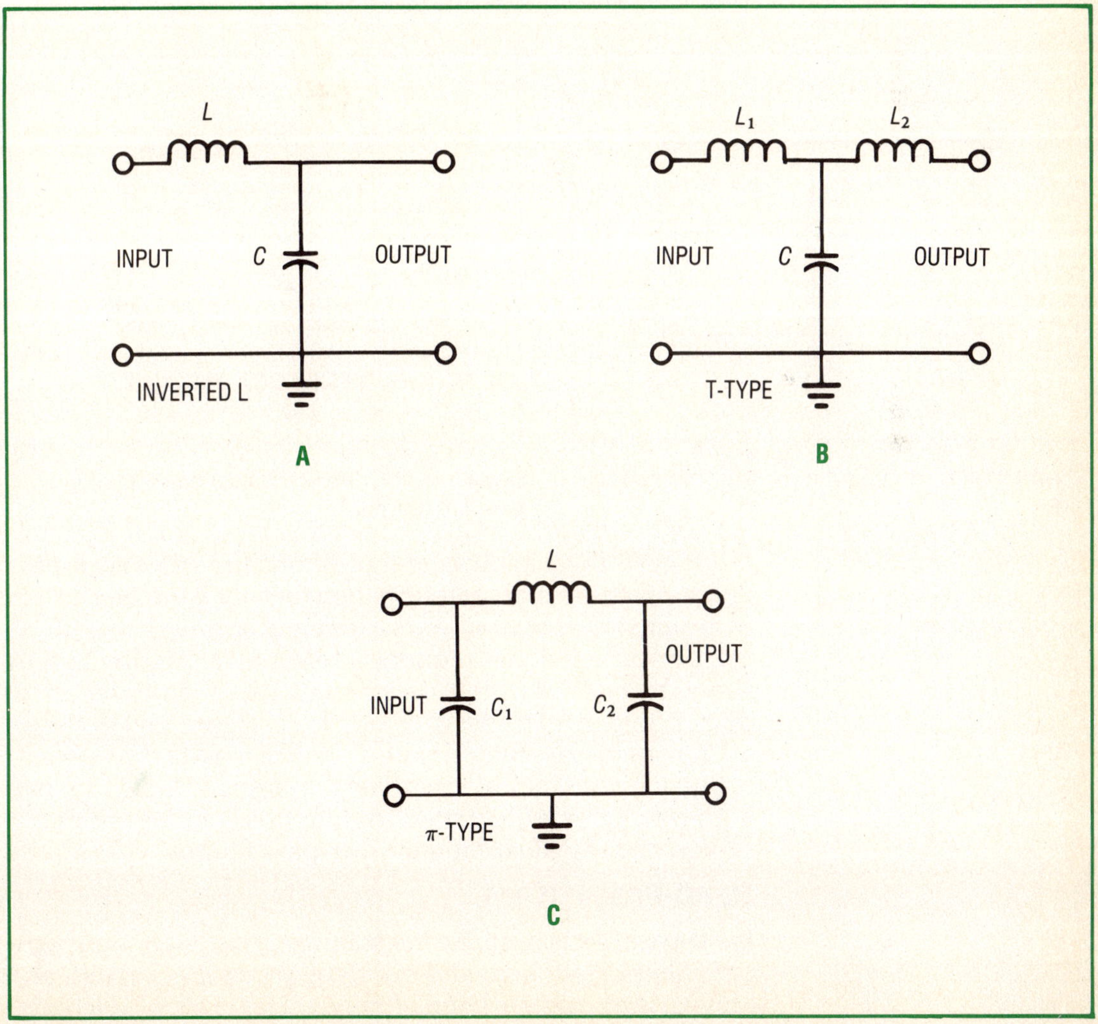

High-Pass Filters

A high-pass filter utilizes components which block, or attenuate, all signal frequencies up to a given level, then pass all frequencies above that cutoff frequency. Many types of components can be used in circuits for high-pass filters. One typical configuration, shown in Figure 11-41, uses a resistor and a capacitor connected opposite to those in the circuits of low-pass filters. This becomes apparent by comparing the schematics of Figures 11-37 and 11-41. In a high-pass filter the resistor is the shunt, or load. Signals below the cutoff frequency are attenuated by the capacitor, which will not pass signals below the cutoff point.

Figure 11-42 shows how a high-pass filter rejects low frequencies and passes high frequencies.

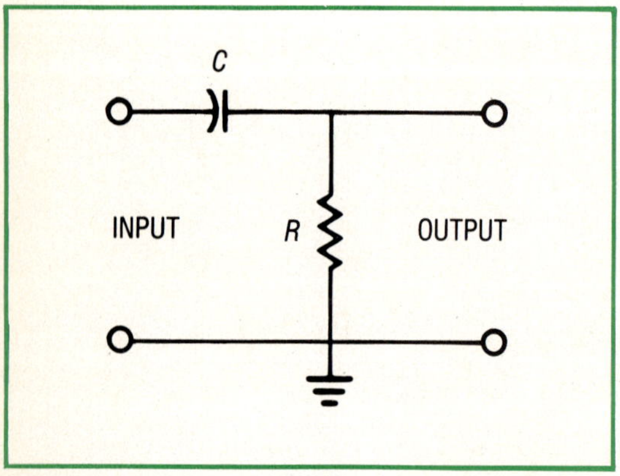

Figure 11-41. Series *RC* circuit for high-pass filter.

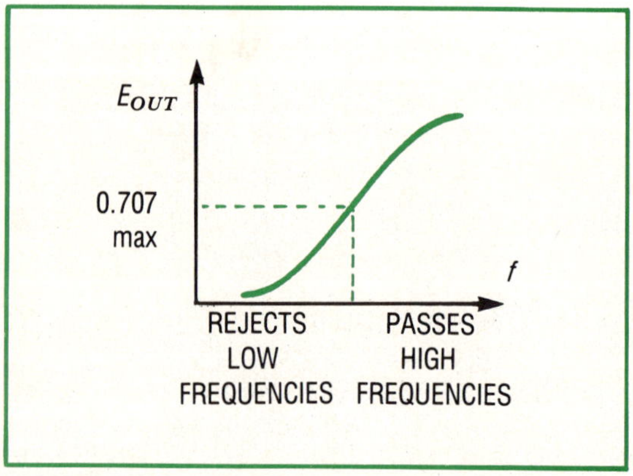

Figure 11-42. Performance curve for high-pass filter circuit.

Figure 11-43 shows configurations for other high-pass filters which are more selective than that of Figure 11-41. Each of the illustrated circuits takes a different approach to improving performance or frequency selectivity. Figure 11-43A shows an *inverted L* high-pass circuit. This circuit uses an inductor, rather than a resistor, as a load. Figure 11-43B shows a *T-type* high-pass filter circuit. This design is an "inversion" of the one used for low-pass filters. In Figure 11-43C, a *π-type* high-pass filter is illustrated. Again, this is an inversion of a design already illustrated for low-pass filters.

Band-Pass Filters

Band-pass filters combine the features of both low- and high-pass filters. Both high and low frequencies are blocked, and only an intermediate band of frequencies is allowed to pass.

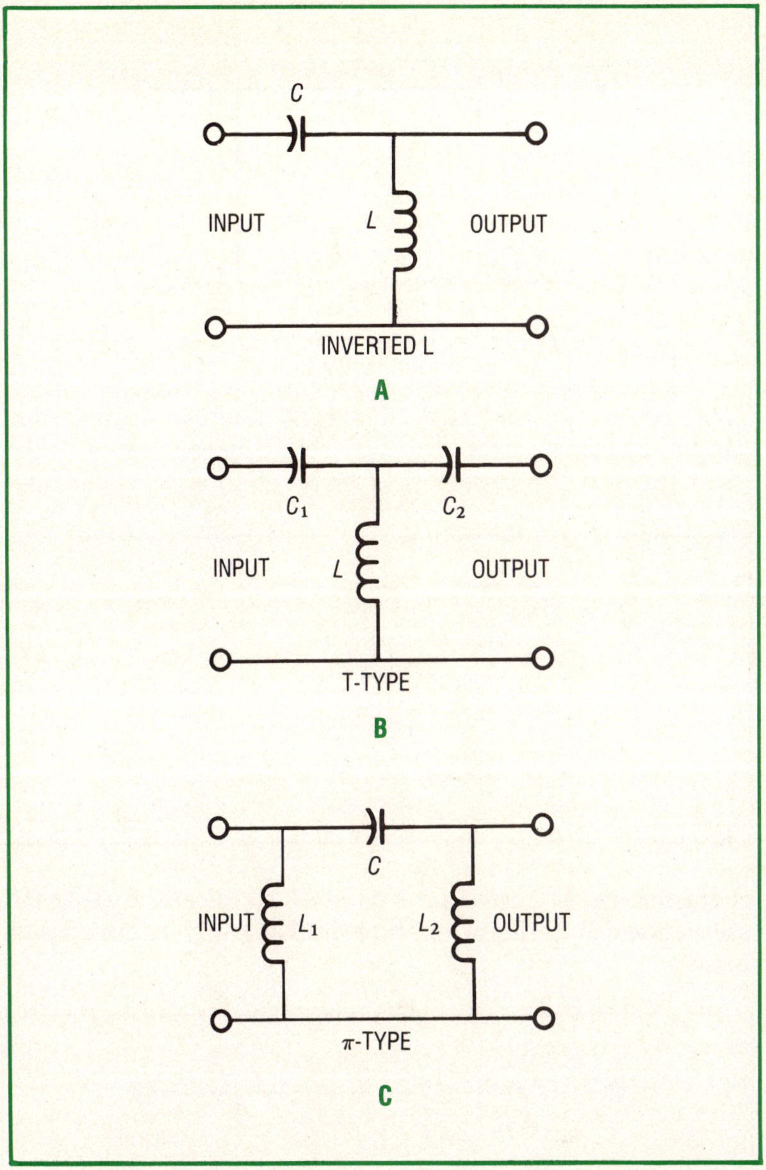

Figure 11-43. Circuit configurations for high-pass filters.

In effect, band-pass filters use resonant circuits. If series resonant circuits are used, they maximize current flow and introduce minimum impedance at their resonant frequencies. Therefore, at the resonant frequency, most of the voltage will be across the load resistor.

Because a band-pass filter depends on resonance, the *Q* factor becomes important. The higher the *Q* factor, the sharper the selectivity will be.

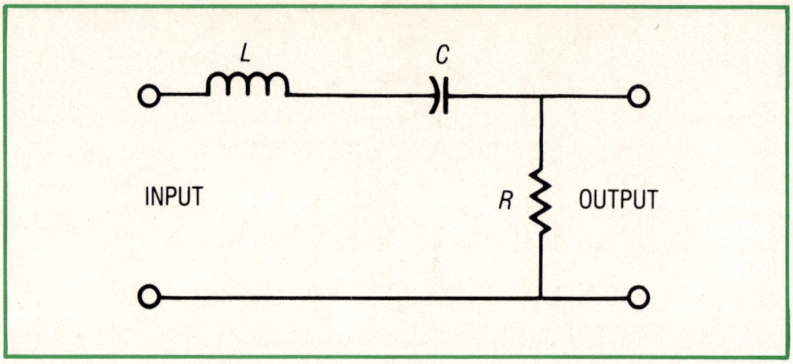

Figure 11-44. *RLC* series circuit for band-pass filter.

Figure 11-44 illustrates a series resonant band-pass filter. This shows a typical *RLC* series circuit in which the resonant frequency can be established by selecting the values of the inductor and the capacitor. Figure 11-45 shows the filter

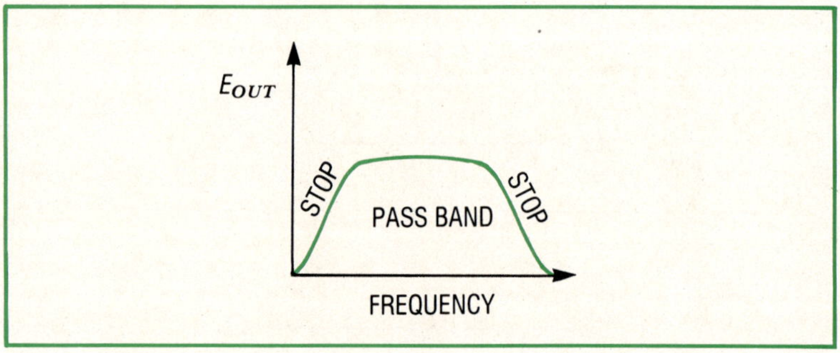

Figure 11-45. Performance curve for series resonant band-pass filter.

characteristics of a band-pass filter. This indicates that signals are stopped at frequencies above and below the desired pass band.

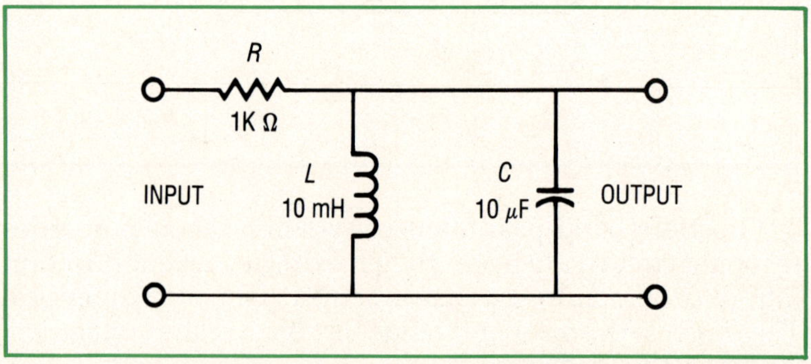

Figure 11-46. *RLC* parallel circuit for band-pass filter.

A circuit for a parallel resonant band-pass filter is shown in Figure 11-46. Again, this is a typical parallel *RLC* circuit like those you have already encountered. This type of circuit develops maximum impedance at the resonant frequency. At that frequency, the circuit will develop maximum output voltage across the load resistor. Component values are given for the drawing in Figure 11-46. When the component values are known, it becomes possible to calculate the center frequency for the pass band, which is the resonant frequency of the *LC* circuit. The formula used for calculating the resonant frequency is as follows:

$$f_r \cong \frac{1}{2\pi\sqrt{LC}}$$

The formula indicates that only approximate, rather than exact, frequencies can be determined for band-pass filters. This is because there are so many variables involved that, in a practical filter, the pass frequencies can never be predicted with complete accuracy. However, filters can be designed to almost any desired level of tolerance.

Applying this formula to the circuit in Figure 11-46, the following approximate center frequency can be calculated:

$$\begin{aligned} f_r &\cong \frac{1}{2\pi\sqrt{LC}} \\ &\cong \frac{1}{2\pi\sqrt{(10\text{ mH})(10\ \mu\text{F})}} \\ &\cong \frac{1}{2\pi \times 0.0003} \\ &\cong \frac{1}{0.0019} \\ f_r &\cong 530\text{ Hz} \end{aligned}$$

Band-Stop Filters

A band-stop filter is intended to be the opposite of a band-pass filter. A band-stop filter blocks a specific center band of frequencies, permitting lower and higher frequencies to pass. This effect can be achieved through either series resonant or parallel resonant circuits. A circuit for a series resonant band-stop filter is illustrated in Figure 11-47. This is a series circuit that inverts the relationship of *RLC* components, as compared with the circuit in Figure 11-44. This circuit has a minimum impedance at the resonant frequency. The signal

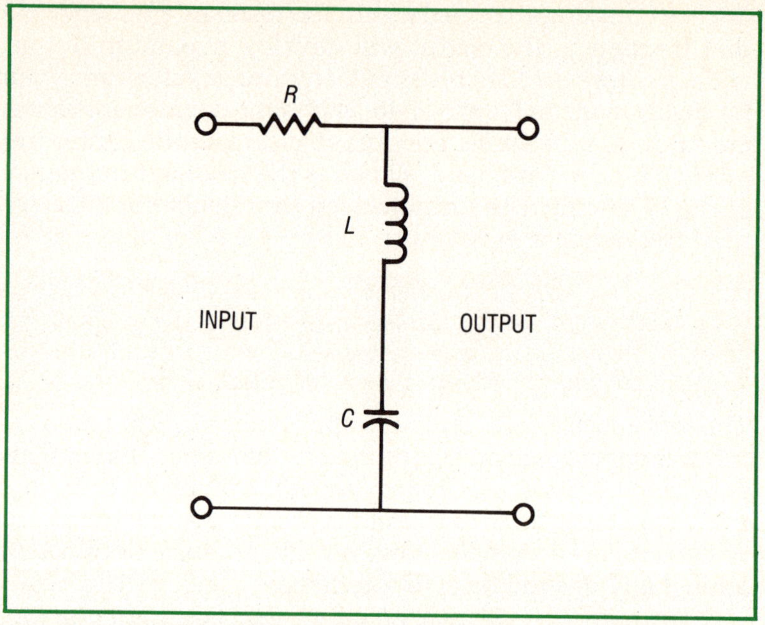

Figure 11-47. Circuit for series resonant band-stop filter.

is effectively shorted to ground. At resonance, output voltage is minimal. At frequencies above and below resonance, impedance increases, permitting the non-blocked frequencies to pass. A graphic representation of this effect is shown in Figure 11-48.

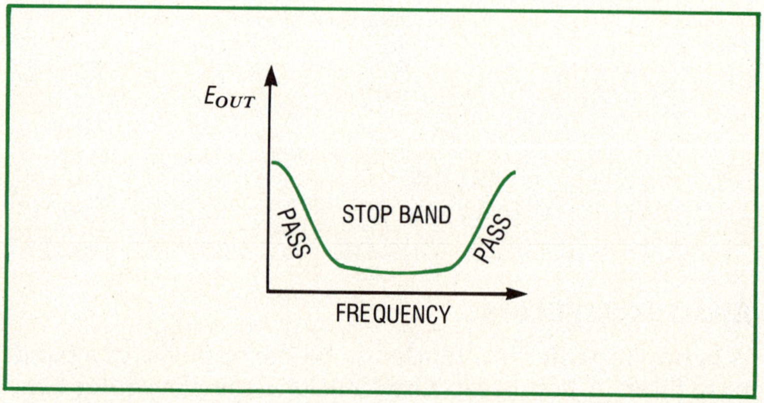

Figure 11-48. Performance curve for series band-stop filter.

Band-stop filters can also be constructed with parallel resonant, or tank, circuits. Figure 11-49 shows a tank circuit used in a band-stop filter. At resonance, the *LC* circuit has maximum impedance and prevents signals from passing to the output. At frequencies other than resonance, an output voltage will exist across the load resistor.

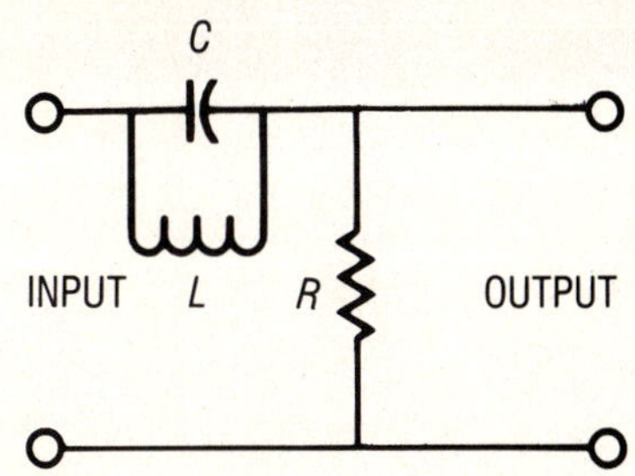

Figure 11-49. Tank circuit for band-stop filter.

Applications for Filters

Some typical applications include:

- Power-line filters, which are used in television sets and electronic instruments. These are band-stop filters that block radio frequency signals which could interfere with television receivers or other electronic devices.
- Power supplies, which convert AC line current to DC, also use band-stop filters to help assure that the DC power delivered to electronic instruments is sufficiently pure.
- Radios, which use band-pass filters to tune out unwanted signals.

Discussion Questions/Problems

1. What is the operating principle of a low-pass filter?
2. What is the operating principle of a high-pass filter?
3. What type of circuit is used for a band-pass filter?
4. How does the circuit of a band-stop filter compare, in the specific use of components, with that of a band-pass filter?

REVIEW

CHAPTERS 10 AND 11

SUMMARY

1. Inductance is an electrical property inherent in coils and is important to their use in circuits. An inductor opposes any change in current and can store electricity in the form of an electromagnetic field around the inductor. It limits the flow of AC in the circuit and allows DC to pass readily. It causes the current through the inductor to lag voltage across the inductor by a phase angle difference of 90 degrees. (10.1)

2. Capacitance is an electrical property inherent in metal plates separated by an insulator. A capacitor opposes any change in voltage and can store energy in the form of charge on the capacitor plates. It limits the flow of AC in a circuit and blocks the flow of DC. It causes the current through the capacitor to lead the voltage across the capacitor by a phase angle of 90 degrees. (10.1)

3. In an inductive circuit, current lags voltage by 90 degrees. In a capacitive circuit, current leads voltage by 90 degrees. (10.2)

4. To determine the impedance of an AC parallel circuit, use Ohm's law as it was used for the series AC circuits. (10.3)

5. Susceptance is the reciprocal of reactance. The symbol for susceptance is *B*. (10.3)

6. Admittance is the reciprocal of impedance. The symbol is *Y*. (10.3)

7. The power loss of any given circuit is a measure of the rate at which energy is dissipated through resistance in the circuit. (10.4)
8. Mathematical techniques can be used to simplify computation for circuits with many reactive parts. One technique is to use a mathematical tool called a *j-operator*. The *j*-operator is the imaginary component of complex numbers in mathematical formulas. (10.5)
9. Mathematical values of the properties of AC circuits can be shown graphically, as well as through complex numbers. For this purpose, a graphic representation known as a *complex plane* is used. (10.6)
10. The complex plane is assumed to be rectangular. The axes of the complex plane would, if all of the value lines were extended, become a graph. When the values of complex numbers are presented in this format, the format is known as *rectangular notation*. This format can also be described as a *rectangular coordinate system*. Another method for presenting the values of complex numbers on the complex plane is through the use of a format called *polar notation*. (10.8)
11. In an *RC* circuit, the time required for the charge on the capacitor to neutralize is an exact function of the capacitance and resistance. This time is called the *time constant*. (11.1)
12. A capacitor charges to 63.2 percent of its eventual steady state charge in the first time constant; to 80 percent in the second time constant; to 95 percent in the third time constant; to 98 percent in the fourth time constant; and to 99 percent in the fifth time constant. At this point, it is considered to be fully charged. (11-2)
13. An exponential form is a curve that shows exponential relationships in the properties of current or voltage. This is the most common way of explaining behavior of components in *RC* and *RL* circuits. (11.3)
14. The logarithm of a number is the exponent to which a base is raised to equal that number. (11.3)
15. The voltage across a capacitor in a series *RC* circuit at any given moment is called the *integrated voltage*. The voltage across a resistor in a series *RC* circuit is called the *differentiated voltage*. (11.5)
16. Inductors oppose changes in the current flowing through them. When a coil is included in a circuit, current flow

cannot be changed instantaneously. Instead, if a current has been flowing through an inductor and the current is interrupted, the inductor causes a time interval to elapse as the current decreases and the magnetic field collapses. This time interval is known as the *time constant* for an L/R circuit. (11.6)

17. An L/R circuit can act as an integrator, and integrated current output is measured across the resistor. When an L/R circuit is used as a differentiator, the output is measured as voltage across the inductor. (11.7)
18. The condition of zero impedance in a series inductor/capacitor circuit is called *resonance*. (11.8)
19. Resonance is a condition when $X_L = X_C$. (11.9)
20. The term *bandwidth* has a specific meaning. It is the range of frequencies for which the current is equal to, or greater than, 70.7 percent of its resonant value. (11.11)
21. Resonant circuits can be rated according to a uniform quality factor (Q). The Q of a circuit is the ratio of its reactive power to its resistive power. (11.12)
22. A filter is a device which incorporates resistors and either or both inductors and capacitors. (11.13)
23. There are many types of electrical and electronic filters. These include low-pass filters, high-pass filters, band-pass filters, and band-stop filters. (11.13)

PROBLEMS

1. What is the phase angle by which voltage lags current across an inductor? (10.1)
2. What is the phase angle by which voltage leads current across a capacitor? (10-1)
3. Given a frequency of 1,000 Hz and an inductance of 10 μH, find X_L. (10.2)
4. Given a frequency of 1,000 Hz and a capacitance of 10 μF, find X_C. (10.2)
5. Find Z if $X_L = 50\ \Omega$, $X_C = 25\ \Omega$, and $R = 12\ \Omega$. (10.2)
6. Find Z if $X_L = 32\ \Omega$, $X_C = 45\ \Omega$, and $R = 12\ \Omega$. (10.2)
7. Find I given $E = 30$ V and $Z = 2\text{K}\ \Omega$. (10.2)
8. Given a series AC circuit, find the power factor when $Z = 20\text{K}\ \Omega$ and $R = 18\text{K}\ \Omega$. (10.2)

9. Find the impedance of an AC parallel circuit given an E of 50 V and total current of 2 A. (10.3)
10. Given a reactance of 100 Ω, what is the susceptance of a coil? (10.3)
11. Find the admittance of an AC circuit given an impedance of 80 Ω. (10-3)
12. Using the circuit in Figure R5-1, find Z. (10.3)

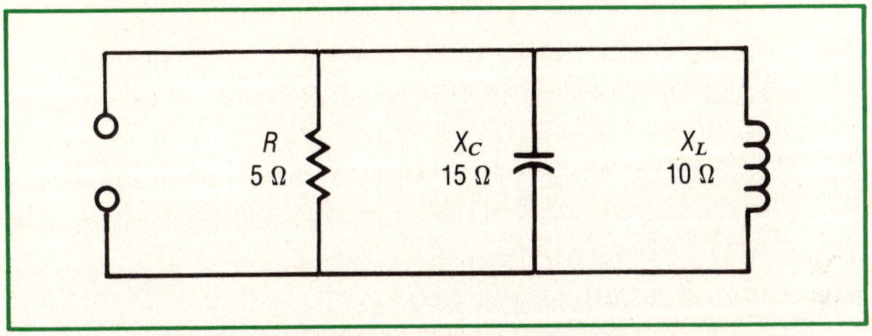

Figure R5-1. Circuit schematic for Problem 12.

13. If a circuit has a current of 0.5 A going through it and a total resistance of 50 Ω, what is the power? (10.4)
14. What is the power factor of a circuit having an R of 75 Ω and a Z of 150 Ω? (10.4)
15. Factor the following: $\sqrt{-12}$. (10.5)
16. Using Figure R5-2, what is the magnitude of the vector line 0A? (10.6)

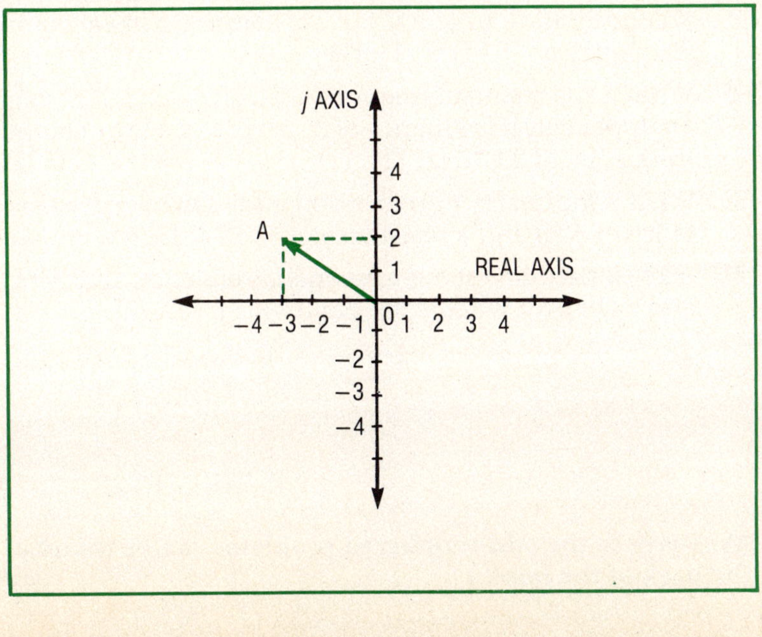

Figure R5-2. Complex plane for Problem 16.

17. Add $(6 + j2) + (5 - j3)$. (10.7)
18. $7\measuredangle 30^\circ \times 3\measuredangle 40^\circ = ?$ (10.8)
19. $5\measuredangle 18^\circ \times 2\measuredangle 24^\circ = ?$ (10.8)
20. Find the time constant of a circuit having a resistance of 10K Ω and a capacitance of 10 μF. (11.2)
21. In a circuit which has a capacitor that has just been switched on, what would be the voltage build-up in the first time constant if the circuit voltage is 50 V? (11.2)
22. What is the instantaneous voltage in a circuit having a charge time of 0.02 seconds, a resistance of 10K Ω, a capacitance of 15 μF, and an E_F of 50 V? (11.3)
23. For a circuit having an E_T of 50 V, an E_C of 41 V across a capacitor, and an R of 50K Ω, what is the instantaneous current value? (11.4)
24. Given a circuit having a coil with a value of 25 mH and a resistance of 2K Ω, what is the time constant? (11.6)
25. What is the resonant frequency of a series circuit having a capacitance of 40 μF and an inductance of 30 mH? (11.8)
26. An *RLC* circuit is at resonance with $X_L = X_C$, and an R of 10 Ω. What is the value of Z? (11.9)
27. What is the bandwidth of a given resonant circuit with an upper cutoff frequency of 50 MHz and a lower cutoff frequency of 40 MHz? (11.11)
28. What is the bandwidth of a given resonant circuit with an upper cutoff frequency of 125 kHz and a lower cutoff frequency of 75 kHz? (11.11)
29. What is the resonant frequency of a given circuit having an upper cutoff frequency of 25 MHz and a lower cutoff frequency of 23 MHz? (11.11)
30. What is the quality (Q) of a resonant circuit having a reactance of 30 Ω and a resistance of 15 Ω? (11.12)
31. If a resonant circuit has a resonant frequency of 20 MHz and a Q of 26, what is the bandwidth? (11.12)

(Answers to the odd-numbered problems can be found at the back of the book.)

12 TESTING AND MEASURING INSTRUMENTS

12.1 TYPES OF INSTRUMENTS

Circuits and their components are tested to determine if they are functioning properly and are meeting their design specifications. It is necessary to identify the types and locations of electrical or electronic malfunctions or problems that may exist or occur within a circuit or component. Testing instruments are used for these checks.

Testing instruments are often incorporated as parts of the circuits or components to be tested. In such cases, the testing instruments are fabricated from various components arranged in specific circuits. Each test instrument, then, must be used according to a specific set of procedures. The testing process consists of the proper combination of procedures and test equipment.

Values for current, voltage, resistance, frequency, and other performance parameters, or factors, are measured to determine how well circuits and components are functioning. Measurement instruments or equipment are used for this purpose. These instruments are often combined with testing devices and may also be part of the circuits, components, or devices being tested and measured. In using any measurement equipment, there are procedures which have to be followed.

There are many types of testing and measuring instruments and equipment available for use with electrical circuits and electronic devices. Some of the most important categories include:

- Meters
- Oscilloscopes
- Signal generators
- Logic probes and similar instruments
- Power supplies.

Discussion Questions/Problems

1. What is circuit testing?
2. What is circuit measurement?
3. What are the basic differences between testing and measurement?

12.2 METERS

Meters are devices that measure the values of the basic properties or characteristics of electricity. For example, there are meters that measure voltage. Others measure current, while still others determine resistance or wattage. The factors, or parameters, measured by meters correspond to the basic nature or properties of electricity. The various types of meters and the properties they measure include:

- Voltmeters measure voltage.
- Ammeters measure current.
- Ohmmeters measure resistance.
- Watt meters measure wattage.

Most electrical measurement tools used in the field or in laboratory situations are general purpose in nature. That is, they measure several characteristics of electricity. Thus, they are called *multimeters*. One typical multimeter is illustrated in Figure 12-1. Note that there are multiple scales, or sets of lines, on the face of the meter movement. These are used to read, or indicate, the different parameters of electricity being measured. This particular meter is a single device that can be used to measure DC current, or DC voltage, or AC voltage, or resistance. Different multimeters are designed for use in

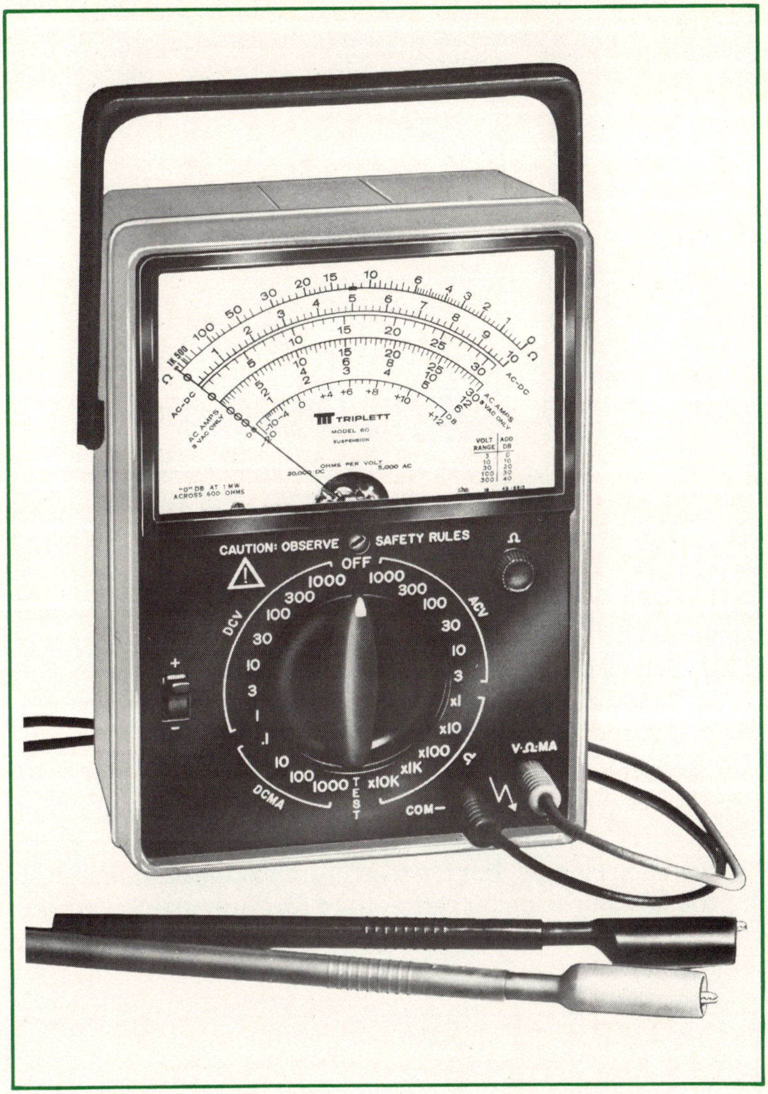

Figure 12-1. Voltohm-meter. (Courtesy Triplett)

different types of measurements. The types of measurements are usually indicated on the control panel, and the particular type of measurement is selected by a function-control switch. The user selects the parameter to be measured, and then uses the corresponding scale on the meter to read values. The device shown in Figure 12-1 is known as a *VOM analog meter*. VOM stands for Volt, Ohm, and Milliampere. The initials identify the parameters it can measure. The designation as an analog meter means that electrical properties or values are continuously displayed through use of electrical components such as coils and electromagnets.

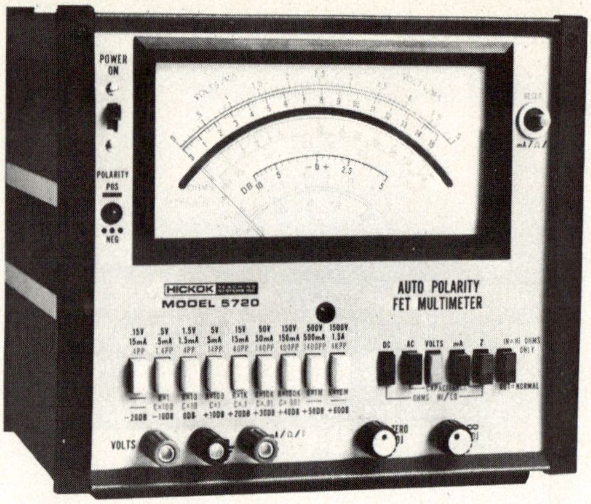

Figure 12-2. Solid-state FET VOM. (Courtesy Hickok Teaching Systems Inc.)

Figure 12-2 shows a different type of VOM. This is known as a *solid-state FET VOM*. The meter in Figure 12-2 measures essentially the same parameters as the one in Figure 12-1. The chief difference lies in the design of the meter circuits. The FET designation stands for *Field Effect Transistor*. This is an electronic device that makes it possible to measure certain parameters at high frequencies which cannot readily be measured with a simple VOM.

Another type of multimeter in common use is the *digital multimeter*, or *DVOM*. A unit of this type is shown in Figure 12-3. This type of meter also measures the flow of electricity with analog circuits, but the results are displayed numerically on the face of the meter rather than through use of a pointer and a series of scales. Some digital multimeters measure with digital circuits.

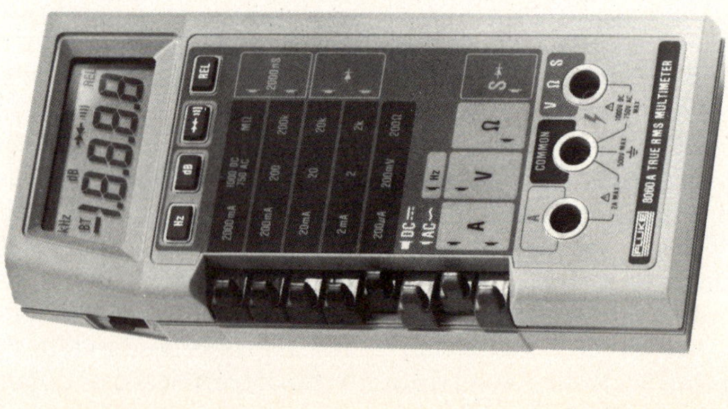

Figure 12-3. Digital multimeter. (Courtesy John Fluke Manufacturing Co.)

Instrument Movements

In an analog electrical meter, the instrument movement is the component of the meter which indicates the measured value. The movement receives an electrical current from the circuit and (usually) translates the electrical signal to a change in position for reading or display.

The d'Arsonval Movement. The most common type of instrument movement uses a moving coil. It is known as the *d'Arsonval meter* movement, named after its inventor, Jacques Arsene d'Arsonval (1851–1940), French physicist. A number of refinements were added to the basic movement by a prominent American electrical engineer, Edward Weston (1850–1936).

The basic version of this type of meter is known as the *d'Arsonval galvanometer*. A galvanometer is a coil that rotates in a magnetic field when electricity is applied. Within the movement, a small, lightweight coil of wire is supported on jeweled bearings that permit free rotation. (A jeweled bearing is a low-friction bearing that carries the weight of a small rotating device.) In a meter, as illustrated in Figure 12-4, the jeweled bearing is known as a pivot point. The pivot-mounted coil is suspended between the poles of a permanent magnet.

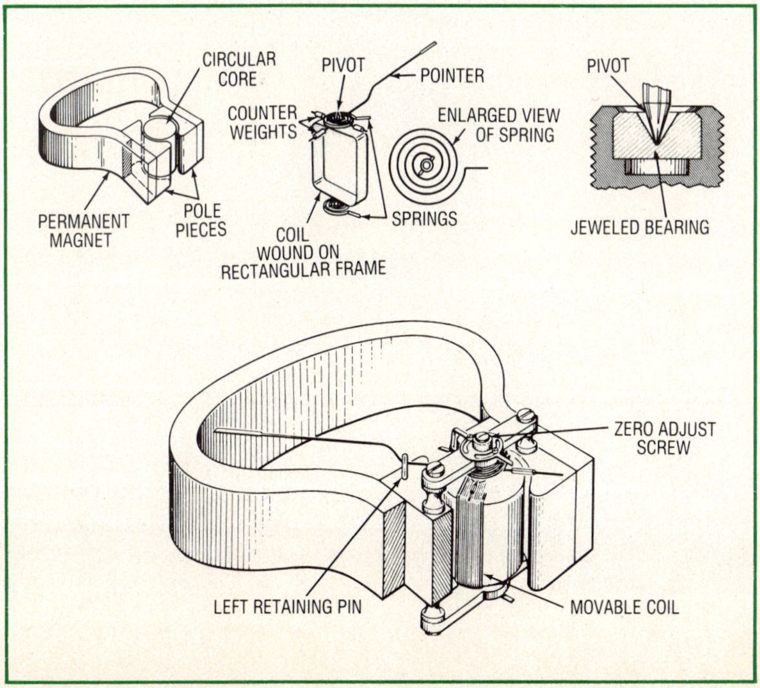

Figure 12-4. Operating principles and construction of d'Arsonval meter movement.

Chapters 6 and 8 discuss the behavior of an electric current in a magnetic field and show that there is a force on a wire carrying a current in a magnetic field. The instrument movement is designed to use that force. When a direct current (from the current to be measured) is applied to the coil, there is a magnetic force on the coil. This force causes the coil to rotate an amount which is related to the current in the coil. A pointer attached to the d'Arsonval galvanometer coil will thus also rotate, and can be used as an indicator.

Iron-vane meter movements. The operating principles of an iron-vane movement are shown in Figure 12-5. The two iron vanes identified in this illustration are permanent bar magnets. They are positioned so that the magnets attract each other and are in a parallel position when no current is applied to the surrounding electric coil. When current is applied to the coil, the polarity of the stationary vane is reversed, repelling the other magnet. The repelled magnet pivots on a spring movement, causing a pointer to move along a scale.

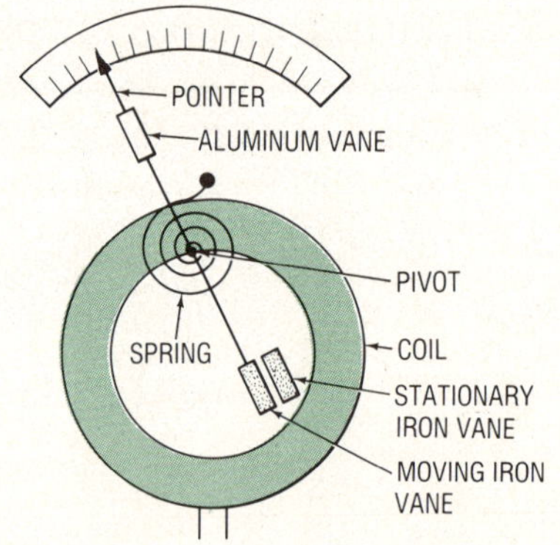

Figure 12-5. Operating principles of iron-vane meter movements.

Movement of the pointer is slowed, or dampened, by two devices. One is a sensitive, low-pressure spring (similar to the hairspring of a watch) that simply restrains the movement of the pointer. The pointer is also dampened by the "wind resistance" of an aluminum vane attached to the pointer.

An iron-vane meter responds the same whether AC or DC is applied.

Figure 12-6 shows how a typical iron-vane meter movement is constructed. This type of movement has a large inductance, so the inductive reactance increases as frequency goes up. Iron-vane meters are therefore inaccurate at higher frequencies, and their use is limited largely to power-line frequencies.

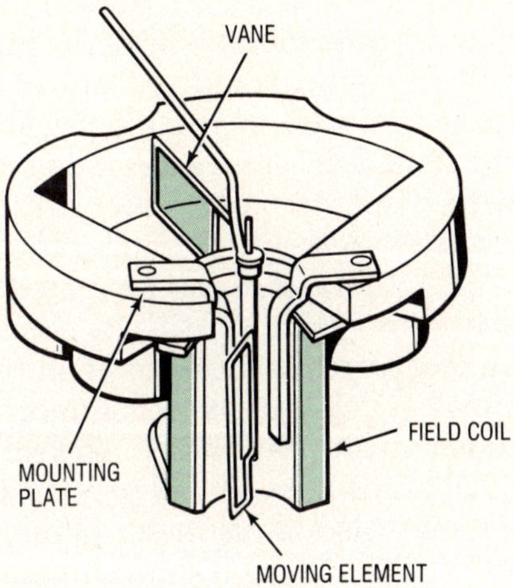

Figure 12-6. Construction of iron-vane meter movement.

The electrodynamometer. This type of movement uses operating principles similar to those of the d'Arsonval movement except that the large field magnet is an electromagnet, rather than a permanent magnet. Figure 12-7 shows a moving coil suspended between two stationary coils to form the basic movement. Figure 12-8 is a cutaway drawing showing a commercial design of an electrodynamometer movement.

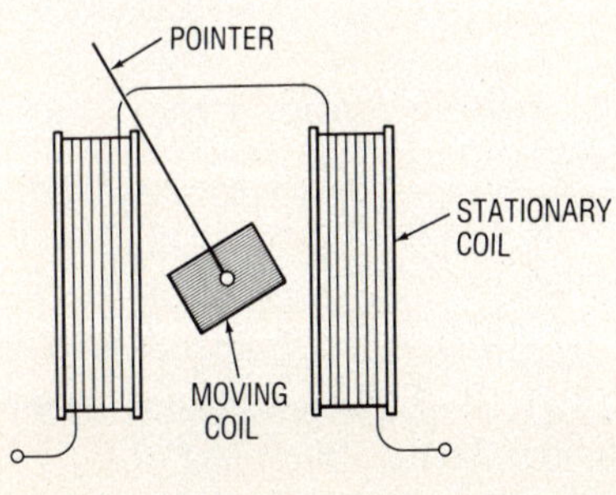

Figure 12-7. Operating principles of electrodynamometer movements.

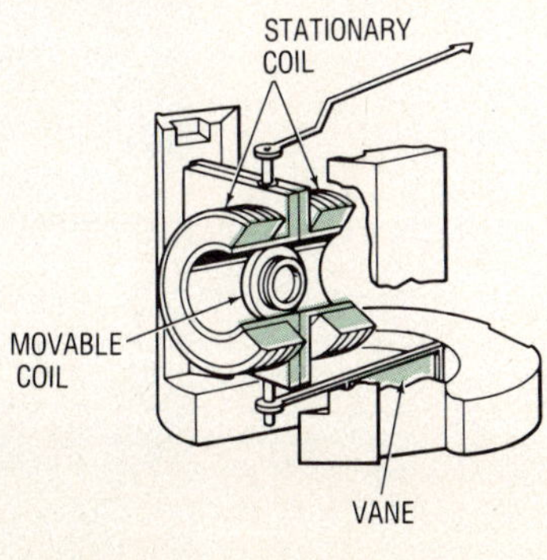

Figure 12-8. Construction of electrodynamometer movement.

Meter movements of this type can be extremely sensitive because the instrument designer can control the magnet field in the stationary coils. Meters of this type are used most commonly as watt meters.

Meter Sensitivity

In a meter, sensitivity refers to how well the meter responds to an input. Sensitivity is measured as a ratio of meter readout response to the magnitude of the electrical signal being measured. Meter sensitivity can be described in a number of ways, including divisions per volt, milliradians per microampere, or other descriptors of the ratio between output and input. Sensitivity can also be stated indirectly by stating a property from which sensitivity can be computed. One example is a measurement in terms of ohms per volt for a given meter. Ohms per volt is the most common measure used to indicate the sensitivity of a voltmeter. An instrument movement rated at 1,000 ohms per volt will present 1,000 ohms to the circuit under test. Also, a voltmeter rated at 1,000 ohms per volt will present a load of 1,000 ohms to the circuit under test when it is set to the 1 volt range. See Figure 12-9.

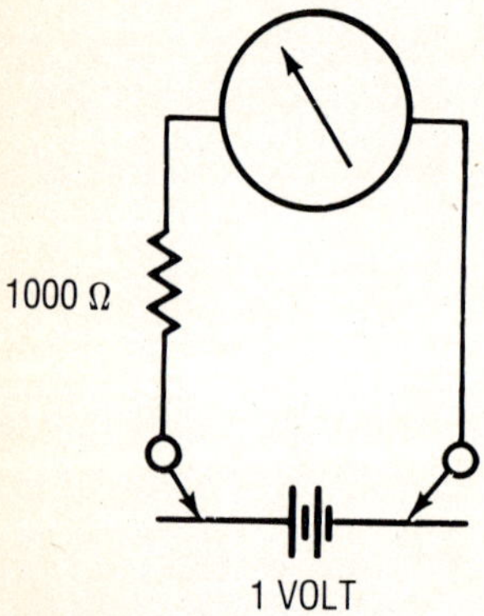

Figure 12-9. Circuit schematic of voltmeter.

Discussion Questions/Problems

1. What is the main electrical component of a d'Arsonval movement?
2. What reaction takes place to cause a reading when electricity is applied to a d'Arsonval movement?
3. What causes an iron-vane meter movement to deflect?
4. What is the function of the hairspring in an iron-vane movement?
5. What is the main difference between a d'Arsonval movement and an electrodynamometer movement?
6. Define sensitivity in meter movements.

12.3 THE AMMETER

The ammeter is an instrument used for measuring current. The way in which an ammeter is used depends upon the magnitude of the current and the type of current involved. To measure DC current, a meter must be connected into the circuit itself. The ammeter must be in series with the flow of current within the portion of the circuit being measured, as indicated in Figure 12-10. Small AC currents may also be measured by connecting the meter into the circuit itself, as shown in Figure 12-11. Note that Figure 12-11 does not have any polarity markings. Otherwise this diagram is the same as the one in Figure 12-10. This difference reflects the fact that AC alternates polarity. Note also that the meter in Figure 12-11 is shown to measure milliamperes. For AC circuits with larger currents, another type of meter connection may be preferred. Figure 12-12 shows a meter used for measuring relatively large AC currents. Instead of being connected into the circuit, meters of this type measure the current indirectly. A detector is placed around the wire, outside the insulation. The current is determined by measuring the magnetic field created by the current flowing through the wire, rather than by passing the current directly through the meter.

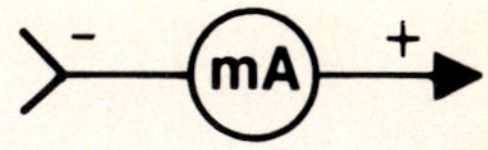

Figure 12-10. Schematic for DC milliammeter.

Figure 12-11. Schematic for AC milliammeter.

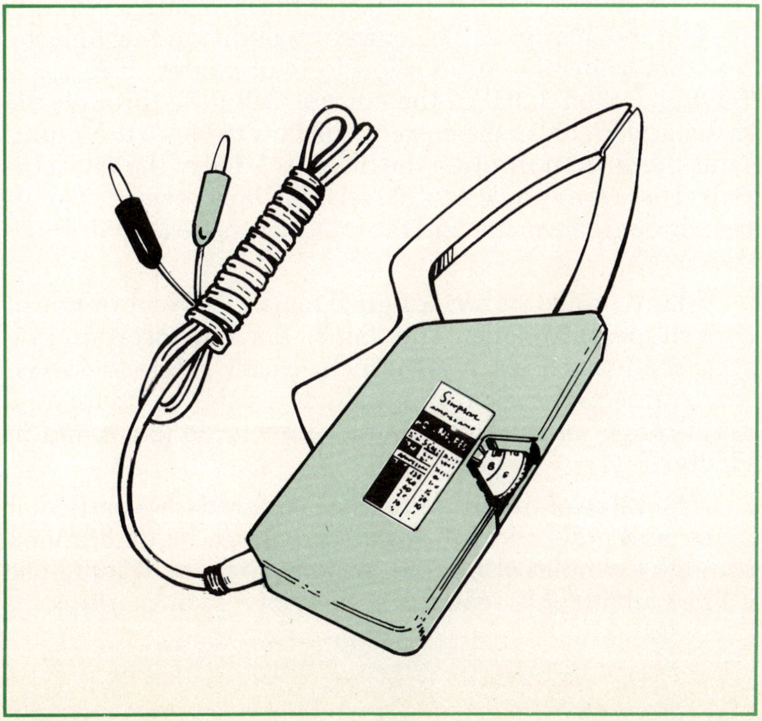

Figure 12-12. Meter used for measuring large AC currents.

Most ammeters use a d'Arsonval movement which, typically, has a sensitivity of 1 mA full scale and an internal resistance of 25 Ω. The internal resistance of a meter is abbreviated R_M.

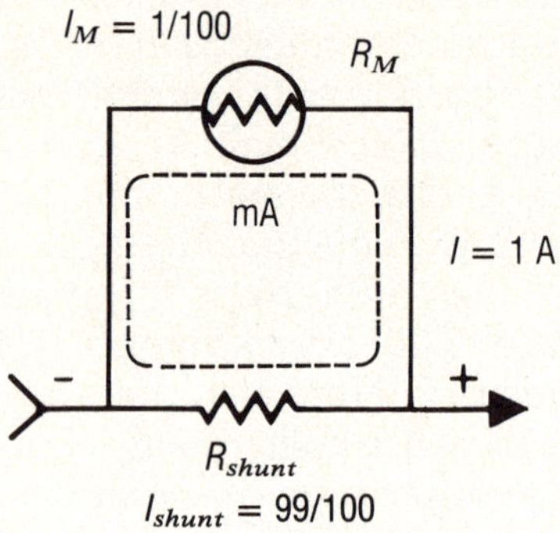

Figure 12-13. Circuit schematic of ammeter with shunt.

Very often, an ammeter with a 1 mA sensitivity is used to measure current that is well beyond this range. This can be done through use of an accessory device, known as a *meter shunt*. In a shunted meter, a resistor is connected in parallel with the instrument movement. A simple milliammeter with a shunt is diagrammed in Figure 12-13. This is a simple circuit with two resistors in parallel, similar to parallel circuits studied in Chapter 2. The current will divide through the resistors in inverse proportion to their values. If $R_{shunt} = 1/99R_{meter}$ then 1/100 of the current will flow through the meter and 99/100 of the current will flow through the shunt. Thus the current through the meter is 1/100 of the total current. There is a simple 1 to 100 relationship between the current through the meter and the total current flowing through the circuit.

Many ammeters are designed to handle a wide range of current measurements. This can be done by incorporating several shunt resistors into the meter circuit. A selector switch then connects the appropriate shunt resistor. A circuit for a multi-range, or multiple-shunt, ammeter is illustrated in Figure 12-14.

The value of the shunt resistor required for a particular meter and a desired full-scale current can be determined through a simple calculation. Assume that you want to use a 1 mA ammeter to read full scale with a 15 mA current:

$$I_S = I_T - I_M = 15 \text{ mA} - 1 \text{ mA} = 14 \text{ mA}$$

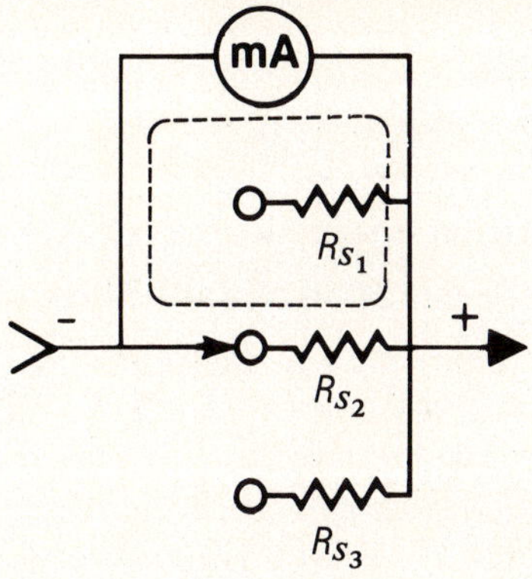

Figure 12-14. Circuit schematic for multiple-shunt ammeter.

Through this simple subtraction, you know that 14 mA of current must go through the shunt. Now, find the voltage drop across the meter (assuming the meter resistance is 25 Ω).

$$E_M = I_M \times R_M$$
$$= 1 \text{ mA} \times 25\ \Omega$$
$$E_M = 0.025 \text{ V}$$

You can now apply the basic Ohm's law formula to find the shunt resistor value:

$$R_S = \frac{0.025 \text{ V}}{14 \text{ mA}}$$
$$R_S = 1.79\ \Omega$$

Note that the value of the shunt resistor, 1.79 Ω, is much smaller than the resistance of the meter itself. The meter resistance is 25 Ω. Current will divide in inverse proportion to the two resistances. Thus, the shunt circuit becomes the primary current carrier. Also, both the internal resistance of the meter movement (25 Ω) and the resistance of the shunt (1.79 Ω) will dissipate power from the source.

Discussion Questions/Problems

1. What is the rule about connecting an ammeter into a DC circuit?
2. How does the magnitude of current affect the measuring technique for current in AC circuits?
3. What is the function of a shunt in an ammeter?
4. Assume you have a 50 μA meter. This meter has an internal resistance of 100 Ω. You want to measure a 1 A current with this meter. What resistance value would be required in a shunt for this meter? What is the power dissipated in the meter movement? In the shunt?

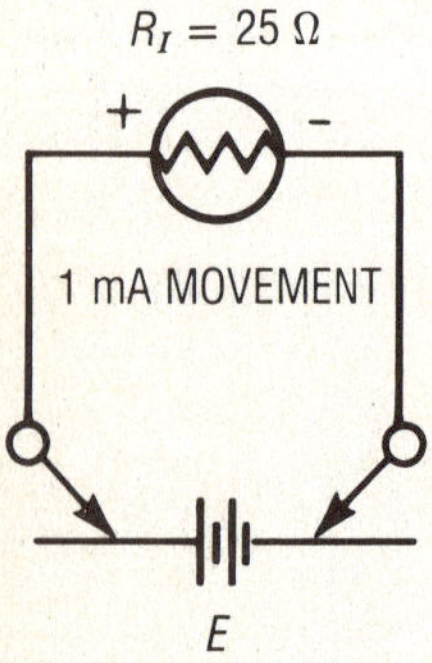

Figure 12-15. Circuit schematic for voltmeter with simple d'Arsonval movement.

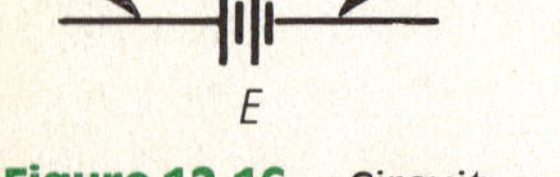

Figure 12-16. Circuit schematic for multiple-resistor voltmeter movement.

12.4 THE VOLTMETER

A voltmeter, as its name implies, measures voltage in an electrical circuit. The most commonly used type of voltmeter is also based on use of the d'Arsonval movement. However, a voltmeter in a circuit behaves differently from an ammeter. The difference is that a voltmeter can use considerably more of the power of the circuit it is measuring. This is much less of a consideration when current is being measured. Ammeter shunts draw very small amounts of power from circuits being measured. However, a voltmeter can consume power in measurable amounts. (Note that a fully deflected movement draws the same power whether it is used as an ammeter or as a voltmeter.)

Suppose a simple 1 mA d'Arsonval instrument movement is applied to a circuit, as shown in Figure 12-15. Assume that the movement has an internal resistance of 25 Ω. In this situation, the full-scale reading would be 0.025 V. In Figure 12-15, a voltmeter would be able to measure only very low voltages.

For measuring larger voltages, voltmeters require what are known as *multiplying resistors*. Figure 12-16 shows the same 1 mA instrument movement with a multiplier resistor R_{mul} connected in series. Chapter 2 shows (Ohm's law) that the voltage across a resistor is proportional to the current through the resistor. Since this is a 1 mA meter, the maximum current the movement can handle is 1 mA. The multiplier resistor must be selected so that, for the voltage to be measured, the total current through R_{meter} and R_{mul} in series will not exceed 1 mA.

To determine the resistance of the multiplier resistor, refer to Figure 12-16 and the following formula:

$$R_{mul} = \frac{E_{full\ scale}}{I_{full\ scale}} - R_M$$

In this example, assume the desired full-scale measurement is 20 V. The full-scale current is the rated value of the meter movement, 1 mA. The internal resistance of the meter (R_M) is 25 Ω. Given these values, the value of the multiplier resistor is determined as follows:

$$E_{full\ scale} = I_{full\ scale} \times R_T$$
$$20\ \text{V} = 0.001\ \text{A} \times R_T$$
$$R_T = \frac{20}{0.001\ \text{A}} = 20{,}000\ \Omega$$
$$R_T = R_M + R_{mul}$$
$$20{,}000\ \Omega = 25\ \Omega + R_{mul}$$
$$R_{mul} = 19{,}775\ \Omega$$

Voltmeters can be designed with multiple voltage measurement ranges by inserting selectable resistors into a circuit, as shown in Figure 12-17. Each of the three resistors would have a different value. Each would create a different measurement range for the meter.

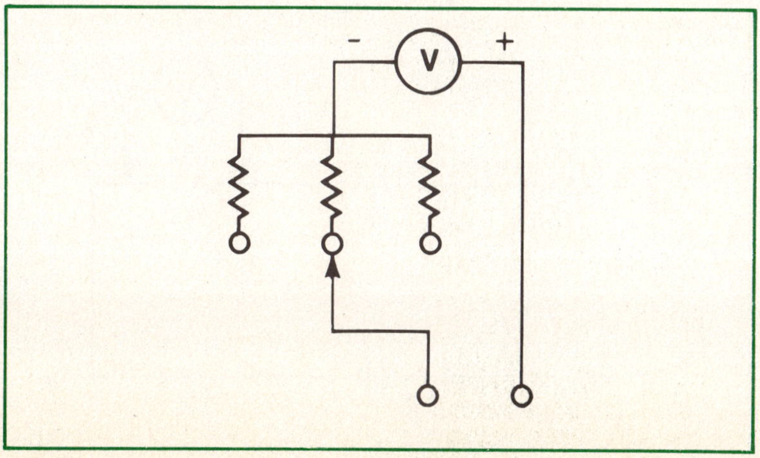

Figure 12-17. Circuit schematic for voltmeter having multiple voltage measurement ranges.

Voltmeter Loading

The beginning of Section 12.4 states that a voltmeter will use some of the power of the circuit it is measuring. In the example in Figure 12-16, the power drain of the meter is:

$$W = EI = 20\ \text{V} \times 10^{-3}\ \text{A} = 20\ \text{mW}$$

Even this low power can be a significant power drain on some circuits.

In general, the more sensitive the meter movement, the less it will load the circuit being measured. This is because a more sensitive meter will read full scale with a lower current through its movement. To reduce the loading effect, many voltmeters are designed with movements which have a sensitivity of 20K Ω per volt. This corresponds to a full-scale current of 1/20,000 A, or 0.05 mA. For extremely sensitive applications, FET meters are used. These meters use solid-state devices which make possible a sensitivity of up to 10 million ohms per volt. This makes FET meters extremely sensitive and useful for low-voltage measurements on low-power circuits.

To illustrate how slight the loading effect of voltmeters with sensitive movements can be, consider the diagram in Figure 12-18. This shows a meter with a sensitivity of 20K Ω per volt connected to read the voltage across a 2K Ω resistor. Assume the full-scale reading of the meter is 10 V. Then the meter would place a resistance of 20,000 Ω/V × 10 V = 200,000 Ω across R_3. (R_{mul} would be about 199,975 Ω.) The

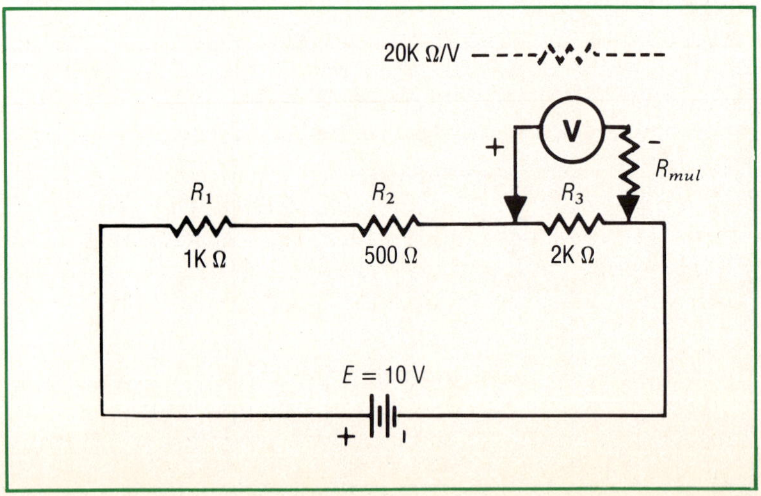

Figure 12-18. Circuit schematic for FET meter.

effect is to give a reading which is in error by 2,000/200,000 or 1 percent. This error is well within the limits to which meter movements are manufactured, and probably less than anyone's ability to distinguish markings on the meter scale. A 1 percent error is thus acceptable for most applications.

Discussion Questions/Problems

1. What are the similarities between ammeters and voltmeters?
2. What is the major functional difference between ammeters and voltmeters?
3. What determines the sensitivity of a voltmeter?
4. What is meant by the loading effect of voltmeters in relation to the circuits being measured?

12.5 THE OHMMETER

An ohmmeter measures resistance. In most instances, this meter is used to measure the resistance, in ohms, of individual resistors. The components being measured are usually not a part of a complete circuit. Most ohmmeters require that the components being measured cannot have power applied to them during measurement.

Ohmmeters are based on Ohm's law. The instrument is an ammeter with a d'Arsonval movement. An ohmmeter, of course, needs a scale that can be read in ohms. In addition, an ohmmeter must have its own internal power source, usually a battery. Also, the ohmmeter must have a built-in resistor to serve as a basis of calibration for the resistance being measured.

A basic ohmmeter circuit and a resistor to be measured are diagrammed in Figure 12-19. Parts of this circuit include a battery, instrument movement, a calibration resistor, and a scale resistor. The circuit is closed when the meter is connected to a part being measured.

Figure 12-20 illustrates the calibration of an ohmmeter scale. Note that this is different from other meter scales presented thus far. The major difference is that the ohmmeter scale is nonlinear. The scale also has larger resistances on the

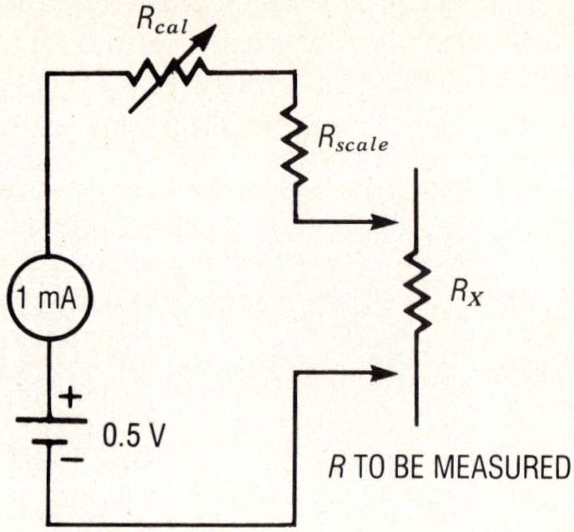

Figure 12-19. Circuit schematic for ohmmeter.

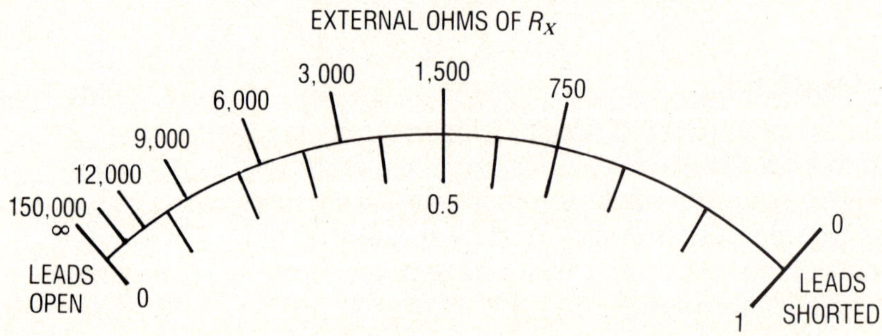

Figure 12-20. Calibration of ohmmeter.

left. This scale is opposite from other meters, such as ammeters or voltmeters. This type of scale is sometimes called a *back-off scale*.

The ohmmeter places a voltage across the unknown resistor R_X. The circuit becomes a series circuit with R_{cal}, R_{scale}, and R_X in series with the battery and the meter movement. Since the internal battery is usually a flashlight cell, the voltage at a given time will depend on how old the battery is. It is essential, therefore, that the meter be "zeroed" before any readings are taken. This adjusts the meter resistance to account for the particular battery voltage. The meter is zeroed by simply putting the two leads together, and adjusting the variable resistor R_{cal} so the meter reads full scale—or zero ohms.

Assume the meter has a 1,000 ohm/volt movement. Then 1 mA will cause full scale deflection when the two leads are touched together. Since the battery is 1.5 V and there is 1 mA of current, the sum of the two resistors (R_{cal} and R_{scale}) must be 1.5 V/0.001 A = 1,500 Ω. If the leads are now placed across R_X and the meter reads half-scale, there must be only 0.5 mA of current flowing. In that case, the total resistance must be 1.5 V/0.0005 A = 3,000 Ω. Since R_{cal} and R_{scale} together are 1,500 Ω, the other 1,500 Ω must be due to R_X. So, R_X = 1,500 Ω.

Changing R_{scale} will make it possible to have a mid-scale reading for resistances much larger or smaller than 3,000 Ω. Most ohmmeters have a switch to select among several different resistors for R_{scale}.

A VOM meter is commonly used for measuring ohms. This single instrument is equipped with selectors that make it possible to measure volts, ohms, or milliamperes. The scale is marked for each type of reading. VOM instruments are used commonly in the electronics field because of their utility. A VOM is illustrated in Figure 12-1.

Discussion Questions/Problems

1. What must be the condition of a circuit or component before it can be measured by an ohmmeter?
2. Why is an internal power source needed for an ohmmeter to function?
3. Why is it necessary that an ohmmeter have a variable resistor within its circuitry?
4. What operating check must be applied before an ohmmeter can be used?

12.6 BRIDGE CIRCUITS

Voltage and current, along with component parameters such as resistance, capacitance, and inductance, can be measured using simple bridge circuits.

Figure 12-21A shows a resistance bridge. If $R_1 = R_2$ and $R_3 = R_4$, then the current through R_1R_3 will be the same as the current through R_2R_4. Also, the voltage at point A will

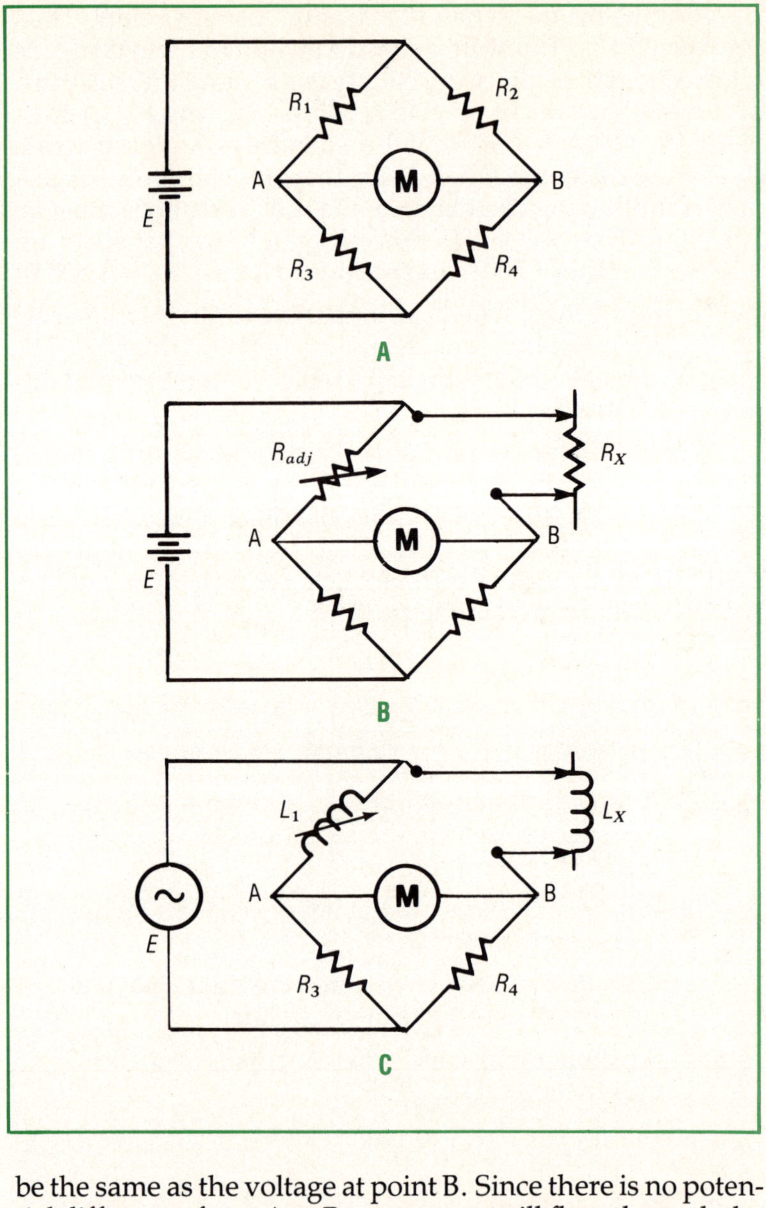

Figure 12-21. Circuit schematics for bridges.

be the same as the voltage at point B. Since there is no potential difference from A to B, no current will flow through the meter (M). The bridge is said to be balanced.

If $R_2 < R_1$, then $R_2 + R_4$ will be less than $R_1 + R_3$, and more current will flow through the right side of the bridge than through the left side. There will be a potential difference between point A and point B, with B being more positive than A. A current will flow from B to A, causing the meter (M) to deflect. The bridge is said to be unbalanced.

With certain changes to this circuit, an instrument of measurement can be produced. Figure 12-21B shows the same basic circuit, with R_1 replaced by an adjustable resistor R_{adj}. Instead of R_2 there are two test leads which are connected to the unknown resistor R_X. To use the bridge, the resistor R_{adj} is adjusted until the meter (M) shows no current is flowing between A and B. (This type of bridge typically uses a meter which is at center scale when there is zero current in the meter. The meter pointer can move to right or left, depending on the direction of the current flow.) When no current flows through M, $R_{adj} = R_X$. The value of R_X can be read from a scale on R_{adj} which, of course, must have been calibrated beforehand.

The same principle can be applied to measuring inductance. In this case, since a property of inductance is its AC impedance, an AC voltage is used. Figure 12-21C shows an inductance bridge. M must be an AC meter, and L_1 must be an adjustable inductor where the inductance has been calibrated. Even though the current through L_1R_3 (and L_XR_4 also) will have both a real and an imaginary component, when the bridge is balanced there will be no current flow between A and B. This is because the complex voltages at A and B are exactly equal when the bridge is balanced.

Bridges like those in Figure 12-21 are sometimes called *substitution bridges*.

Discussion Questions/Problems

1. What is the purpose of a bridge circuit?
2. How can a bridge circuit be used to measure capacitance?
3. What is meant by an unbalanced state in a bridge circuit?

12.7 THE OSCILLOSCOPE

An oscilloscope is a versatile tool that makes it possible to display and visualize signals that are being measured. Oscilloscopes measure instantaneous values of one or more electrical quantities. These quantities can be either steady or varying. The rate of variation can be extremely rapid. Oscilloscope measurements are based on a scale of either time or another known electrical or mechanical quantity.

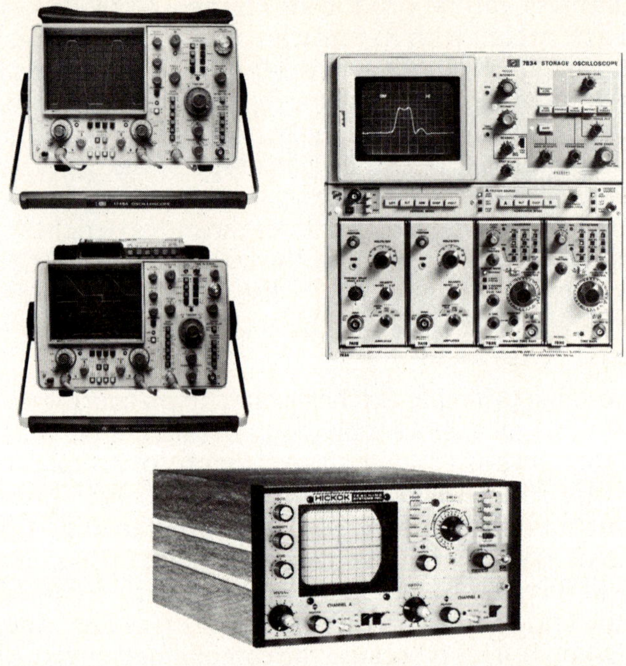

Figure 12-22. Oscilloscopes. (Courtesy Hewlett-Packard, Hickok Teaching Systems Inc., and Tektronix)

Figure 12-22 shows several different types of oscilloscopes. Instruments of this type are often used with electronic circuits. The main feature of an oscilloscope is a cathode ray tube, or CRT. The basic elements of a CRT are shown in Figure 12-23. A CRT includes an electron gun that generates and shoots an electron beam upon the face of a screen coated with a phosphorescent material. The phosphors in the coating make the image glow. They also retain the image long enough to make it visible. There are controls to adjust the position and focus of the electron beam. The images displayed on an oscilloscope tube are replicas of the signals being measured.

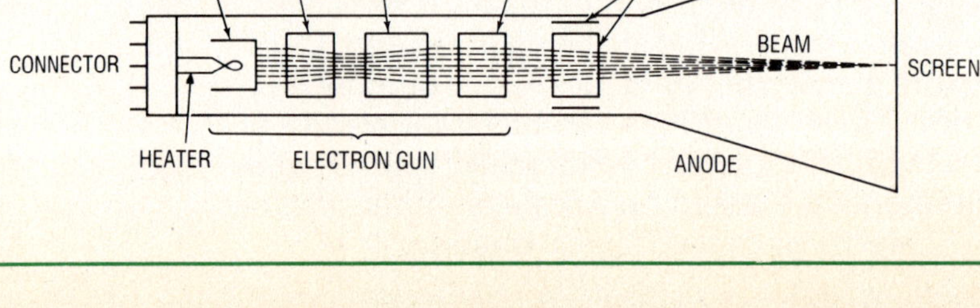

Figure 12-23. Basic elements of cathode ray tube (CRT).

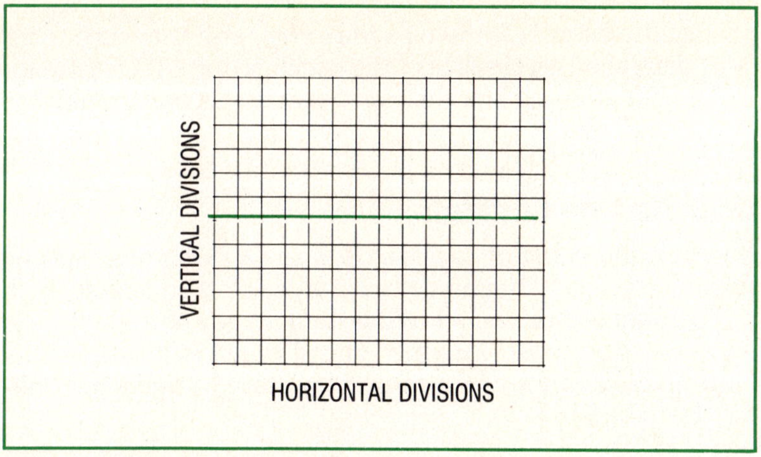

Figure 12-24. Rectangular grid of CRT screen.

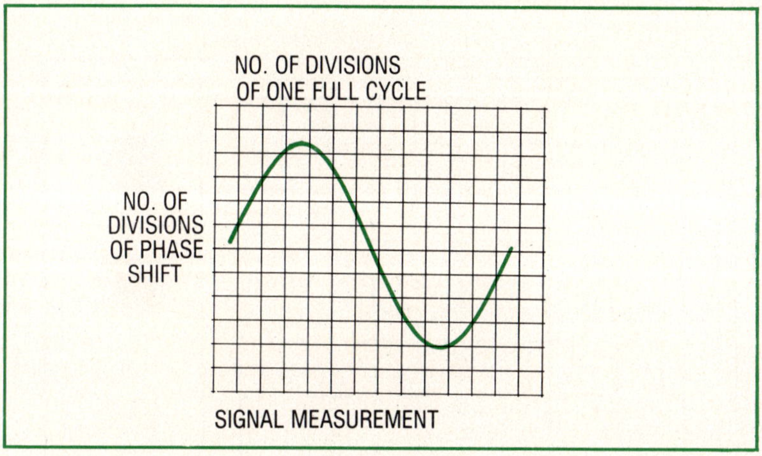

Figure 12-25. Sine wave displayed on oscilloscope screen.

The CRT screen of an oscilloscope is marked with a rectangular grid (recticule) set up in a series of vertical and horizontal divisions, as shown in Figure 12-24. Signal measurements are in terms of deflections on this grid. Figure 12-25 shows a typical sine wave displayed on an oscilloscope screen. This type of waveform display is the main purpose of an oscilloscope. With this display, a signal can be measured according to several parameters:

- Vertical deflection of the electron beam measures the *amplitude* of the signal.
- The horizontal deflection of the oscilloscope display measures the *time* and *frequency* of waveforms, and can be used to measure phase shift. The horizontal motion

of the beam is known as the *sweep*, and the wave form displayed on the face of the oscilloscope CRT is usually called a *trace*. (The electron beam traces out a replica of the signal.)

Oscilloscope Controls

Much of the value of oscilloscopes as working tools lies in their versatility. They usually have controls to adjust their performance features, allowing them to adapt to the job at hand. A typical oscilloscope control panel is illustrated in Figure 12-26. Some of the more important of these controls are described below.

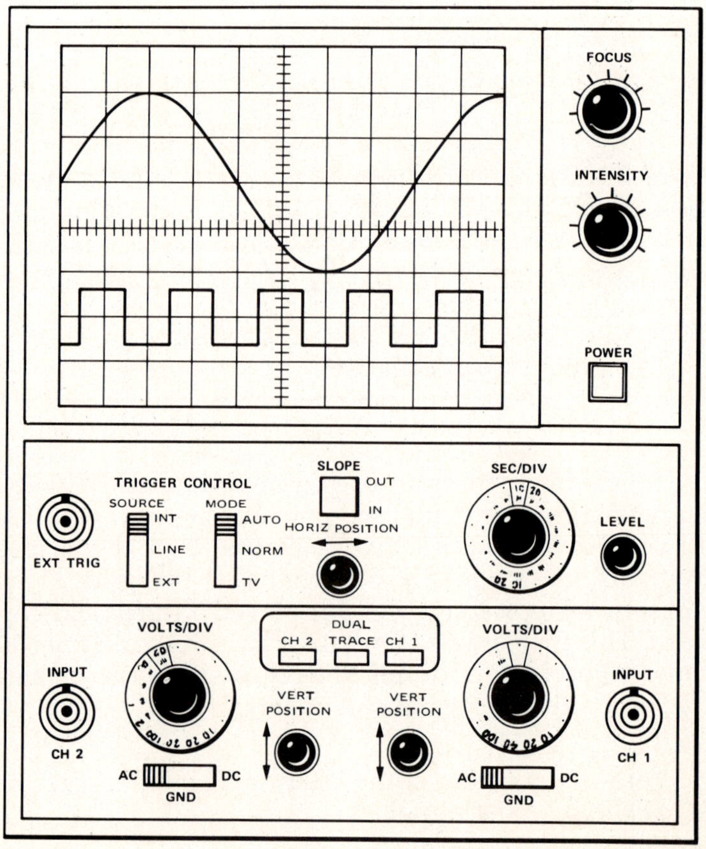

Figure 12-26. Control panel and screen of oscilloscope.

Power switch and lamp. A switch for turning the instrument on and off is essential. The useful lifetime of a CRT screen is limited. If the instrument is left on when not needed, the continuous electron beam could burn the phosphorescent material off the point on the face of the CRT screen where the beam is aimed. There is usually a pilot lamp that indicates when primary power is on.

Intensity. The term *intensity* is a synonym for the brightness of the trace displayed on the screen. This control should be set so that the signal is visible to the user. Setting the intensity too high can damage the phosphorescent coating on the screen.

Focus. This adjustment controls the sharpness, or exactness, of the wave image on the screen. When the electron beam is in *focus*, the electrons forming the beam are properly aligned when they strike the phosphor. This is an exact analog to focus in a camera or projector. The more exact the focus, the more exact the reading will be.

Vertical/horizontal position. There are usually two separate knobs controlling the position of the trace on the screen. One directs the electron beam up or down, the other from side to side.

Voltage/division. This control makes it possible to select the vertical sensitivity of the oscilloscope. This control makes it possible either to amplify or attenuate the incoming signal. By using this switch, the operator is able to keep the image of the wave on the screen, and of observable size, no matter what the incoming amplitude might be.

Seconds/division. This switch controls the sweep rate. It is known as a *time base control*. This adjustment makes it possible to select the time interval to be represented by each horizontal division on the oscilloscope screen. For some tests it may be useful to set the horizontal adjustment so that a full cycle of the signal can be observed on the screen; other times, many cycles are preferred.

Not all oscilloscopes use time base controls. Some are limited to the 60 cycle AC source that powers the instrument.

Trigger control. The *trigger*, as its name implies, initiates a horizontal sweep operation. This switch activates the beginning of a sweep across the face of the screen. The oscilloscope then cuts off after a single cycle. This makes it possible to freeze the image of a single cycle on the screen for closer observation. If the single cycle actually is one which recurs, then the sweep will recur.

Probes. Probes are the devices that connect oscilloscopes to their input signals. The operator touches the probes to the circuit being tested. The probe is the adaptor between the circuit and the oscilloscope. Probes can adapt high voltage to a low-voltage oscilloscope. Probes can include capacitors, filters, or attenuators. See Figure 12-27.

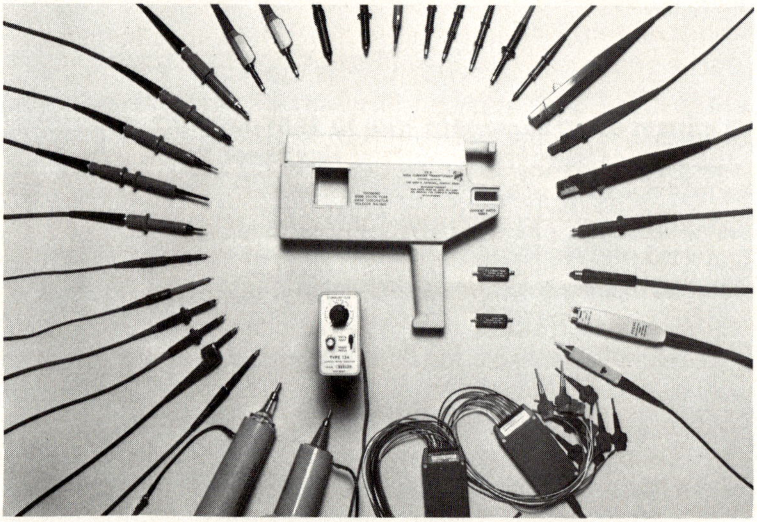

Figure 12-27. Oscilloscope probes. (Courtesy Tektronix)

Special-Purpose Oscilloscopes

General-purpose oscilloscopes have the functions described above. In addition, there are a number of oscilloscopes tailored to special applications. These include:

- Storage-type oscilloscopes
- Spectrum analyzers
- Logic analyzers.

Storage-type oscilloscopes. These units have electronic memories that store sweep patterns for display. The waves displayed from storage can then be compared with live signals. The stored wave is a standard against which other inputs can be compared. In some cases the signal occurs only once, rather than repetitively. The storage feature allows the signal to be displayed continuously, long after it has gone. A storage-type oscilloscope is illustrated in Figure 12-28.

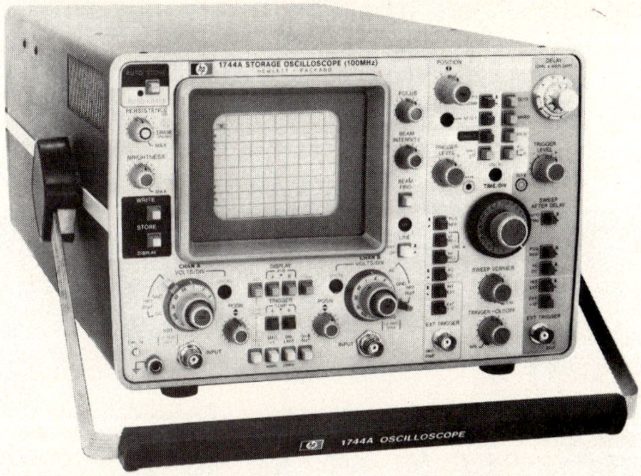

Figure 12-28. Storage oscilloscope. (Courtesy Hewlett-Packard)

Spectrum analyzers. A spectrum analyzer is used for diagnosing relatively complex signals. For example, consider a normal television signal. It includes separate components for the video picture, the audio portion, and several control signals. With a spectrum analyzer, all of the components of the single signal can be displayed individually. This makes certain types of troubleshooting and analysis possible. A spectrum analyzer is illustrated in Figure 12-29.

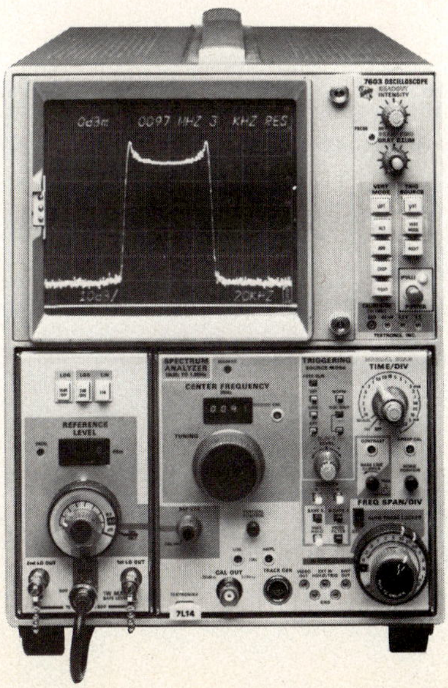

Figure 12-29. Spectrum analyzer. (Courtesy Tektronix)

Logic analyzer. This is a specialized instrument used for digital circuits. Digital signals are represented by their digital components. The display generated on the CRT screen of a logic analyzer is in the form of binary coding—the digits 0 and 1. A logic analyzer breaks down a digital signal into its binary components. See Figure 12-30.

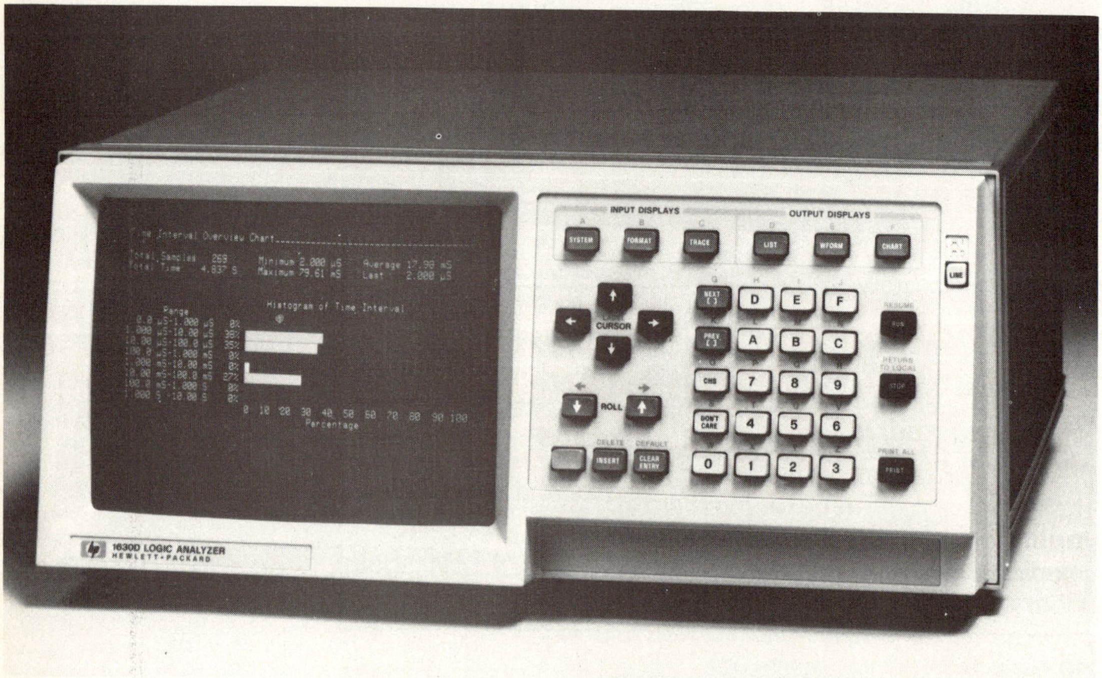

Figure 12-30. Logic analyzer. (Courtesy Hewlett-Packard)

Discussion Questions/Problems

1. What are the main functions of an oscilloscope?
2. What does an oscilloscope show?
3. Why do oscilloscope screens have display grids?
4. How are the factors of amplitude, time, and frequency related in an oscilloscope display?
5. What is the function of the trigger control on an oscilloscope?
6. What is a spectrum analyzer and what does it do?
7. What is a logic analyzer and what does it do?

12.8 SIGNAL GENERATORS

A signal generator is an electronic unit that supplies selected output signals to be used for testing purposes. Signals from a signal generator are used to test the circuit behavior of other electronic circuits. The main requirement for a signal generator is that it must provide a signal of known amplitude, known frequency, and also known waveform. Usually the operator can select the amplitude and frequency and, often, the waveform as well.

Types of signal generators used regularly in electronic testing include:

- Audio frequency generator
- Radio frequency generator
- Sweep generator
- Television pattern and color bar generator
- Pulse generator
- Function generator
- Programmable waveform generator.

Audio Frequency Generator

Audio signal generators are used to test amplifiers, cassette players or other devices that generate outputs in the audible signal range. This range, generally, is from 20 Hz to 20 kHz. Some audio devices produce outputs at frequencies higher than the audible range. This is particularly true for devices that use inaudible signals as trigger mechanisms. For example, many devices used for presenting sound-slide films can be triggered to change pictures on the basis of signals that are outside the audible range. Primary functions of audio frequency generators include testing speakers, amplifiers, or other components of sound systems. An audio frequency generator is illustrated in Figure 12-31.

Radio Frequency Generator

These units, often called *RF generators*, cover the AM, FM, and TV ranges of the radio frequency spectrum. Typical signal output ranges are from 250 kHz to 250 MHz. These units are used to test such devices as radio transmitters and receivers. In addition to generating AM and FM waves, some RF generators also produce continuous wave (CW) signals. An RF generator is shown in Figure 12-32.

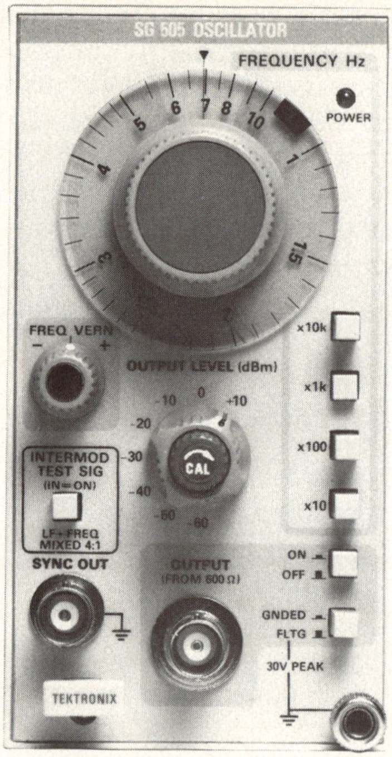

Figure 12-31. Audio frequency generator. (Courtesy Tektronix)

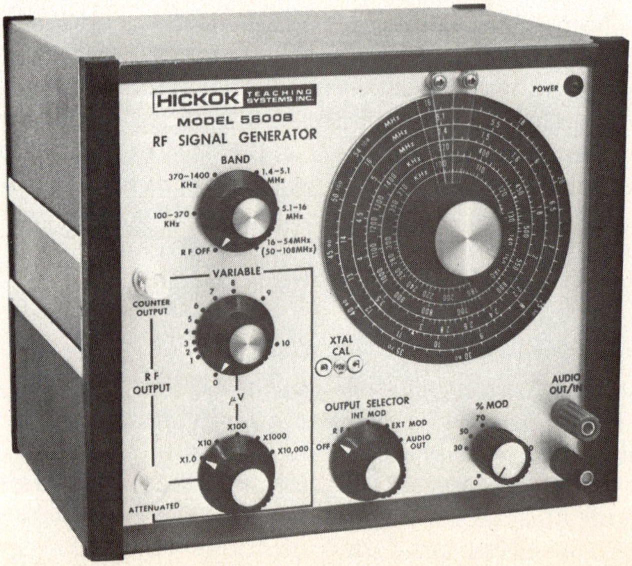

Figure 12-32. Radio frequency (RF) signal generator. (Courtesy Hickok Teaching Systems Inc.)

Sweep Generator

A sweep generator is a special type of signal generator. Output from a sweep generator is a signal of continuously varying frequency. These devices are used for testing television and other communication receivers. The RF sweep signal is particularly useful in testing signal filters. By covering a broad range, beyond the cutoff points of filters, an RF sweep generator used with an oscilloscope can show the filter characteristics of band-pass, low-pass, and high-pass filters. Audio sweep generators are useful to measure characteristics of audio filters. A synthesized sweep generator is illustrated in Figure 12-33.

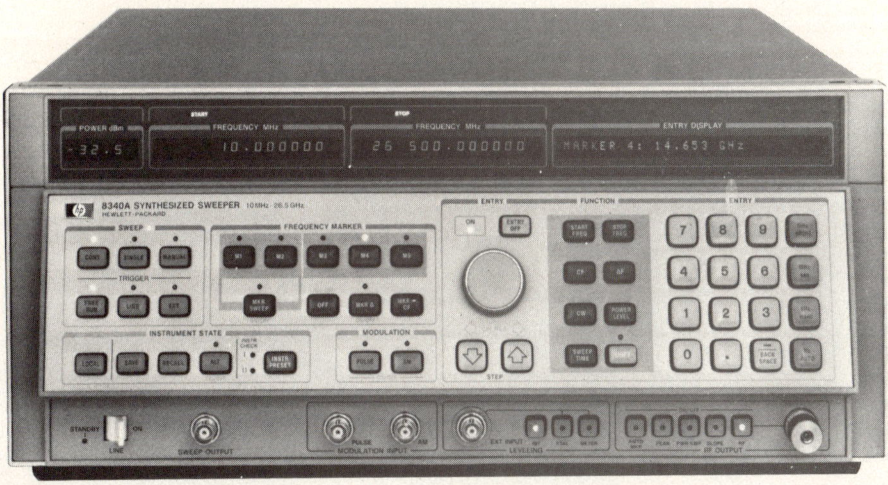

Figure 12-33. Sweep generator (Courtesy Hewlett-Packard)

Television Pattern and Color Bar Generator

These units produce video picture signals. Patterns displayed include dot, line, cross-hatch, and other patterns used to check and repair television receivers. Also generated are sets of vertical bar images covering the range of the color spectrum. These are used to check and adjust color television receivers. A television pattern and color bar generator is shown in Figure 12-34.

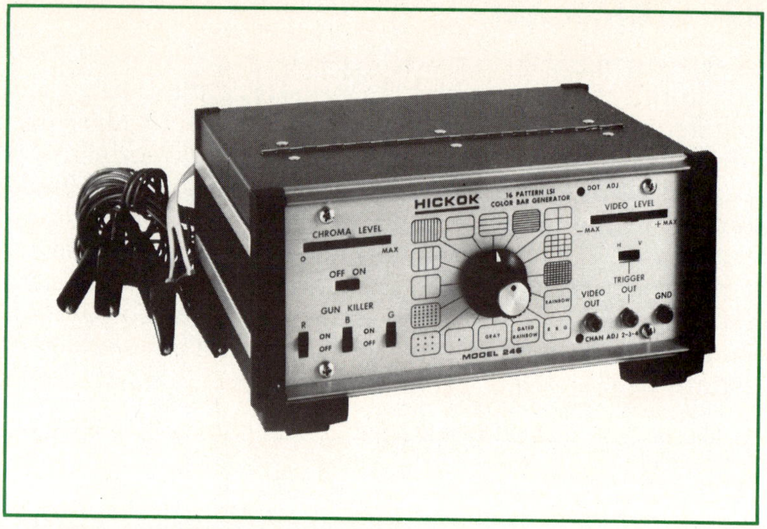

Figure 12-34. Television pattern and color bar generator. (Courtesy Hickok Teaching Systems Inc.)

Pulse Generator

The output of a pulse generator is a continuing series of pulses. This is useful to adjust the timing of digital circuits. The periodicity and the rise time or fall time of the pulses can usually be adjusted, as well as the amplitude and duration of each pulse. Computers and other digital devices operating on nonsinusoidal signals are usually tested and adjusted with a pulse generator. A pulse generator is shown in Figure 12-35.

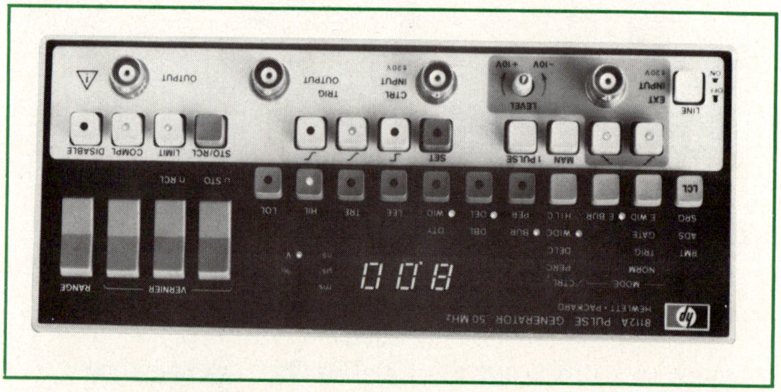

Figure 12-35. Programmable pulse generator. (Courtesy Hewlett-Packard)

Function Generator

Function generators are versatile instruments that can produce a variety of waveform outputs. Included are square, triangular, sine, and ramp waveforms. Examples of these waveforms, along with a photograph of a function generator, are shown in Figure 12-36.

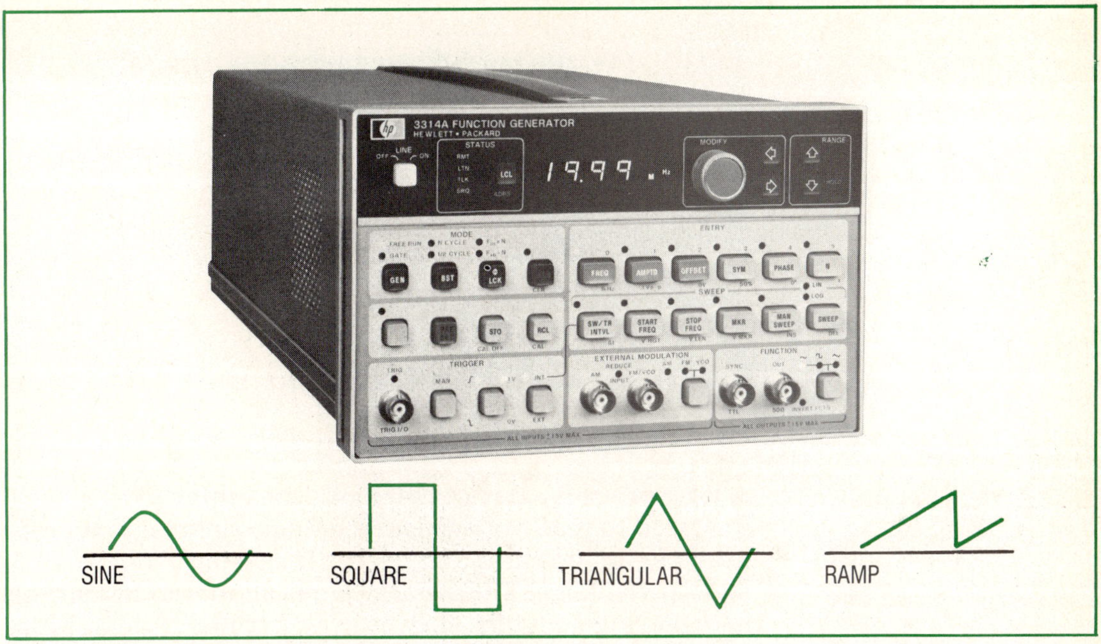

Figure 12-36. Function generator (courtesy Hewlett-Packard) and examples of wave forms produced.

Programmable Signal Generators

A programmable signal generator is a device which can be "programmed," or preset, to generate almost any desired signal. The signals generated by this test instrument can be programmed to meet specific application needs. This means that a programmable generator can be used for virtually any testing application requiring a signal input. Nonstandard signals can be created for use in testing special devices or instruments, or for checking the error-detection capabilities of filter or testing units. A programmable signal generator is shown in Figure 12-37.

Figure 12-37. Programmable signal generator. (Courtesy Hewlett-Packard)

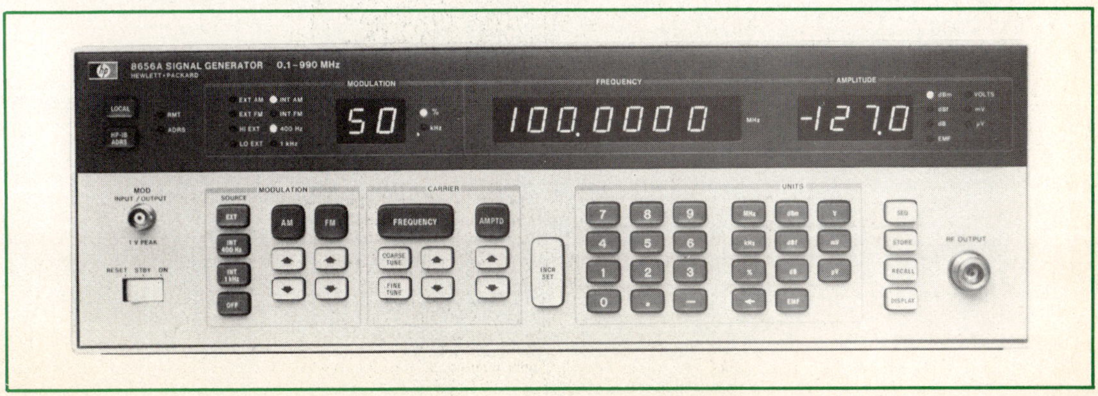

Discussion Questions/Problems

1. Which of the types of generators mentioned above might you use for work on a television receiver?
2. What type of signal generator would you use to test a tape recorder?
3. What type or types of signal generators might you use if you were working on a computer?

12.9 LOGIC PROBES

A logic probe is a device that looks something like a pencil. It is used to read voltage levels in logic circuits. The probe itself has an indicator to show if it is sensing a logic high or logic low voltage. The primary application is to test signals in digital logic circuits, such as in a computer. A logic probe is shown in Figure 12-38.

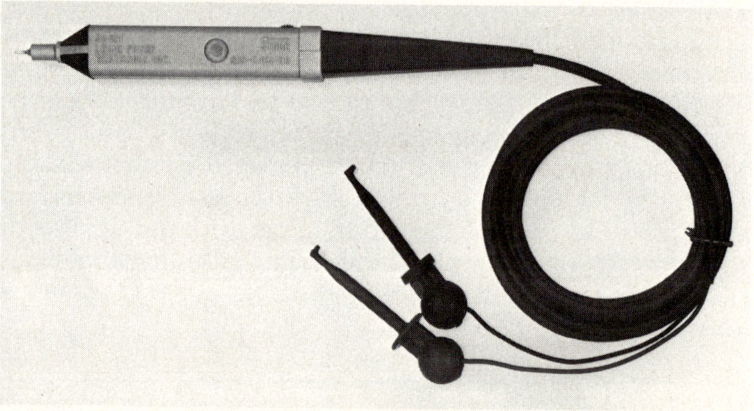

Figure 12-38. Logic probe. (Courtesy Tektronix)

Discussion Questions/Problems

1. Why is a logic probe well suited for digital circuits?
2. From your other experience, why is it important that logic circuits deliver voltages within tolerances?

12.10 POWER SUPPLIES

A power supply can be an important tool in testing circuits and devices. For example, suppose you are testing an instrument's power supply to see if it is functioning properly. A power supply is shown in Figure 12-39. One test is to substitute an external power supply, which is known to be good, for the suspect supply. This would then allow you to determine whether the power supply was faulty or whether there was a problem elsewhere in the circuit.

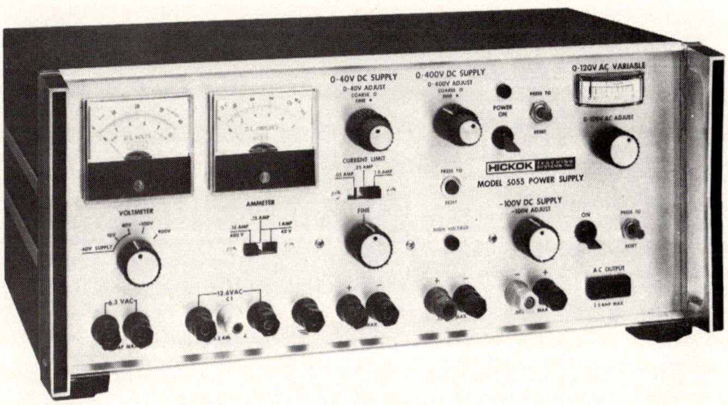

Figure 12-39. Power supply. (Courtesy Hickok Teaching Systems Inc.)

Power supplies are a necessity when working on mobile communications equipment, such as car radios or CB units. Both these devices rely on a 12 V power source from the vehicles in which they are installed. When you take out a radio or CB transmitter for repair on a test bench, the power supply stays in the car. Thus, you will need a power supply in the shop in order to work on the unit. Similar problems make a power supply necessary when working on equipment associated with aircraft, ships, etc.

Discussion Questions/Problems

1. Why should an auto electronics store selling mobile radios have a 12 V power supply on hand?
2. Under what conditions might you use a power supply as a test instrument in checking a device that has its own internal power supply?

13

ELECTRONIC DEVICES

13.1 ELECTRONICS

Electronics is a field dealing with the behavior of free electrons, particularly within environments such as vacuums, gases, special conductors, or semiconductors.

It is important to emphasize the difference between electronics and electricity. Although no simple definition is adequate, one way to describe the difference is to say that electricity deals with the flow of electrons in metal conductors. In other words, electricity is a limited, highly proscribed field of study. Electronics has ramifications far beyond those of electricity.

The field of electronics deals mainly with active components—those that either undergo or stimulate changes when electron flow is applied. Examples of active components include electron tubes and semiconductor devices. By contrast, passive components include resistors, inductors, and capacitors. Passive components are used as integral parts of electronic circuits. It is the *active components*, however, that make a circuit electronic, rather than electrical.

Discussion Questions/Problems

1. What is the primary difference between the fields of electronics and electricity?
2. What are active components? Give examples.
3. What are passive components? Give examples.

13.2 VACUUM TUBES

Vacuum tubes were the first electronic devices put into practical use. Development of the vacuum tube built upon a discovery in 1883 by Thomas Edison. Edison's basic discovery was the thermal emission of electrons. As a by-product of his main work (seeking materials for use in light bulbs), Edison discovered that the electrons in the outer shells of some metal atoms were bound very loosely to their nuclei. When sheets or strips of these metals were heated in a vacuum, electrons were emitted. The effect was as though they had boiled off, or evaporated, from the base metal elements.

Subsequently, researchers placed strips of these electron-emitting metals in evacuated, or vacuum, tubes. When these metals were connected to a source, current or free electrons moved along the metal strips, known as *filaments*. The released electrons collided with other free electrons within the atoms of the filament, breaking loose other electrons. In addition to causing heat, the electron collisions generate sufficient levels of energy so that some of the free electrons overcome the attractive force of the base material and break away from the surface of the filament. These electrons travel freely through surrounding space, propelled by the energy of the collisions. The electrons that break away in this way are called *emitted electrons*. The heated metal that emits electrons is called an *emitter*.

To build vacuum tubes, an emitter filament is placed in close proximity to a metal device known as a *plate*, or *anode*. When a potential difference is applied between the plate and the emitter, with the plate more positive than the emitter, electrons will flow from emitter to plate. The plate thus attracts free negative electrons from the emitter. When electrons flow from an emitter to a plate within the same vacuum tube, this phenomenon is called the *Edison effect*. It is also called the *Richardson effect*. Figure 13-1 illustrates the electron flow within a vacuum tube when the emitter is negative and

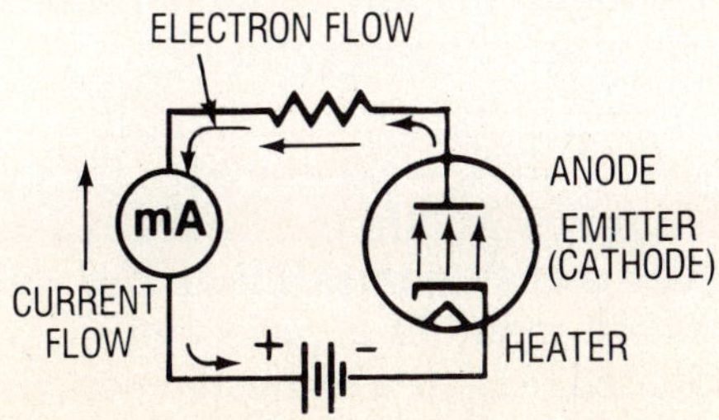

Figure 13-1. Circuit schematic of vacuum tube with negative emitter and positive plate.

the plate, or anode, is positive. In this diagram, the wire device beneath the emitter is a filament that serves as a heater. The emitter itself is called a *cathode*.

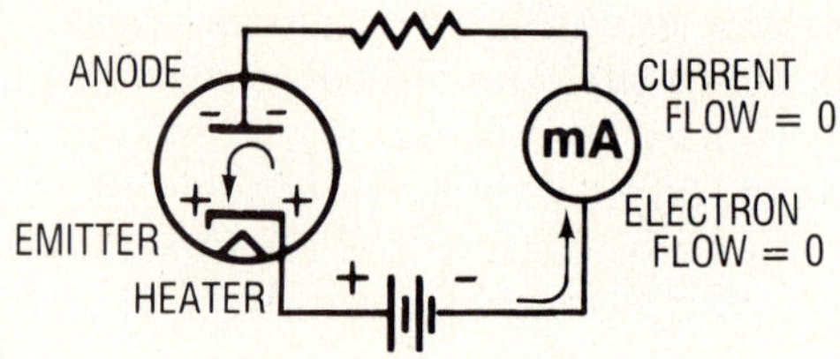

Figure 13-2. Effect in vacuum tube when negative plates repel electrons from heated filament.

Figure 13-2 illustrates the effect that takes place when electrons from the heated filament are repelled by the negative plates. In this case, the flow of current is stopped entirely. Note that the potential source in Figure 13-2 is the inverse, or opposite, of the one in Figure 13-1.

Discussion Questions/Problems

1. What is the Edison effect?
2. How is heat generated in a filament?
3. What is an emitter and what is the source of its name?
4. What is a plate and what is its function within an electron tube?

13.3 TUBE CONSTRUCTION

Figure 13-3. Vacuum tube.

A vacuum tube consists of a series of metal parts built into either a glass or metal vacuum container, called an *envelope*. A vacuum tube is shown in Figure 13-3.

The metal parts within vacuum tubes are all known as *electrodes*. Figure 13-4 shows the symbol for a vacuum tube, together with a drawing of the electrodes in a simple vacuum tube, giving their specific names. Electrodes within electronic vacuum tubes are made of a variety of metals, including tungsten, thoriated tungsten (tungsten with thorium added to improve emissions), and other oxide-coated metals.

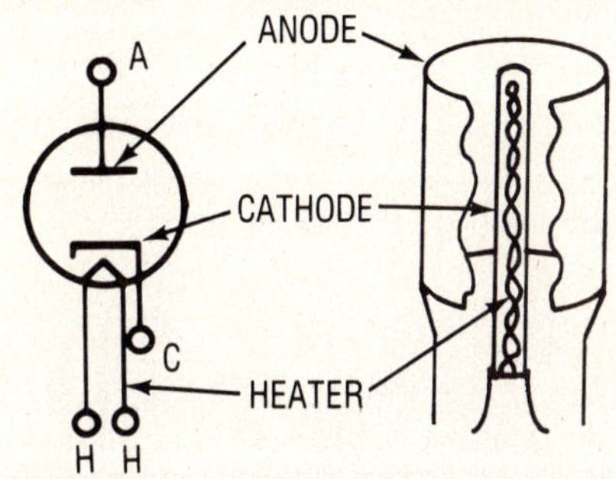

Figure 13-4. Electrodes in vacuum tube and symbol for vacuum tube.

As indicated earlier, the flow of electrons is a result of heating an emitter within a vacuum tube. Two methods can be used for heating an emitter—direct or indirect. Figure 13-5 illustrates the design and configuration of tubes with both directly and indirectly heated emitters. When emitters are heated directly, as shown in Figure 13-5A, they are known as *filaments*. When vacuum tubes are heated indirectly, as shown in Figure 13-5B, the emitter unit is called the *cathode*. In these tubes, the wire generating the heat is simply called the *heater*.

The power supply for the heater or filament of either directly or indirectly heated tubes can be either AC or DC. However, AC-heated tubes tend to produce an objectionable hum for some applications, particularly audio or video circuits. The hum is simply the 60-cycle AC superimposed on the desired signal.

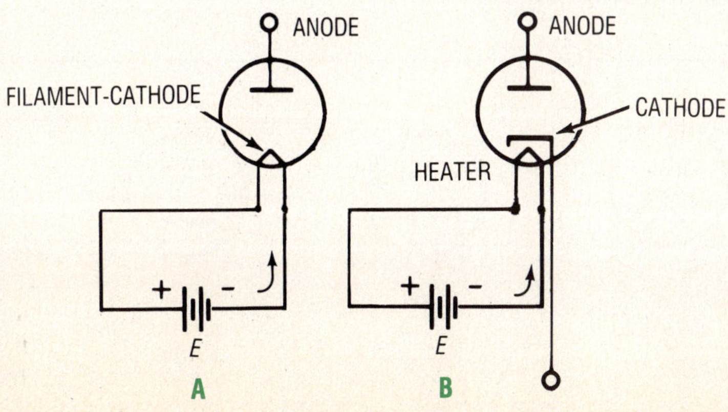

Figure 13-5. Schematics of (A) filament cathode and (B) cathode.

The voltage and current needed to heat the tube to an operational temperature are determined by the tube designer. All tube manufacturers provide extensive manuals. These are used as references listing filament voltage, plate voltage, current rating, and other significant characteristics. Common filament voltage ratings for vacuum tubes include 2.5 V, 6.3 V, 12.6 V, as well as many others.

Discussion Questions/Problems

1. What is a vacuum tube envelope and from what materials can it be made?
2. What is an electrode?
3. What methods can be used to develop heat for emitters in vacuum tubes?
4. What types of currents can be used in vacuum tubes?

13.4 DIODES

The term *diode* indicates that a tube has two electrodes, usually a cathode and an anode. These are the simplest types of vacuum tubes. The cathode, or emitter, can be either directly or indirectly heated, as shown in the diagrams and schematic symbols in Figure 13-6.

When a positive voltage is applied to a diode plate, electrons flow from the cathode to the plate and then return through an external circuit to the cathode. If negative voltage is applied to the plate, free electrons from the cathode will not flow to the plate.

Figure 13-6. Symbols for diodes.

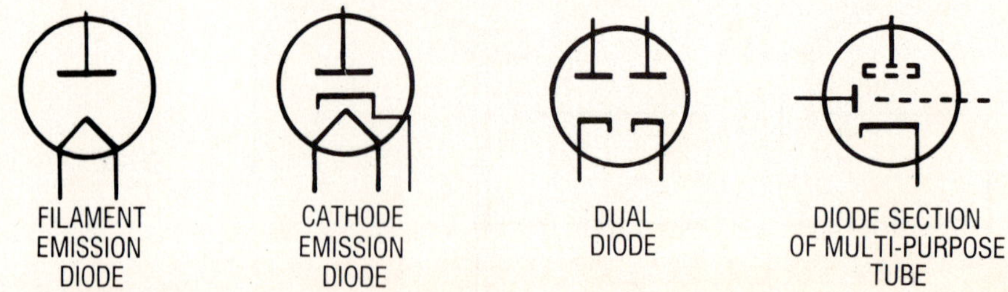

Rectification

A major use for a diode is as a *rectifier*. A rectifier changes, or converts, AC to DC. This is done by accepting the bidirectional input of AC but allowing only a unidirectional current, or DC, to flow in the output. A schematic of a rectifier is shown in Figure 13-7, along with waveforms indicating the nature of input and output signals.

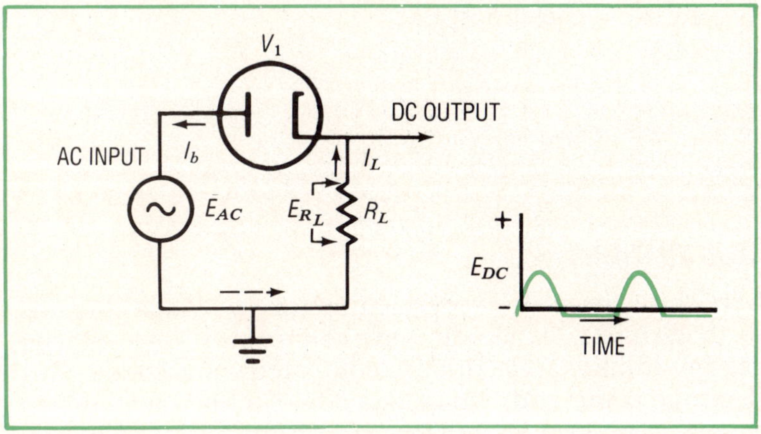

Figure 13-7. Circuit schematic of rectifier with waveforms of input and output signals.

Emission Saturation

There are limits to the rate at which electrons can be emitted by an emitter. The electron emission can become saturated. Saturation simply means that the emitter cannot release any more electrons. When the saturation point is reached, increasing the plate voltage will not increase the cathode emission.

Space Charge

A condition known as *space charge* occurs when the plate cannot accept any more electrons. The electrons have no place to go, but are suspended in the space between the emitter and the plate. This condition occurs only at low-plate voltages.

A space charge exerts a repelling force upon other electrons being emitted. When space charge occurs, the circuit is not operating effectively. Thus, operating voltages should be selected so that the plate voltage is large enough to continue to attract the electrons from the emitter.

Plate Saturation Point

When the plate reaches a condition at which it cannot absorb any more electrons, even if a higher voltage is applied, it is said to be at its *saturation point*. The factors that affect the plate saturation point include the specific structure of the plate, the vaporization factor of the cathode, and the metallurgical structure of the plate materials. None of these factors relates to the electrical properties of the plate or cathode.

Discussion Questions/Problems

1. How many electrodes are in a diode?
2. What is rectification?
3. How do the conditions of emission saturation and plate saturation differ?
4. What is space charge?

13.5 TRIODES

A *triode* is a three-electrode vacuum tube. A triode contains an anode (plate), a cathode, and a control grid. Thus, a triode has one additional electrode as compared with a diode. This electrode is the grid, which is made of a fine metal mesh, or wire. The grid is usually made from a relatively low-conductivity material, such as molybdenum, iron, or nickel. Electrons must pass through the grid on their path from the cathode to the anode, as shown in Figure 13-8. The triode is usually operated with a low negative voltage on the grid, as compared to that of the cathode. This *bias voltage* is usually called a *C-voltage*. The voltage applied to the heater is known as the *A voltage*. The plate voltage is designated by the symbol *B*. These designations come from the early days of radio, when separate batteries, called the *A*, *B*, and *C* batteries, were used to power radios.

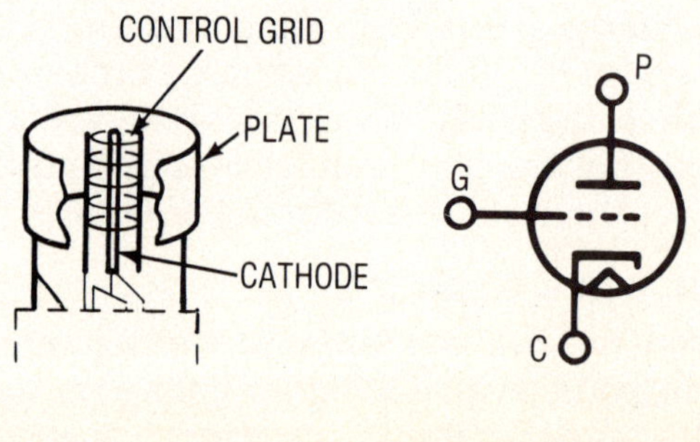

Figure 13-8. Construction (A) and Symbol (B) of triode.

Early experimenters, the most well-known of whom is Lee DeForest, discovered that an electrode between the cathode and anode of a vacuum tube could be used to control the flow of electrons to the anode. A large flow of current could be controlled by a very small voltage on this intermediary grid electrode. The effect is much like using a small amount of force to turn a water valve which controls a very large flow of water. For this reason, in England and some other countries, an electron tube is called a *valve*.

The function of the control grid is to control the plate current. Adjusting the negative bias voltage on the grid will adjust the plate current. The more negative the grid voltage, the less the plate current. If the grid voltage becomes sufficiently negative, there will be no plate current at all, even for a highly positive plate. When the current to the plate is reduced to zero, this condition is known as *grid cutoff*.

The major application for a triode is for use as an *amplifier*. This means that the triode accepts an input signal and amplifies it to a large output signal. The output signal is greater than the input. An amplifier circuit has these important performance characteristics:

- The smaller the (negative) grid voltage, the greater the plate current.
- The phase of the output signal is inverted. The alternation pattern of the output AC waveform is reversed when compared to the input. A functional diagram of a triode amplifier circuit, together with input and output waveforms, is shown in Figure 13-9. Notice that the

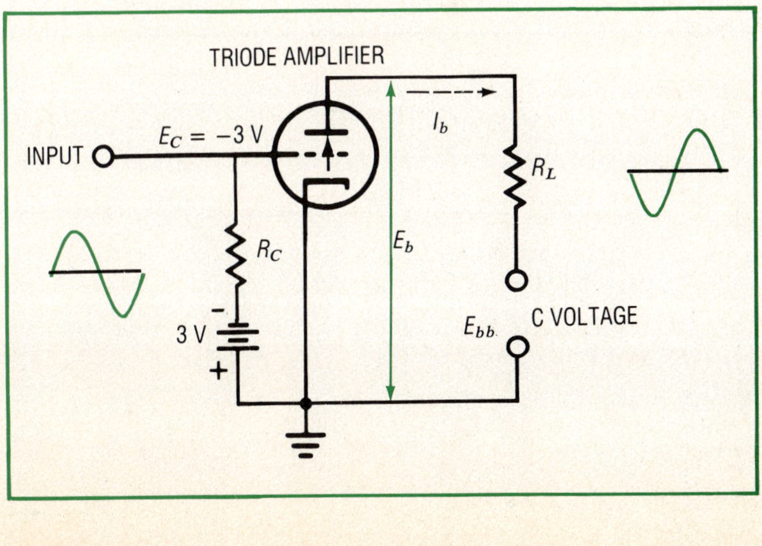

Figure 13-9. Functional schematic of triode amplifier circuit with input and output waveforms.

grid is given a negative DC bias voltage that will maintain an average negative grid voltage, even with an AC signal input.

- An amplifier triode produces a voltage gain. The peak-to-peak signal output voltage is greater than the peak-to-peak signal input voltage.
- The plate current is not a linear function of the grid voltage. To select the desired plate current, it is necessary to use reference materials that give the characteristics of individual tubes. Figure 13-10 shows behavior patterns for a variety of grid voltages and their effects upon plate voltage and current for a typical triode tube.

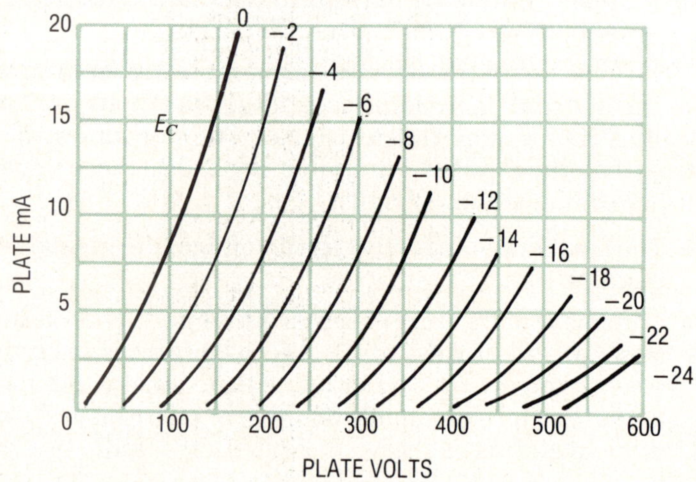

Figure 13-10. Behavior patterns for grid voltages of triode tube and effects upon plate voltage and current.

Discussion Questions/Problems

1. How does a triode differ from a diode?
2. What is a grid and how does it affect electrical properties of a vacuum tube?
3. What is the relationship between grid voltage and plate current?
4. What is the primary application for triodes?

13.6 OTHER TYPES OF TUBES

A number of other types of vacuum tubes are manufactured. Each of these includes some sort of special design feature to improve performance over triodes or to achieve a special type of performance. Some of the more popular of these other types of tubes are identified and described below.

Tetrodes. A tetrode contains a fourth electrode. This is known as a *screen grid*. The function of this additional electrode is to reduce the likelihood that the metal parts within a tube will function as capacitors. Any two metal electrodes can, of course, behave as a capacitor. This is called the *interelectrode capacitance*. This capacitance can become a serious problem for high-frequency circuits, thus the tetrode was developed to reduce interelectrode capacitance. The symbol for a tetrode is shown in Figure 13-11.

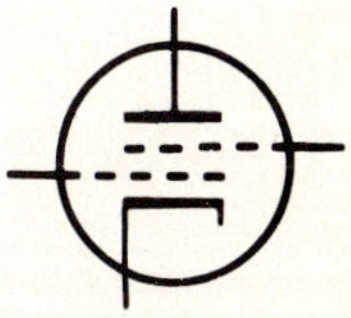

Figure 13-11. Symbol for tetrode.

Pentodes. As the name implies, a *pentode* is a five-element vacuum tube. A pentode symbol is shown in Figure 13-12. The fifth element in a pentode is known as a *suppressor grid*. The suppressor grid is positioned between the screen grid and the plate. Electrons traveling from the cathode to the plate move through three separate grids, in order. These are the control grid, the screen grid, and the suppressor grid. The purpose of these grids is to control electron flow between the cathode and the anode. The function of the suppressor grid is to control a phenomenon known as *secondary emission*. Secondary emission occurs when electrons, traveling at extremely high speeds, bounce back off the plate into the space within the vacuum tube. The suppressor grid holds back, or prevents, the secondary emission of electrons to make sure the plate absorbs most of the electrons.

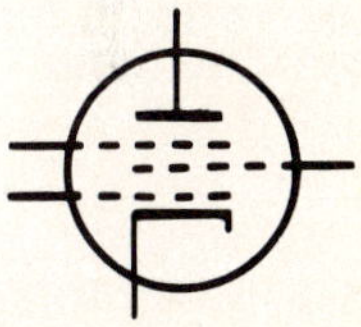

Figure 13-12. Symbol for pentode.

Beam power tubes. A beam power tube is a special type of tetrode. Its characteristics are similar to those of a pentode tube, but without a suppressor grid. There are special beam-forming plates in the tube which help concentrate the electron flow from cathode to plate into well-defined beams. The effect is to cause secondary electrons to be repelled back toward the plate, as if there were a supressor grid. Also, a beam power tube can handle large currents and thus larger power levels. Figure 13-13 shows the symbol for a beam power tube.

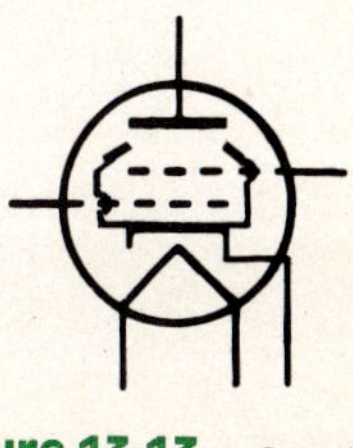

Figure 13-13. Symbol for beam power tube.

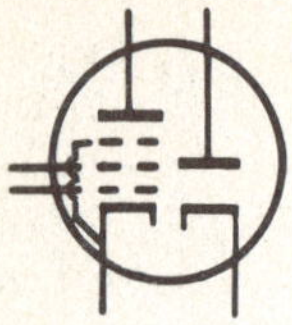

Figure 13-14. Symbol for multipurpose tube with rectifier and beam power amplifier in single unit.

Multipurpose tubes. As the name implies, these tubes are actually several sets of elements in a single envelope. For example, consider the tube illustrated in Figure 13-14. This tube includes both a rectifier and a beam power amplifier in a single unit. Other types of multipurpose tubes are also available.

Gas-filled tubes. Rather than having a complete vacuum, these tubes include a small amount of inert gas. Certain properties of inert gases can be an advantage for tubes. Primarily, gases can ionize, and the ions in the tube become part of the current flow. Tubes of this type can be designed to maintain a fixed voltage drop across the circuit and are used to regulate a power supply. Other tubes of this type, because of the ability of the ionized gas to conduct a current, can be used in applications which require very large currents. Some gas-filled tube configurations are shown in Figure 13-15.

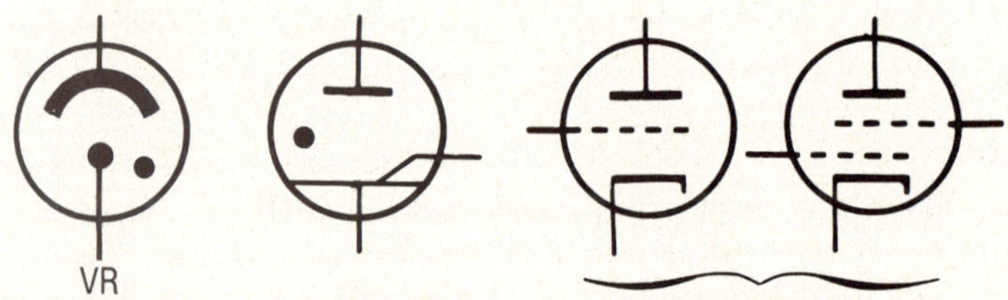

Figure 13-15. Symbols for gas-filled tubes.

Important Operational Parameters for Tubes

In order to design circuits that incorporate vacuum tubes, you should be able to use some relatively simple formulas to determine performance characteristics of the circuits. Three of these characteristics are identified and described below.

Amplification factor. An equation can be applied to show the effect of a change in control grid voltage on the resulting plate voltage. The ratio of these is the *amplification factor*, indicated by the lowercase Greek letter *mu* (μ). Note in the formula below that this is a ratio with no unit of measure.

For constant i_b:

$$\mu = \frac{\Delta E_b}{\Delta E_c}$$

Plate resistance. The plate resistance of a tube is the ratio of the change in plate voltage caused by a change in plate current. The equation is:

For constant E_c:

$$r_p = \frac{\Delta E_b}{\Delta i_b}$$

Note that the unit of measure for plate resistance is ohms, and is directly related to Ohm's law: $R = E/I$.

Transconductance. Transconductance compares the effect of changes in grid voltage upon plate current. The formula is as follows:

For constant E_b:

$$gm = \frac{\Delta i_b}{\Delta E_c}$$

Note that the unit of measure for transconductance is the mho or the micromho. The mho was adopted before the siemens, and is still used in working with tubes.

Discussion Questions/Problems

1. What is the difference between a tetrode and a triode?
2. What is the main application of a pentode?
3. Describe the operation of a beam power tube.
4. What are the primary differences in characteristics and advantages of gas-filled tubes?
5. What is the amplification factor and how can it be determined?
6. What are plate resistance and transconductance? What are the formulas for determining them?

13.7 CATHODE RAY TUBES

The *cathode ray tube* is a special type of vacuum tube that displays an image on its front surface. In electronic systems, the cathode ray tube (CRT) is used to display a waveform or other pictorial representation.

The CRT plays a unique role in the electronics field in that it is one of the few types of vacuum tube enjoying an increasing utilization. In electronics work, you will encounter CRTs as display screens on test units such as oscilloscopes. You are no doubt familiar with the role of the CRT as the picture tube in a television set. In addition, CRTs are gaining widespread use for video games and as display screens for computer terminals.

Internal Elements of the CRT

The primary internal elements of the CRT are the electron gun, the deflection plates, and the phosphor screen (see Figure 13-16). The electron gun is used to create and focus a stream of electrons which are directed by the deflection plates to a specific point on the phosphor screen. The electron gun is made of parts similar to those of a triode or pentode. It includes a filament, cathode, accelerating grid, and an annular plate. The voltage on the plate focuses the beam, and the hole in it allows the beam to continue on to the phosphor screen. The deflection plates are set up in pairs, one pair to deflect the beam vertically and one pair to deflect the beam horizontally. The phosphor screen is a chemical coating on the inside face of the CRT glass envelope. The phosphor glows when the electrons strike it.

Figure 13-16. Primary internal elements of CRT.

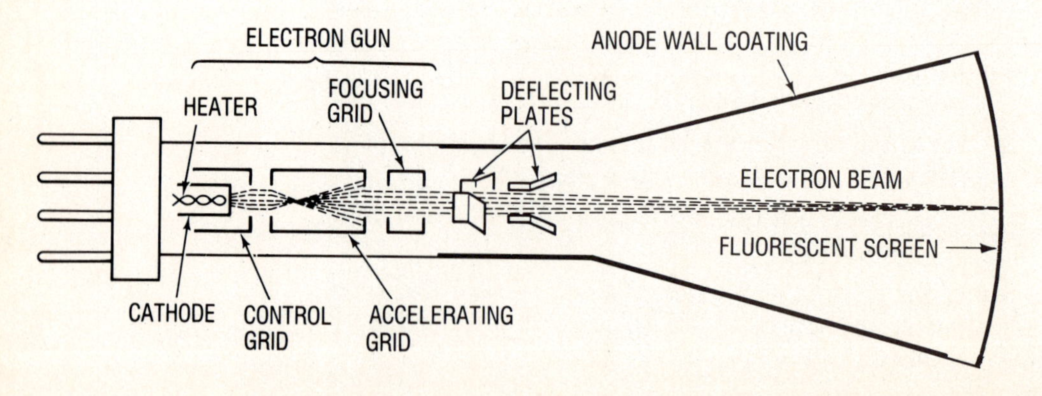

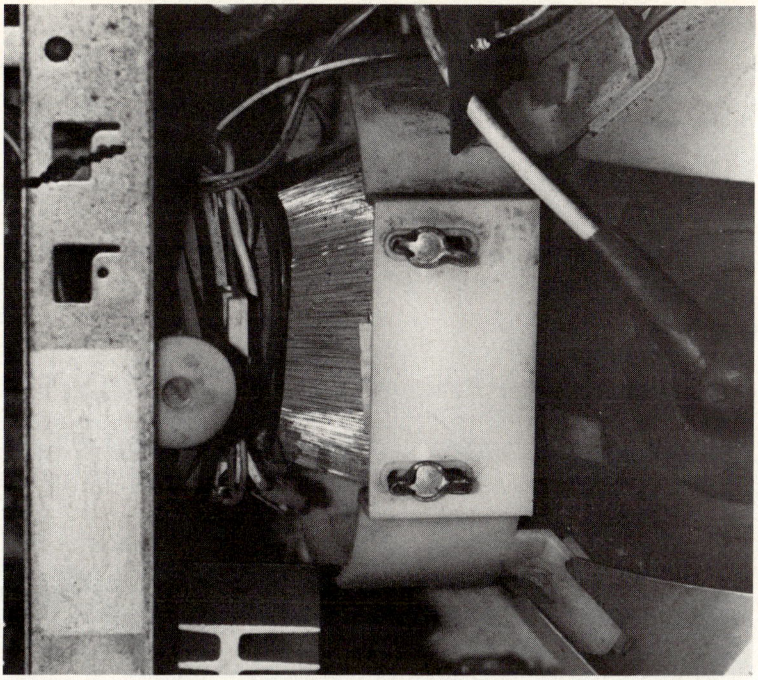

Figure 13-17. CRT with external yoke.

Some CRTs are made without the internal deflection plates, and instead use an external *yoke*, as shown in the illustration in Figure 13-17. Yokes are used to deflect the electron beam for applications requiring scanning patterns, such as television sets, computer terminals, or electronic games.

Phosphor Coating

The actual picture or waveform display results from the effect of the electron beam upon a *phosphor* coating on the inside of the face of the CRT. A phosphor is a chemical that has fluorescent properties. When excited by a stream of electrons, the phosphor sets up a glowing image that is retained for a fraction of a second. It is the glowing effect of the phosphor that you see on an instrument screen.

Several kinds of phosphor coatings are available, depending upon the application of a particular CRT. Some of the more common phosphor coating designations and their applications include:

- A P1 phosphor, which displays a green image
- A P4 phosphor coating, which causes a white image
- A P22 phosphor, which is used for color picture tubes, displaying, as required, red, green, and blue images.

SAFETY NOTE:

The envelopes for CRTs are made of glass and have a high vacuum. When the glass is cracked or broken, the CRT will *implode*. This means that the broken parts of the envelope collapse inwardly toward the vacuum. Because of the vacuum, considerable force is exerted. After imploding, pieces of glass can be scattered widely.

When a CRT is inside a rigid housing, such as a television cabinet, much of the implosion effect is contained by the cabinet. However, there can be considerable danger if a CRT implodes in an open area while it is being worked on or carried. Shards of glass from CRTs have been known to embed themselves in cement walls following an implosion. Be aware that handling an exposed CRT can involve considerable prospective danger. These devices should be handled and stored accordingly.

Discussion Questions/Problems

1. What is the function of a CRT?
2. Name at least three grid devices included in a typical CRT.
3. What is the function of the deflecting plates on a CRT within an oscilloscope?
4. What is the function of the phosphor coating on the inside surface of a CRT?
5. What inherent danger exists in handling and storing CRTs?

13.8 SEMICONDUCTOR DEVICES

The term *semiconductor* describes a family of materials or elements, either solid or liquid, with intermediate current conductivity characteristics. The way a semiconductor behaves with an electrical current is somewhere between the behaviors of conductors and insulators. Semiconductors might be called *partial* or *intermediate* conductors of electricity.

Semiconductor materials are also used to make electronic devices that serve as active components in electronic circuits. A basic characteristic of most semiconductors is that resistance decreases as heat increases. This means that as a semiconductor is heated its ability to conduct current increases.

The molecular structures of semiconductor devices incorporate two mechanisms for current flow. One is a surplus of electrons in some atoms. These electrons are available to flow toward positive materials. The other electrical mechanism inherent in semiconductor materials is the *positive hole*. This is a position within an atom where a negative electron has been dislodged. The atom or atoms that have lost electrons take on a positive charge. Thus, free electrons are attracted to these holes in these atoms.

Semiconductor materials that are to be used in making active electronic devices are modified to improve their electrical conductivity. This is done while the material is being prepared by adding impurities known as *dopants* to the basic semiconductor substances. This chemical modification serves to increase the number of free charges within the semiconductor materials, thereby promoting the movement of electrons when external voltages are applied. The type of impurity added determines the type of semiconductor device that will be created from the material.

- When a dopant increases the number of free electrons, the semiconductor device will be negative, known as *N-type*.
- When the number of free electrons in the basic semiconductor material is reduced, the device will be positive, or *P-type*.

Semiconductor Materials

A number of materials can be used as semiconductors. However, two are the most common. Most semiconductors are constructed from silicon (Si). Next in popularity is germanium (Ge). Either silicon or germanium can be used for N-type or P-type semiconductor devices, depending upon the type of dopant applied.

Forming Semiconductor Devices

When N-type and P-type materials are joined, the resulting device is referred to either as an *NP* or a *PN junction*. The structure of such a junction is shown in Figure 13-18. This establishes a boundary at the junction point, with the N material on one side and the P material on the other. A small internal contact potential, or separation, keeps the N and P charges apart.

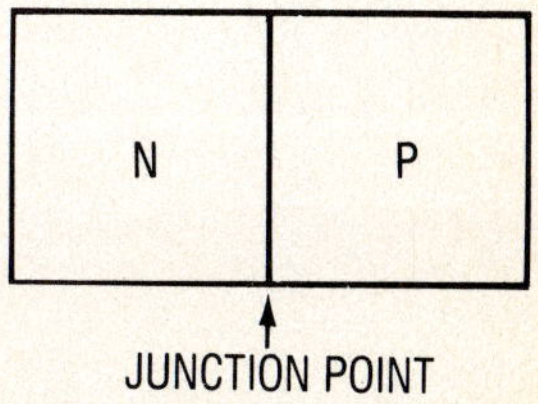

Figure 13-18. Diagram of NP junction structure.

It is this junction point that is useful in devising semiconductors for specific electronic applications. This is because control can be exerted at this relatively small, concentrated junction region.

For example, a PN junction is potentially useful because it can be connected into a circuit as a semiconductor diode. This semiconductor diode, shown by the symbol illustrated in Figure 13-19, can be used as a rectifier. Such semiconductor devices can be substituted for diode tubes.

Figure 13-19. Symbol for semiconductor diode.

As another example, a three-terminal device, either NPN or PNP, as illustrated in Figure 13-20, can be used to form a transistor to be used instead of a tube triode in amplifier applications. In this three-terminal construction, one unit of semiconductor material functions as an emitter. Another serves as the controlling device, or base. The final part of the transistor functions as the collector.

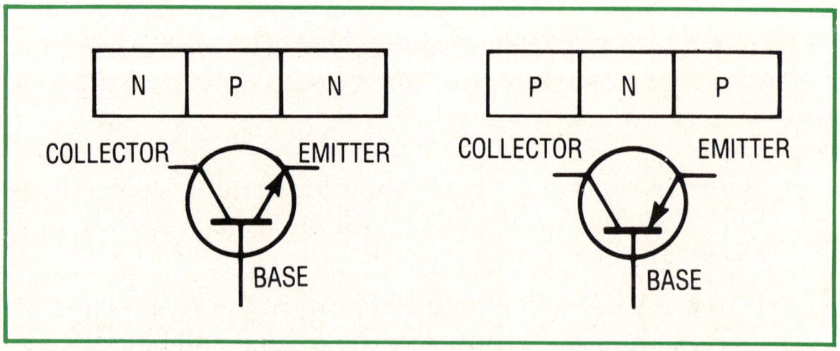

Figure 13-20. NPN and PNP structures of transistors.

These brief examples illustrate the principle of semiconductor construction. There are many other types of semiconductors. These are described and their applications explained in subsequent chapters.

Discussion Questions/Problems

1. What is a semiconductor?
2. How does response to heat affect semiconductor performance?
3. What are the two basic types of semiconductor materials and what are their primary characteristics?
4. What type of semiconductor device is indicated by the symbols *NP* or *PN*?
5. What type of semiconductor device is indicated by the symbols *NPN* or *PNP*?

13.9 SEMICONDUCTOR CHARACTERISTICS

To understand what semiconductors are and how they are applied, it is important to understand the basic atomic and molecular structures of these materials. Though this information is presented in Chapter 5, it is reviewed here as a basis for further discussions of semiconductors.

Both free electrons and positive holes exist in the outer electron shells of semiconductor atoms. This outer shell is called the *valence shell*, or orbit. The electrons of this outer shell are known as *valence electrons*.

When a positive hole develops in the outer shell of one atom, it is possible for that atom to share a valence electron with another atom. This joining of atoms through the sharing of valence electrons is called a *covalent bond*. Atoms joined through a covalent bond form a multi-atomic structure known as a *molecule*. Semiconductors derive their structural and performance characteristics from the nature of their molecular structures. For an illustration of how molecules are formed, consider Figure 13-21. This is a drawing showing the most common molecule found in nature—water. Water is formed through covalent bonding of two atoms of hydrogen and one atom of oxygen.

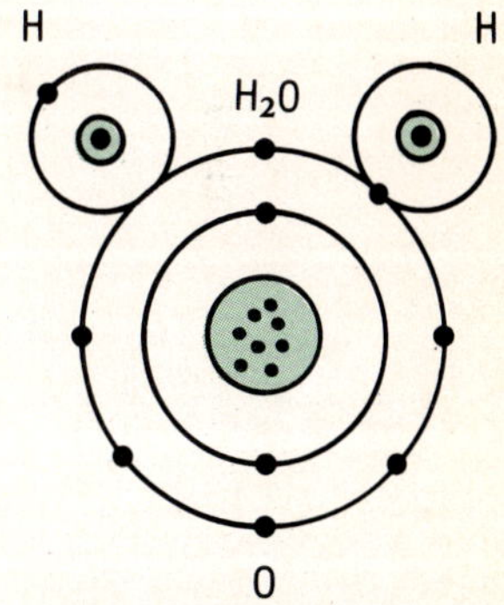

Figure 13-21. Diagram of water molecule structure.

Both silicon and germanium lend themselves to molecular bonding. In their molecular state, both of these materials form what are known as *crystal lattice* structures. These are cubes of solid crystal materials. A typical crystal lattice structure for silicon is shown in Figure 13-22. This type of solid crystal lattice structure is the reason why semiconductor devices are referred to as *solid state*, in contrast to the vacuum or gas in electron tube devices.

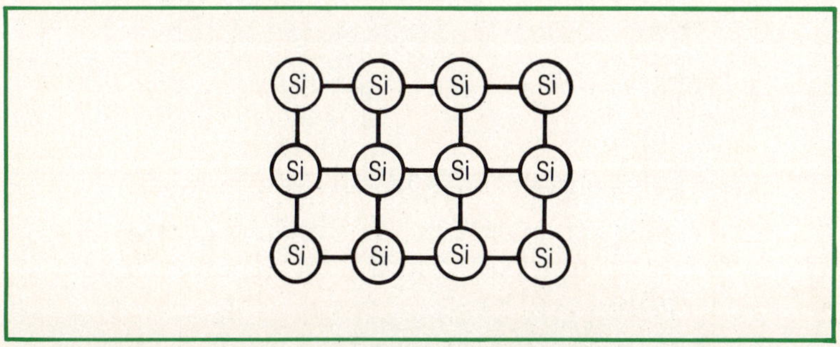

Figure 13-22. Typical crystal lattice structure for silicon molecule.

When pure semiconductor materials are used—with no dopent added—the result is known as an *intrinsic* semiconductor. When dopant is added, the material is known as an *extrinsic* semiconductor. In either case, the fabrication of semiconductors begins by "growing" pure silicon crystals in a laboratory. Continuous rods of silicon material are formed by a special process. These rods are then cut into thin wafers of silicon material. Figure 13-23 shows silicon wafers.

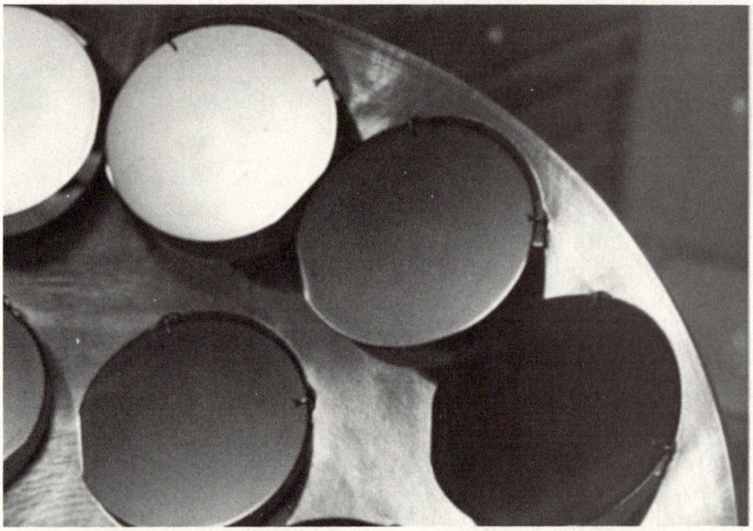

Figure 13-23. Silicon wafers.

Intrinsic and extrinsic semiconductors have opposite thermal conductivity characteristics. An intrinsic semiconductor has a negative temperature coefficient. That is, resistance decreases (and conductivity increases) when heat is applied. Extrinsic semiconductors exhibit opposite behavior. Resistance increases with rising temperature.

Discussion Questions/Problems

1. What is a valence electron?
2. What is a covalent bond?
3. What is a molecule and how is it formed?
4. What is a crystal lattice structure and how is this structure important in the formation of semiconductor devices?
5. What is an intrinsic semiconductor?
6. What is an extrinsic semiconductor?

ELEMENT	SYMBOL	ATOMIC NUMBER	VALANCE ELECTRONS
ANTIMONY	Sb	51	5
ARSENIC	As	33	5
PHOSPHORUS	P	15	5
ALUMINUM	Al	13	3
BORON	B	5	3
GALLIUM	Ga	31	3
INDIUM	In	49	3

Figure 13-24. Properties of common doping elements.

13.10 DOPING

Doping materials are selected for their ability to promote the formation of covalent bonds. This makes it possible for the dopants to integrate easily into the lattice structures of silicon or germanium. Most dopant materials have either three or five valence electrons. Depending upon the characteristics of the doping material, either N- or P-type semiconductors can be formed. Figure 13-24 is a chart showing the properties of many commonly used doping elements.

A basic characteristic of N-type semiconductors is that they have large numbers of free electrons. This imparts a negative charge to the semiconductor molecules. The electrons thus become what are called the *majority charges*. The positive holes within negative semiconductor materials are then called *minority charges*.

P-type semiconductors are the opposite. The positive holes represent the majority charge while the negative electrons are the minority charge.

When majority charges are moved by an outside voltage, this movement is called the *forward current*. The forward current in an N-type semiconductor material is an electron flow. In a P-type material, the forward current is caused by a flow of positive holes. Molecular structures promoting these two types of forward current are illustrated in Figure 13-25.

The forward current in a semiconductor is determined by the nature of the majority charge. The majority charge moves in the normal direction for its charge. If the majority charge is positive holes, they will flow toward the negative terminal of an applied voltage. At the same time, the minority charges within a semiconductor are affected by the majority

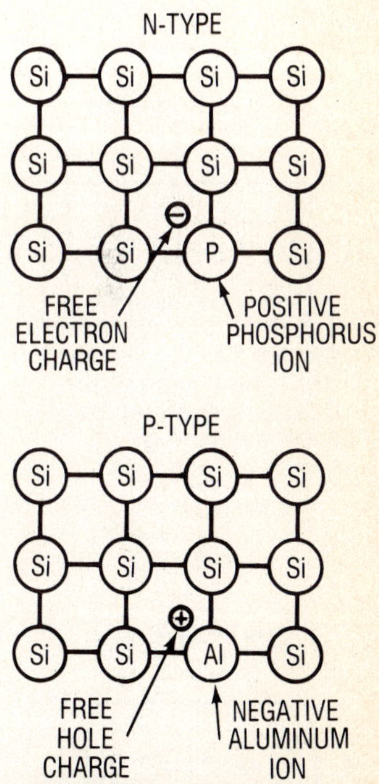

Figure 13-25. Diagrams of molecular structures for (A) N-type and (B) P-type semiconductor materials.

flow. The minority charges move in a direction opposite to the majority charges. This minority current is called *leakage current*. Leakage current increases with temperature. The higher the temperature, the greater the leakage current will be.

Discussion Questions/Problems

1. What is characteristic of the valance shells of doping materials?
2. What are the basic differences in electrical properties between N- and P-type semiconductor materials?
3. What does the term *majority current* mean?
4. What does the term *minority current* mean?

13.11 THE SEMICONDUCTOR DIODE

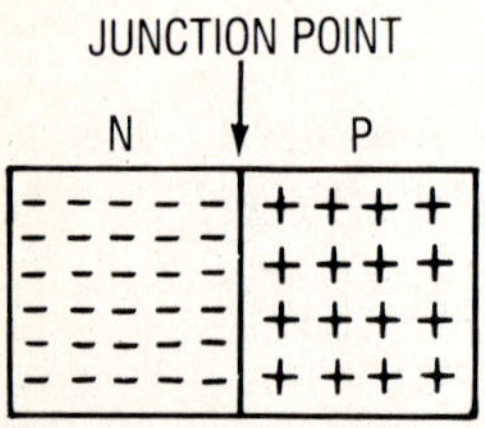

Figure 13-26. Diagram of N- and P-Type semiconductors joined to form junction diode.

A semiconductor diode is formed by joining N- and P-type semiconductors at a junction. The resulting diode is known as a *junction diode*. As shown in Figure 13-26, electrons and positive holes combine at the junction point, with initial movement of free electrons from the N side of the diode to the P side. As these free electrons cross the junction to P, the N-type material loses free electrons. Ultimately, the supply of electrons on the N side becomes depleted. At the same time, the positive holes on the P side are being filled. Thus, the availability of holes to be filled by the moving electrons is also depleted. This depletion is greatest, obviously, in the areas nearest the junction point, with depletion occurring on both the N and P sides. The areas adjacent to the junction form what is known as a *depletion region*, as illustrated in Figure 13-27. This depletion region extends only a short distance from the junction.

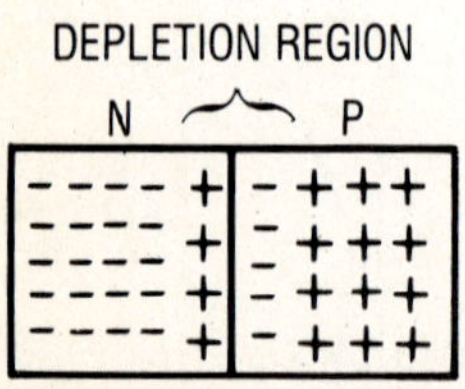

Figure 13-27. Diagram of NP junction with depletion region created.

The Depletion Region

The electrical properties in the depletion region ultimately change. Free electrons are the majority charges or carriers in the N-type material. In the P-type material, the positive holes are the majority charges or carriers. In the depletion region, electrical properties are changed in that there are no majority carriers.

As shown in Figure 13-27, the electrical charges of the affected materials in the depletion region are reversed. As free electrons leave the material on the N side of the diode, positive holes remain. Ultimately, these positive holes outnumber the negative electrons, imparting a positive charge value. The opposite effect occurs in the depletion region on the P side of the diode. The positive holes acquire so many negative atoms that the charge value is reversed. An area of negative charge is formed on the P side of the junction point. Opposing charges ultimately exist on each side of the junction.

The Potential Barrier

The depletion action at the junction cannot go on indefinitely. The reversal of charges in the depletion region ends pretty quickly. This is largely because the area itself is relatively small. The depletion action is brought to a close because each side accumulates enough charges to repel opposing charges. On the P side, for example, there is a negative area of charges between two positive areas, as shown in Figure 13-28. The same happens on the N side, where a positive area of charges is accumulated. Ultimately, these fields with reversed charges repel opposing charges, establishing a difference in potential which forms what is known as a *potential barrier* or *barrier voltage*. The term *barrier* indicates that the drift of charges has been halted, limiting the size of the depletion region within the diode.

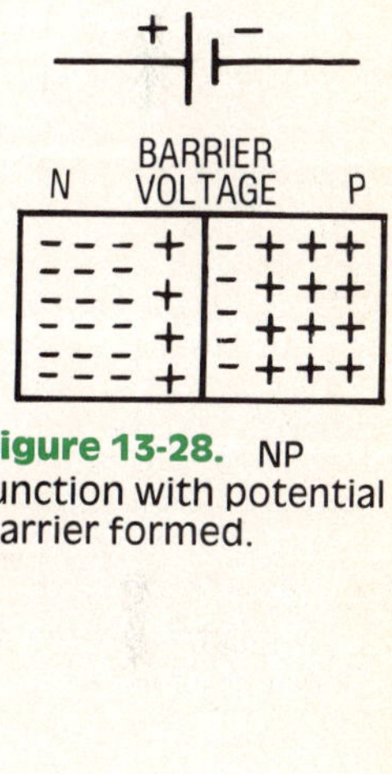

Figure 13-28. NP junction with potential barrier formed.

The potential difference in the barrier area serves to generate a very small voltage—usually measured in tenths of a volt. For germanium diodes, the barrier voltage (E_b) is 0.3 V. For silicon diodes, the barrier voltage is 0.7 V. Both these values are based on an ambient temperature of 25°C—or room temperature.

Diode Bias

The barrier voltage, it is important to remember, is generated internally within the diode. The barrier voltage exists when no external voltage is applied to the diode. When an external voltage is applied, the diode changes its functional patterns.

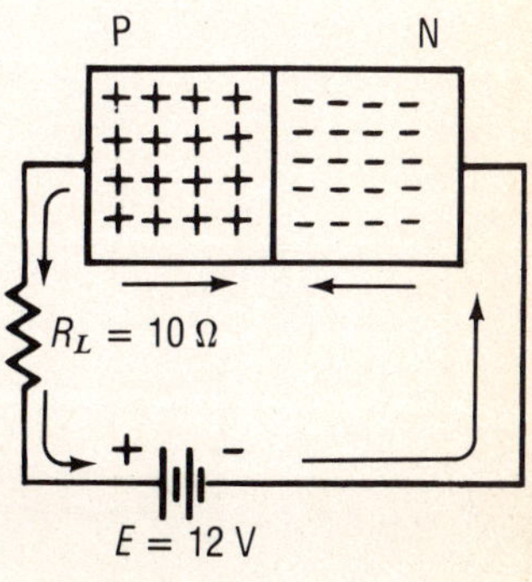

Figure 13-29. Circuit schematic with diode forward bias.

For example, consider the circuit in Figure 13-29. The diode has been incorporated in a circuit in which the negative pole of the battery is connected to the N side of the diode. In addition, a resistor has been positioned between the P side of the diode and the battery. In this circuit configuration, the electrons from the negative battery terminal repel the free electrons on the N side of the diode. This forces them to the junction, where they neutralize the positive ions in the depletion region. At the same time, the negative ions that had accumulated on the P side of the barrier are attracted toward the positive battery terminal. This causes neutralization of positive ions on the P side of the diode.

As a result of these effects, the ions that form the depletion region barrier voltage are neutralized. When this happens, the diode can conduct current. In the configuration shown in Figure 13-29, there is a flow of electrons through the diode toward the positive pole of the battery. However, this effect can occur only if the battery voltage is greater than the barrier voltage in the barrier area. When an external source is applied as shown in Figure 13-29, the effect indicated is that negative electrons and positive holes both flow from their respective sides of the diode toward the junction. Eventually, the exchange of ions that takes place results in the barrier being overcome by the negative electrons, causing a current flow. This condition creates what is known as a *forward-biased diode*. A forward-bias current flows within the diode from the negative to the positive side. When forward biased, the diode will offer only a small resistance to the current flow. This resistance is approximately equivalent to the effect of the barrier voltage: 0.3 V for germanium and 0.7 V for silicon.

Except for this small resistance, the forward-bias current (I_F) is a function of the amount of source voltage E, the forward-bias voltage (E_F), and the load resistance within the circuit. Calculations to determine forward-bias current apply Ohm's law. In this case, the battery voltage is reduced by the amount of the forward-bias voltage. The formula is as follows:

$$I_F = \frac{E - E_F}{R}$$

Computations to determine forward-bias current for the circuit in Figure 13-29 (assuming that a silicon diode is used) are as follows:

$$I_F = \frac{12 - 0.7}{10}$$

$$= \frac{11.3}{10}$$

$$I_F = 1.13 \text{ A}$$

When the polarity of source voltage is reversed, a condition known as *reverse-bias* occurs. This condition is illustrated in Figure 13-30. In this circuit, the positive pole of the battery is connected to the N side of the diode. The positive pole thus attracts the free electrons on the N side of the diode, leaving a growing number of positive holes on the N side. As this electron flow gains momentum, the width of the depletion region is increased. At the same time, a flow of electrons from the negative battery terminal enters the P side of the diode. These fill the holes in the positive ions, creating a flow condition in which the positive holes appear to move toward the junction. The ultimate effect is a flow of electrons toward the N side of the diode: a reverse-bias current.

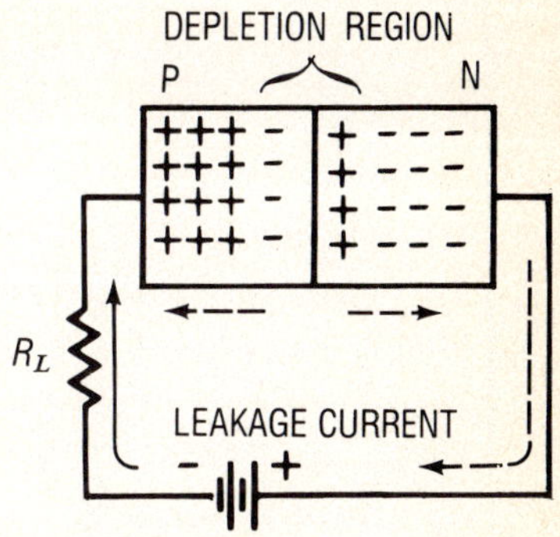

Figure 13-30. Circuit schematic with diode reverse bias.

There will always be some current leakage in a semiconductor device. This is a small current that flows in the direction opposite to the movement of majority carriers. This leakage current is set up by minority carriers and must be considered in evaluating the performance of a semiconductor diode.

Discussion Questions/Problems

1. What is a junction diode?
2. What is the depletion region of a junction diode?
3. What is the barrier voltage and how is it formed?
4. What is a forward-bias current?
5. What is a reverse-bias current?

13.12 SEMICONDUCTOR DIODE CHARACTERISTICS

A graphic representation of the forward- and reverse-bias characteristics for silicon diodes is shown in Figure 13-31. Note that the forward current remains at zero until the applied voltage level approaches the barrier level. There is, then, a breakthrough point at which the forward-bias and barrier voltages are equal. After that, current flow increases rapidly, in almost a linear manner.

A critical point, in terms of diode performance, is the breakdown point. This is the point on the graphic diagram at which the internal resistance within the diode breaks down due to an external applied reverse-bias voltage. The normal

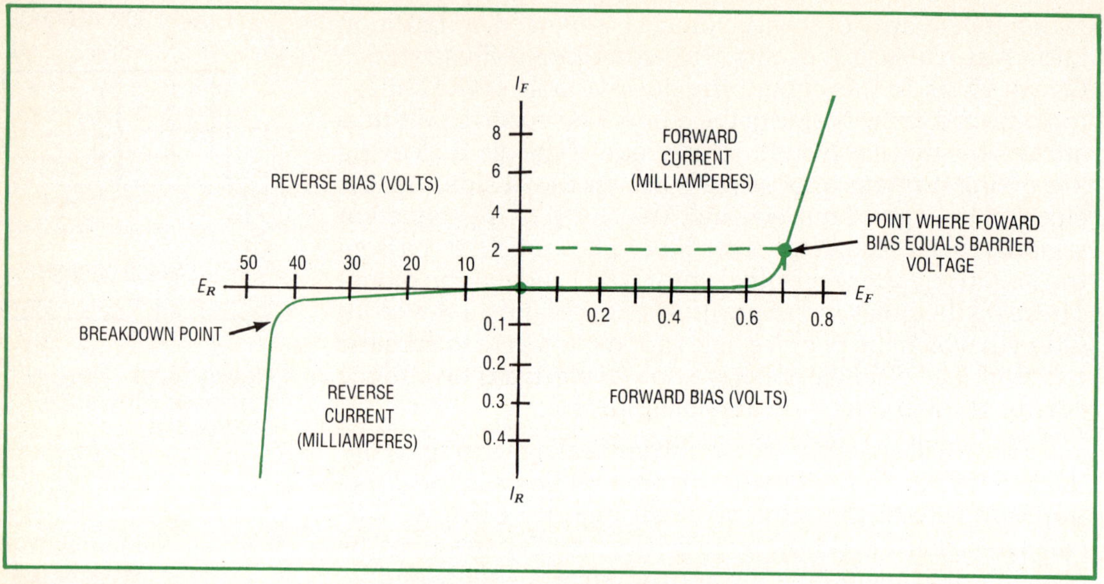

Figure 13-31. Forward and reverse bias for silicon diodes.

performance characteristics of the diode break down and large amounts of current are able to flow in the reverse direction. Junction diodes should not be permitted to operate beyond their breakdown points.

Figure 13-32 provides similar data for a germanium diode. Note that, for voltages greater than the barrier voltage, the slope of forward current versus forward voltage is more gradual than for silicon diodes. This is a basic difference in performance characteristics between germanium and silicon semiconductors.

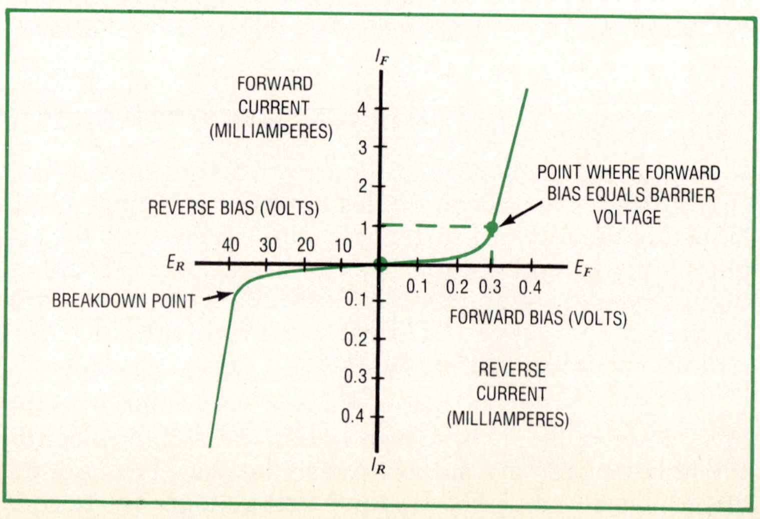

Figure 13-32. Forward and reverse bias for germanium diode.

Diode Ratings

The primary parameters used in rating diode performance are the forward-bias current characteristics and the peak inverse voltage (*PIV*) prior to the breakdown point. The *PIV* is the maximum reverse-bias voltage that can safely be applied to a diode without damaging it. These rating points vary widely according to types of diode and the practices of manufacturers.

Temperature Considerations

Comparatively speaking, semiconductor diodes are temperature-sensitive devices. The performance of a semiconductor diode is affected more by temperature than is the performance of a vacuum tube. Further, the reverse-bias current is more adversely affected by temperature than the forward-bias current. The degradation of service quality is straightforward: The higher the temperature, the higher the leakage current.

Diode Symbol and Numbering System

The symbol used to represent diodes is shown in Figure 13-33A. The straight line touched by the arrowhead in the symbol is the cathode. The arrowhead itself is the anode. An external pointer is usually used with this symbol to indicate the direction of current flow. (Electron flow is in the opposite direction.) In production diodes, the arrowhead is often printed on the case. Figure 13-33B shows the positions of the P and N semiconductor materials within the junction diode. Figure 13-34 shows a number of actual diodes.

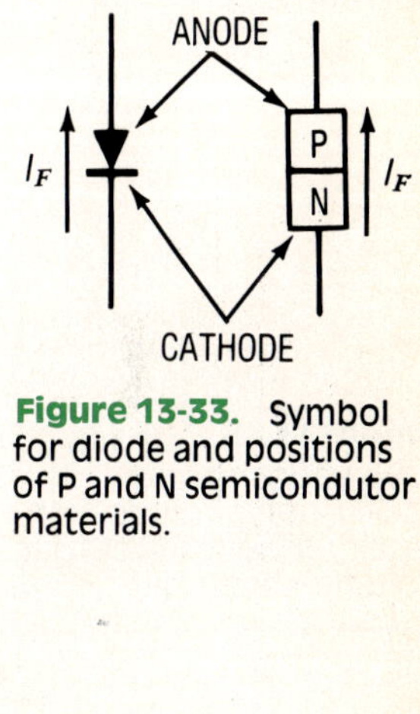

Figure 13-33. Symbol for diode and positions of P and N semicondutor materials.

The numbering system used for diodes usually has a prefix 1N. For example, a typical model number might be

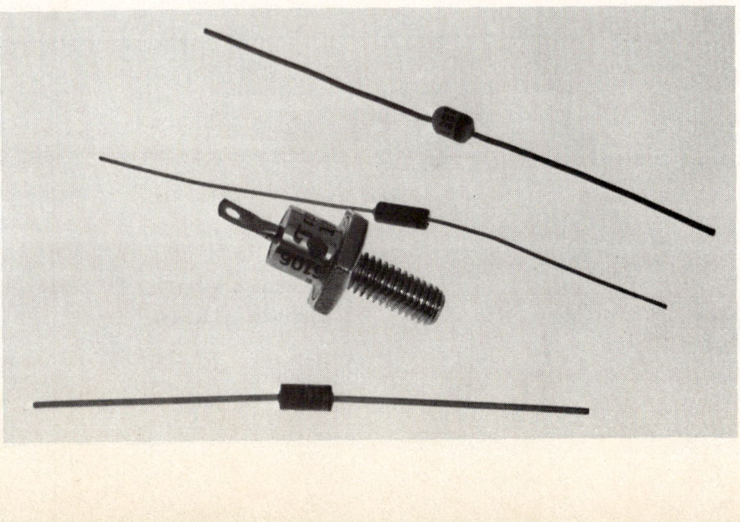

Figure 13-34. Typical diodes.

1N4001. The numbers after the prefix are assigned sequentially for models by the manufacturer. These numbers are used in cataloging junction diodes. With this number, you can identify detailed performance characteristics for the unit.

Discussion Questions/Problems

1. What is meant by breakdown point for diodes and how is it determined?
2. What does the abbreviation *PIV* stand for and what is its meaning?
3. How does temperature affect diode performance?
4. In which direction does current flow through a forward-bias diode?

13.13 DIODE APPLICATIONS

Junction diodes can be used within circuits to convert AC signals to single-directional DC current. This important, typical application stems from the fact that junction diodes conduct current in one direction only when they are functioning properly. This process and a typical rectifier circuit are shown in Figure 13-35. The unidirectional diode simply blocks the flow of current on the negative alternation of the AC current. Only the positive alternation is carried through the diode, producing a pulsating DC signal. This type of circuit is commonly used in DC power supplies that operate from AC current.

Figure 13-35. Circuit schematic showing application of diode with resulting input and output voltage waveforms.

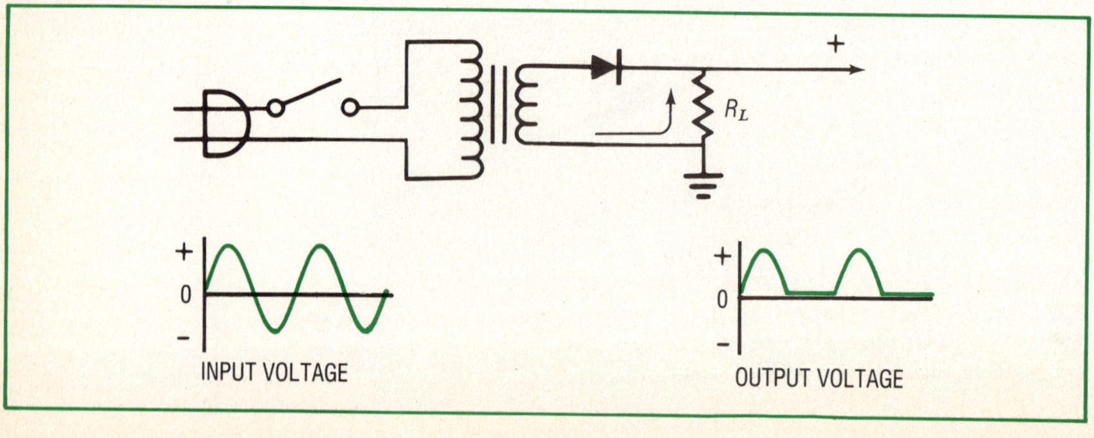

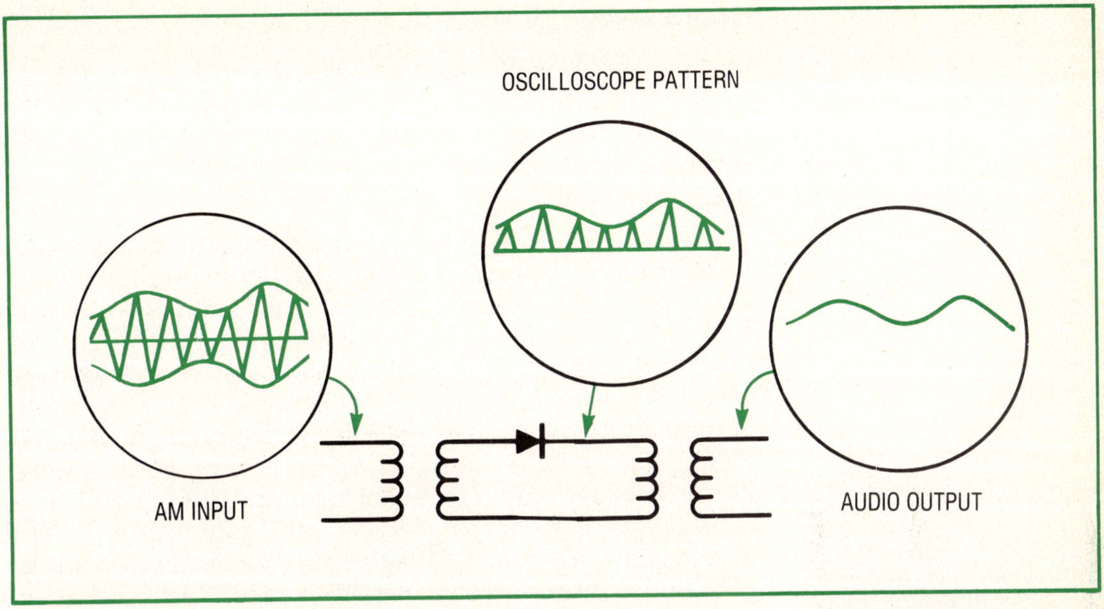

Figure 13-36. Detector circuit for AM radio receiver with typical oscilloscope signal patterns.

Another important application for junction diodes is to demodulate radio signals. Figure 13-36 shows a detector circuit for a typical AM radio receiver. The AM, or amplitude modulation, signal received by a radio is not audible in its transmitted form. This is because the transmitted signal contains the audio signal as modulation on a carrier signal. To produce an audible signal in the radio, it is necessary to rectify the modulated signal. Again, the negative alternations are eliminated. The circuit passes through only the positive alternations that carry the audio portion of the signal. These positive pulses are at a very high frequency and are filtered by a capacitor so only their average value is available at the output. This average value is the audio signal.

Other applications for junction diodes include waveshaping, switching, digital logic, and protecting electronic devices against random occurrences of inverse voltage.

Discussion Questions/Problems

1. How does a junction diode convert an AC signal to a DC signal?
2. How does a junction diode function in demodulating radio signals?

13.14 ZENER DIODES

Ordinary junction diodes, as explained above, are considered to be malfunctioning and in danger of damage when the reverse breakdown voltage is reached. This is because high reverse current produces heat that can destroy the semiconductor material.

However, a special type of diode is made that takes advantage of the properties of the breakdown point to deliver assured, reliable voltage sources. These diodes, known as *zener diodes*, can thus be used as voltage regulators. They are used in situations in which a stable, predictable voltage source is required. Figure 13-37 illustrates how a zener diode is specially designed to operate at a specific, single, reverse-bias voltage. Unlike other diodes, the zener unit operates at its rated voltage only. Zener diodes commonly available have ratings of 4.7 V, 5 V, 9 V, 12 V, and many others. The abbreviation for zener voltage is E_Z. Zener diodes can be obtained with a power dissipation capability from a few milliwatts to as high as 75 W.

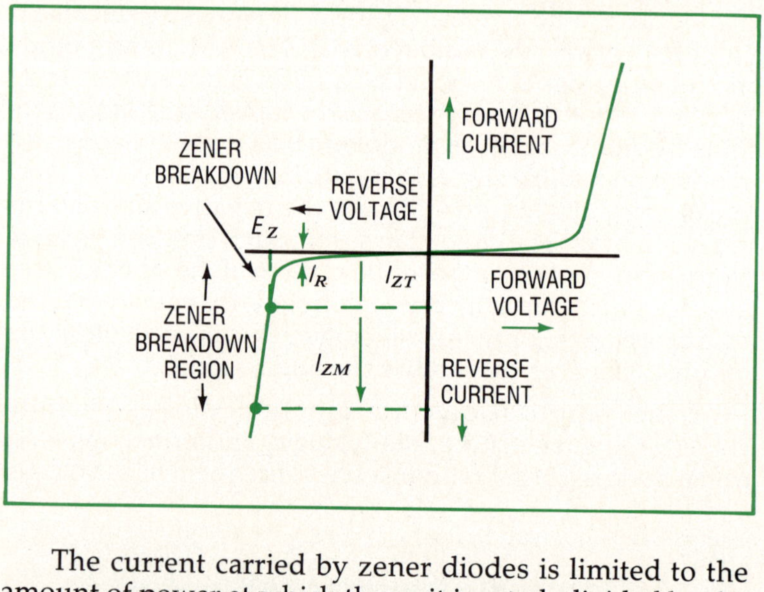

Figure 13-37. Reverse and forward bias of zener diodes.

The current carried by zener diodes is limited to the amount of power at which the unit is rated, divided by the zener voltage. The formula is as follows:

$$I_{ZM} = \frac{\text{Power rating}}{\text{Zener volts}}$$

Suppose, for example, a zener diode had a power dissipation rating of 1 W and a zener voltage of 9 V. The current capacity of this unit would be computed as follows:

$$I_{ZM} = \frac{1\ \text{W}}{9\ \text{V}} = 0.11\ \text{A}$$

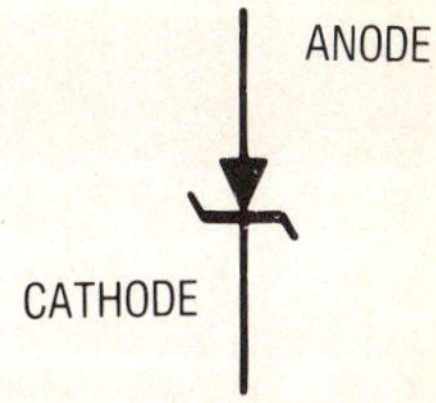

Figure 13-38. Symbol for zener diode.

The symbol for a zener diode is shown in Figure 13-38. Zener diodes also carry identification prefixes of 1N.

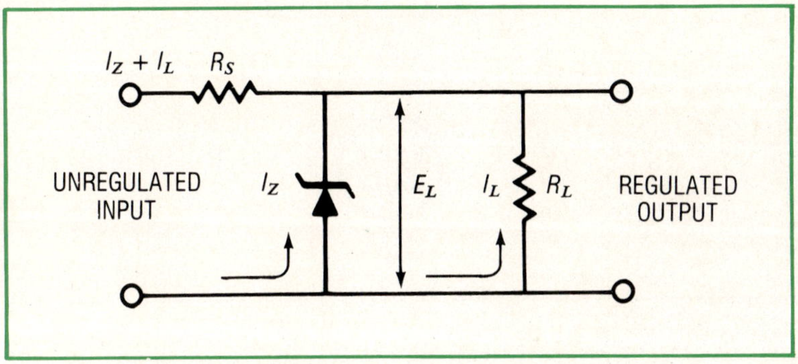

Figure 13-39. Circuit with zener diode.

Figure 13-39 demonstrates the operation of a zener diode as a voltage regulator. The circuit in this illustration is designed to accept unregulated input and produce a regulated output at a reliable voltage level. This is possible because the zener diode varies its current I_Z so the voltage E_Z remains constant.

Discussion Questions/Problems

1. What is the zener point?
2. Why can a zener diode operate at a steady, even voltage level?
3. How is the zener voltage determined?
4. What factors limit current through a zener diode?
5. What is the principal application for zener diodes?

13.15 THE BI-POLAR TRANSISTOR

A transistor is a dual-junction device with three semiconductor regions. There are two basic types of transistors. The NPN transistor, as illustrated in Figure 13-40, has two areas of N-type semiconductor material with an enclosed base of P-type material.

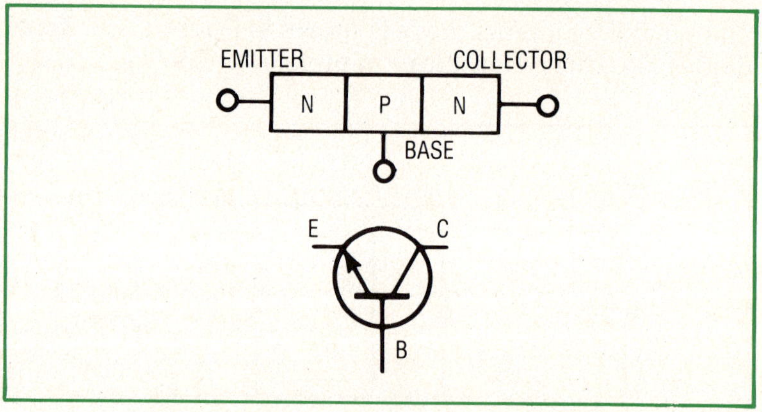

Figure 13-40. NPN transistor.

A PNP transistor has, as shown in Figure 13-41, two P-type regions and one N-type region.

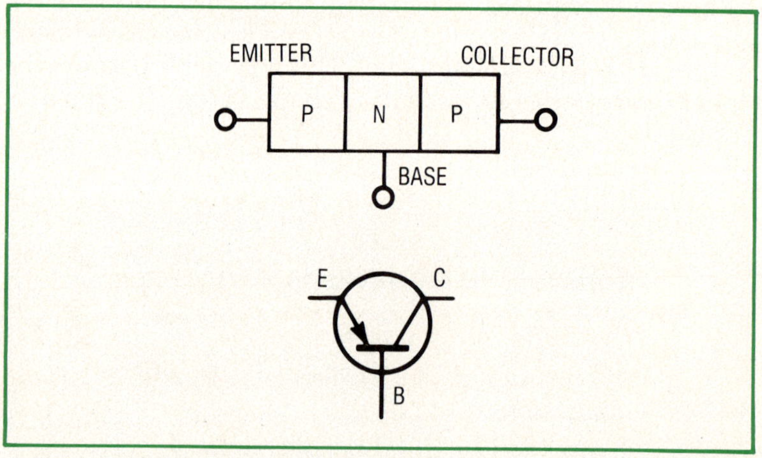

Figure 13-41. PNP transistor.

Several commercial transistors are illustrated in Figure 13-42.

As with diodes, transistors can be made from either silicon or germanium that has been doped for its specific application. The most important uses for transistors are as amplifiers.

In many applications, transistors take the place of triode tubes. However, the functional principle is somewhat different. Tubes are voltage devices and can be simulated in circuits as voltage sources. Bi-polar transistors control the flow of current through their terminals and are simulated as current sources.

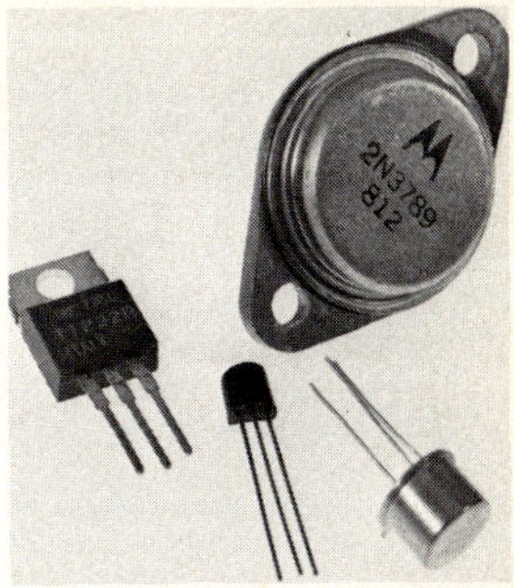

Figure 13-42. Commercial transistors. (Courtesy TRW, Inc.)

Biasing Transistors

To perform properly as amplifiers, transistors must be properly biased. Biasing techniques for transistors will naturally be more complex than for diodes because each transistor has two NP junctions, rather than one. The two junctions lie between the emitter and base on one side and between the collector and base on the other. The emitter-base junction is usually called the *emitter junction*. The collector-base junction is known as the *collector junction*. An emitter junction is shown schematically in Figure 13-43. Note that the emitter junction in this drawing is forward biased. This is the normal bias for the emitter junctions of NPN transistors.

Figure 13-44 shows the electron flow for the collector junction of an NPN transistor. Note that this is reverse biased, which is the normal bias for collector junctions of NPN transistors. As was the case for diodes, reverse-biased

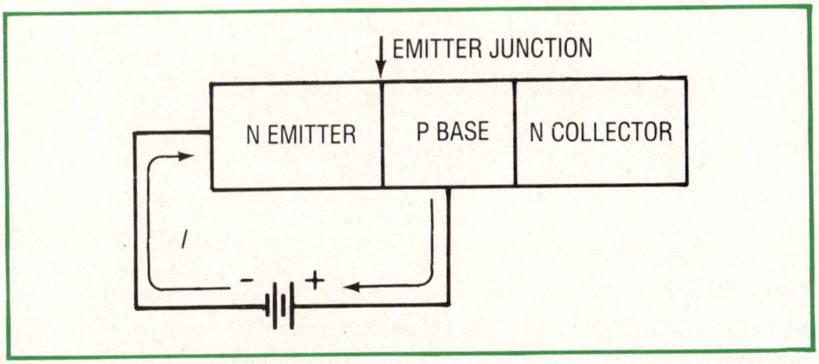

Figure 13-43. Circuit schematic of NPN transistor with emitter junction.

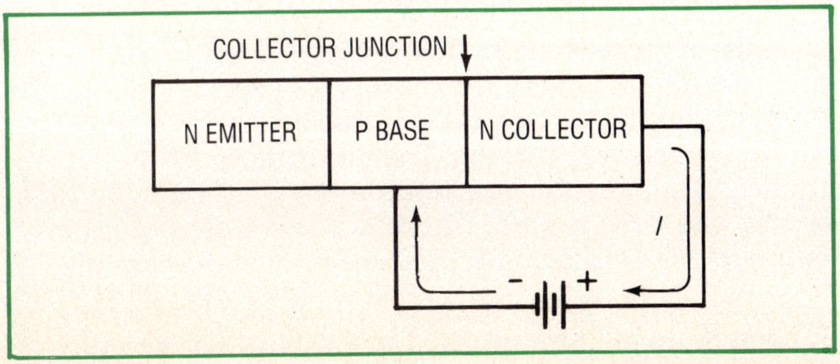

Figure 13-44. Circuit schematic of NPN transistor with collector junction.

junctions will always have some small amount of current leakage. For an NPN transistor to function in an amplifier circuit, the junctions must be connected so that the emitter junction is forward biased and the collector junction is reverse biased, as shown in Figure 13-45. A forward-biased PNP transistor is shown in Figure 13-46.

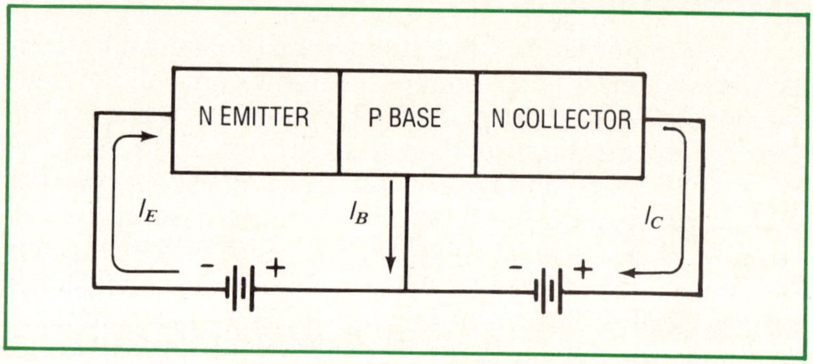

Figure 13-45. Schematic for current flow of NPN transistor.

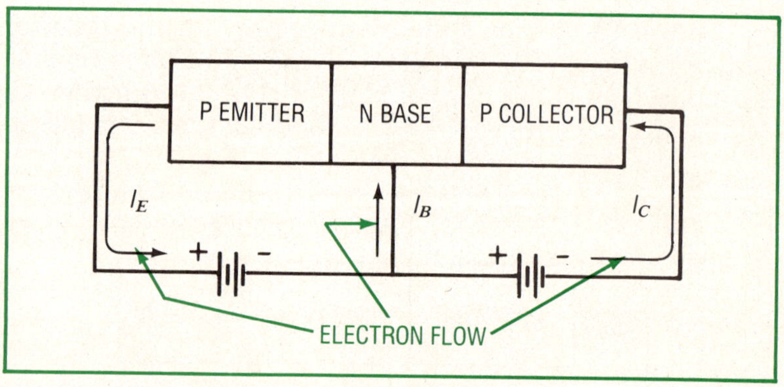

Figure 13-46. Schematic for current flow of forward-biased PNP transistor.

A transistor should never be thought of as consisting of two diodes, fixed back-to-back. A transistor has performance characteristics that are markedly different from those of a diode or a pair of diodes. This is because of the way a transistor is constructed, both physically and electrically.

One of the reasons for the way transistors perform is that the base region is very thin and is also lightly doped. The emitter region, by contrast, is thicker and has a greater number of majority carriers. Therefore, most of the electrons that cross the emitter junction do not combine with holes within the base region. There is comparatively little accumulation of electrons in the P material in the base region.

Because the base region is so thin, the barrier voltage is comparatively unimportant in establishing resistance to the flow of electrons. Rather, the attraction for electrons by the source battery is sufficiently great so that electrons flow through the entire transistor. Electrons move from the emitter, through the base, through the collector and out to the circuit. Some 95 percent of the current that flows through a transistor moves along this path to the positive voltage terminal. This is the majority current. It is known as the *collector current*.

Only about five percent of the current from a transistor, the minority current, flows from the base. This is called the *base current*. The formulas below are used to determine emitter current, collector current, and base current (in order) for bi-polar transistors.

$$I_E = I_B + I_C \qquad I_C = I_E - I_B \qquad I_B = I_E - I_C$$

The same type of behavior is characteristic for a PNP transistor, although in that case the voltage polarity must be opposite the polarity for an NPN transistor.

Note that these performance characteristics hinge upon a thin base region. Should the base region not be thinner than those of the collector and emitter, the transistor action described above would not occur. The barrier action of the base would be too large to overcome.

Each of the junctions in a transistor has the same barrier voltage and barrier functions as the single junction in a diode. For silicon transistors, the barrier voltage at each junction is 0.7 V; for germanium materials, the barrier voltage is 0.3 V.

Discussion Questions/Problems

1. How many junctions are there in a bi-polar transistor?
2. What is the most important function, or application, for bi-polar transistors?
3. What is the normal electron flow pattern for an NPN transistor?
4. What is the normal electron flow pattern for a PNP transistor?
5. How do barrier voltages for two-junction transistors compare with those for single-junction diodes?

13.16 TRANSISTOR AMPLIFIERS

Amplifier circuits utilizing transistors can be connected in three basic ways. These are known as the *common base (CB)*, the *common emitter (CE)*, and the *common collector (CC)* configurations. All of the illustrations presented previously were for common base circuits.

The term *common* when used in connection with transistor circuits can have two meanings. One is the common connection point for all of the current paths in a circuit. In a common-base circuit, the base is connected to this common point. In many instances, the word *common* can also refer to the circuit ground. This is true, for example, in Figure 13-47, which shows a common-base transistor circuit. E_{EB} is the forward-bias voltage applied to the emitter; E_{CC} is the collector supply. Note that the lead from the base is connected directly to both a common point and a ground. A common-base circuit of this type has forward biasing at the emitter junction and reverse biasing at the collector junction, as indicated in the circuit diagram. As another characteristic, the input signal is received at the emitter, with the output at the collector. It is important to note that, as shown in the waveforms at the input and output points, this type of circuit has no phase inversion.

Another characteristic of a common-base circuit is that it has a low-impedance input and a high-impedance output. This combination of characteristics means that a circuit of this type is considered to have good stability. Undesired transient signals in the input will have little effect on the output.

A circuit with a common emitter is shown in Figure 13-48. This is the most usual type of amplifier circuit using transistors. Note that the signal input is to the base (E_{BE}) and output is from the collector. The emitter is connected to ground. This makes for the highest possible current gain. This type of circuit inverts the phase of the signal, a characteristic which may be useful to the designer.

Figure 13-49 illustrates a common-collector configuration for an amplification circuit. In this circuit, the signal is also input to the base (E_{BB}). However, the output is through the emitter, while the collector is connected to ground. An advantage of this configuration is that there is a high input impedance. This can be valuable in matching characteristics of devices in which the circuits are used. A common-collector circuit does not invert phase.

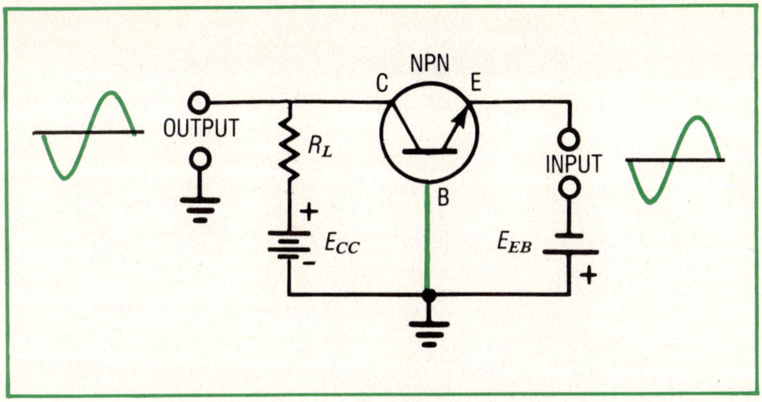

Figure 13-47. Circuit schematic for common-base transistor, showing input and output waveforms.

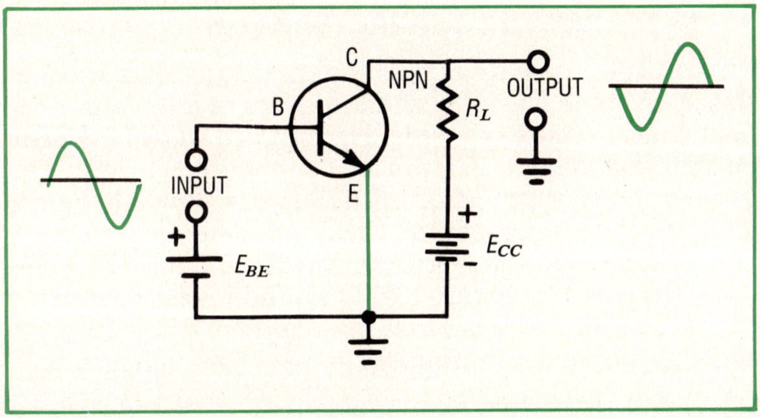

Figure 13-48. Schematic of amplifier circuit with common emitter, showing input and output waveforms.

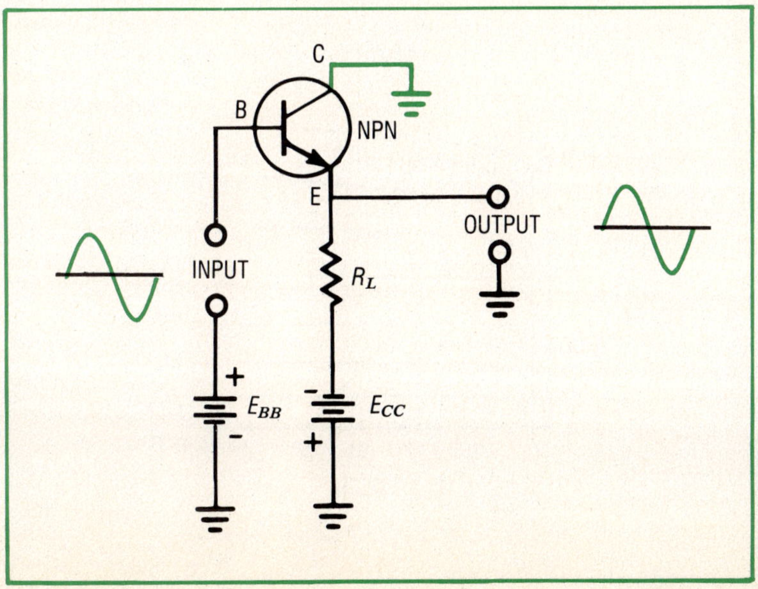

Figure 13-49. Schematic for amplifier circuit with common collector, showing input and output waveforms.

Discussion Questions/Problems

1. What are the meanings of the word *common* in electrical circuits?
2. Where is input applied in a common-base circuit?
3. Where is output derived from a common-emitter circuit?
4. Where is input applied in a common-collector circuit?

13.17 TRANSISTOR CHARACTERISTICS

Electrical performance characteristics of transistors are nonlinear. In some ways the characteristics of a transistor are analogous to those of a tube. Figure 13-10 shows the plate current and voltage as a function of the grid current. The analogous collector characteristic curve is shown in Figure 13-50. For a common-emitter circuit, the amount of collector current generated varies according to the amount of base current. The curves in Figure 13-50 show how collector current varies for a number of different base currents. These base currents are nonlinear in comparison with base voltages.

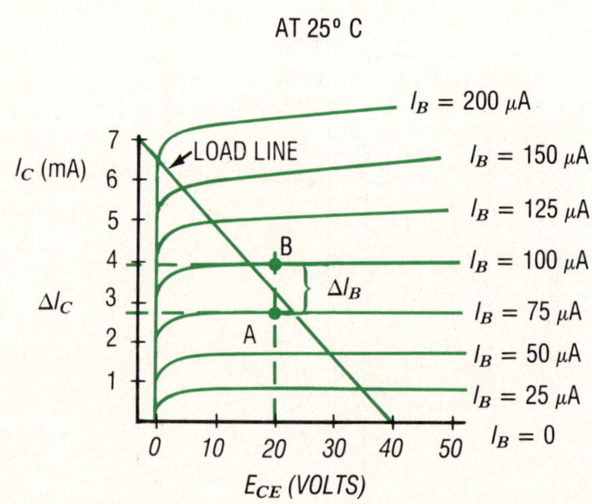

Figure 13-50. Performance characteristics of transistors.

The current gain, or amplification, of a common-emitter transistor is the ratio (indicated by the Greek letter *beta*, β) of the change in collector current divided by the corresponding change in base current. The formula is as follows:

$$\beta = \frac{\Delta I_C}{\Delta I_B}$$

As an illustration, suppose the collector current is 6 mA and the base current is 30 μA. The beta ratio, as calculated below, would be 200.

$$\beta = \frac{6 \text{ mA}}{30 \ \mu\text{A}} = 200$$

Note that the ratio is given as a numeric value, without electrical dimensions. In most transistors, beta values range from ten to over 200. This ratio is called the *transistor forward-current transfer ratio*. Different notations are used to identify these ratios for AC and DC measurements. For AC the notation is h_{fe}. For DC the notation is h_{FE}.

When a transistor is applied in a common-base circuit, the forward-current transfer ratio is called the *alpha* (α) *gain*. The formula for determining alpha gain is as follows:

$$\alpha = \frac{\Delta I_C}{\Delta I_E}$$

If the alpha gain for a given transistor is known, the beta value can be developed by applying the following equation:

$$\beta = \frac{\alpha}{1 - \alpha}$$

Conversely, if the beta is known, the alpha can be developed by applying the following equation:

$$\alpha = \frac{\beta}{\beta + 1}$$

Beta Cutoff Frequency

As the signal frequency applied to a transistor increases, the current gain decreases. Ultimately, the current gain of a transistor reaches a cutoff point. This is known as the *beta cutoff frequency*. This is the frequency at which the collector current of a common-emitter transistor will fall to $\sqrt{2}$ times its low-frequency value. In other words, no matter how much additional input signal current is applied, the output current of a common-emitter transistor will be limited to 141 percent of the base low-frequency current. The symbol for the base low-frequency value is f_{ae}.

Performance at frequencies above the beta cutoff level will be seriously reduced for a common-emitter transistor circuit. The notation for the cutoff frequency of a common-base transistor circuit is f_{ab}. The cutoff frequency for a common-base transistor circuit is usually substantially higher than for a common-emitter transistor circuit. In terms of applications, this means that a common-base transistor circuit can amplify voltage at much higher frequencies. Conversely, however, a common-base transistor circuit cannot provide current gain.

Discussion Questions/Problems

1. What is the meaning of the term *nonlinear device*?
2. What is beta?
3. What is alpha?
4. What is the meaning of transistor forward-current transfer ratio?
5. What is the beta cutoff frequency?

13.18 TRANSISTOR RATINGS

Transistors are rated by their manufacturers. These ratings are designed to establish operating parameters and conditions which will protect the transistor from overload. Some of the important rating parameters are described below.

Emitter Breakdown Voltage

The emitter breakdown voltage (V_{EBO}) is measured with the transistor's collector open. That is, the collector lead is not connected into the circuit. The emitter breakdown voltage is the amount of reverse-bias voltage required to break down the emitter junction. Voltage applied to the emitter junction should never exceed this rating.

Collector Breakdown Voltage

The collector breakdown voltage (V_{CBO}) is the level of the reverse bias on the collector junction required to produce a sharp increase in collector-to-base current. This rating is measured with the emitter open. Voltage applied to the collector lead should not exceed this rating.

Maximum Collector and Emitter Currents

Standard transistor specifications set limits on current that should be applied to the collector and emitter. These specifications should not be exceeded.

Collector-to-Emitter Leakage Current

The collector-to-emitter leakage current (I_{CEO}) can be an important factor in the design of some circuits. It is desirable to select transistors with the lowest available collector-to-emitter leakage current because excessive leakage can cause considerable problems in certain types of circuits. This measurement is made with the base lead disconnected.

Maximum Collector Dissipation

This rating measures the heat dissipation of transistors. A heat sink like those shown in Figure 13-51 can be used to help dissipate the heat and avoid transistor overheating. The transistor should be designed so that the heat generated internally is effectively transmitted to some external surface, which, in turn, is in contact with the heat sink. The measurement itself should be taken at room temperature, 25°C. The maximum collector dissipation rating is usually stated in watts.

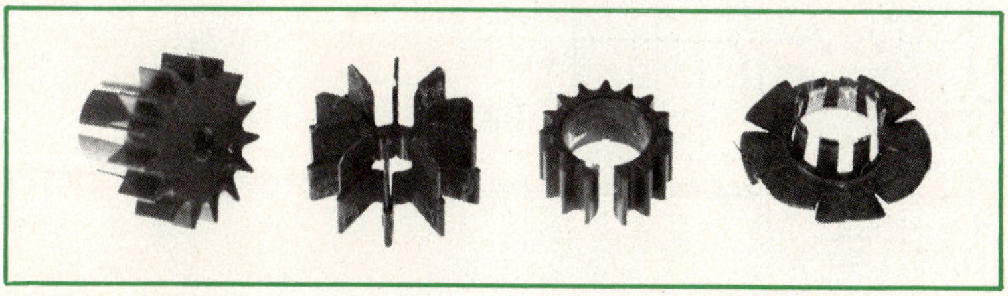

Figure 13-51. Heat sinks for transistors.

Temperature Ratings

These ratings, provided by manufacturers, provide operating temperature ranges for the transistors covered. Again, heat sinks can be used to dissipate excess heat and to extend the operating temperature range for individual transistors.

Discussion Questions/Problems

1. What is emitter breakdown voltage?
2. How is a transistor connected in measuring collected breakdown voltage?
3. Why is collector-to-emitter leakage current an important rating for transistors?
4. What is the application for heat sinks in transistor circuits?

13.19 TRANSISTOR AMPLIFIER CIRCUITS

An amplifier is a device that generates at its output terminals a signal which is essentially the same as the input signal, but of greater magnitude. Input and output signals are essentially the same in shape and timing. The magnitude, however, is increased. One of the most commonly used amplifiers today is the common-emitter amplifier circuit. A circuit of this type is illustrated in Figure 13-52. Input to this circuit is to the base and emitter.

Figure 13-52. Schematic of common-emitter amplifier circuit, showing input and output waveforms.

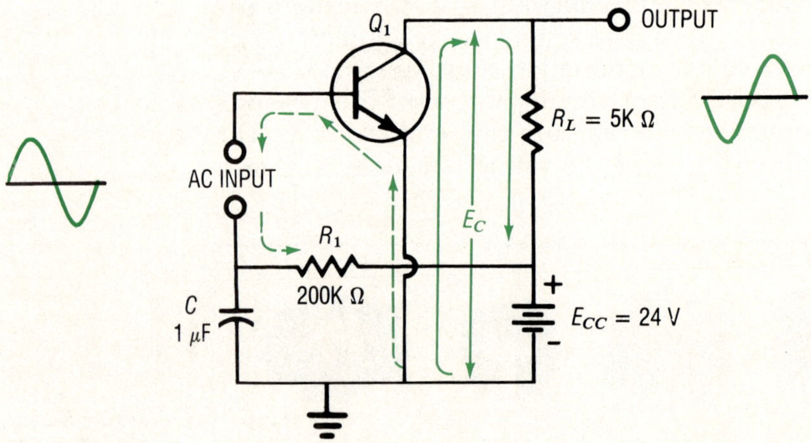

The capacitor (C) is selected so that its reactance at the signal frequencies is very low. C acts as a short circuit at signal frequencies, so the signal is thus applied between base and ground. The resistor (R_1) is selected to provide a bias current for the base, from the battery (E_{CC}). Notice that the AC input must have some means to complete the DC circuit from R_1 to the base. The collector current includes both the input and the output signal.

In this particular circuit, the output voltage across the load resistor (R_L) will be greater than the voltage of the input signal. This circuit has amplified the input signal. There is also a substantial current amplification in this circuit. Since both output current and output voltage are much greater, there is an overall increase in power output for a circuit of this type.

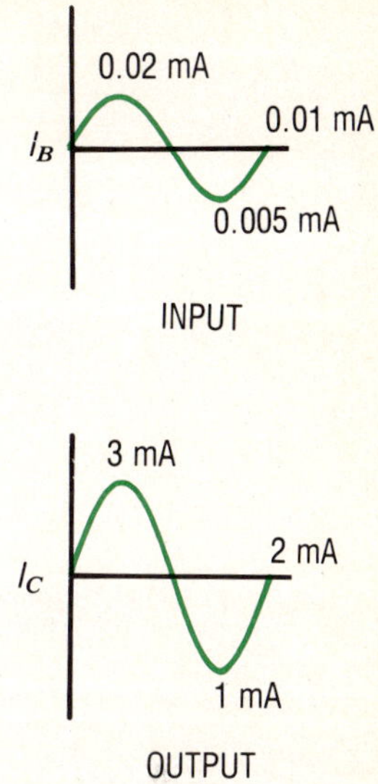

Figure 13-53. Input and output curves for circuit of Figure 13-52.

One way to evaluate a common-emitter amplifier is to review a typical collector characteristic curve for input and output. Curves of input and output current for the circuit in Figure 13-52 are shown in Figure 13-53. These curves demonstrate that a small amount of input current serves to control a large amount of output current. In this circuit, then, variations in input current cause corresponding—and amplified—variations in output current.

As the input signal varies in the circuit shown in Figure 13-52, a varying voltage drop is developed across the load resistor (R_L) which is the collector load. This varying load voltage can be described with the following formula:

$$E_C = E_{CC} - I_C R_L$$

To illustrate, suppose the collector current for the circuit in Figure 13-52 is 2 mA. Corresponding base current is 0.01 mA. Collector voltage would be computed as follows:

$$2 \text{ mA} \times 5\text{K } \Omega = 10 \text{ V}$$
$$E_C = 24 \text{ V} - 10 \text{ V} = 14 \text{ V}$$

However, when base current is at a level of 0.02 mA, and collector current is correspondingly at 3 mA, the values can be computed as follows:

$$3 \text{ mA} \times 5\text{K } \Omega = 15 \text{ V}$$
$$E_C = 24 \text{ V} - 15 \text{ V} = 9 \text{ V}$$

As base current increases, collector voltage decreases. In this situation, output voltage is 180 degrees out of phase with the input signal. In other words, there is a phase inversion.

Transistor Gain in an Amplifier Circuit

Transistor circuits can be designed with either voltage amplification or current amplification. Together, the factors of voltage gain and current gain can be used to determine power amplification.

Voltage gain is determined, as shown below, by dividing voltage out by voltage in:

$$\text{Voltage gain} = \frac{\Delta E_{out}}{\Delta E_{in}}$$

Current gain is determined by dividing current output by current input. The formula is:

$$\text{Current gain} = \frac{\Delta I_{out}}{\Delta I_{in}}$$

Given these values, power amplification is determined by multiplying voltage gain by current gain:

$$\text{Power amplification} = \text{Voltage gain} \times \text{current gain}$$

Discussion Questions/Problems

1. What is an amplifier?
2. Which transistor connection configuration is most commonly used for amplifier circuits?
3. How does the voltage across the load resistor in a common-emitter amplifier circuit compare with the input voltage?
4. What is the phase relationship between base current and collector voltage in a common-emitter amplifier circuit?

SPECIAL SEMICONDUCTOR DEVICES

14.1 TYPES OF DEVICES

In addition to the devices reviewed in the previous chapter, you will also encounter a number of other types of semiconductors in electronic circuits. Many of these are designed to meet special needs or for special applications. These special devices fall into a series of groups, or families:

- Special diodes
- Special transistors
- Thyristors
- Optoelectronic semiconductors.

This chapter describes these devices and some of the special applications for which they are designed.

Discussion Questions/Problems

1. Give reasons why special semiconductor devices are necessary.
2. What are some groups of special semiconductor devices?

14.2 VARACTOR DIODES

One of the devices produced to meet special application needs is the *varactor diode*. Several different symbols are used to represent varactor diodes in electronic drawings. These are shown in Figure 14-1.

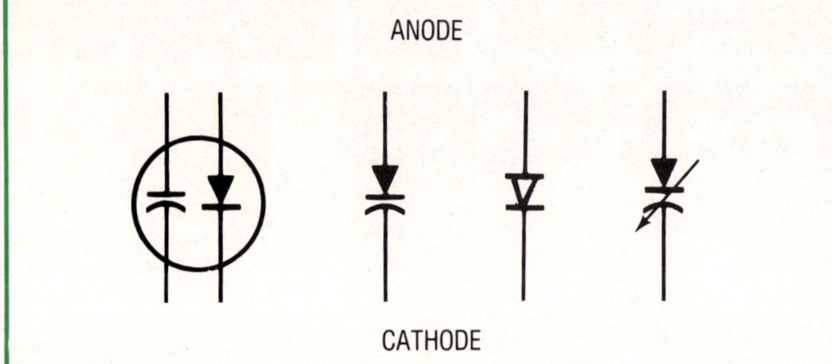

Figure 14-1. Symbols for varactor diodes.

A varactor diode is a semiconductor device designed with special capacitance characteristics. Varactor design takes advantage of characteristics that are basic to any diode junction. When a diode is connected so that it is reverse biased, the majority carriers are moved away from the junction. This leaves a wide depletion area. As reverse-bias voltages are varied, there are also variations in the extent of the depletion area. The depletion area becomes wider or narrower. As the depletion area becomes wider, it acts as an insulator, becoming a zone through which no current can flow.

Under these conditions, the N and P sections develop a relationship similar to that of the electrodes of a capacitor. The depletion area of the diode functions in the same manner as the dielectric separating the electrodes of a capacitor. Changes in the depletion area are similar to the changes in capacitance that take place when the properties of the dielectric in a capacitor are varied.

Normally, diodes have only a very small amount of internal junction capacitance. However, in the case of varactor diodes, it is possible to design them with substantially more internal capacitance. They are used in circuits just as variable capacitors are used.

To operate as capacitors, varactors are reverse biased, with a voltage less than the breakdown voltage rating. As reverse voltage is increased, the depletion area widens.

Because capacitance varies inversely with the thickness of dielectrics, the capacitance of a varactor diode decreases as the reverse voltage increases. Therefore, the capacitance of varactor diodes varies inversely with reverse-bias voltage. The variation in capacitance as a function of reverse-bias voltage varactor diodes is nonlinear. A curve representing capacitance vs. voltage for a varactor diode is illustrated in Figure 14-2.

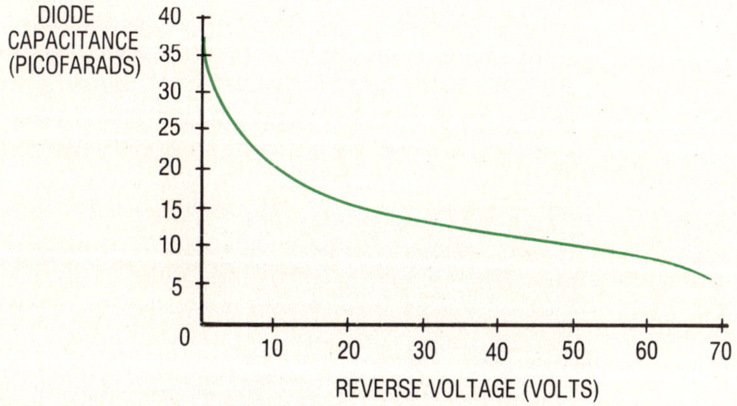

Figure 14-2. Relationship of capacitance and voltage for varactor diodes.

A significant performance difference between varactor diodes and ordinary capacitors lies in temperature sensitivity. As a general characteristic, varactor diodes are much more sensitive to temperature than ordinary capacitors. Each manufacturer provides a temperature coefficient (*TC*) indicating the effect of temperature upon capacitance for each model of varactor diode. The *TC* is expressed as a percentage of capacitance change for each degree C of temperature.

Varactor Diode Characteristics

A varactor diode attains its maximum Q at a level just before the breakdown voltage is reached. This is the same quality factor discussed in Chapters 8 and 9.

Since Q is X_C/R, Q will vary with frequency. Therefore, the Q factor is usually indicated by manufacturers as a value that applies at a specific frequency. This specific frequency is called the cutoff frequency (F_{CO}).

Because leakage current can seriously degrade the Q factor of a varactor diode, it can be important to know the maximum leakage current for any given device. Leakage for varactor diodes is usually in the microampere range.

Construction of Varactor Diodes

Varactor diodes are usually made of silicon. However, certain high-frequency units—designed to operate in microwave ranges—are made from gallium arsenide.

Applications for Varactor Diodes

Varactor diodes are electronic capacitors whose capacitance can be adjusted by an applied voltage. They are extremely useful in tuning circuits because they provide better reliability and accuracy of adjustment than traditional capacitors. Varactor diodes are particularly useful as the tuning component in the tank circuits described in earlier chapters. The primary application for varactor diodes is in tuners for broadcast receivers, including television sets, radios, citizens band units, and other radio frequency (RF) range devices.

Discussion Questions/Problems

1. What is the most unusual characteristic of a varactor diode?
2. What is the function of the depletion area at the junction of a varactor diode?
3. What is the principal application of varactor diodes?
4. What is the *Q* factor as it applies to varactor diodes?

14.3 TUNNEL DIODES

The chief characteristics of a *tunnel diode* are its ability to operate at high frequencies and its negative resistance region. Thus, tunnel diodes are included in many signal generators, as part of the *oscillator circuit*.

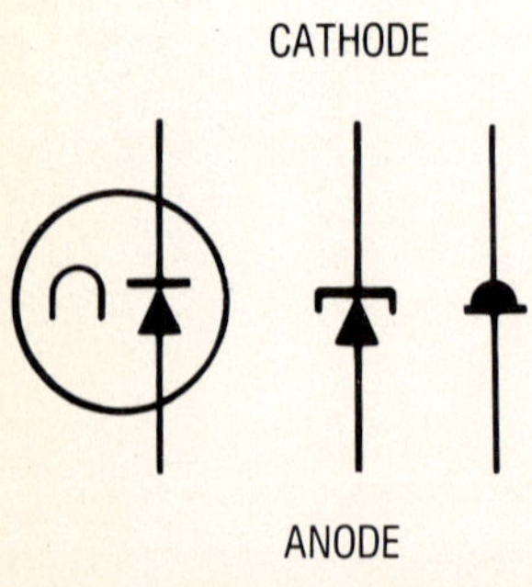

Figure 14-3. Symbol for tunnel diodes.

In their design, tunnel diodes have characteristics that are notably different from those of ordinary diodes. Tunnel diodes are heavily doped and have extremely small depletion areas. Symbols used to represent tunnel diodes in electronic diagrams are shown in Figure 14-3.

Applications for tunnel diodes are built around the fact that they have an extremely low reverse breakdown voltage. Therefore, under reverse-bias conditions, tunnel diodes conduct large amounts of current. However, at low forward-bias voltages, electron flows through tunnel diodes are at extremely high velocity. This is because of the high concentration of charges on both sides of the junction. The name for this device stems from this behavior: Electrons appear to tunnel through the junction point. As this tunneling action takes place, a level is reached at which an increase in forward voltage causes a decrease in forward current. This is an effect of the heavy doping applied when fabricating the materials.

Tunnel Diode Characteristics

A graphic representation of the operating parameters of tunnel diodes is shown in Figure 14-4. As indicated, the tunnel diode conducts high levels of reverse currents when reverse voltage is applied. However, a more important characteristic is that, when forward-bias voltage is applied, current increases as voltage is increased. This continues up to the point identified in Figure 14-4 as the peak voltage (E_P). At this point, voltage/current characteristics change. Continued increase in forward voltage actually decreases forward current.

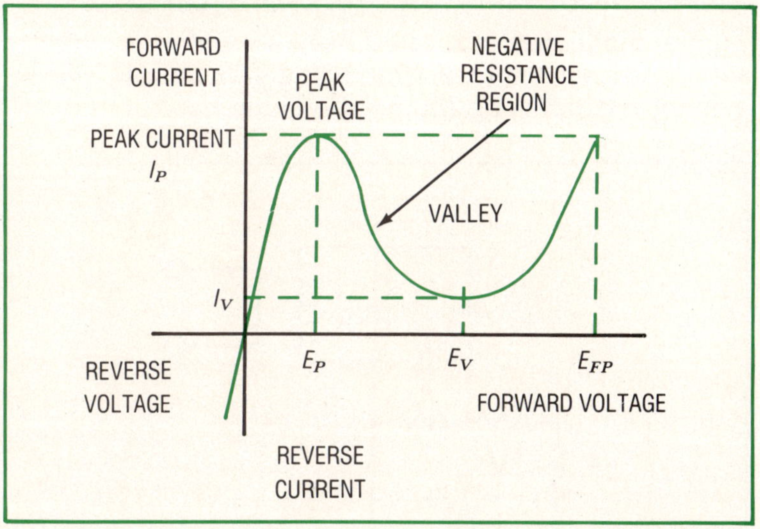

Figure 14-4. Operating parameters or characteristics of tunnel diodes.

After some continued increase in voltage, current will again increase. The portion of the cycle at which the current is at a minimum is called the *valley current* (I_V). Similarly, the voltage at this point is called the *valley voltage* (E_V). The left side of this valley area is identified as the tunnel diode's *negative resistance region*. When the low point in the valley current is reached, tunnel action ends. This clearly defined negative resistance region serves to make the tunnel diode useful for its major application as an oscillator.

The negative resistance characteristics of tunnel diodes are unique. According to Ohm's law, an increase in voltage should lead to a corresponding increase in current through a resistor. Because the current decreases with an increase in voltage along this segment of the tunnel diode's characteristic, it must be a negative resistance.

If a tank circuit is made up of a capacitor and an inductor, the inherent resistance in the inductor will quickly damp out any tendency to oscillate. However, when a tunnel diode is used, the negative resistance of the diode can be made to balance out the positive resistance of the inductor. The tank circuit will oscillate at its resonant frequency.

To bias a tunnel diode properly for an oscillator application, all values should be known, including peak current, peak voltage, valley current, and valley voltage. This information is provided by manufacturers of tunnel diodes. Specifications usually include negative conductance and negative resistance characteristics, stated in ohms.

Tunnel Diode Construction Characteristics

Tunnel diodes are fabricated from a number of types of semiconductor materials, including germanium, silicon, gallium arsenide, or gallium antimonide.

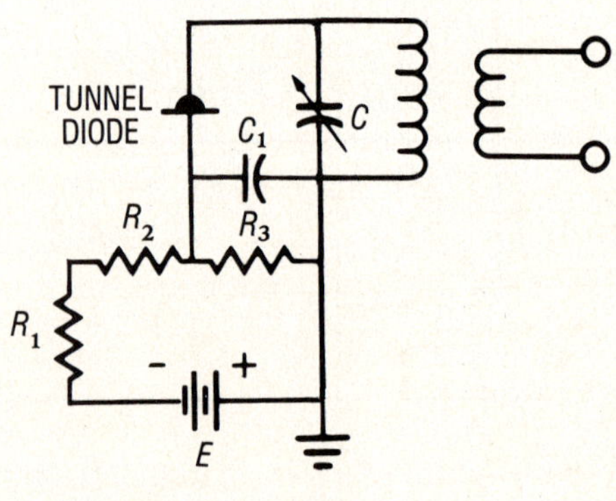

Figure 14-5. Circuit schematic for tunnel-diode oscillator.

Tunnel Diode Applications

The major applications for tunnel diodes are in oscillator circuits, typically in television and radio receivers and transmitters. Figure 14-5 shows a typical tunnel-diode oscillator.

Discussion Questions/Problems

1. What is the chief performance characteristic of a tunnel diode?
2. What happens to a tunnel diode when forward-bias voltage is applied?
3. What happens within a tunnel diode when the peak voltage is reached?
4. How do current and voltage behave in the valley area of a tunnel diode's performance cycle?
5. In what way, specifically, is the behavior of a tunnel diode counter to Ohm's law?

14.4 TUNNEL RECTIFIERS (BACK DIODES)

As indicated above, the performance characteristics of tunnel diodes are achieved through high levels of doping. If the dopent levels of semiconductor materials are reduced, a device can be developed with excellent high-frequency response characteristics. This is the *tunnel rectifier* or *back diode*. Tunnel rectifiers are identified by the same symbols as those for tunnel diodes.

A tunnel rectifier has an extremely low peak-to-valley ratio. However, the device keeps its reverse-bias characteristics. The performance parameters of back diodes emphasize reverse conduction capabilities and de-emphasize negative resistance. Figure 14-6 shows the low peak-to-valley ratio as

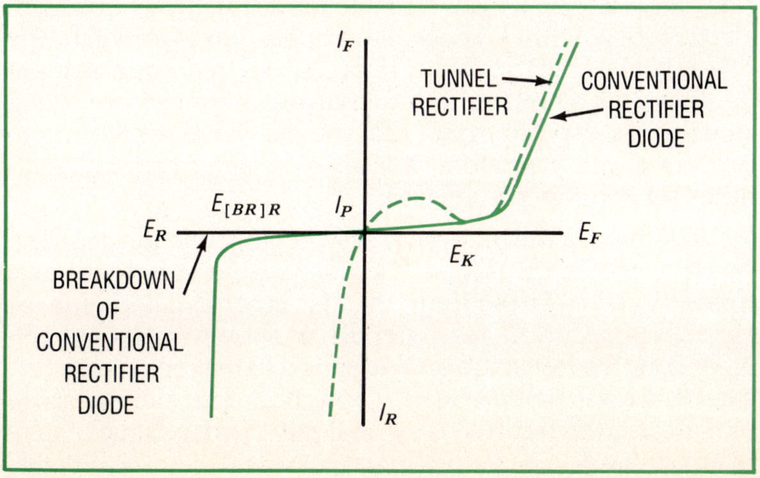

Figure 14-6. Performance characteristics of tunnel rectifiers.

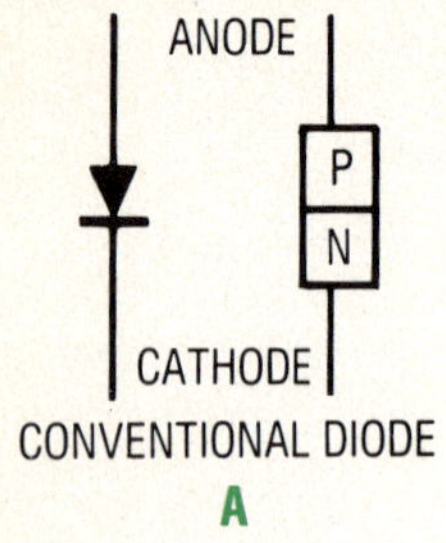

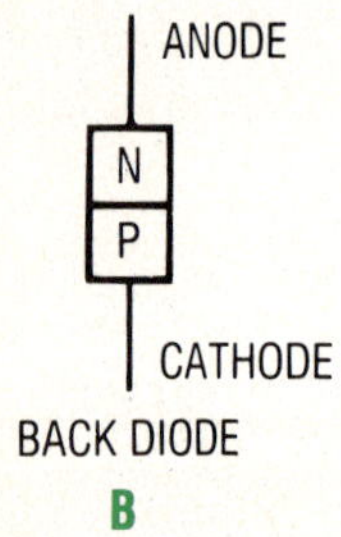

Figure 14-7. Diode symbol and configurations of (A) conventional diode and (B) back diode or tunnel rectifier.

well as the high reverse current that results from a very low reverse voltage. This illustration also compares performance characteristics of tunnel rectifiers and conventional diodes.

Because of these characteristics, tunnel rectifiers can be used as low-voltage rectifiers, operating on the reverse portion of the characteristic curve. For this reason, the polarity of a rectified signal is opposite to that of a signal produced by a conventional rectifier. (Figure 14-7 shows relative configuration for a conventional diode and a back diode.) The back diode gets its name from this performance characteristic.

Because of its excellent high-frequency response characteristics, the tunnel rectifier is used in many high-frequency applications. Examples include tuning circuits and oscillators.

Discussion Questions/Problems

1. How does the dopant level of tunnel diodes compare with that for tunnel rectifiers?
2. What are the peak-to-valley characteristics of a tunnel rectifier?
3. Why is a tunnel rectifier well suited for use in a low-voltage rectifier?

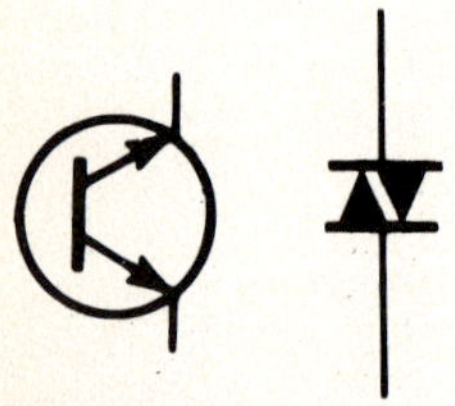

Figure 14-8. Symbols for diacs.

14.5 THE DIAC

The diac is a specially constructed diode that functions as an electronic switching device. For application as a switch, the diac must be able to provide bidirectional operation. It must be operated by either positive or negative current. As a switch, it is also important that the diac have something to turn on or off. Therefore, it is always used in circuits with other semiconductor devices.

To illustrate the bidirectional switching function of a diac, special symbols are used, as shown in Figure 14-8. To achieve this bidirectional operation, the diac is constructed of three layers of semiconductor materials, as shown in Figure 14-9. Note that this construction is identical to that of a transistor, except there is no base lead. Though there are three layers of material, the fact that there are only two leads makes the diac a diode.

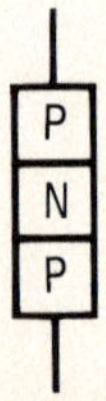

Figure 14-9. Diac construction.

Because of its specialized application, the diac is also known as a *trigger diode*, or *bidirectional trigger diode*. To assure uniformity of bidirectional operation, the junctions of a diac are different from those of transistors. In addition, doping materials must be closely uniform around both junctions. This makes it possible for the diac to respond uniformly to currents flowing from either direction. Regardless of the direction of current flow, one junction of a diac will always be forward biased. The other will always be reverse biased. For flow in either direction, there must be sufficient voltage to overcome the junction barriers. Until this voltage level is achieved, the diac switch is open, and it is in an *off* state. When sufficient voltage is applied, this is known as the *breakover voltage* (E_{BO}). The breakover voltage can be either negative or positive. Remember, the diac remains in an off state until the breakover point is reached. Once the breakover voltage is reached, the diac can conduct relatively high currents. Performance characteristics for a diac are shown graphically in Figure 14-10.

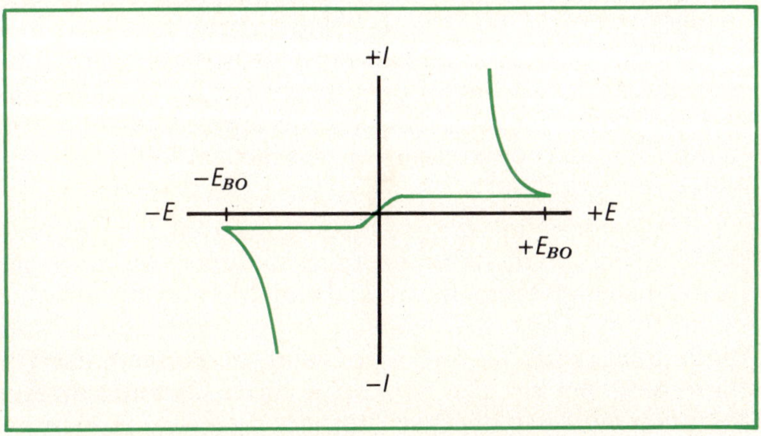

Figure 14-10. Diac performance characteristics.

Discussion Questions/Problems

1. Why is the diac called an electronic switch?
2. What gives the diac its bidirectional capabilities?
3. How does the construction of a diac resemble and differ from that of a transistor?
4. What is the breakover voltage and how does it affect the performance of a diac?

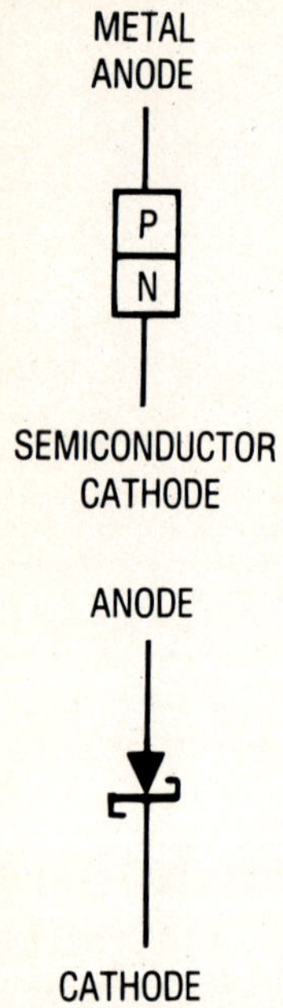

Figure 14-11. Construction configuration and symbol for hot carrier diodes.

14.6 HIGH-FREQUENCY DIODES

Special types of diodes are used in equipment that operates at unusually high frequencies, including very high frequency (VHF), ultra-high-frequency (UHF), and microwave systems. Ordinary diodes cannot operate in these frequency ranges.

Among the many types of high-frequency diodes, some of the most important are:

- Hot carrier diodes
- PIN diodes
- IMPATT diodes.

Hot Carrier Diodes

Hot carrier diodes are constructed with N-type semiconductor material (usually silicon) that is placed in contact with a metallic P-type material. The P-type material is usually fabricated with gold, aluminum, or silver. This construction forms a semiconductor-to-metal junction point. The N-type material serves as the cathode. Diodes of this type have low barrier voltages—approximately 0.3 V. Also, hot carrier diodes have very low levels of minority carriers. This configuration makes for rapid response to applied current. The symbol for, and construction of, hot carrier diodes are illustrated in Figure 14-11.

The name of hot carrier diodes stems from a similarity of these devices to diode tubes. In diode tubes, cathodes are heated to a high temperature and emit high-energy streams of electrons. Thus, the behavior of hot carrier diodes and diode tubes with hot cathodes is similar. Both are used in high-frequency circuits. Hot carrier diodes can turn on and off much more rapidly than ordinary diodes.

Hot carrier diodes are sometimes called *Schottky diodes*, after the German scientist who discovered the operating principles behind these units.

PIN Diodes

A PIN diode is, in effect, a solid-state resistor for which the resistance is controlled by the amount of current applied. The result is an electronic variable resistor.

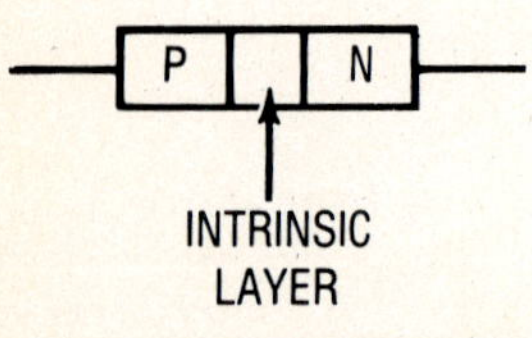

Figure 14-12. Construction configuration of PIN diodes.

This performance is achieved through the design illustrated in Figure 14-12. P and N semiconductor materials are separated by a layer of intrinsic (undoped) semiconductor material. In fabrication, the dopant materials are applied from the P and N poles, leaving less semiconductor material at the center.

As a result of this design, the PIN diode has a storage time that is sufficiently long so the device cannot rectify at microwave frequencies. Instead, a PIN diode functions as a variable resistor, with its resistance value controlled by a DC bias current. Because of these characteristics, PIN diodes are well suited for application as microwave attenuators.

IMPATT Diodes

The name *IMPATT* is an acronym for *IMPact Avalanche Transit Time*. These units are formed around a PN junction and are operated with a heavy back bias that produces an avalanche breakdown in the active region. Considerable heat is generated by this design, illustrated in Figure 14-13. To prevent

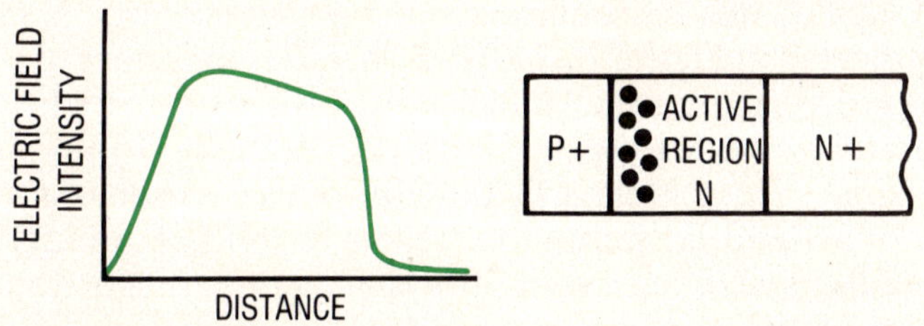

Figure 14-13. Performance characteristics and construction configuration of IMPATT diodes.

burnout, IMPATT diodes are constructed so that their active regions are close to efficient heat sinks. As another means of preventing burnout, a constant current must be used as the bias supply.

Though it is activated by reverse breakdown voltage, the IMPATT diode cannot be used in the same way as the zener diode. An IMPATT diode cannot be used as a voltage regulator. The IMPATT diode is one of the best methods available to produce frequencies in the microwave band.

Gunn-Effect Diode

Another type of high-frequency diode used for the same type of application as the IMPATT diode is the Gunn-effect diode. These units are not true diodes because they do not have a PN junction. Gunn-effect diodes have only an N-type gallium arsenide area.

Discussion Questions/Problems

1. What kinds of jobs are high-frequency diodes designed to do?
2. What are the main differences between hot carrier diodes and ordinary diodes?
3. What special fabrication techniques are used to produce PIN diodes?
4. What are the primary applications of PIN diodes?
5. What is the primary application for IMPATT diodes?

14.7 SHOCKLEY (FOUR-LAYER) DIODES

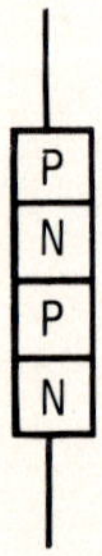

Figure 14-14. Four-layer construction configuration of Shockley diodes.

Shockley diodes are constructed in four semiconductor layers as indicated in Figure 14-14. Note that the alignment of P and N regions are consistant throughout the four layers. Therefore, the Shockley diode passes current in a single direction. With this construction, the device is usable as a trigger, or switching, diode that reacts to unidirectional flow. Symbols used to represent Shockley diodes are illustrated in Figure 14-15.

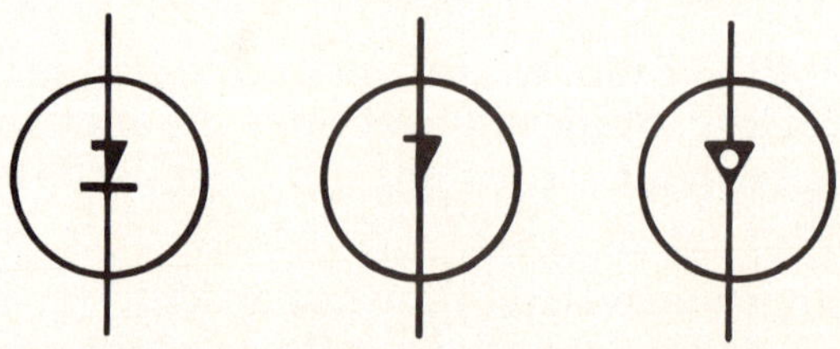

Figure 14-15. Symbols for Shockley diodes.

Operating characteristics for the Shockley diode are shown graphically in Figure 14-16. As indicated by this diagram, the Shockley diode presents a relatively high level of resistance to forward-bias voltage. This is because of the four-layer construction. Considerable voltage must be applied before the breakover point is reached. With the four-layer construction, it is possible to vary the materials so that the breakover point can be predicted with relative precision.

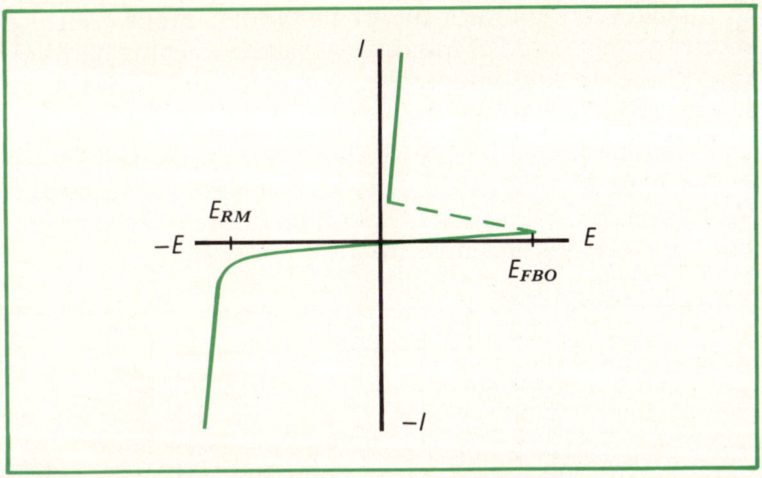

Figure 14-16. Performance behavior of Shockley diodes.

Once the breakover voltage is reached, current flow accelerates rapidly, as illustrated in Figure 14-16. This behavior leads to the description—and to the application—of Shockley diodes as trigger switches. This behavior pattern also leads to the classification of Shockley diodes as part of the thyristor family of electronic devices. Thyristors are electronic switching units. Once the breakover voltage is reached and conduction begins, the flow of current maintained within the Shockley diode is called the *holding current* (I_H).

In the reverse-bias mode, the Shockley diode presents an extremely high impedance. Therefore, these units allow only very small leakage currents.

Discussion Questions/Problems

1. How does construction of a Shockley diode differ from ordinary diodes?
2. Why is a Shockley diode unidirectional in current flow?
3. What are the characteristics of a Shockley diode that cause it to function as an electronic switch?

14.8 LIGHT-EMITTING DIODES

As the name *light-emitting diodes (LEDs)* implies, these diodes emit electronic energy in the visible range. Therefore the emissions can be seen. This makes these devices ideal for use

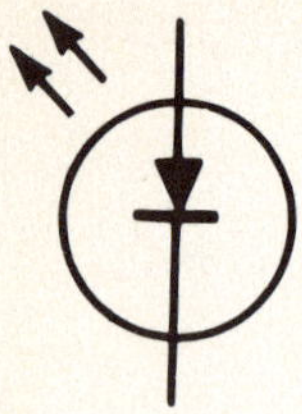

Figure 14-17. Symbol for light-emitting diode (LED).

in instruments as information readouts. Light-emitting diodes are also used as indicator or safety-warning lights on many types of equipment. The symbol for an LED is shown in Figure 14-17.

The operating principle of a light-emitting diode is shown in Figure 14-18. LEDs are PN diodes made from semiconductor materials that promote the combination, or recombination, of electrons and holes to form photons.

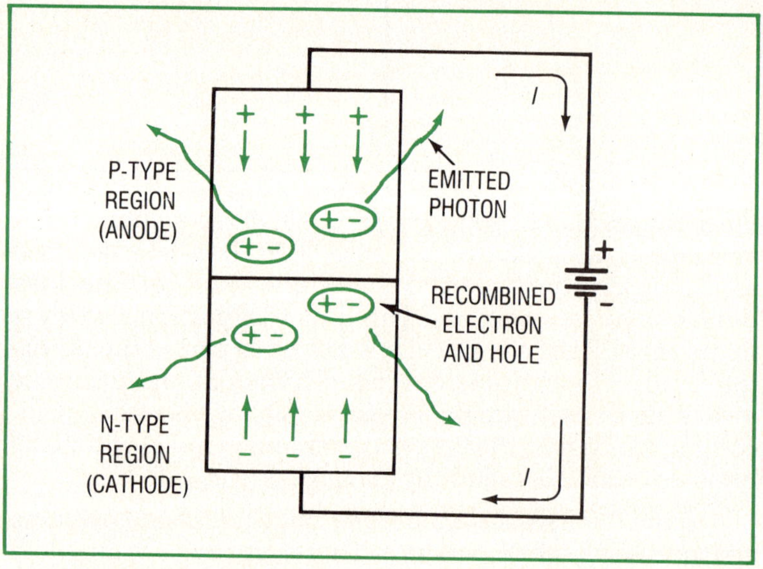

Figure 14-18. Operating principles of LEDs.

Photons are high-energy atomic particles, each consisting of one proton or positive hole, and one electron. It takes certain types of materials and conditions to form photons. Although photons are formed in the silicon deposited upon germanium materials normally used for diodes, these materials are quite opaque and screen out the light. By contrast, the materials used for LEDs permit a relatively free passage of light. This means that an LED does not use popular semiconductor materials such as silicon or germanium. Rather, special materials are used. These include gallium arsenide, gallium-arsenide phosphide, or gallium phosphide. With these materials, photons are formed in the barrier region of the diode.

The materials used in light-emitting diodes lend themselves well to recombining electrons and holes to form photons. The photons release so much energy when they are formed that the resulting energy waves fall into the visible-light spectrum. Photon emissions from LEDs are at different wavelengths. Depending upon the frequency of the wavelength emitted, the light will appear blue, yellow, red, or green. The color of light emission varies with the type of semiconductor used.

The construction of a typical LED is shown in Figure 14-19. This package is assembled to allow maximum light visibility from emitted photons. Light emission begins as soon as forward-bias voltage is high enough to stimulate the flow of current. In a gallium-arsenide phosphide (red) LED, this is 1.2 V. At a forward voltage of 1.6 V, current applied to the LED remains constant.

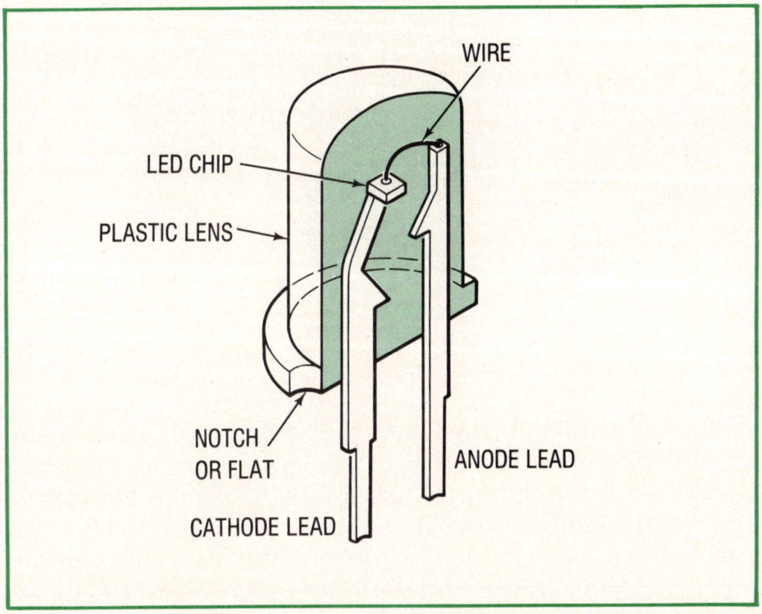

Figure 14-19. Cutaway of LED.

Another characteristic of LEDs is that output intensities vary with the color of the light generated. Red light has the highest output. Much of the light produced by an LED never reaches the surface to become visible. The light waves are simply absorbed within semiconductor materials. Thus, as lighting devices, LEDs are not as efficient as normal light bulbs.

There are advantages in favor of using LED display devices. One of these is ruggedness. Another is extremely rapid response times. Response times are much faster than for conventional incandescent lamps. Still other advantages include low cost for the devices themselves and low operating voltages for the circuits that support them.

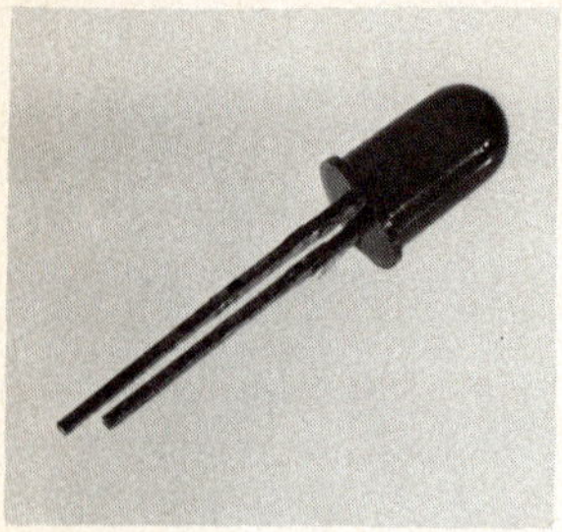

Figure 14-20. LED unit.

Offsetting disadvantages for LEDs are that light outputs are low and that these units are damaged relatively easily if excess heat or voltage is applied.

Typical LED units are shown in Figure 14-20. Units of this type are easily installed in display or control panels for use as indicators. Figure 14-21 shows how LED units can be arranged for displays of alphabetic or numeric information.

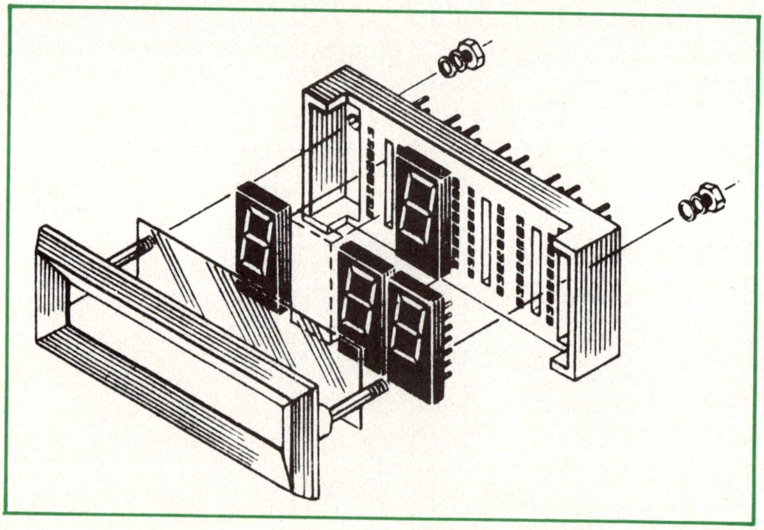

Figure 14-21. LEDs arranged for alphabetic or numeric displays.

Some LED units produce infrared radiation that can be applied in conjunction with other light-sensitive devices. Together, these light emitters and sensors can be used to form optical couplers like those shown in Figure 14-22. Units

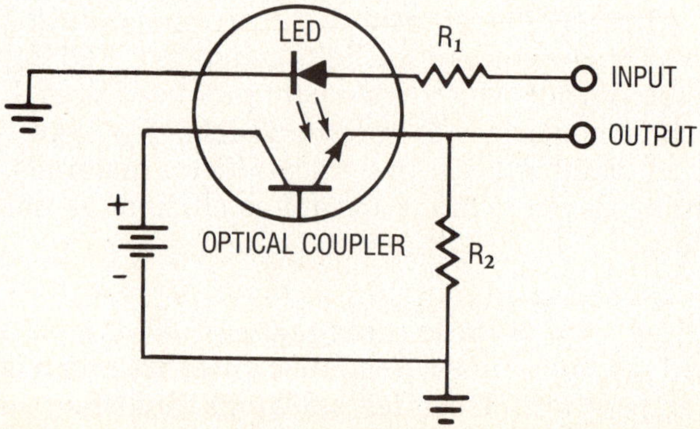

Figure 14-22. LEDs used in optical couplers.

such as these can be incorporated in automatic door-opening or burglar-alarm equipment.

Discussion Questions/Problems

1. Why is it necessary to use semiconductor materials other than silicon and germanium when fabricating LEDs?
2. What is the source of light radiation from LEDs?
3. Which color of LEDs is brightest?
4. How is the visible energy within an LED generated?

14.9 PHOTODIODES

Figure 14-23. Symbol for photodiode.

While some diodes such as LEDs emit light, others are sensitive to light. These are called *photodiodes*. The diode materials respond to light by producing an electron flow. Photodiodes, which are used in many devices that respond automatically to light stimuli, are represented by the symbol in Figure 14-23. A typical photodiode is pictured in Figure 14-24.

Photodiodes are made from silicon semiconductor materials. As shown in Figure 14-24, the case of a photodiode is built to permit light beams to enter. When light strikes the surface of the semiconductor material, the photons are excited. This energy generates a combining action for electrons and holes in pairs. The pairs move to opposite sides of the junction area. This creates a small reverse-bias current flow. The higher the light intensity, the more electron-hole pairings result. The more pairings, then, the higher the currents generated. Because of this responsiveness to light, photodiodes are also called *optoelectronic* devices.

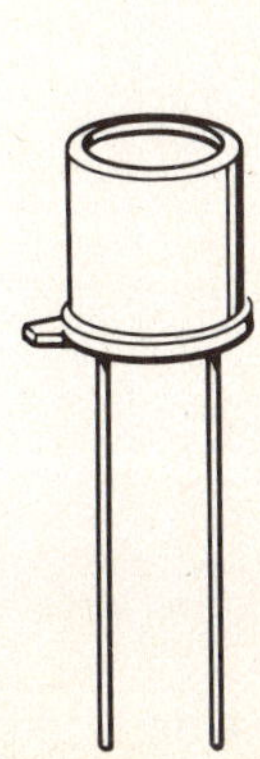

Figure 14-24. Photodiode.

Some photodiodes are fabricated with an intrinsic layer between the N and P junctions. This type of fabrication enlarges the depletion area within the diode. The resulting device is called a *PIN photodiode*. PIN photodiodes have better responses to lower light frequencies than ordinary photodiodes. Also, PIN diodes have substantially faster response times.

Discussion Questions/Problems

1. How is energy generated when the semiconductor material in a photodiode is exposed to light?
2. What is the relationship between exposure to light and the generation of current within a photodiode?
3. What special design features are used to create PIN photodiodes?

14.10 FIELD-EFFECT TRANSISTORS

A field-effect transistor (FET) is a three-terminal device used primarily as an amplifier. The three terminals of an FET are called the *source*, *drain*, and *gate*. Outwardly, an FET looks something like an ordinary transistor. The designations of the leads as the *source*, *drain* and *gate*, however, do mark a difference from the traditional identifications of *emitter*, *collector* and *base*. FETs are often compared with vacuum tubes because their operating characteristics are somewhat more typical for tubes than for transistors. In some ways, an FET behaves more like a vacuum tube than like a transistor.

As shown in Figure 14-25, the construction of an FET holds the key to its ability to function as an amplifier. The

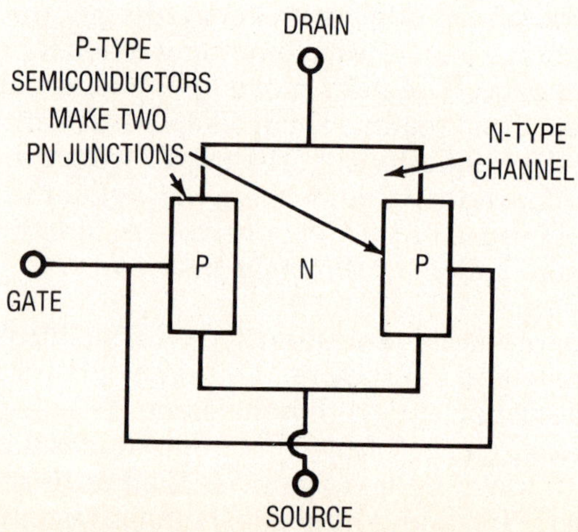

Figure 14-25. Construction configuration of field-effect transistor (FET).

simplest form of FET, illustrated in Figure 14-25, is an N-type semiconductor that is diffused on two sides by P-type materials. The center part of this kind of device is called the *channel*. FET channels may be either P-type or N-type.

In use, an FET requires two bias voltages, as illustrated in Figure 14-26. The first voltage source is connected between the source and the drain, causing current to be pushed through the channel. The second bias voltage is connected between the gate and the source, to establish a voltage gradient.

As with any PN junction device, application of reverse bias causes the buildup of a depletion region between the P and N junctions. The higher the reverse bias, the wider the depletion region becomes, as illustrated in Figure 14-26A. A reverse-bias depletion widens the depletion region around the junctions. As the depletion regions grow wider, the width of the conductivity channel grows narrower. If the reverse bias that is applied becomes great enough, the flow of electrons through the channel can be cut off altogether. The point at which this occurs is called the *gate-to-source cutoff voltage*.

When the reverse bias is low, as indicated in Figure 14-26B, depletion areas are small and the channel becomes wider, permitting the passage of larger currents. The current flow itself is stimulated because the drain lead is positive with respect to the source lead. These connections cause the flow of current.

Figure 14-26. Circuit schematics showing (A) high reverse bias and (B) low reverse bias for FETs.

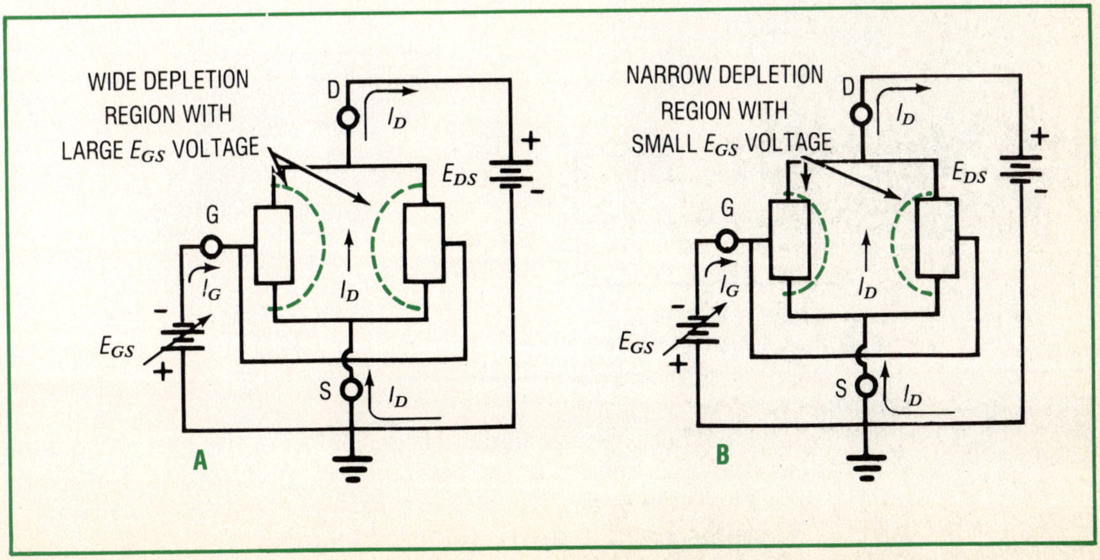

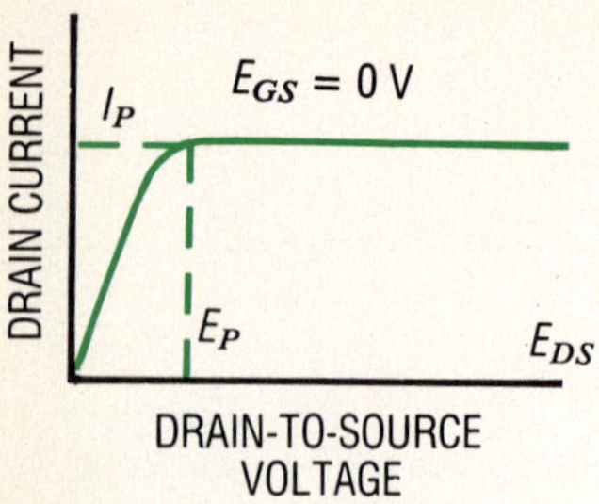

Figure 14-27. Drain current and drain-to-source voltage relationships for FETs.

If voltages increase between the drain and source, this causes a higher drain current (I_D). The drain current increases only up to a certain point. After that, the voltage applied to the drain and source leads rises a point and is called the *pinch-off voltage* (E_P). This means that any further increase in voltage will not cause increases in current flow. These electrical characteristics are shown graphically in Figure 14-27. This type of performance is not achieved with the type of construction shown in Figure 14-25. Rather, this diagram is used as a convenient way to explain an FET. An actual device of this type, with diffusion materials at the sides, would be difficult to fabricate and extremely difficult to control in actual use. A more practical construction approach for an FET is shown in Figure 14-28. This type of FET is built upon a supporting layer called a *substrate*. The substrate serves as an electrically neutral support. On top of the substrate, an N-type diffusion material is spread and a P-type material is diffused into this. The effect is as shown in the cutaway drawing, Figure 14-28A. Electrical conductivity is increased or decreased in proportion to the depletion activity around the P-type material.

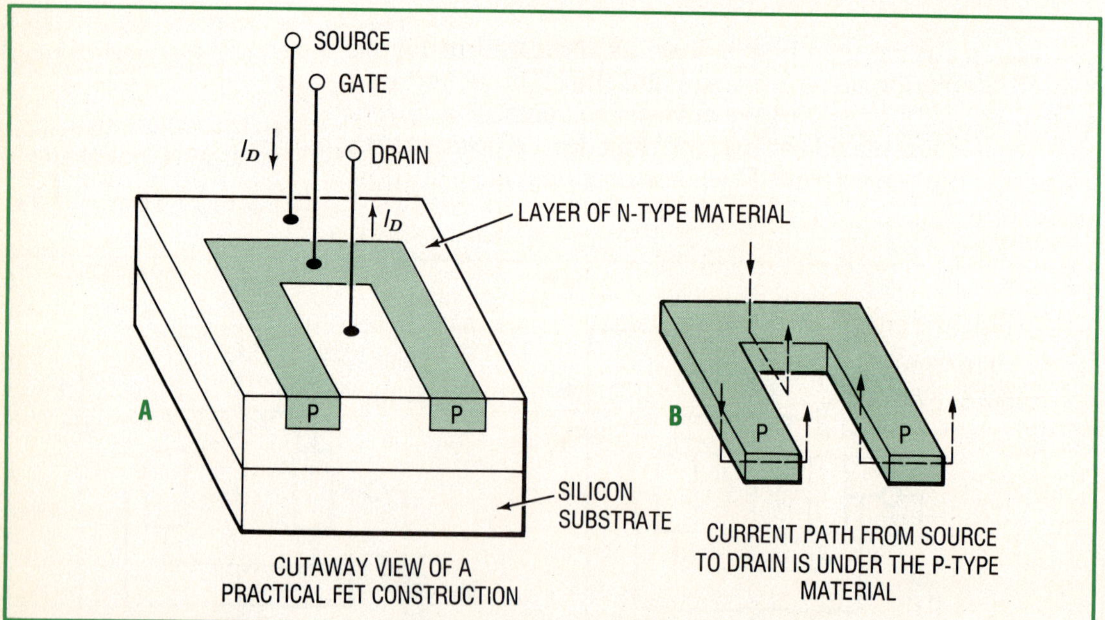

Figure 14-28. Construction configuration and cutaway of FET.

There are two general types of FETs. The first, described above, is known as the *junction field-effect transistor*, or *JFET*. The other type, *insulated-gate field-effect transistor*, or *IGFET*, is described in the section that follows.

The symbols used to represent FETs in circuit diagrams depend on the characteristics of individual units. Several types of FET symbols are shown in Figure 14-29.

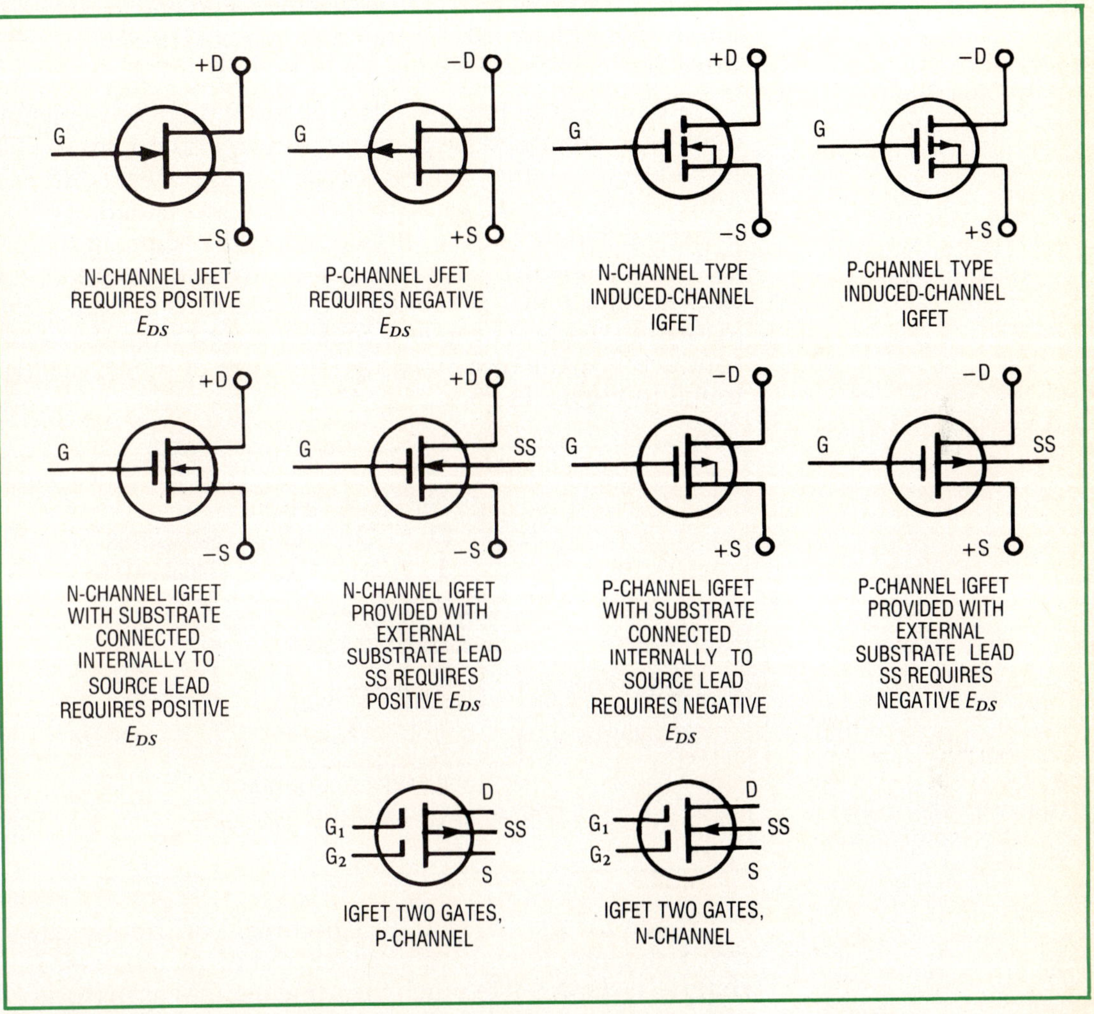

Figure 14-29. Symbols for FETs.

Discussion Questions/Problems

1. What is the primary application for an FET?
2. What are the terminals of an FET called?
3. What is the function of a channel in an FET?
4. What is the gate-to-source cutoff voltage?

14.11 INSULATED-GATE FIELD-EFFECT TRANSISTORS

The insulated-gate field-effect transistor (IGFET) is also known as a *metal-oxide semiconductor field-effect transistor*, or *MOSFET*. On an IGFET, the gate is insulated with a metal oxide semiconductor (MOS) material. Symbols used for IGFET units are included in Figure 14-29.

An IGFET can operate in two different modes—depletion or enhancement. For depletion-mode operation, an IGFET is made with a lightly doped N-type or P-type layer diffused into a substrate to act as a channel. Then an insulating oxide is formed over the entire surface area of the semiconductor. Holes are made in the surface, and source and drain connections are attached to these holes. The gate is connected to the insulating oxide. The gate is thus insulated from the semiconductor by the oxide coating. This type of IGFET is illustrated in Figure 14-30.

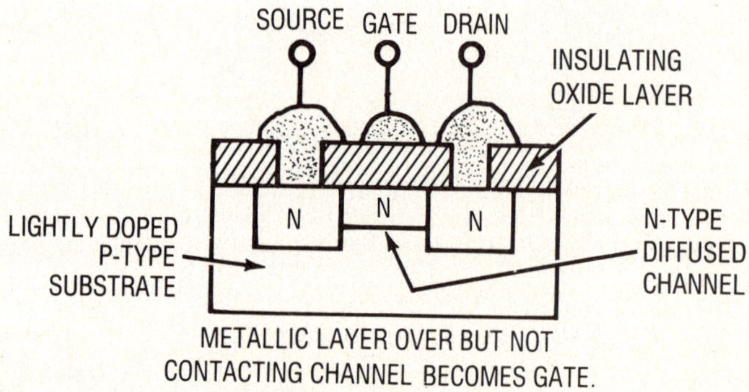

Figure 14-30. Construction configuration for IGFET.

When the gate of an IGFET becomes sufficiently negative, it depletes the majority carriers out of the channel. The channel is depleted in much the same manner as in the normal operation of a JFET. The channel is narrowed and, thus, reduces or cuts off drain current.

When the gate of an IGFET is made more positive, this helps to attract majority carriers.

There are some differences between the operation of depletion-mode IGFETs and JFETs. If a JFET has a positive bias on the gate, this will forward bias the PN junction and serve to lower the input resistance. In an IGFET this does not happen because the gate is insulated. The gate input resistance is unaffected by polarity of the gate voltage.

An induced-channel IGFET, as shown in Figure 14-31, is constructed somewhat differently from a regular IGFET. The lightly doped layer diffused into a substrate is not included in an induced-channel IGFET. There is only substrate material between the source and the drain. The substrate then acts as the channel. The substrate material incorporates very few charge carriers, so it takes a positive gate-to-source voltage to develop a drain current. The positive gate pulls electrons into the channel. It then induces them to travel through the channel. This induced electron flow gives the induced-channel IGFET its name. An induced-channel IGFET works best in the enhancement mode, meaning that current is being pulled through the channel.

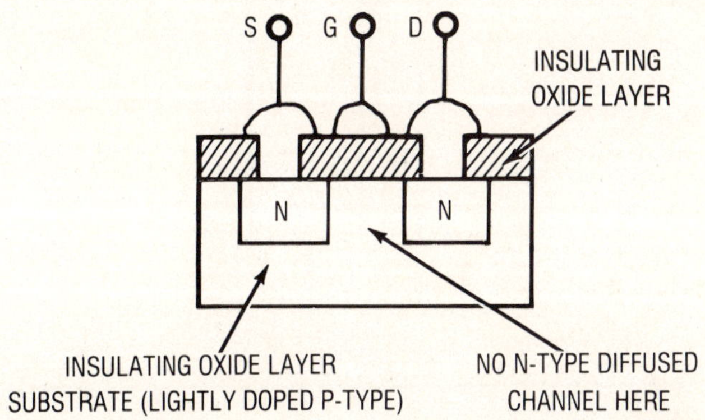

Figure 14-31. Construction configuration for induced-channel IGFET.

Discussion Questions/Problems

1. What special design feature gives the insulated-gate field-effect transistor its name?
2. What force stimulates the flow of electrons through the channel of an IGFET?
3. What is the purpose of coating an IGFET with metal oxide insulating material?
4. What is the main design difference between depletion-mode and induced-channel IGFETs?

14.12 UNIJUNCTION TRANSISTORS

The unijunction transistor or UJT is a special type of semiconductor device. As shown in Figure 14-32, a UJT is composed of N-type semiconductor material and has only one junction. The semiconductor material itself is lightly doped.

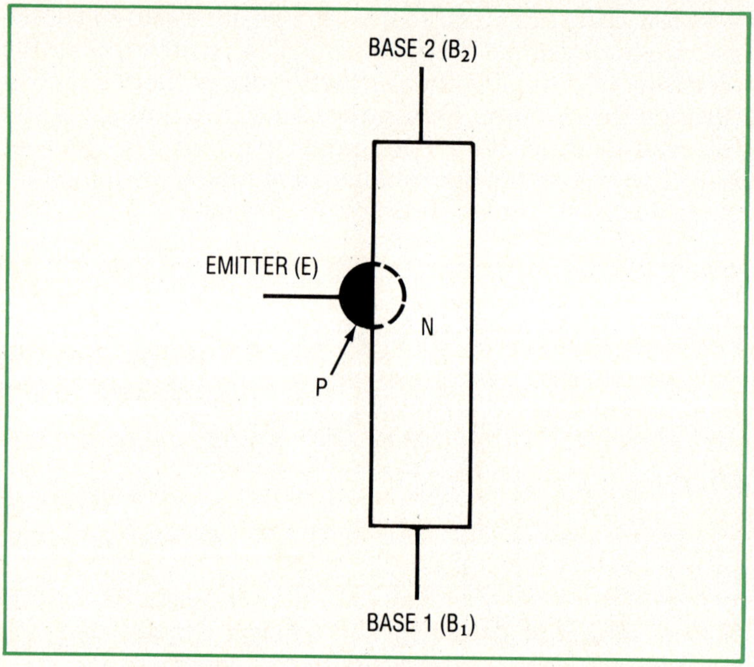

Figure 14-32. Construction configuration for UJT.

The P-type material is fused to the bar that forms the main body of the UJT. Within a UJT there are one emitter and two bases. The symbol used to represent this structure is shown in Figure 14-33. These are identified, as shown, as Base 1 and Base 2. Because the bar is lightly doped, there are few majority carriers to support the current flow. The resistance between Base 1 and Base 2 is high.

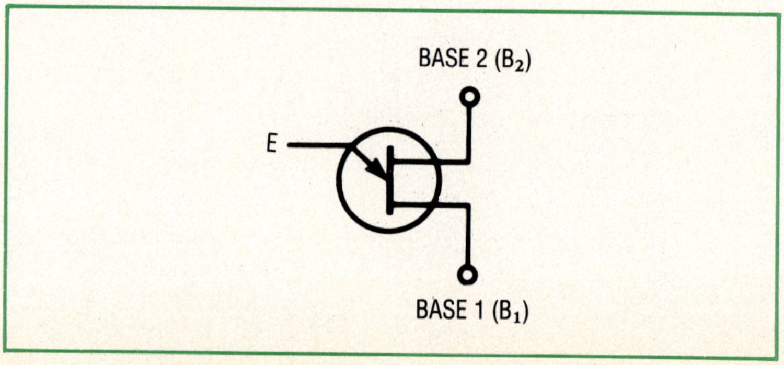

Figure 14-33. Symbol for UJT.

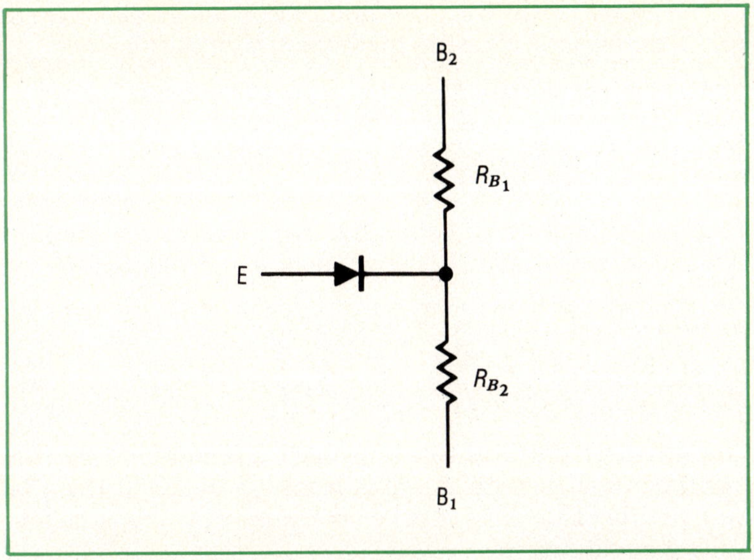

Figure 14-34. Circuit representing UJT.

A UJT can be represented by an equivalent circuit as shown in Figure 14-34. Within an equivalent circuit, electrical components are used to represent electronic properties. Analysis of the equivalent circuit helps build an understanding of the functions of a UJT. Figure 14-35 shows an equivalent circuit linked into power sources that provide the proper bias for normal operaton. If the voltage applied to this circuit (E_S) is not high enough to forward bias the diode, the two resistors will severely restrict the amount of current that can flow. The supply voltage (E_{BB}) is distributed across both resistors. The voltage across resistor R_{B_1} is called E_{B_1}. The ratio between E_{BB} and E_{B_1} is called the *intrinsic standoff ratio*. The

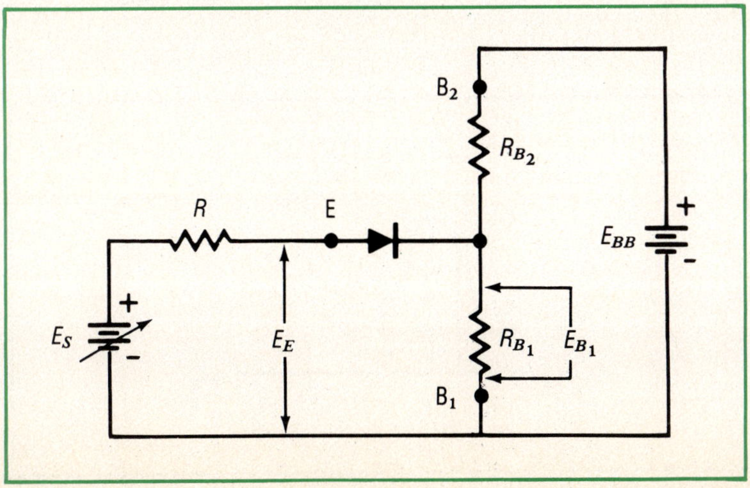

Figure 14-35. Circuit schematic of UJT linked to power source.

abbreviation for intrinsic standoff ratio is the greek letter *eta* (η). This ratio is expressed mathematically as follows:

$$\eta = \frac{E_{B_1}}{E_{BB}}$$

Because voltage drops across the two resistors are proportional to their resistance, the value of η is also equal to:

$$\eta = \frac{R_{B_1}}{R_{B_1} + R_{B_2}}$$

The intrinsic standoff ratio for each UJT is given by the manufacturer. Typical range is from 0.5 to 0.8. If the η is known, the value of E_{B_1} can be determined by using the following equation:

$$E_{B_1} = \eta E_{BB}$$

When the voltage (E_S) of a UJT is high enough to establish a forward bias, the UJT will have a high conductivity. To reach this voltage, a peak voltage is needed:

$$E_p = \eta E_{BB} + E_F$$

In this equation the value E_F is the voltage required for breakover of the diode junction. At this point, represented by the peak of the curve in Figure 14-36, the diode will conduct current, following the pattern shown. UJTs have a negative resistance characteristic, making them useful as

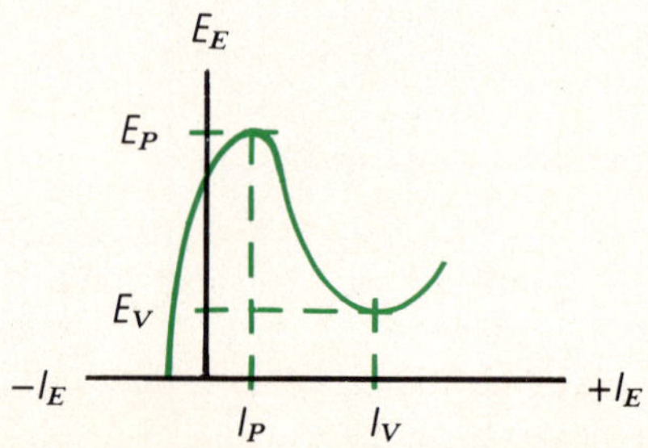

Figure 14-36. Performance characteristics of UJT.

oscillators and trigger devices. This type of application is illustrated in Figure 14-37.

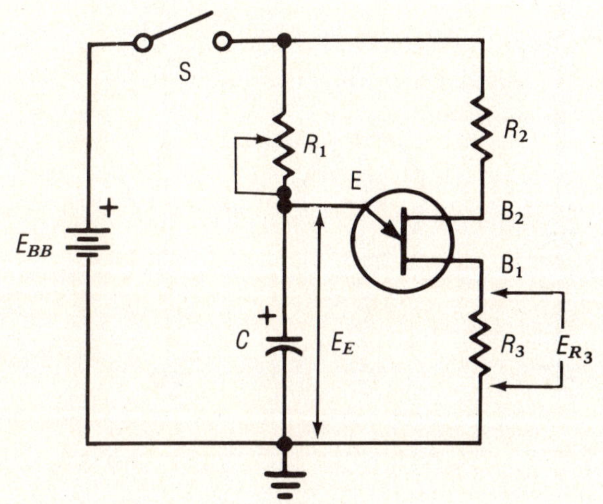

Figure 14-37. Oscillator circuit with UJT.

UJTs can also be produced with programmable characteristics. A programmable UJT is called a *PUT*, for *programmable unijunction transistor*. The symbol for a PUT is shown in Figure 14-38.

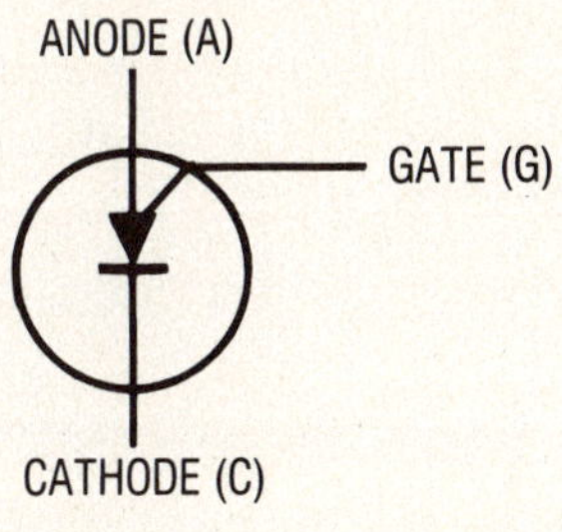

Figure 14-38. Symbol for PUT.

A PUT performs essentially the same way as a UJT. The main difference is that the PUT is programmable, meaning that it is possible to select the peak voltages that function as trigger points. This programmability results partly from a difference in construction between the UJT and the PUT. A PUT is fabricated with four semiconductor layers.

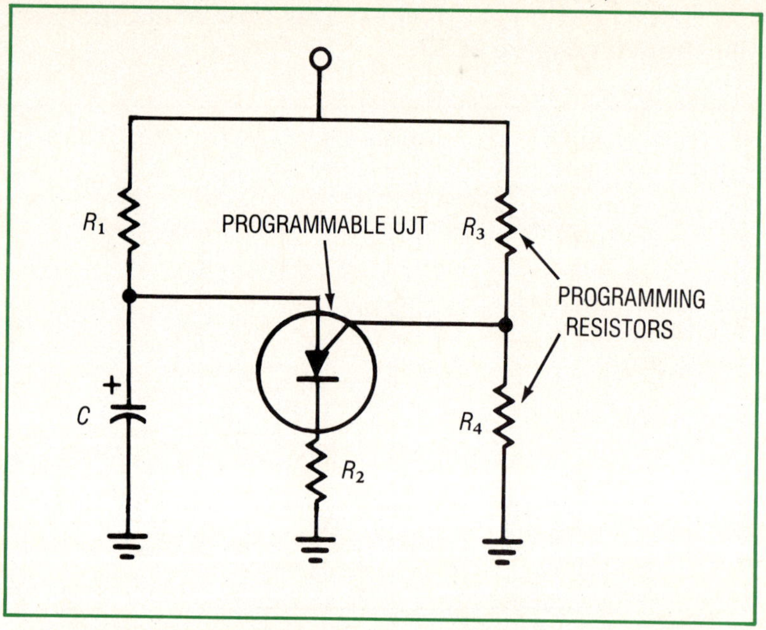

Figure 14-39. Oscillator circuit with PUT.

A circuit in which a PUT is applied is shown in Figure 14-39. This is an oscillator circuit that is controlled by varying voltages between the gate and cathode terminals in a circuit in which the gate is positive with respect to the cathode.

Discussion Questions/Problems

1. What design feature gives the unijunction transistor its name?
2. What are the leads of a UJT called?
3. What is the intrinsic standoff ratio?
4. What are the differences between a UJT and a PUT?

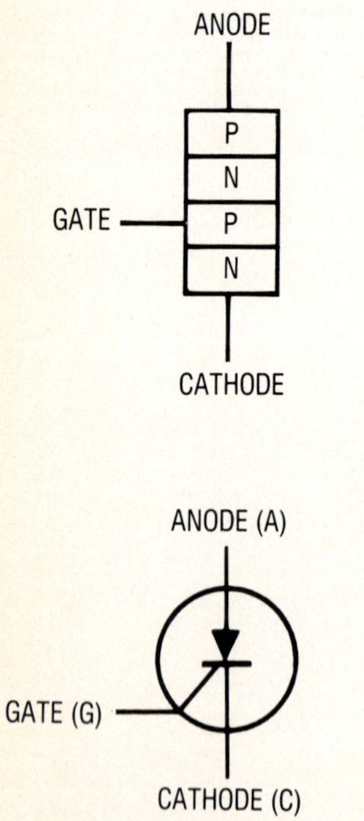

Figure 14-40. Construction configuration and symbol for SCR.

14.13 SILICON-CONTROLLED RECTIFIERS

A silicon-controlled rectifier (SCR) is a device with four semiconductor regions and three leads. Figure 14-40 shows the construction and symbol for the SCR. SCRs are part of the thyristor family of electronically controlled switches. However, unlike bipolar transistors used for switching, the SCR does not amplify the signals it processes.

An SCR is basically a rectifier that conducts current in one direction only. Though there are four regions of semiconductor material, leads are connected to only three of these regions, as shown in Figure 14-40.

One effective way to describe operation of SCRs is through use of equivalent circuits as shown in Figure 14-41. The equivalent relationship established in these circuits is to separate the four-layer SCR into two interconnecting transistors. One of the equivalent transistors is PNP; the other is NPN. The equivalent circuit is forward biased in the same way that a unidirectional SCR would be. The anode must be positive with respect to the cathode. The gate is left open.

Figure 14-41. Operation of SCR illustrated with (A) construction configuration and (B) equivalent circuit.

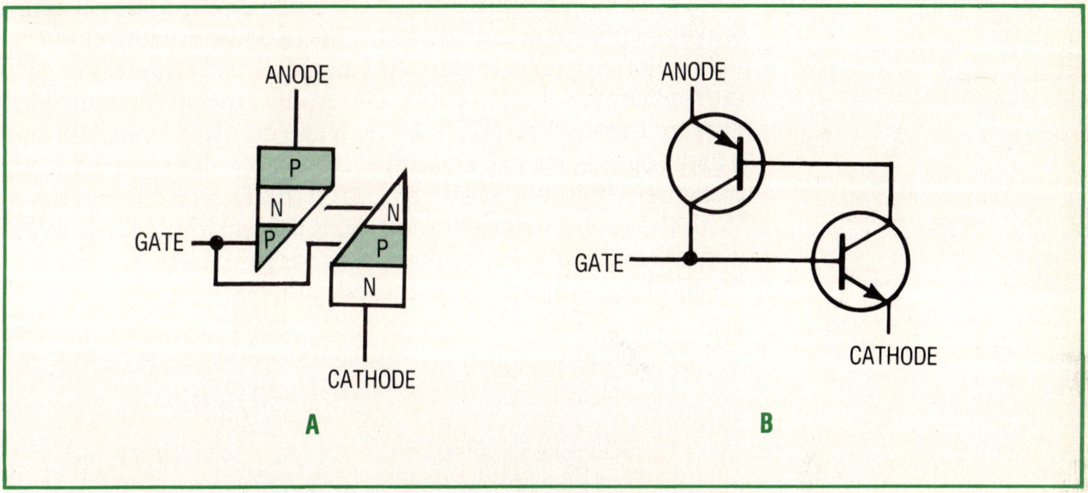

Under these conditions, the NPN transistor will not conduct because the emitter junction is not forward biased. If the emitter junction were forward biased, a base current would be produced. However, because the NPN transistor is not conducting, it will not allow the base current to flow through the emitter junction of the equivalent PNP transistor. Therefore the PNP transistor will not conduct.

This condition changes if the gate is made positive—for a moment only—with respect to the cathode. When the gate becomes positive, a forward bias is applied to the emitter junction of the NPN transistor. The emitter junction then begins to conduct and causes a base current to flow through the PNP transistor, causing this transistor to conduct. At this point, the PNP transistor can cause a base current to flow through the NPN transistor. The combined transistors keep each other switched on, or in a conducting mode. As long as

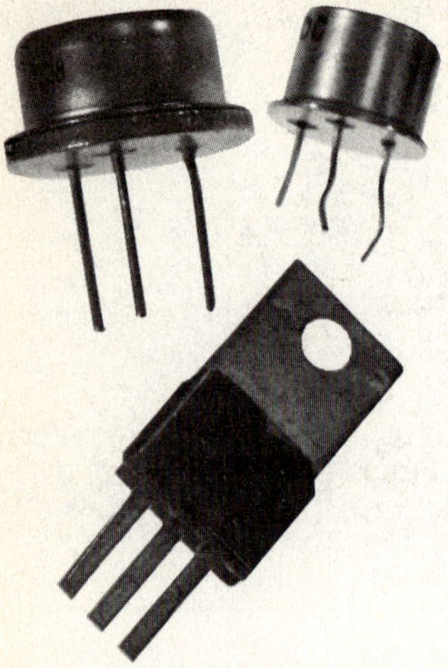

Figure 14-42. SCR units.

the gate remains positive, the current keeps flowing continuously—from cathode to anode—in the equivalent circuit.

An interesting principle is applied in this equivalent circuit: Current continues to flow even after the gate voltage is removed. The only way to turn off the equivalent circuit would be to reduce or turn off the anode-to-cathode voltage. An SCR operates in exactly the same manner as the equivalent circuit in Figure 14-41.

Several SCR units are shown in Figure 14-42. When these three-lead devices are connected and turned on, they deliver high cathode-to-anode current. This is called the forward current. If an SCR is connected in a reverse-bias manner, a small leakage current is introduced. This leakage current flows in a direction opposite to the bias voltage. Small amounts of gate current can be used to control large amounts of forward-bias current through an SCR. An SCR can conduct even without a gate voltage. However, without the gate voltage introduced, extremely high voltages are needed to trigger a current flow. With these characteristics, the SCR is ideal for use in power-control applications. Typical uses for SCRs are in dimmers for lighting systems or as motor controls.

Discussion Questions/Problems

1. How many semiconductor regions are included in a silicon-controlled rectifier?
2. Why is it feasible to show an SCR as the equivalent of two separate transistors?
3. What is the function of a current applied to the gate in an SCR?
4. Why is it more efficient to control an SCR through gate current than through foward-bias current?

14.14 BIDIRECTIONAL TRIODE THYRISTORS

A bidirectional triode thyristor, also known as a *triac*, is part of the thyristor family. Again, this means that the devices are usable as electronic switches.

An SCR, as explained above, is typically usable to control current flow in one direction only. This makes SCRs ideal for use in DC applications. When electronic switching is required in AC circuits, a single SCR cannot be used since it would cut off half of the AC signal. Triacs serve the same purpose in AC circuits as SCRs do in DC circuits: Triacs serve as electrically controlled switches in AC circuits. The symbol and construction principles of a triac are shown in Figure 14-43.

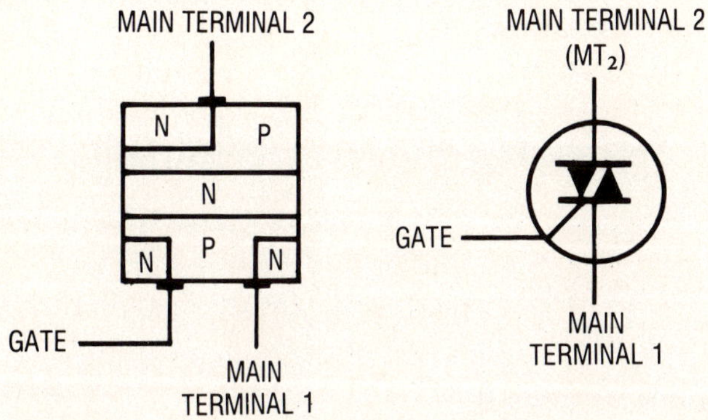

Figure 14-43. Construction configuration and symbol for triac.

A triac is basically two SCRs that are interconnected for control as a single unit. To illustrate operation of a triac, consider the equivalent circuit in Figure 14-44. As is true for a single SCR, the triac will not conduct until a trigger current

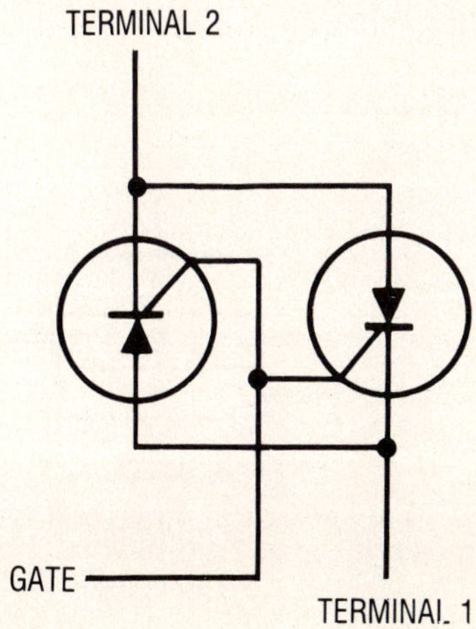

Figure 14-44. Circuit schematic for triac.

is applied to the gate. However, in a triac, the gate trigger can be either positive or negative with respect to terminal 1, which is the reference terminal of the equivalent circuit in Figure 14-44. Once the triac is conducting, it cannot be turned off by removing the gate voltage. As with the SCR, voltage across the terminals must be reduced or turned off to stop the flow of current through a triac. Triacs are used in application situations similar to those for SCRs, except that these applications handle AC loads. Typical jobs include controls over lights, motors, and so on.

Discussion Questions/Problems

1. What are the differences in current-carrying characteristics between SCRs and triacs?
2. Why is a triac called part of the thyristor family?
3. What is the function of the voltage applied to the gate of a triac?
4. Once a triac is operational and carrying current, how is it turned off?

14.15 SILICON-CONTROLLED SWITCHES AND SILICON BILATERAL SWITCHES

A silicon-controlled switch (SCS) is a thyristor constructed with four silicon semiconductor regions. As shown in Figure 14-45, the arrangement of these regions is PNPN. The symbol for an SCS is also shown in Figure 14-45.

Figure 14-45. Symbol and construction configuration for SCS.

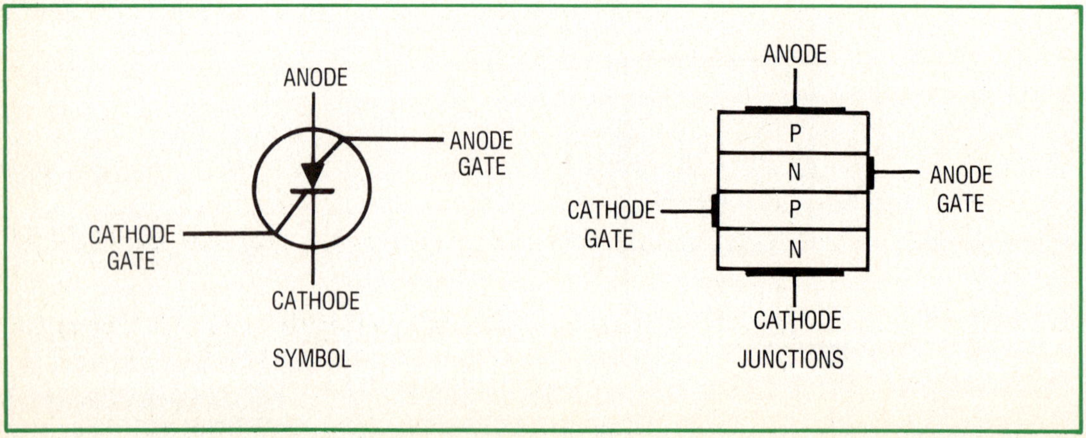

An SCS is similar to an SCR except that there are two gate connections: an anode gate and a cathode gate. An SCS can be triggered into conduction through application of either positive or negative pulses, making these units usable for a wide range of special applications.

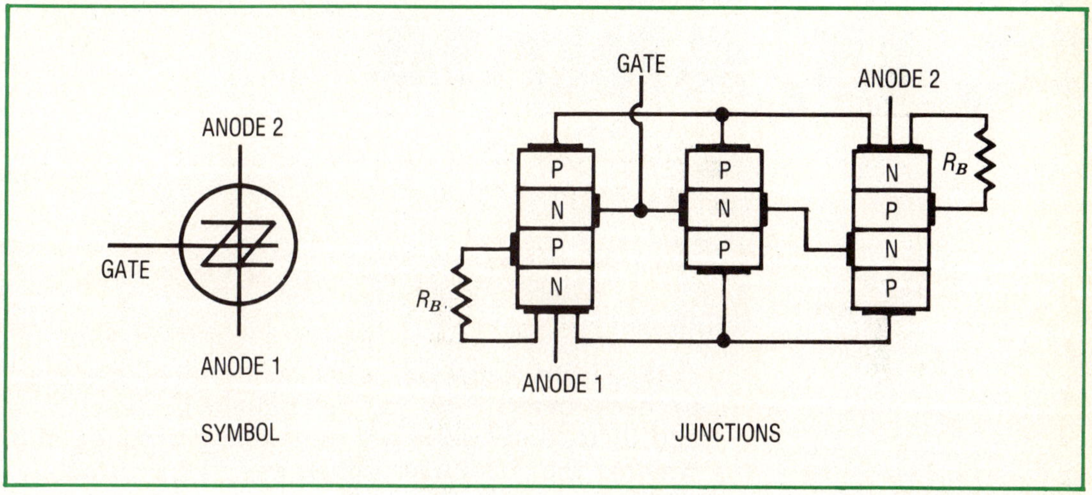

Figure 14-46. Symbol and construction configuration for SBS.

A silicon bilateral switch (SBS) is also a thyristor. As shown in Figure 14-46, an SBS has a construction similar to that of a triac. Operationally, an SBS is similar to an SCS, with one major exception: An SBS can allow current flow in either direction while an SCS permits current flow in one direction only.

Discussion Questions/Problems

1. What is the main structural difference between an SCS and an SCR?
2. What are the differences in triggering characteristics between an SCS and an SCR?
3. What is the main functional difference between an SBS and an SCS?

CHAPTERS 12, 13, AND 14

SUMMARY

1. Testing instruments are, as the name implies, designed to check circuits. They can determine whether circuits and their components are functioning properly and whether the circuits meet their design specifications. There are many types of testing and measuring instruments available to persons working with electrical circuits and electronic devices. These include meters, oscilloscopes, signal generators, logic probes, and power supplies. (12.1)
2. Meters are devices that measure the values of the basic properties or characteristics of electricity. (12.2)
3. The ammeter is an instrument used for measuring electrical current. (12.3)
4. A voltmeter is an instrument that measures voltage in an electrical circuit. (12.4)
5. An ohmmeter measures resistance. (12.5)
6. Voltage and current, along with component parts such as resistors, capacitors, and inductors, can be measured using simple bridge circuits. (12.6)
7. An oscilloscope is a versatile tool that makes it possible to display and visualize signals being measured. (12.7)
8. There are a number of oscilloscopes tailored to special applications. These include storage-type oscilloscopes, spectrum analyzers, and logic analyzers. (12.7)
9. A signal generator is an electronic unit which supplies selected output signals used for testing purposes. (12.8)

10. A logic probe is a device that looks something like a pencil. It is used to read voltage levels in logic circuits. (12.9)
11. A power supply can be an important and useful tool in testing circuits and devices. (12.10)
12. Electronics is a field dealing with the behavior of free electrons, particularly within environments such as vacuum, gas, special conductors, or semiconductors. (13.1)
13. Vacuum tubes were the first electronic devices put into practical use. (13.2)
14. A vacuum tube consists of a series of metal parts built into either a glass or metal vacuum container called an *envelope*. (13.3)
15. The term *diode* indicates that a tube has two electrodes, usually a cathode and an anode. A major use of diodes is as rectifiers. A rectifier converts AC to DC. (13.4)
16. A triode is a three-electrode vacuum tube. A triode contains an anode, a cathode, and a control grid. (13.5)
17. A number of other types of vacuum tubes are manufactured. Each of these includes some sort of special design feature to improve performance over triodes or to achieve a special type of performance. (13.6)
18. The cathode ray tube (CRT) is a special type of vacuum tube that displays an image on its front surface. (13.7)
19. The term *semiconductor* describes a family of materials or elements, either solid or liquid, with intermediate current conductivity characteristics. (13.8)
20. To understand what semiconductors are and how they are applied, it is important to understand the basic atomic and molecular structures of these materials. (13.9)
21. Doping materials are selected for their ability to promote the formation of covalent bonds. This makes it possible for the dopants to integrate easily into the lattice structures of silicon or germanium. (13.10)
22. A semiconductor diode is formed by joining N- and P-type semiconductors at a junction. The resulting diode is known as a *junction diode*. (13.11)
23. The primary parameters used in rating diode performance are the forward-bias current characteristics and the peak inverse voltage (PIV) prior to the breakdown point. (13.12)

24. Junction diodes can be used within circuits to convert AC signals to DC. (13.13)
25. A special type of diode is made that takes advantage of the properties of the breakdown point to deliver assured, reliable voltage sources. This diode is known as a *zener diode*. (13.14)
26. A transistor is a dual-junction device with three semiconductor regions. There are two basic types of transistors. The NPN transistor has two areas of N-type semiconductor material with an enclosed base of P-type material. A PNP transistor has two P-type regions and one N-type region. (13.15)
27. Amplifier circuits utilizing transistors can be connected in three basic ways. These are known as the *common base (CB)*, the *common emitter (CE)*, and the *common collector (CC)*. (13.16)
28. Electrical performance characteristics of transistors are nonlinear. In some ways, the characteristics of a transistor are analogous to those of a tube. (13.17)
29. An amplifier is a device that generates, at its output terminals, a signal that is essentially the same as the input signal but of a greater magnitude. (13.19)
30. A varactor diode is a semiconductor device designed with special capacitance characteristics. (14.2)
31. The chief characteristics of a tunnel diode are its ability to operate at high frequencies and its negative resistance region. (14.3)
32. If the dopant levels of semiconductor materials are reduced, a device can be developed with excellent high-frequency response characteristics. This is the tunnel rectifier, or back diode. (14.4)
33. The diac is a specially constructed diode that functions as an electronic switching device. (14.5)
34. Special types of diodes are used in equipment which operates at high frequencies, including very high-frequency (VHF), ultra high-frequency (UHF), and microwave systems. There are many types of high-frequency diodes. Some of the most important of these are hot carrier diodes, PIN diodes, and IMPATT diodes. (14.6)
35. A PIN diode is a solid-state resistor for which the resistance is controlled by the amount of current applied. The result is an electronic, variable resistor. (14.6)

36. The term *IMPATT* is an acronym for *IMPact Avalanche Transit Time*. The IMPATT diode is one of the best methods available to produce frequencies in the microwave band. (14.6)
37. Shockley diodes are constructed in four semiconductor layers. This construction makes the Shockley diode usable as a trigger, or switching, diode that reacts to unidirectional flow. (14.7)
38. *Light-emitting diodes,* as the name implies, emit electronic energy in the visible range. (14.8)
39. While some diodes, such as LEDs, emit light, other diodes are sensitive to light. These are called *photodiodes*. The diode materials respond to light by producing an electron flow. (14.9)
40. A field-effect transistor (FET) is a three-terminal device used primarily as an amplifier. The three terminals of an FET are called the *source, drain,* and *gate*. (14.10)
41. The insulated-gate field-effect transistor (IGFET) is also known as a *metal oxide semiconductor field-effect transistor,* or *MOSFET*. On an IGFET, the gate is insulated with a metal oxide semiconductor (MOS) material. (14.11)
42. The unijunction transistor (UJT) is a special type of semiconductor device. UJTs are useful as oscillators or trigger devices. (14.12)
43. A silicon-controlled rectifier (SCR) is a device that has four semiconductor regions and three leads. SCRs are part of the thyristor family of electronically controlled switches. (14.13)
44. A bidirectional triode thyristor, also known as a *triac,* is part of the thyristor family. This means that the device is usable as an electronic switch. (14.14)
45. A silicon controlled switch (SCS) is a thyristor constucted with four semiconductor regions. A silicon bilateral switch (SBS) is also a thyristor. (14.15)

PROBLEMS

1. Assume that you want to use a 1 mA ammeter to read full scale with a 25 mA current. How much current would flow through the required shunt? (12.3)
2. What would be the voltage drop across the ammeter in the above problem, assuming the meter resistance is 25 Ω? (12.3)

3. What would be the value of the shunt resistor discussed in Problem 1? (12.3)
4. In a given voltmeter, the desired full-scale measurement is 25 V. The full-scale current is 1 mA. The internal resistance of the meter is 25 Ω. Given these values, what is the resistance value of the multiplier resistor? (12.4)
5. What is the amplification factor of a tube having a ΔE_b of 100 V and a ΔE_c of 5 V with i_b of 1 mA? (13.6)
6. What is the plate resistance of a tube having a ΔE_b of 200 V and a Δi_b of 1 mA with an E_c of 5 V? (13.6)
7. What is the transconductance of a tube having a Δi_b of 1 mA and a ΔE_c of 5 V with an E_b of 200 V? (13.6)
8. What is the forward-bias current of a diode having a source voltage (E) of 22 V and a forward-bias voltage of 0.7 V with a 15 Ω load resistance? (13.11)
9. What is the current capacity of a zener diode rated at 2 W and with a voltage of 12 V? (13.14)
10. What is the beta of a transistor having a ΔI_C of 8 mA and a ΔI_B of 4 μA? (13.17)
11. What is the alpha gain of a transistor having a ΔI_C of 8 mA and a ΔI_E of 20 μA? (13.17)
12. A given transistor has an alpha of 400; what would be its beta? (13.17)
13. What is the collector voltage of a transistor having a collector current of 2 mA, a load resistance of 10K Ω, and a supply voltage of 30 V? (13.18)
14. What is the Q of a varactor having an X_C of 2K Ω and an R of 100 Ω? (14.2)
15. What is the intrinsic standoff ratio of a UJT having an E_B of 20 V and an E_{BB} of 18 V? (14.12)

(Answers to the odd-numbered problems can be found at the back of the book.)

INTEGRATED CIRCUITS

15.1 TYPES OF INTEGRATED CRCUITS

The integrated circuit (IC) is a complete, functional circuit manufactured as a single unit. All of the components and connections of the circuit are built in a unit so that the entire circuit can be plugged in and used as a single entity. Figure 15-1 shows different types of ICs. These units are designed to serve a number of different purposes. However, the thing they have in common is that all of the circuit components and connections are complete—manufactured as a single unit.

Figure 15-2 shows a close-up of the actual circuit surface inside an IC package. This illustrates some of the basic characteristics of ICs. All of the components and circuit carriers are affixed to or deposited upon a nonconductive surface called a *substrate*. The substrate is simply the structural base for an IC. To build semiconductors in place, it is often necessary to deposit multiple layers of materials upon a substrate.

Figure 15-1. Integrated circuits.

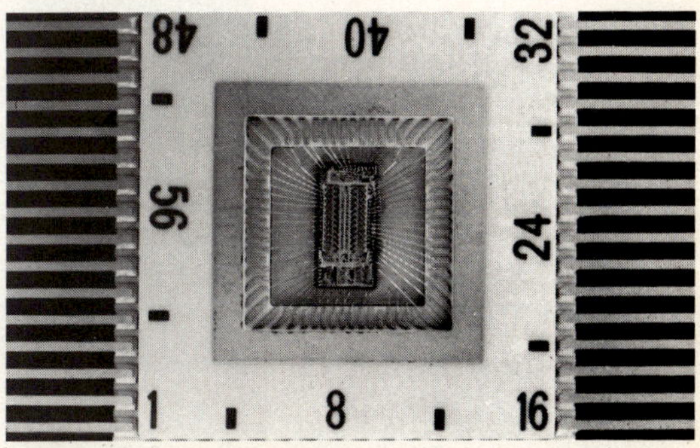

Figure 15-2. Close-up of integrated circuit surface. (Courtesy TRW, Inc.)

Construction techniques serve to separate integrated circuits into some basic, distinctive groups, or families, of products. These include:

- Monolithic integrated circuits
- Film (thin, thick) integrated circuits
- Hybrid integrated circuits.

Discussion Questions/Problems

1. What do all integrated circuits have in common?
2. What is the substrate of an IC?

15.2 MONOLITHIC INTEGRATED CIRCUITS

Monolithic circuits are manufactured through use of certain photographic and lithographic techniques. The name *monolithic* comes from the fact that all components are formed in one element, applied to one surface.

Development of a monolithic circuit begins with a drawing of a comparatively large-size image of the circuit. Such drawings were once done entirely by hand. Increasingly, however, computer-operated plotters have replaced hand drafting. Figure 15-3 shows a full-size drawing of circuit components being examined as a preliminary step in the manufacture of integrated circuits.

Figure 15-3. Negative of integrated circuit being inspected. (Courtesy TRW, Inc.)

After the design has been approved, the large drawing is photographed in miniature. A negative is then made which is composed of many copies of the miniature photograph. The process is similar to the preparation of printing plates for postage stamps. A single, large drawing of the postage stamp is made. Then this is reduced and used to set up the images for full sheets of stamps.

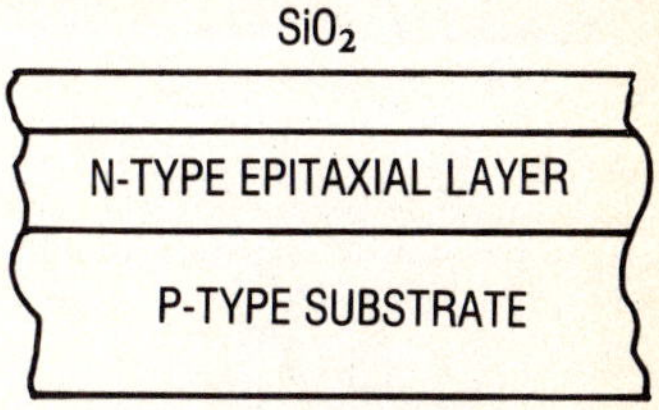

OXIDIZED SLICE

Figure 15-4. Cross section of monolithic IC wafer with unexposed top silver oxide layer.

A carefully prepared ultrathin sheet of silicon material is used to produce monolithic integrated circuits. These silicon sheets, or wafers, are cut from solid bars of silicon materials that are "grown" in laboratories through continuous deposit, or buildup, of the basic crystals. The pure silicon wafers are coated with one or more layers of semiconductor materials that will form the actual electrical circuit. The semiconductor layer is covered with a coating called an *epitaxial* layer of material. This epitaxial layer has both electrical and protective properties. The epitaxial layer is a semiconductor.

A cross section of one type of monolithic IC wafer is shown in Figure 15-4. In this illustration, the substrate is made of P-type material. The epitaxial layer is made from N-type material. In other configurations, the substrate might be neutral and two or more epitaxial layers of semiconductor material might be added.

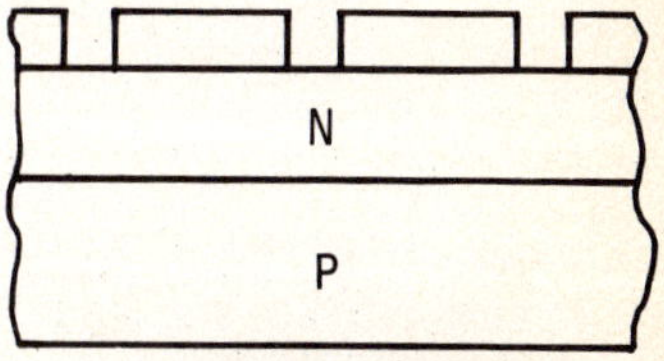

DEVELOPED IMAGE

Figure 15-5. Cross section of IC wafer after exposure and chemical development of top layer.

Figure 15-4 also shows that a final layer of material is added above the epitaxial layer. This consists of silver oxide, a light-sensitive material. Once the silver oxide coating is in place, the entire wafer can be exposed photographically. This is done by placing the film negative against the surface of the wafer and exposing the wafer to light. The photosensitive layer, which is the top layer of the wafer, is then developed chemically. The result, as shown in Figure 15-5, is an exact image of the photograph used for the design.

The photographic material used for the top coating of the wafer is known as a *resist*. This means that the material is plastic based and will resist the action of acid applied to the surface during the next step. This process is known as *etching*. The acid, or etching, solution, when applied to the surface, cuts through the epitaxial layers, down to the substrate. This etching process actually forms electronic components that are produced in place on the substrate. Components that can be formed in this manner include diodes, transistors, resistors, capacitors, and other electronic components. The etching process is illustrated in Figure 15-6.

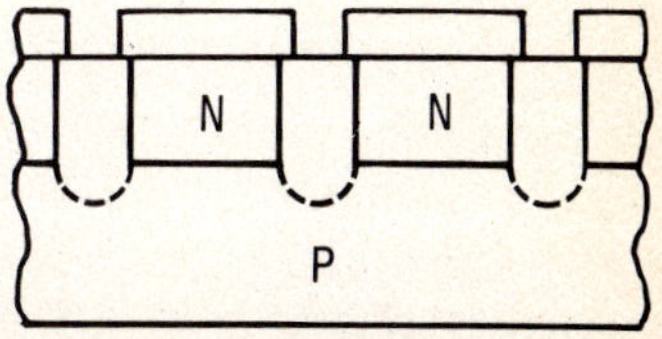

ETCHING

Figure 15-6. Cross section of IC wafer after etching process.

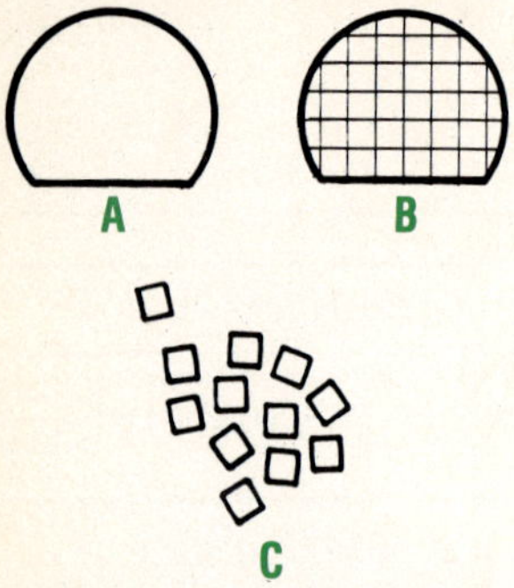

Figure 15-7. Integrated circuits formed on surface of single substrate wafer (A), segregated into individual ICs (B), and cut apart as separate chips (C).

Multiple integrated circuits are formed on the surface of a single substrate wafer during the etching process. Following photographic coating and etching, the wafers are segregated into individual chips and are cut apart, or *diced*, as shown in Figure 15-7. The result is a large number of relatively small, integrated semiconductor circuits. The diced circuits are usually called *chips*.

Once they are separated into individual units, the chips are placed in packages that include protective covering and leads to be used to link the chips into overall circuits of the devices in which they will be used. In the process of packaging, a series of small wires are welded to connection areas printed on the surface of the chips. These welded wires are connected to the leads of the various packages, as shown in Figure 15-8. Figure 15-9 shows a number of typical IC packages.

Figure 15-8. Integrated circuits or chips and lead wire connections. (Courtesy TRW, Inc.)

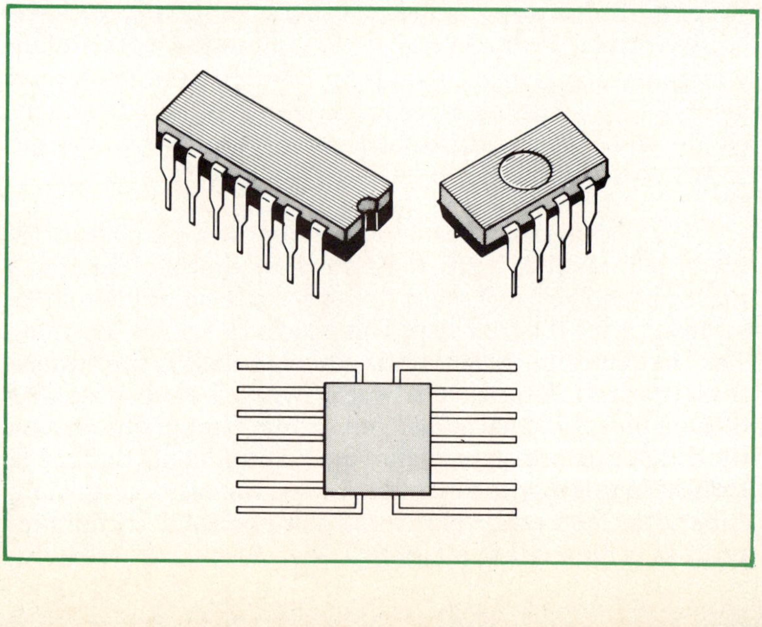

Figure 15-9. Integrated circuit packages.

Monolithic circuits come in a number of sizes and serve a number of purposes. The common denominator for circuits of this type is that they are produced lithographically through photographic processes.

Discussion Questions/Problems

1. What is an integrated circuit?
2. What procedure is used in the initial design of monolithic integrated circuits?
3. Why are the surfaces of integrated chip wafers exposed photographically?
4. What is the function of etching solution in the manufacture of monolithic integrated circuits?

15.3 FILM INTEGRATED CIRCUITS

In film fabrication techniques, the photographic image is placed on the surface of a wafer and developed before the coating of semiconductor materials takes place. That is, the silver oxide resist material is applied first. Then the photographic image is developed in place. For film-type integrated circuits, the substrate materials (usually glass or ceramic) coated with the photo-resist material are all excellent insulators.

The photographically coated substrates are then placed in a vacuum chamber. Two techniques can be used for coating the substrates with the semiconductor materials.

One technique is *evaporation*. Metallic materials, such as tantalum, nichrome, or others, are vaporized by applying heat. The vapors settle on the surface of the substrates, coating the areas where the photographic resist materials have been etched away.

The second coating process is known as *sputtering*. Under this approach, high voltage is applied to a metallic material that has been placed in a gas-filled chamber. The voltage ionizes the gas, bombarding the material with ions. The atoms of the material become dislodged and drift toward the substrate. The areas where the resist has been etched away are then coated with the material, and the remaining area is

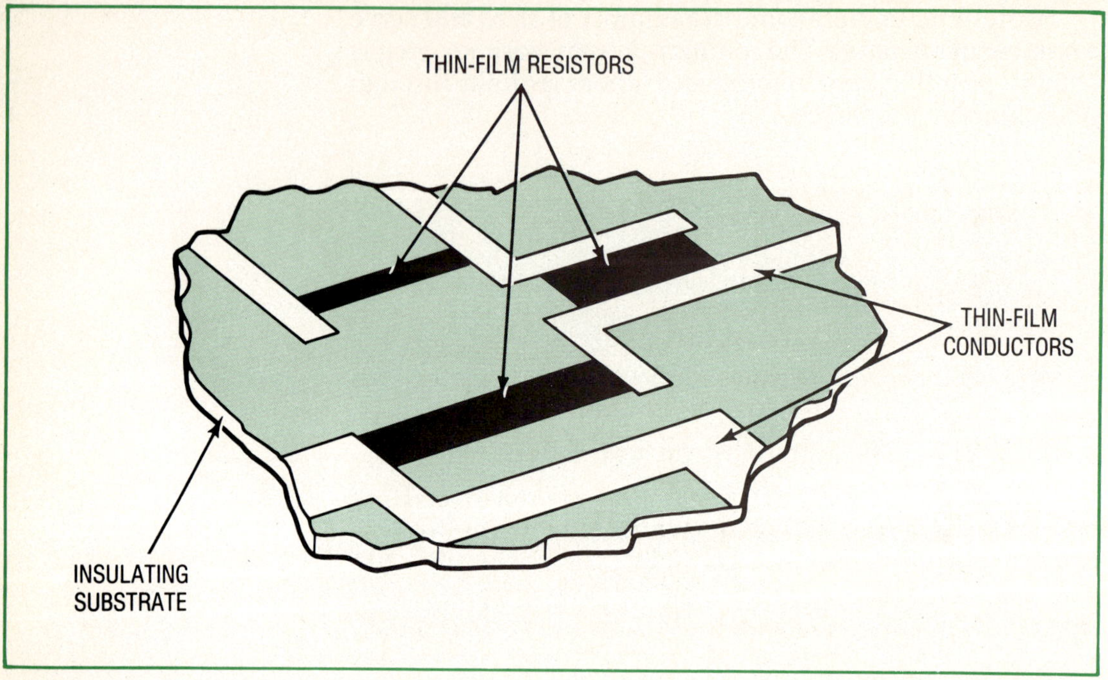

Figure 15-10. Cutaway of film integrated circuit.

protected by the resist. Figure 15-10 illustrates the way in which vaporized or ionized thin-film material is coated onto substrates.

Thick-film processes use a different approach. Photographic materials are used to form a printing screen. The procedures are those of the screen printing process. A finely meshed screen is coated so that inks or other liquidized materials can be pushed through the screen in places where there is no photographic coating. These materials are forced onto the printed surface with a rubber blade, known as a *squeegee*, and deposited beneath the screen. In thick-film integrated circuit processes, metalized inks are used to deposit semiconductor materials on substrates.

Comparision of Film and Monolithic Techniques

Different results are produced with film and monolithic techniques. Under monolithic methods, coatings must be uniform across the surface of the wafer. Usually, the coatings are extremely thin. Under film techniques, the thicknesses of the coatings can be varied. Densities can be increased as necessary for specific integrated-circuit components. Thus, it is easier with film methods to increase deposits in areas that will serve as resistors and capacitors. The thickness cannot be varied readily with monolithic methods.

Discussion Questions/Problems

1. What is the meaning of the word *film* as it applies to integrated circuits?
2. What are the basic differences between thin-film and thick-film integrated-circuit production methods?
3. What is the main advantage of film-type integrated circuits as compared with monolithic-type circuits?

15.4 HYBRID INTEGRATED CIRCUITS

As the name implies, hybrid integrated circuits are combinations. They are formed by using all of the fabrication techniques described above—monolithic, thin film, and thick film. A hybrid integrated circuit consists of a mosaic, or composite formation from a number of small chips. These miniature chips are mounted together, side by side, to produce an effect that takes advantage of the special features of each of the three types of fabrication.

Additional Fabrication Techniques

In addition to the basic methods of monolithic and film, there are still other methods of fabrication. The number of fabrication techniques used for integrated circuits is growing rapidly. Research continues to find better methods, higher performance, or ways to increase yields (reducing the reject rate during manufacturing). One alternate method is to form integrated circuits through the same methods used to create insulated-glass field-effect transistors, or IGFETs. This technology uses metal-oxide semiconductor (MOS) construction techniques. Instead of forming separate, or discrete, semiconductor components, deposits on glass substrates are used to create entire integrated circuits. One of the advantages of MOS technology is that these units take up less space than traditional bipolar components. Thus, MOS techniques lead to higher density integrated circuits. More circuits and components can be built in less space, comparatively, than with other fabrication techniques. However, there is a corresponding drawback: MOS integrated circuits do not operate at as high a frequency as other types of integrated circuits.

A number of semiconductor fabrication techniques have been adapted to the integrated circuit field. These include the

Schottky integrated circuit, high-speed MOS integrated circuits, low-power integrated circuits, and complementary MOS (CMOS) integrated circuits.

Discussion Questions/Problems.

1. What is a hybrid integrated circuit and how is it formed?
2. What types of substrates are used for IGFET integrated circuits?
3. What are the major advantages and disadvantages of MOS integrated circuits?

15.5 INTEGRATED CIRCUIT CONSTRUCTION

Construction of integrated circuits is based on the size of the package itself, the function to be performed, and the type of application to be supported.

Size of Integrated Circuit Packages

Broadly, integrated circuit packages are separated into these categories:

- Small-scale integration (SSI)
- Medium-scale integration (MSI)
- Large-scale integration (LSI).

Size indications for integrated circuit packages are broad and, unavoidably, inexact. The breakdowns of SSI, MSI, and LSI are used almost universally. However, these terms do not have specific, accepted meanings. For example, an SSI package is generally considered to be one with fewer than 200 components. This means that any individual package could, conceivably, have anywhere from a dozen to 200 or a few more integrated components and fit into the same category. At higher levels, the spread is even greater. MSI packages are generally classed as having between 200 and 1,000 components. Then, anything above 1,000 components falls under the broad umbrella of LSI packages. A device with 1,000 mixed components would have the same classification as a memory chip with 256,000 diodes built in place.

Classification according to size is not completely definitive. Rather, the idea is to examine individual applications and to find packages that fit specific needs.

In choosing semiconductor packages, be aware of another term. Diodes or transistors packaged individually are called *discrete* components. Integrated circuits, which can have hundreds or even thousands of diodes or transistors, are classified as SSI, MSI, or LSI.

Function

Integrated circuits are classified according to the way they operate. There are two broad categories of IC functions:

- Linear
- Digital.

The term *linear* means simply that there is a direct relationship between input and output signals for an integrated circuit. This direct relationship classifies linear ICs as *analog* devices. Linear integrated circuits are used for such applications as voltage regulators, signal comparators, sensing amplifiers, operational amplifiers, signal drivers, and linear amplifiers. Linear circuits can also be adapted to other applications through the addition of nonlinear devices or components.

Digital integrated circuits act as though they are a special combination of switches. The output signal of these units can be turned on and off electronically in such a way as to represent information.

Applications

Linear and digital integrated circuits have markedly different uses within electronic circuits. The remainder of this chapter and Chapter 16 are organized around these application patterns. The remainder of this chapter deals with applications for linear integrated circuits. Chapter 16 is about digital integrated circuits.

Discussion Questions/Problems

1. If an integrated circuit had 64 components, how would it be classified in terms of package size?
2. What is the major difference between discrete and integrated circuit components?
3. What are the major differences between linear and digital integrated circuits?

15.6 OPERATIONAL AMPLIFIERS

Operational amplifiers (OP AMPs) are the most popularly used type of linear integrated circuit. The symbol for the operational amplifier is shown in Figure 15-11.

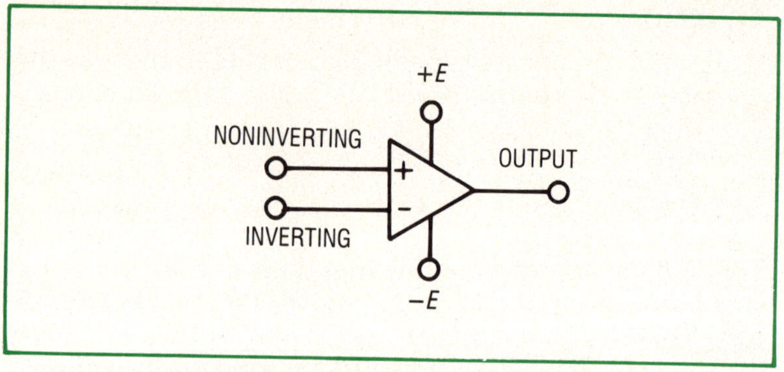

Figure 15-11. Symbol for operational amplifier.

Functionally, operational amplifiers produce very high gains, typically in a frequency range from 0 to approximately 1 MHz. Operational amplifiers can be used in two broad types of circuit configurations:

- Open-loop
- Closed-loop.

Open-Loop Circuits

A typical open-loop circuit using an operational amplifier is shown in Figure 15-12. Note that the signal moves directly through the OP AMP from input to output. There is no feedback, or signal returned from the output side back to the input side. This is the basis for describing this circuit as an open loop. The gain from this type of circuit comes entirely from the OP AMP. It is called an *open-loop gain*. An open-loop gain

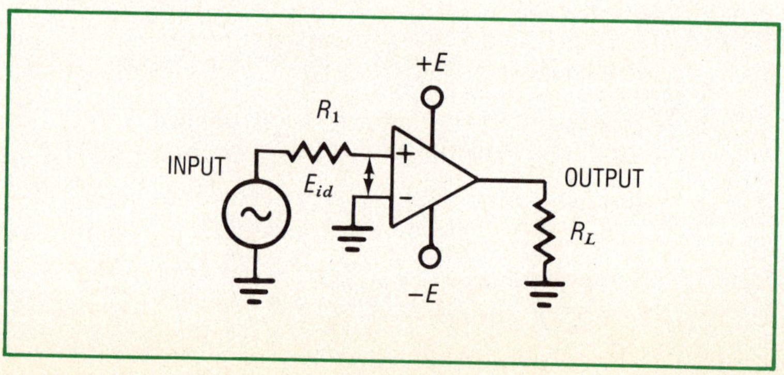

Figure 15-12. Open-loop OP AMP circuit.

can be defined as a ratio of the output signal voltage (E_o) to its differential input (E_{id}). The equation for this ratio can be expressed as follows:

$$A_{VOL} = \frac{E_o}{E_{id}}$$

Closed-Loop Circuits

A closed-loop OP AMP circuit is shown in Figure 15-13. Note that, as compared with an open-loop, an extra connection is established around the OP AMP circuit. This is called the *feedback loop*. In circuitry, a feedback loop is a connection that returns a portion of the output signal back to the input. The feedback loop gives the closed-loop circuit its name. Feedback can be either positive or negative. Positive feedback reinforces the input signal. The effect is analogous to continually pushing someone on a swing. The push reinforces the swinging. Properly adjusted for phase and amplitude, positive feedback can cause the amplifier to oscillate.

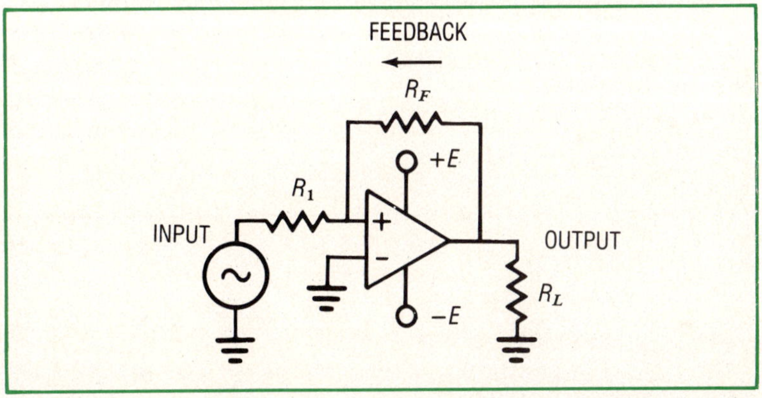

Figure 15-13. Closed-loop OP AMP circuit.

A positive feedback loop is connected to the non-inverting side of the input signal, as in Figure 15-13. A positive feedback loop serves to increase the output of a circuit. This is because the input and output are in phase, thereby aiding signal power.

The gain of the closed-loop OP AMP circuit is a ratio of output to input voltage. This is shown in the following equation in which E_s is the input voltage signal:

$$A_{VOL} = \frac{E_o}{E_s}$$

Negative feedback occurs when the closed loop delivers part of the output signal back to the input signal, and the phases of the two signals are opposite. The feedback loop would be connected to the inverted side of the input signal. Negative feedback is 180 degrees out of phase with the input signal. Therefore, the overall amplification is decreased by the feedback signal. At the same time, the feedback signal also serves to reduce distortion of the output of this circuit and increase circuit stability.

Operational Amplifier Applications

Operational amplifiers are used in many diverse applications, including industrial, medical, commercial, and military instruments. A large measure of their appeal in these applications is due to their sensitivity. OP AMPs can amplify to a useful level relatively small AC or DC signals.

These units are used as general amplifiers and as amplifiers for sensors in many types of instruments and in wave generators.

The first practical integrated circuits used as operational amplifiers were produced in 1965 by Fairchild Semiconductors. The first successful OP AMP IC was the μA709, produced by Fairchild. Since then, many advanced versions of the OP AMP have been introduced. A popular OP AMP IC in current production is the 741. The internal schematic is shown in Figure 15-14.

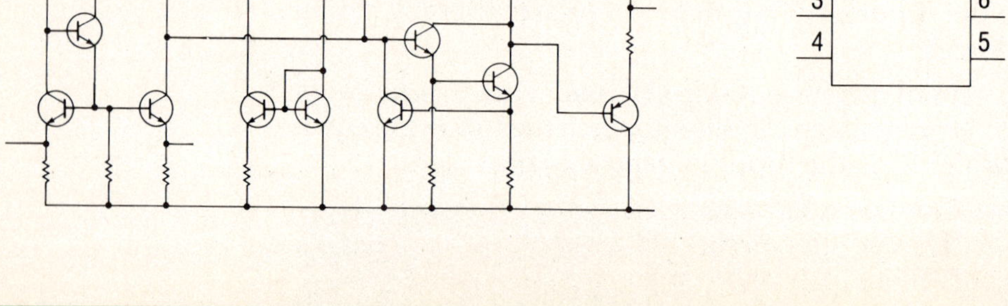

Figure 15-14. Circuit schematic and connection diagram for 741 OP AMP IC.

Discussion Questions/Problems

1. What is a feedback loop in an operational amplifier circuit?
2. What is the difference between open-loop and closed-loop operational amplifier circuits?
3. What is the difference between negative and positive closed-loop operational amplifier circuits?
4. What is the main effect of a negative feedback loop in an operational amplifier circuit?
5. What is the main effect and value of a positive feedback loop in an operational amplifier circuit?

15.7 DIFFERENTIAL AMPLIFIERS

A differential amplifier (DIFF AMP) is a linear integrated circuit that does not include capacitors within its components. The symbol for a DIFF AMP is shown in Figure 15-15.

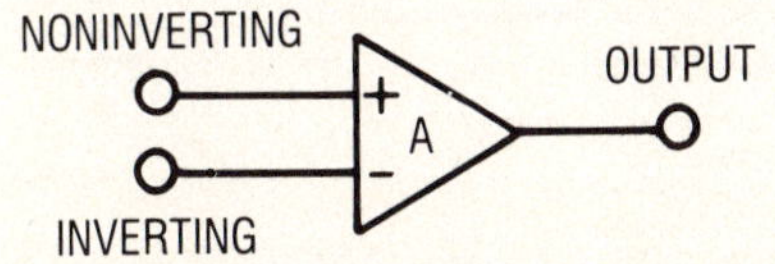

Figure 15-15. Symbol for DIFF AMP.

The chief characteristic of a differential amplifier is that it has two input signals from which it generates a difference output signal. Components within the differential amplifier integrated circuit respond to differences between the voltages of the two input signals. Like signals are effectively suppressed. Thus, the output is a signal representing the differential between the inputs. One type of differential amplifier circuit includes seven individual circuits. Different makes or models of differential amplifier integrated circuits can have varying numbers of individual components, ranging from just a few up to dozens.

Differential amplifier circuits can be constructed from discrete components. However, this is one situation in which the miniaturization and symmetry possible in integrated circuits make for better balancing of components and carriers. For example, an effective differential amplifier circuit requires a close, exact balancing of the two transistors used. An IC can be manufactured to very close tolerances, resulting in better balance and more exact filtering than circuits constructed from discrete components.

In an integrated circuit like the one shown in Figure 15-16, the transistors are biased in the common emitter mode. Note the symmetry of the circuit, which shows why the two transistors must be as close to identical as is commercially feasible.

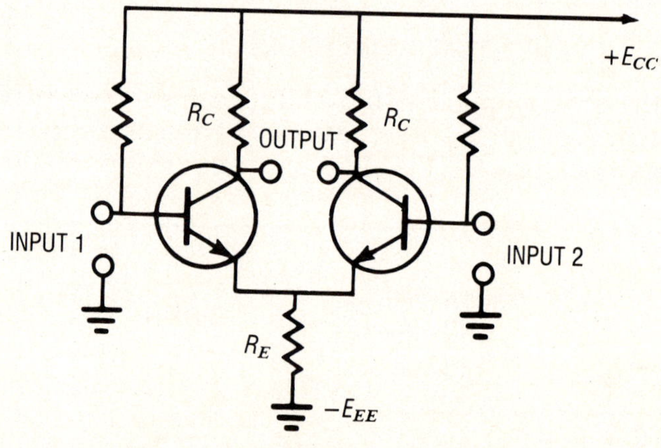

Figure 15-16. DIFF AMP circuit.

Many circuits of the type shown in Figure 15-16 can be fabricated on a single integrated circuit chip. One of the possible configurations is to have a number of differential amplifier circuits connected to one another in sequence. This is known as *cascading* of circuits. A configuration of this type is shown in Figure 15-17. Connecting circuits in this way makes it possible to use one circuit to trigger another. This amplifies the signal.

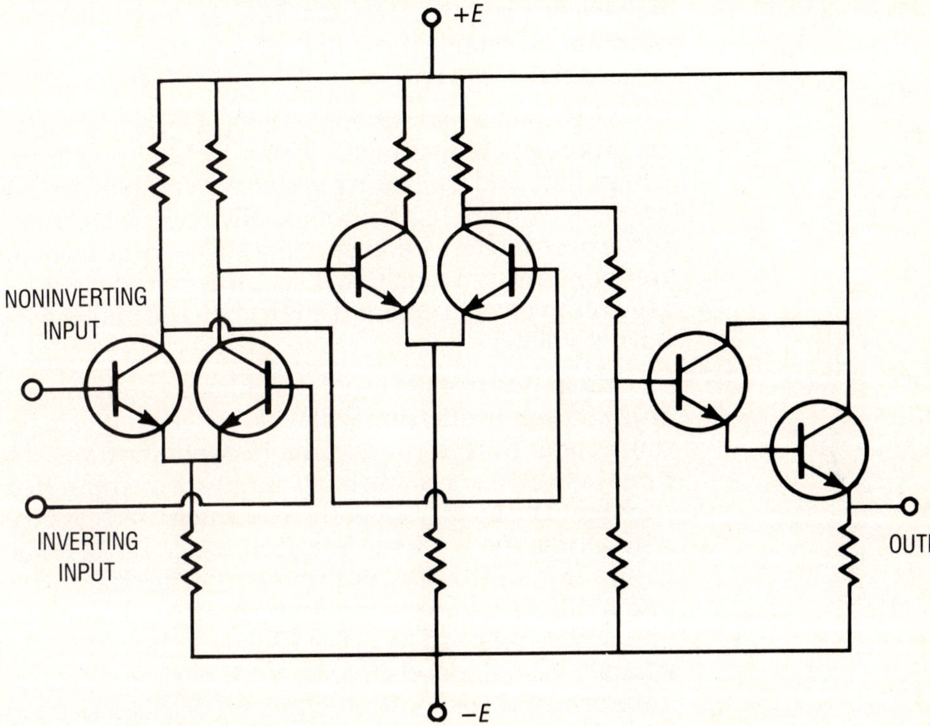

Figure 15-17. Three DIFF AMP circuits connected in sequence.

Discussion Questions/Problems

1. What is the relationship between input and output signals of a differential amplifier integrated circuit?
2. What are the special advantages over discrete components for an integrated circuit differential amplifier?
3. What is cascading and how are circuits cascaded within an integrated circuit differential amplifier?

15.8 VOLTAGE REGULATING INTEGRATED CIRCUITS

A voltage regulator is a circuit that produces an output voltage of a predetermined value. The regulated voltage is constant and stable regardless of any changes in circuit conditions that might occur.

This kind of performance is critical for applications which require regulated voltages. These include power supplies, computers, and measuring systems. Within these devices or systems, voltage regulators usually receive DC that has already been filtered. Because of the ability to incorporate many components in a small package, integrated circuit voltage regulators are a frequent design choice for delivering fixed output voltages.

Integrated circuits usable as voltage regulators are available with many different voltages and current values. These units come in fixed and variable configurations. Usually, three or four leads (some have more) are incorporated in the package. The most frequently used integrated circuit voltage regulator is the National Semiconductor LM309, shown in Figure 15-18. (This unit is also manufactured by other firms under different names, but for the purposes of this text, the unit will be referred to as the LM309.) The LM309 includes a family of integrated circuits, each designed to produce a different regulated output. One unit of this series produces a regulated output of 5 V. The most common application for the 5 V LM309 integrated circuit voltage regulator is in computer applications that require regulated 5 V power. A power supply circuit using a typical LM309 IC package is shown in Figure 15-19. A number of integrated circuit voltage regulators are pictured in Figure 15-20.

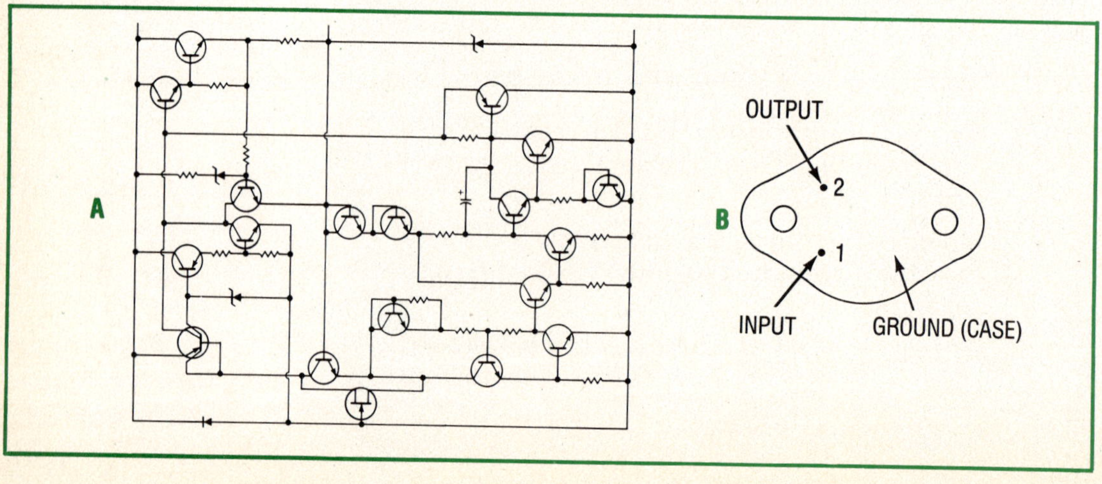

Figure 15-18. Circuit schematic (A) and bottom view (B) of LM309.

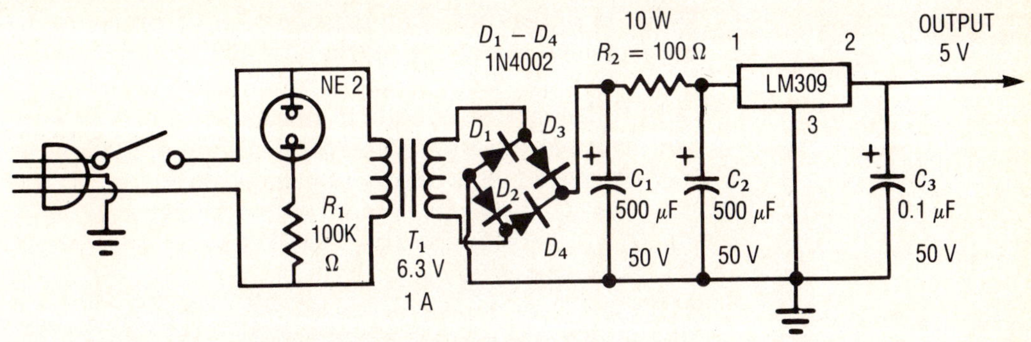

Figure 15-19. Circuit schematic of power supply with LM309 IC package.

Discussion Questions/Problems

1. What is the most important performance characteristic of an integrated circuit voltage regulator?
2. What are some of the common uses for integrated circuit voltage regulators?
3. Name three types of components built into integrated circuit voltage regulators.

Figure 15-20. Integrated circuit voltage regulators.

15.9 INTEGRATED CIRCUIT AUDIO AMPLIFIERS

Audio amplifiers are devices that accept signals in the audio frequency (AF) range and increase the magnitude of output. The range of audio frequencies is considered to be from 20 Hz to 20,000 Hz. Most audio amplifiers are used in devices such as radios, television sets, stereo units, or public address systems. Several types of devices are used in combination to produce desired signals:

- Preamplifiers
- Audio power amplifiers
- Dual amplifiers used either for preamplification or power amplification of parallel signals, as is necessary in stereo systems.
- Integrated amplifiers that are both preamplifiers and power amplifiers.

Amplifiers are designed as either preamplifiers or power amplifiers. This is because it is usually first necessary to amplify the signals and then provide enough power to drive a loudspeaker. Many devices that generate audio signals, such as phonograph cartridges or microphones, produce only weak output. Integrated circuits in devices of this type improve performance and also reduce cost, as compared with circuits that do these jobs with discrete transistors.

Discussion Questions/Problems

1. What is the main function of a preamplifier?
2. Describe the relationship between preamplifiers and power amplifiers.

15.10 OTHER LINEAR INTEGRATED CIRCUITS

Linear integrated circuits are now produced in many configurations to handle a number of special jobs. Figure 15-21 shows a typical television receiver chassis. This incorporates

Figure 15-21. Television receiver chassis.

an array of integrated circuits for signal conditioning. Additional functions for linear integrated circuits include such applications as:

- Wideband amplifiers
- RF amplifiers
- IF (intermediate frequency) amplifiers
- AM amplifiers
- FM amplifiers.

The initials or designations for these various linear integrated circuits describe the frequencies they process. Wideband amplifiers are used in devices that must process signals with wide frequency bands, such as oscilloscopes, television receivers, and FM radios.

RF amplifiers deal with radio frequencies only and are usually included in radio receivers. Intermediate frequency amplifiers are used in both AM and FM radio receivers. These circuits receive signals from the RF/converter section of a receiver and select signals from only that frequency to which they are tuned. A typical AM receiver uses an IF amplifier set specifically for 455 kHz. This selected signal is then passed on to the detector in the radio.

Linear integrated circuits are used in applications that call for amplification or other conditioning of signals in audio, radio, and television devices.

Discussion Questions/Problems

1. What is a common use for wideband amplifiers?
2. What relationship might typically exist between an RF and an IF amplifier?

16 DIGITAL INTEGRATED CIRCUITS

16.1 BINARY NOTATION

The term *digital* describes a class of devices working in pulsating cycles. Signals in digital circuits can be used to represent numbers, and the circuits can be connected to represent the usual mathematical operations such as Add, Subtract, Multiply, and Divide. Digital circuits are the basis for many useful devices such as watches, timers, calculators, and computers. Most digital devices are based on the binary numbering system.

The common numbering system is the decimal system—ones, tens, hundreds, thousands, etc.—based on the number *10*. This system probably developed as a natural extension of having 10 fingers for counting. Other numbering systems are possible, based on a number other than 10. One system, based on the number *2*, is called the *binary system*. The binary system and the decimal system both have the same structure. That is, in the decimal system, a number is made up of a digit in the units (10^0) column, a digit in the tens (10^1) column, a digit in the hundreds (10^2) column, etc. Similarly, a number in the binary system is made up of a digit in the units (2^0) column, a digit in the twos (2^1) column, a digit in the fours (2^2) column, a digit in the eights (2^3) column, etc. Figure 16-1 shows this similarity between the decimal and the binary systems.

The binary system is particularly useful in electronics, because just as a digit in binary can be either zero or one, an electric current or voltage can be either on or off. Thus, there

DIGITAL						
	N^5	N^4	N^3	N^2	N^1	N^0
BASE 10	100,000	10,000	1,000	100	10	0 1 2 3 4 5 6 7 8 9

BINARY						
BASE 2	32	16	8	4	2	0 1

Figure 16-1. Digital and binary numbering systems.

can be a direct correspondence between binary notation and the electrical signals in a circuit. The usual notation is:

- If a switch or circuit has an *off* value, the meaning or numerical value of the binary digit is 0.
- If the switch or circuit condition is *on*, the binary numeric value is 1.

Discussion Questions/Problems

1. How are the decimal and binary systems similar?
2. On what is the binary system based?
3. How do binary notations correspond to electrical signals?

16.2 BINARY PROCESSING

As mentioned above, digital electronic or electrical devices are two-state devices. They process only in values of 0 and 1. Therefore, to accept, process, and deliver information usable to people, computers must have a capability to convert decimal input to binary values for processing. Then, binary equivalents must be reconverted to decimal values for output.

Binary-Decimal Conversion

Figure 16-2 shows the decimal values for binary places in a numbering system. The illustration shows how the binary number 100101 is converted into a decimal value. On the bottom line of Figure 16-2, the values of the 1 entries are added to establish the equivalent value: 32 plus 4 plus 1 = 37.

DECIMAL	32	16	8	4	2	1
BINARY	1	0	0	1	0	1

$$\overline{32} + \overline{0} + \overline{0} + \overline{4} + \overline{0} + \overline{1} = 37$$

Figure 16-2. Decimal values for binary places to use for binary-decimal conversions.

Decimal-Binary Conversion

To convert from the decimal to the binary numbering system, use a successive series of divisions by 2, as illustrated in Figure 16-3. The decimal number to be converted to binary values is 23. This number is divided by 2, producing a quotient of eleven with a remainder of 1. The 1 becomes part of the binary number. This process is then followed for each of the answers, in succession, until the value of 0 is reached. The numbers divided are 11, 5, 2, and 1. The results read from the least to the highest positions as 10111.

23 ÷ 2 = 11 with a remainder of 1
11 ÷ 2 = 5 with a remainder of 1
5 ÷ 2 = 2 with a remainder of 1
2 ÷ 2 = 1 with a remainder of 0
1 ÷ 2 = 0 with a remainder of 1

Figure 16-3. Conversion of decimal to binary numbers.

Discussion Questions/Problems

1. Convert the decimal number 17 to a binary value.
2. Convert the binary number 1101 to a decimal value.
3. Convert the decimal number 1523 to a binary value.

16.3 LOGIC

For circuits to process digital values, a system of logic is needed to define their interconnection and the meaning of signals. The logic system commonly used in digital circuitry is *Boolean algebra*. This is a system of mathematical notation for common mathematical functions such as addition and subtraction. Boolean algebra is particularly well suited for work with digital signals and circuits. This system of mathematics was introduced first in 1847 by George Boole (1815–1864), a famous English mathematician and logician.

Using Boolean algebra, which uses the binary system, it is possible to establish notations for mathematical operators such as AND, OR, NOT, EXCEPT, IF . . . THEN, and other, similar terms. Circuits designed to work with Boolean algebra have made it possible to compute and display mathematical information automatically.

AND Truth Table		
Inputs		Output
A	B	C
0	0	0
0	1	0
1	0	0
1	1	1

Figure 16-4. Two-input AND truth table.

Truth Tables

A standard method for demonstrating how Boolean algebra is used in electronic circuits is through use of *truth tables*. These are numeric tables which describe a number of inputs and show the output generated by each set of inputs. Truth tables for binary digital logic use binary values, with the numbers 0 and 1. In a truth table, the digit 1 symbolizes a true condition. The 0 indicates an untrue condition. Digital integrated circuits can be assembled as building blocks for a computer system in such a way as to represent the logic of truth tables. Therefore truth tables are used to represent different functions of Boolean algebra, as described below.

AND Truth Table			
Inputs			Output
A	B	C	D
0	0	0	0
0	0	1	0
0	1	1	0
1	1	1	1
1	0	0	0
1	1	0	0
0	1	0	0
1	0	1	0

Figure 16-5. Three-input AND truth table.

AND truth table. In an AND statement, an output is true only when all inputs are true. (There can be two or more, up to a virtually infinite number.) In an AND truth table, any untrue condition results in an untrue output. Figure 16-4 is an AND truth table for two inputs. Figure 16-5 is an AND truth table for three inputs.

OR truth table. If any input to an OR circuit has an on or 1 value, the output will have an on or 1 value. This on or 1 condition can exist in any number of inputs, from two upward. An OR truth table is illustrated in Figure 16-6.

OR Truth Table		
Inputs		Output
A	B	C
0	0	0
0	1	1
1	0	1
1	1	1

Figure 16-6. Two-input OR truth table.

NOT Truth Table	
Input	Output
A	B
0	1
1	0

Figure 16-7. One-input NOT truth table.

NAND Truth Table		
Inputs		Output
A	B	C
0	0	1
1	0	1
0	1	1
1	1	0

Figure 16-8. Two-input NAND truth table.

NOR Truth Table		
Inputs		Output
A	B	C
0	0	1
1	0	0
0	1	0
1	1	0

Figure 16-9. Two-input NOR truth table.

XOR Truth Table		
Inputs		Output
A	B	C
0	0	0
0	1	1
1	0	1
1	1	0

Figure 16-10. Two-input XOR truth table.

XNOR Truth Table		
Inputs		Output
A	B	C
0	0	1
1	0	0
0	1	0
1	1	1

Figure 16-11. Two-input NXOR truth table.

NOT truth table. A NOT circuit puts a negative value on the input it receives. NOT circuits accept a single input and invert the signal that is received. If a positive signal is input, the output is negative. If a negative signal is received, the output is positive. NOT circuits process single inputs and produce single outputs, directly inverting the values received. A NOT truth table is shown in Figure 16-7.

NAND truth table. NAND simply means negative AND. The interpretations and outputs of a NAND circuit are exactly opposite from those of an AND circuit. A truth table for NAND conditions is shown in Figure 16-8.

NOR truth table. NOR circuits produce outputs that are opposite of, or negative to, OR circuits. Thus, a NOR truth table indicates that any on or 1 condition in a circuit produces a 0 output. Following the same logic, a circuit that has only 0 values would produce an output equal to a 1 signal. This is shown in the truth table in Figure 16-9.

Exclusive OR (XOR) truth table. XOR circuits get their name from the fact that they respond to only a single condition, an on or 1 condition. This on or 1 condition may be present in only one of two inputs. Any 1 input will produce a 1 output. Two zeros will produce a 0 output. Two 1 signals will also produce a 0 output. Normally, XOR circuits are used with two inputs only. A truth table for this type of circuit is shown in Figure 16-10.

Exclusive NOR (XNOR) truth table. Exclusive NOR circuits function exactly opposite from exclusive OR circuits. That is, the sensing of a single on or 1 condition among two inputs produces a 0 output. A single 0 in either input signal produces a 0 output. Two zero inputs produce a 1 output. Two 1 input values also produce a 1 output. This is demonstrated in the truth table in Figure 16-11.

Truth tables describe circuit conditions. These circuit conditions are built into a wide variety of digital integrated circuits, some of which are described later in this chapter. By combining elements of the truth tables described above, it is possible to establish such circuit functions as logic gates, timers, counters, flip-flop circuits, and others.

Discussion Questions/Problems

1. What is the logical function of an AND circuit?
2. What is the logical function of an OR circuit?
3. What is the functional response of a NOT circuit?
4. What is the meaning, in terms of circuit performance, of the function Exclusive OR?
5. What does an Exclusive NOR circuit do?

16.4 LOGIC GATES AND INVERTERS

Logic gates and inverters are the tools for building digital processing devices. These are digital integrated circuits that perform the switching functions needed to apply digital logic. These integrated circuits are connected to correspond to the logic diagram the designer has developed. The designer has worked with truth tables in deciding how to meet the design objectives. Integrated circuits are used to calculate, store, or retrieve information, or for logical functions. All digital computers use these types of integrated circuits.

The symbols for logic-gate integrated circuits are shown in Figure 16-12. Most of the integrated circuits used in computers or other digital devices are designed to work with negative logic. This is because negative logic structures are easier and less expensive to fabricate. Thus, most integrated circuits are NAND, NOR, and XNOR. Some of the most commonly used types of integrated circuits are described below.

Figure 16-12. Symbols for logic-gate integrated circuits.

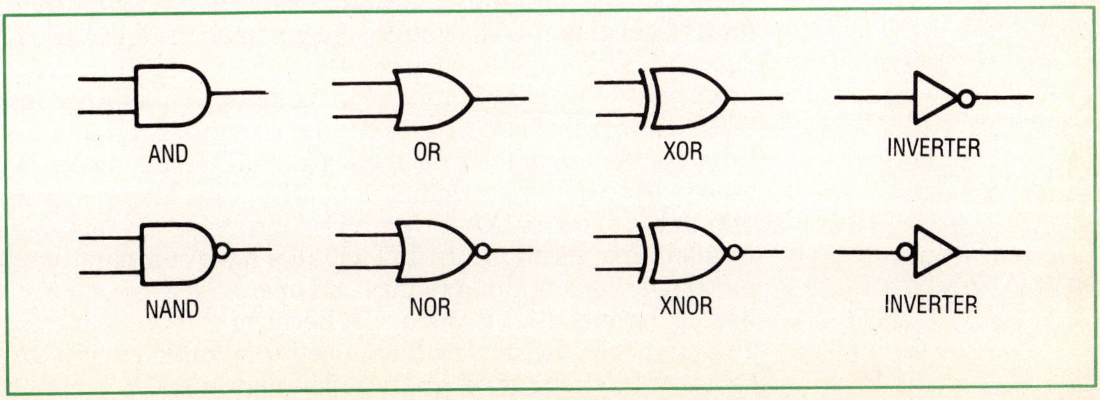

Resistor Transistor Logic (RTL) integrated circuits. These circuits get their name from the fact that the leads for all of the base-connected transistors in the circuit include resistors. Each connected transistor receives its input through a resistor that has the effect of controlling current flow. Through this design, RTL circuits are able to stand comparatively high levels of abuse from current fluctuations. However, these units tend to be slower, to produce more noise, and to consume more power than other types of units. A circuit for an RTL integrated circuit is illustrated in Figure 16-13.

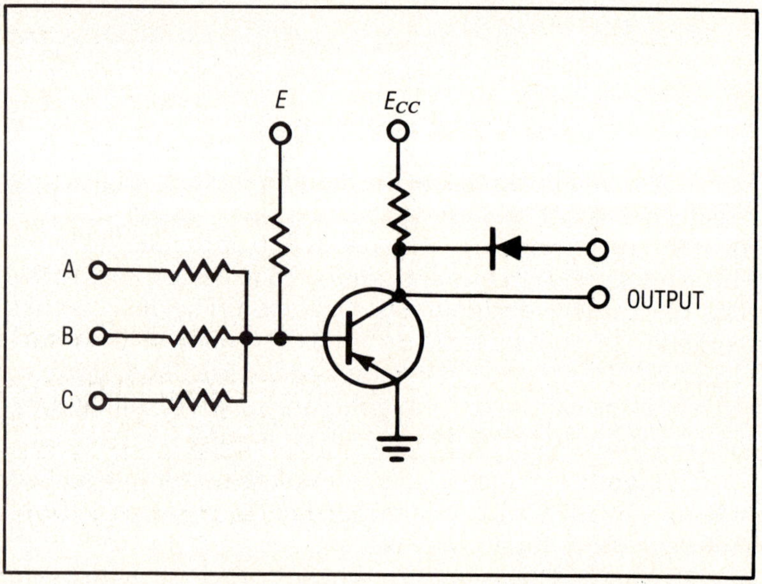

Figure 16-13. RTL integrated circuit schematic.

Diode Transistor Logic (DTL) integrated circuits. This type of circuit includes a diode in series with each input lead. DTL circuits are adaptable to many design situations and have moderate speed ratings. Power consumption is lower than for RTL circuits. A DTL circuit is diagrammed in Figure 16-14.

Transistor Transistor Logic (TTL) integrated circuits. This type of integrated circuit uses emitter or multiple emitter inputs. As a result of their design features, TTL circuits can be built in a smaller space than RTL or DTL units. They can provide far greater packing density—more logic circuits in a smaller area. In addition, TTL circuits have comparatively low power consumption ratings and operate at a low voltage, usually between 4.5 V and 5.5 V. Because of these features, TTL units are the most widely used integrated circuits. A typical TTL circuit is diagrammed in Figure 16-15.

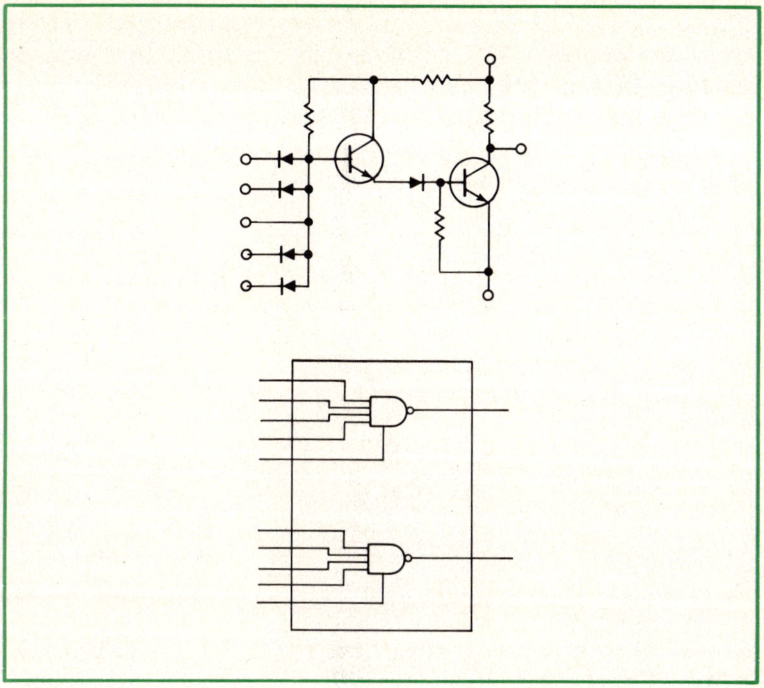

Figure 16-14. DTL circuit schematic and connection diagram.

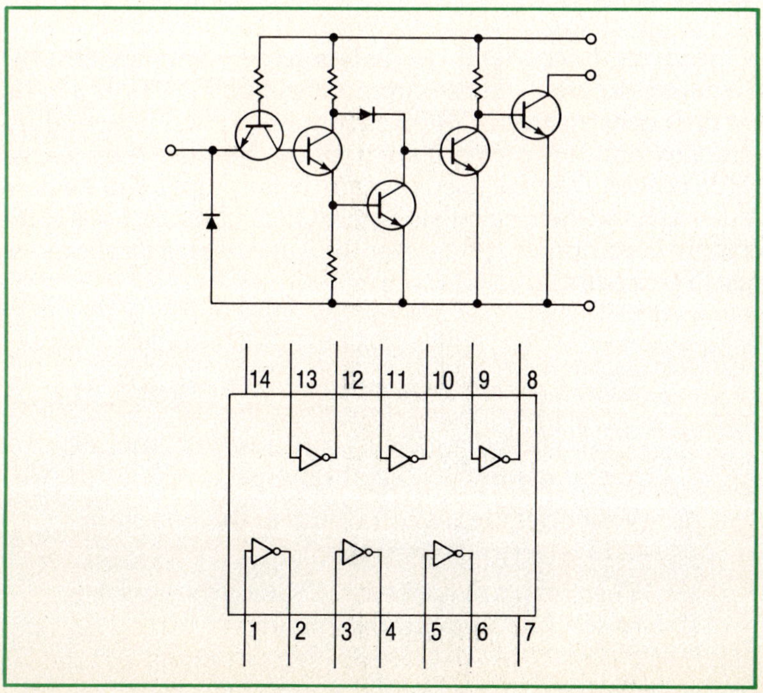

Figure 16-15. TTL circuit schematic and connection diagram.

Emitter Coupled Logic (ECL) integrated circuits. These units are similar to TTL circuits, with the main difference in biasing. ECL circuits can be described as emitter-follower configurations. They have excellent output power capabilities and also offer extremely high-speed operation. A diagram of an ECL circuit is shown in Figure 16-16.

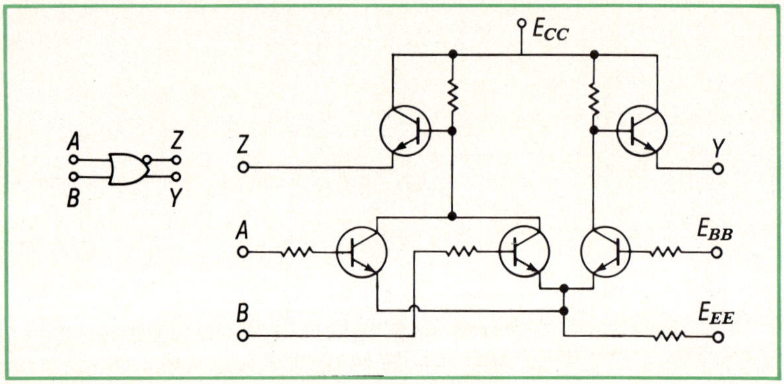

Figure 16-16. ECL circuit schematic and connection diagram.

Metal Oxide Semiconduction (MOS) integrated circuits. These are digital integrated circuits that use insulated-gate field-effect transistors (IGFETs). Power consumption is extremely low in comparison with most other digital integrated circuits.

Complementary Metal Oxide Semiconductor (CMOS) integrated circuits. These circuits also use IGFETs as a basis of their construction. CMOS circuits have good temperature stability and low power consumption. Because they use so little power, they are used in many battery-operated devices such as hand-held calculators. CMOS circuits are used commonly in computer systems for the same type of applications as TTL circuits.

Discussion Questions/Problems

1. What types of logic-gate integrated circuits are most commonly used and why?
2. What advantages contribute to the popularity of TTL integrated circuits?
3. What design features contribute to the popularity of CMOS integrated circuits?

16.5 TTL-TYPE INTEGRATED CIRCUITS

Complete families of devices are available for each of the types of digital integrated circuits described in the previous section. Each device in a family is designed for a special purpose. Many of these units are now identified by standard names or numbers that are used regularly in cataloging and ordering integrated circuits. An excellent example is the 7400 series of TTL integrated circuits. To illustrate, a few of these are described below.

Regular TTL integrated circuits. These units are usually designated by the number 7400. They are the most widely available and lowest priced of digital integrated circuits. Regular TTL units were the first in this family to be developed. They have been used as a basis for developing a whole family of TTL integrated circuit products.

One widely used type of TTL is diagrammed in Figure 16-17. This is known as a 7400 Quad-NAND gate circuit. The Quad-NAND gate circuit can also be made up in any of the types described below.

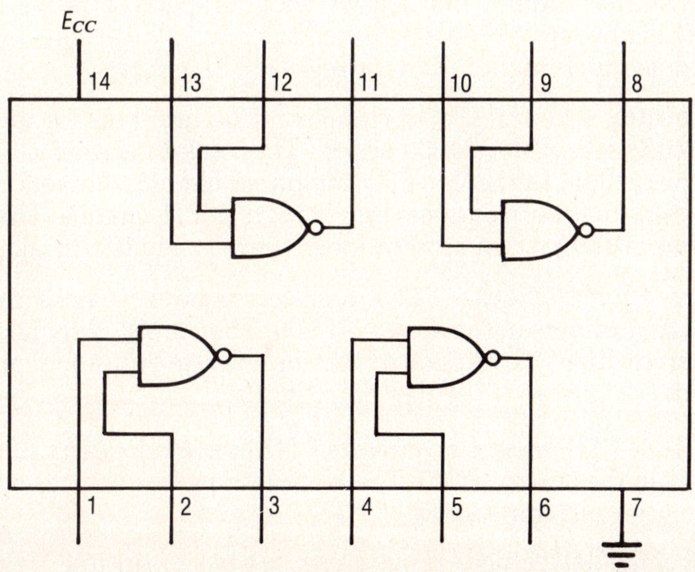

Figure 16-17. Connection diagram for 7400 Quad–NAND gate circuit.

Another typical TTL circuit package, shown in Figure 16-18, is the Quad two-input NOR gate. The model number for this unit is 7402. Again, any of several fabrication techniques can be used.

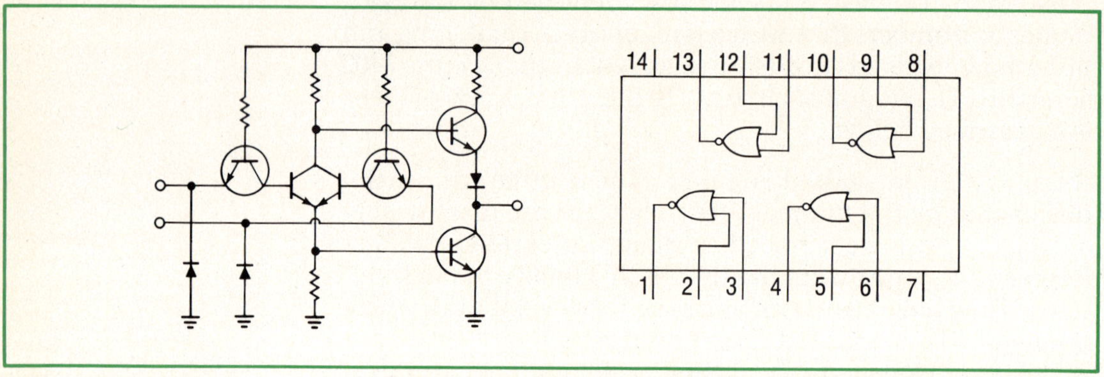

Figure 16-18. 7402 Quad two-input NOR gate circuit schematic and connection diagram.

Low-Power TTL integrated circuit. A typical unit of this type is the 74L00. This is a low-power circuit that uses one-tenth the power of a regular TTL. Its cycle time, or operating speed, is ten times longer than that of the regular TTL.

High-Power TTL integrated circuit. The model number sequence for this type of unit is 74H00. Units in this series have cycle times running into at least twice the speed of the regular TTL. However, speed is achieved at a price: Power consumption is much higher than for regular TTL circuits.

Schottky TTL integrated circuits. The numbers for these circuits are in the 74S00 series. They are three-and-a-half times as fast as regular TTL integrated circuits. Power consumption is about double that of regular TTL circuits. These circuits are used in applications where speed is critical.

Low-Power Schottky TTL integrated circuits. These units carry numbers in the series 74LS00. They are slightly faster than regular TTL integrated circuits and use only about one-fifth the power of regular TTL circuits.

CMOS TTL integrated circuits. These circuits carry numbers in the series 74C00. These are low-power devices with a high input impedance.

All of the descriptions above illustrate a basic principle of electronics. With integrated circuits, as with most components, there are many types and many choices of performance specifications. The choice of the proper unit depends on the individual application.

Discussion Questions/Problems

1. What are the advantages and disadvantages of low-power TTL integrated circuits?
2. What are the advantages and disadvantages of high-power TTL integrated circuits?
3. What are the advantages and disadvantages of the Schottky TTL integrated circuits?
4. What are the major differences, advantages, and disadvantages of the low-power Schottky TTL integrated circuit as compared with the regular Schottky TTL integrated circuit?

16.6 OTHER DIGITAL INTEGRATED CIRCUITS

For the purposes of illustration, some of the other types of digital integrated circuits are described below. These are:

- Astable multivibrator integrated circuits
- Timers
- Flip-flop integrated circuits.

Astable Multivibrator Integrated Circuits

This type of integrated circuit produces a synchronizing digital signal that can be used for some timing applications. As shown in Figure 16-19, the astable multivibrator is a transistorized oscillator circuit in an IC package. This circuit

Figure 16-19. Circuit schematic for astable multivibrator.

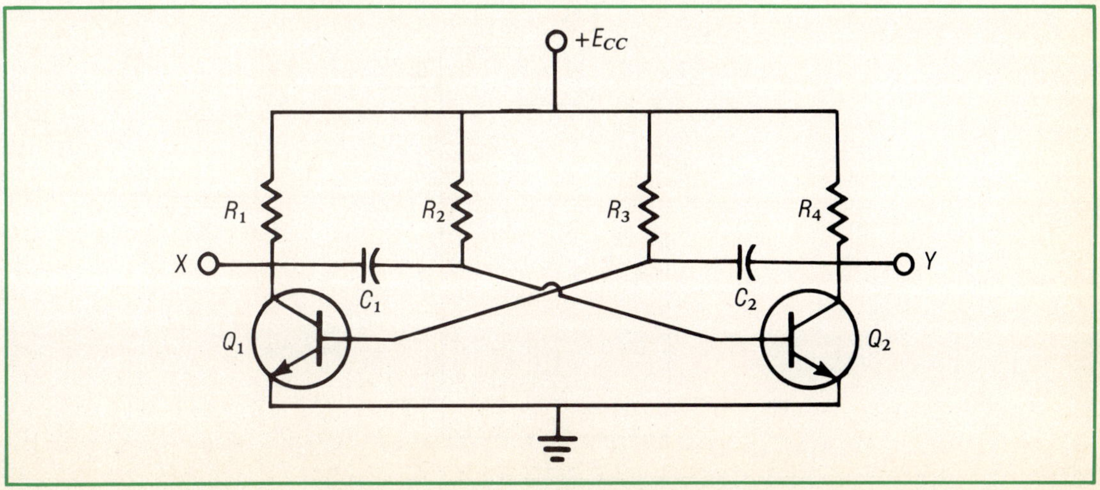

generates a *square wave* output voltage. No input signal is required. Therefore, the astable multivibrator is also called a *free-running multivibrator*. An important application for this type of circuit is to generate a train of pulses used to synchronize electronic devices that must function in unison.

Timers

These are integrated circuits designed specifically for timing applications. They are used for variable-frequency clock functions. The 555 timer, shown in Figure 16-20, is a free-running astable multivibrator which is the most commonly used timing circuit. It is applied widely to produce timing signals in computers or other digital devices.

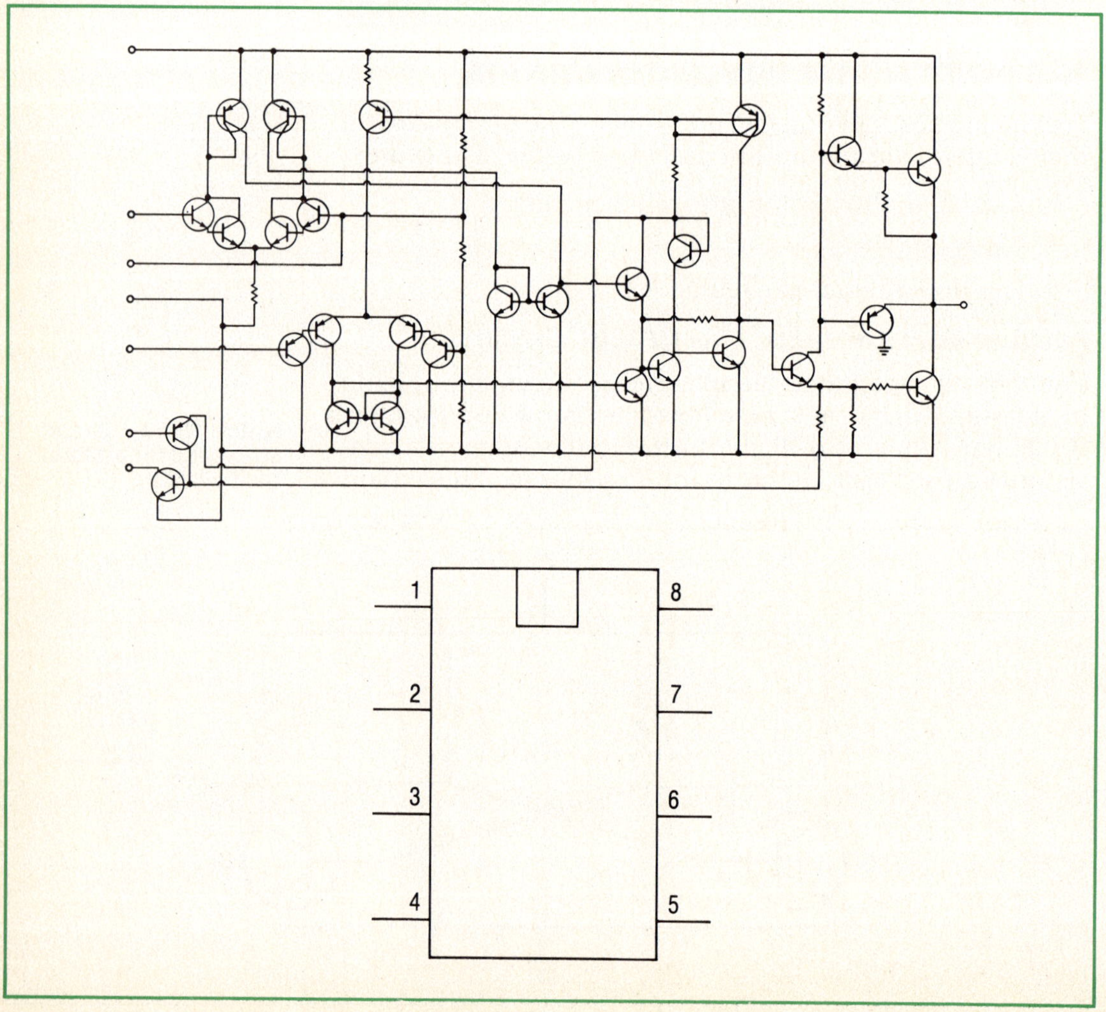

Figure 16-20. Circuit schematic and connection diagram for 555 timer.

Flip-Flop Integrated Circuits

Flip-Flop (FF) circuits are used for a variety of digital applications, including counters. Counters, as the name implies, count digital impulses. Counters can also generate frequencies and wave forms in clock circuits.

The flip-flop circuit symbol shown in Figure 16-21 is the dual JK level-triggered flip-flop. The letters used in flip-flop circuits to designate functions are also shown in this Figure.

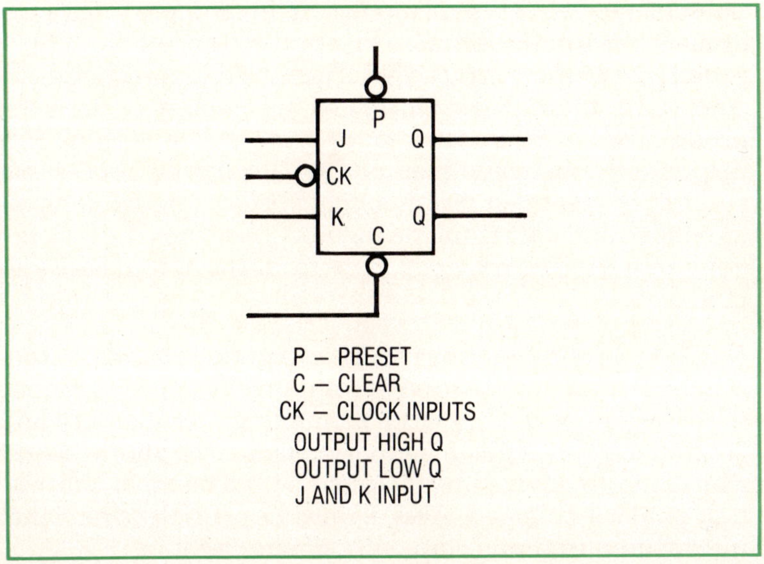

Figure 16-21. Symbol for FF circuit.

Many other types of integrated circuits are used in computers. Just a few of the available units are full adders, shift registers, counters, latchers, interfacing units, and memory chips.

Discussion Questions/Problems

1. Name an important characteristic of astable multivibrators and explain why this feature is valuble.
2. What is the principal application for flip-flop circuits?
3. In what types of devices are timing circuits most commonly used?

16.7 COMPUTER APPLICATIONS OF INTEGRATED CIRCUITS

Computer systems use most of the integrated circuits described in this chapter. In addition, other types of digital integrated circuits are used. Special computer integrated circuit chips include memory and microprocessor devices.

Memory Integrated Circuits

A number of types of integrated circuits are used for computer memories. A typical memory chip can store thousands of binary bits of information by applying current to set the conditions of the various cells. The speed of integrated circuits is important. The cycle time of the memory controls the productivity of a computer since there is a one-to-one correspondence between memory and computer processing cycles. The computer processor reads both programming instructions and data from the memory, executes the instructions with the data provided, then reads the output back into memory.

Random Access Memory (RAM) integrated circuits. Random access memory is used in computer systems where it is necessary to be able to write something into memory and later read it out. A given memory cell must be able to accept a bit of information, either a 0 or a 1, and retain that information until it is changed. Also, it must be possible to read that information from the chip without altering it.

Read Only Memory (ROM) integrated circuits. ROM circuits provide fixed files of information that can be referenced by computers. ROM memory circuits are permanent. Once data are recorded in manufacture, the information cannot be changed. This is a marked difference from the RAM memory chips which are volatile. This means that RAM circuits are continually changing while ROM memories are fixed, or permanent. Typically, computer software and special operating instructions are stored on these memories which direct the basic configuration and operation of computer equipment.

Programmable Read Only Memory (PROM) integrated circuits. These devices are similar to ROM, but they can be pre-programmed by the user, rather than the manufacturer. PROMs make it possible for computer users to insert their own fixed programs into a computer system. This gives the user increased flexibility while still providing fixed program instructions to the computer.

Microprocessor Integrated Circuits

Mircroprocessors are the basis for many low-cost high-productivity business machines, particularly microcomputers. A microprocessor is often described as a "computer on a chip." A microprocessor integrated circuit provides complete logic capabilities in a miniature integrated circuit. A microprocessor is the heart of a microcomputer. A microprocessor chip is mounted on a circuit board along with memory and other integrated circuit devices to form the basic microcomputer, as shown in Figure 16-22.

Microprocessors have come into wide use for many computing or control functions. For example, microprocessors are used in modern automobiles to control fuel, for emissions control, and in other operational systems.

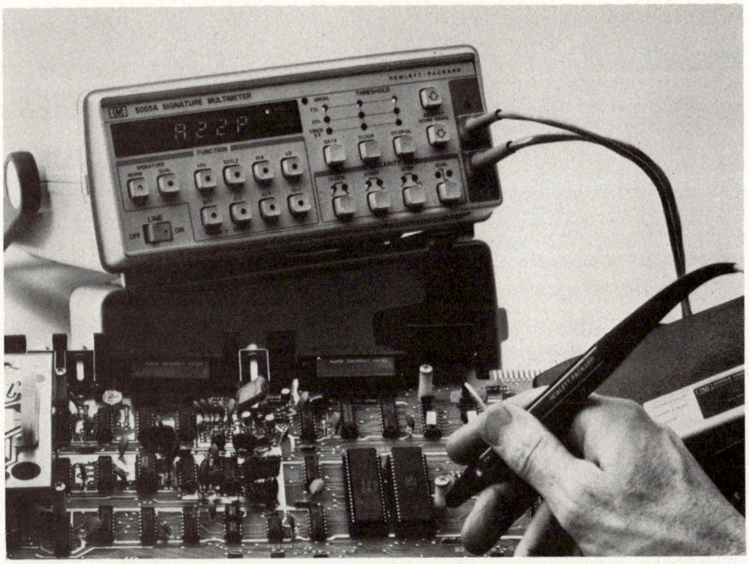

Figure 16-22. Microprocessor circuit board being tested. (Courtesy Hewlett-Packard)

Discussion Questions/Problems

1. How do RAM and ROM integrated circuits differ in purpose and function?
2. How do ROM and PROM integrated circuits differ in purpose and function?
3. How are the functions of microprocessors and memory chips integrated?

REVIEW

CHAPTERS 15 AND 16

SUMMARY

1. The integrated circuit (IC) is a complete, functional circuit manufactured as a single unit. (15.1)
2. Monolithic circuits are manufactured by adopting certain photographic and lithographic techniques. (15.2)
3. In film fabrication techniques, the photographic image is placed upon the surface of a wafer and developed before the coating of semiconductor materials takes place. Silver oxide resist material is applied first. Then the photographic image is developed in place. (15.3)
4. Hybrid integrated circuits are combinations of circuits. They are formed with all of the fabrication techniques that are used in making monolithic, thin film, and thick film ICs. (15.4)
5. Construction of integrated circuits is based on the size of the package itself, the function to be performed, and the type of application to be supported. Broadly, integrated circuit packages are separated into three catagories:
   1. small-scale integration (SSI)
   2. medium-scale integration (MSI)
   3. large-scale integration (LSI). (15.5)
6. Integrated circuits are classified according to the way they operate. There are two broad catagories of IC functions; they are linear and digital. The term *linear* means simply that there is a direct relationship between input and output signals for an integrated circuit. This direct relationship classifies linear ICs as analog devices. Digital

integrated circuits act as though they are special combinations of switches. The output signals of these units can be turned on and off electronically in such a way as to represent information. (15.5)

7. Operational amplifiers are the most popular types of linear integrated circuits. These units are used as general amplifiers, as amplifiers for sensors, in many types of instruments and wave generators. (15.6)
8. A differential amplifier (DIFF AMP) is a linear integrated circuit that does not include capacitors within its components. The chief characteristic of a differential amplifier is that it has two input signals from which it generates a difference output signal. (15.7)
9. A voltage regulator is a circuit that produces an output voltage of a predetermined value. (15.8)
10. Audio amplifiers are devices that accept signals in the audio frequency (AF) range and increase the magnitude of output. (15.9)
11. Linear integrated circuits are now produced in many configurations to handle a number of special jobs, including such applications as wideband amplifiers, RF amplifiers, IF amplifiers, AM amplifiers, and FM amplifiers. (15.10)
12. The common numbering system is the decimal system (ones, tens, hundreds, thousands, etc.) based on the number 10. This system was probably developed as a natural extension of having ten fingers for counting. Other numbering systems are possible, based on a number other than 10. One system, based on the number 2, is called the *binary system*. The binary system and the decimal system both have the same structure. The binary system is particularly useful in electronics because, just as a digit in binary can be either zero or one, an electronic current, or voltage, can be either on or off. (16.1)
13. Computers must have a capability to convert decimal input into binary values for processing. (16.2)
14. For circuits to process digital values, a system of logic is needed to define their interconnection and the meaning of signals. The logic system commonly used in digital circuitry is Boolean algebra. A standard method for demonstrating how Boolean algebra is used in electronic circuits is through the use of truth tables. (16.3)
15. Logic gates and inverters are the tools for building digital processing devices. These are digital integrated circuits

that perform the switching functions needed to apply digital logic. (16.4)

16. There are many other types of integrated circuits used for special purposes such as timers and flip-flops. (16.6)
17. Computer systems use most of the integrated circuits described. Special computer integrated circuit chips include memory and microprocessor devices. (16.7)

PROBLEMS

1. What is the open-loop gain of an operational amplifier having an output signal voltage of 25 V and a differential input voltage of 0.5 V? (15.5)
2. What is the closed-loop ratio of an OP AMP that has an output voltage of 10 V and an input voltage of 0.2 V? (15.5)
3. Convert the binary number 1001 to a decimal value. (16.2)
4. Convert the decimal number 20 to a binary value. (16.2)
5. Convert the decimal number 32 to a binary value. (16.2)
6. Write a truth table for an OR circuit that has three inputs. (16.3)
7. Write a truth table for a NAND circuit that has three inputs. (16.3)
8. Write a truth table for a NOR circuit that has three inputs. (16.3)

(Answers to the odd-numbered problems can be found at the back of the book.)

ELECTRONIC SCHEMATIC SYMBOLS

The ability to read schematic drawings is essential in electrical or electronics work. Symbols for the various components in circuits have been standardized. When used in a circuit schematic, these symbols indicate the components needed for a project. Such drawings also show how all components within a circuit or device are connected to each other. This appendix covers those electronic symbols most commonly found in circuit schematics.

At the end of this appendix is a list of some of the recommended practices for drawing symbols on a circuit schematic.

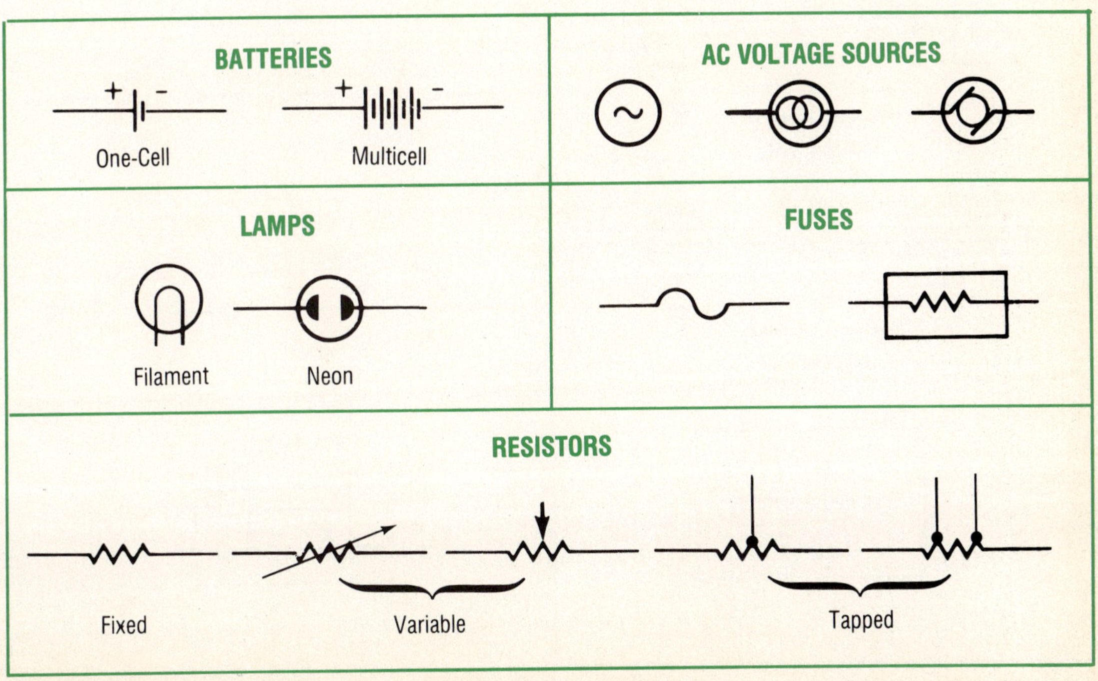

CAPACITORS

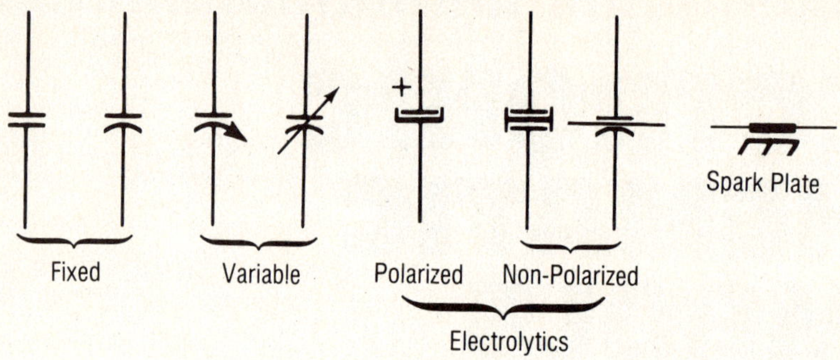

JACKS

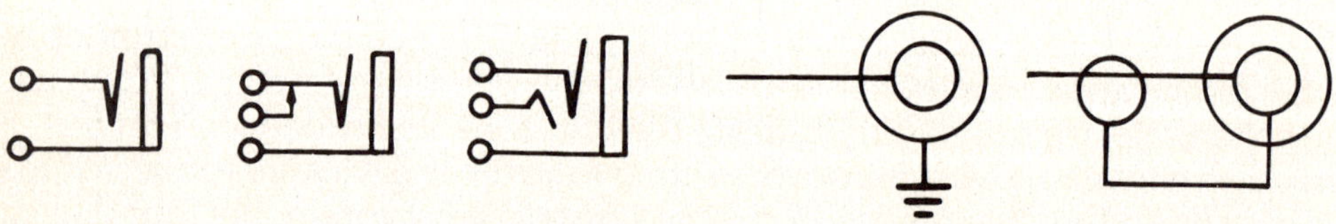

SHIELDS

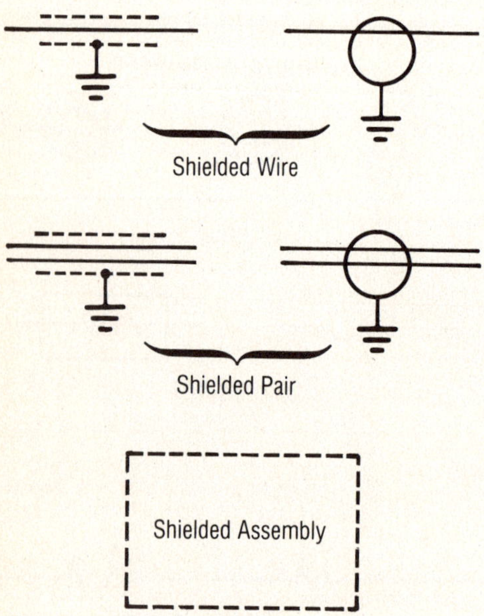

CIRCUIT BREAKERS

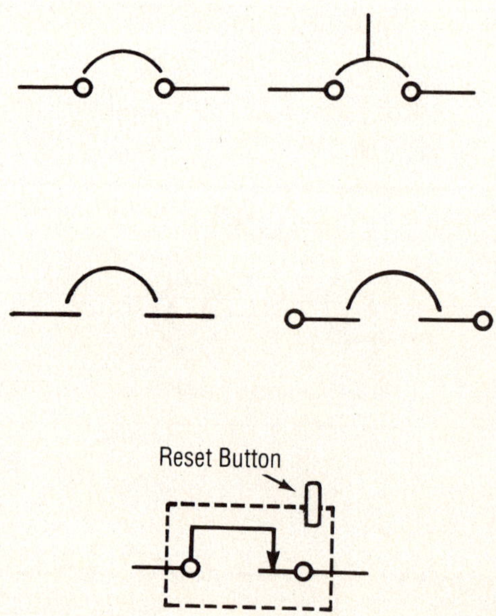

INDUCTORS

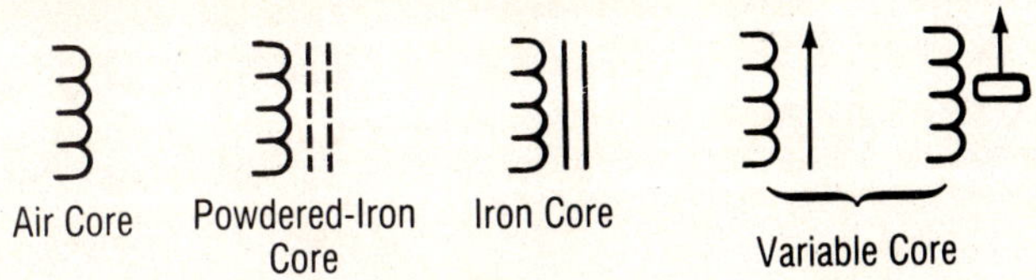

TRANSFORMERS

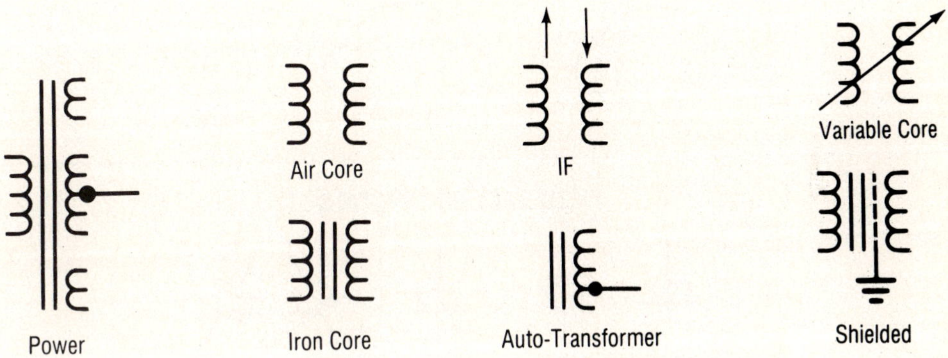

WIRING

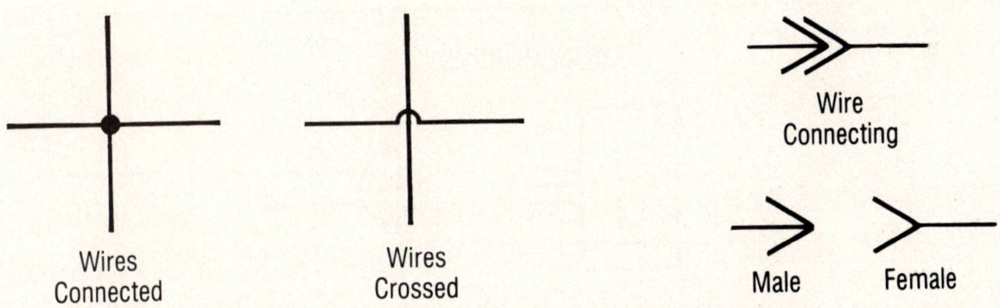

AC RECEPTACLES

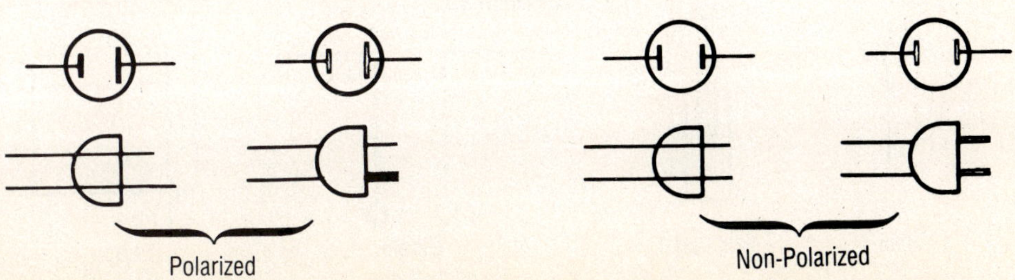

GROUNDS

SPEAKERS

HEADPHONES

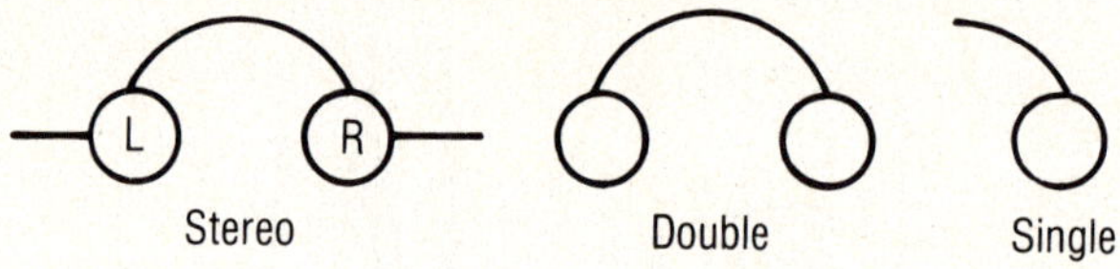

PIEZOELECTRIC CRYSTALS

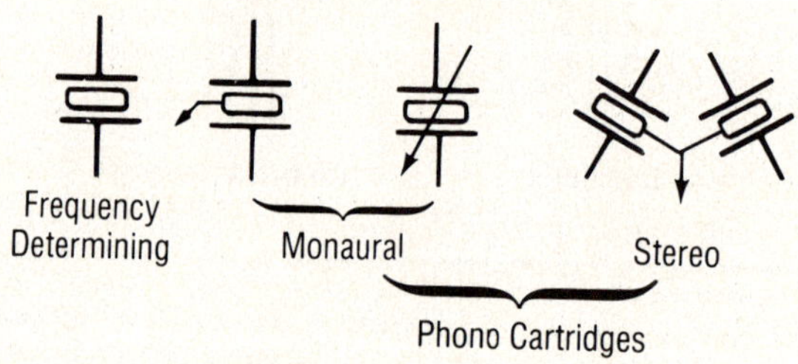

MICROPHONES

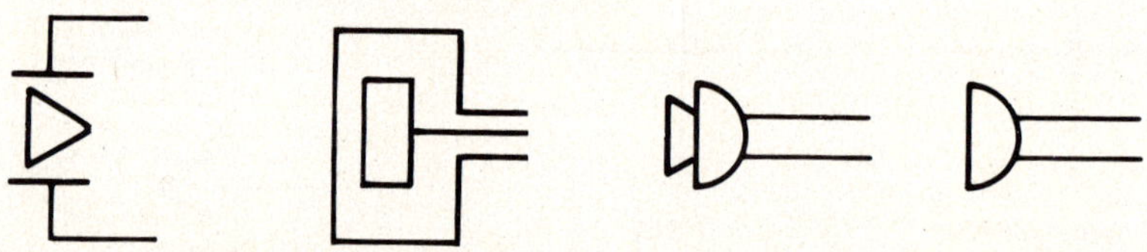

Type indicated by note: Ceramic, Crystal, Dynamic, etc.

ANTENNAS

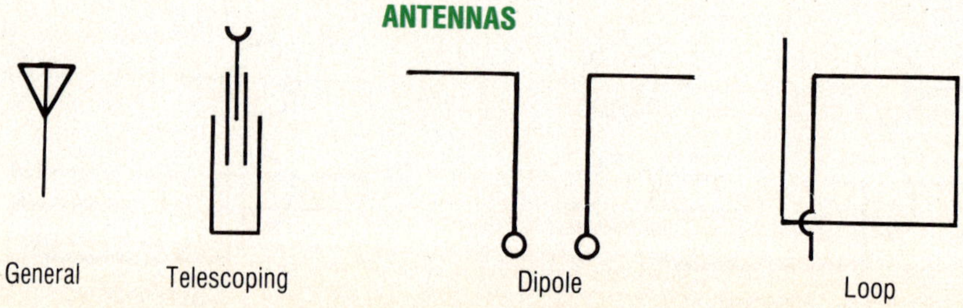

TUBES

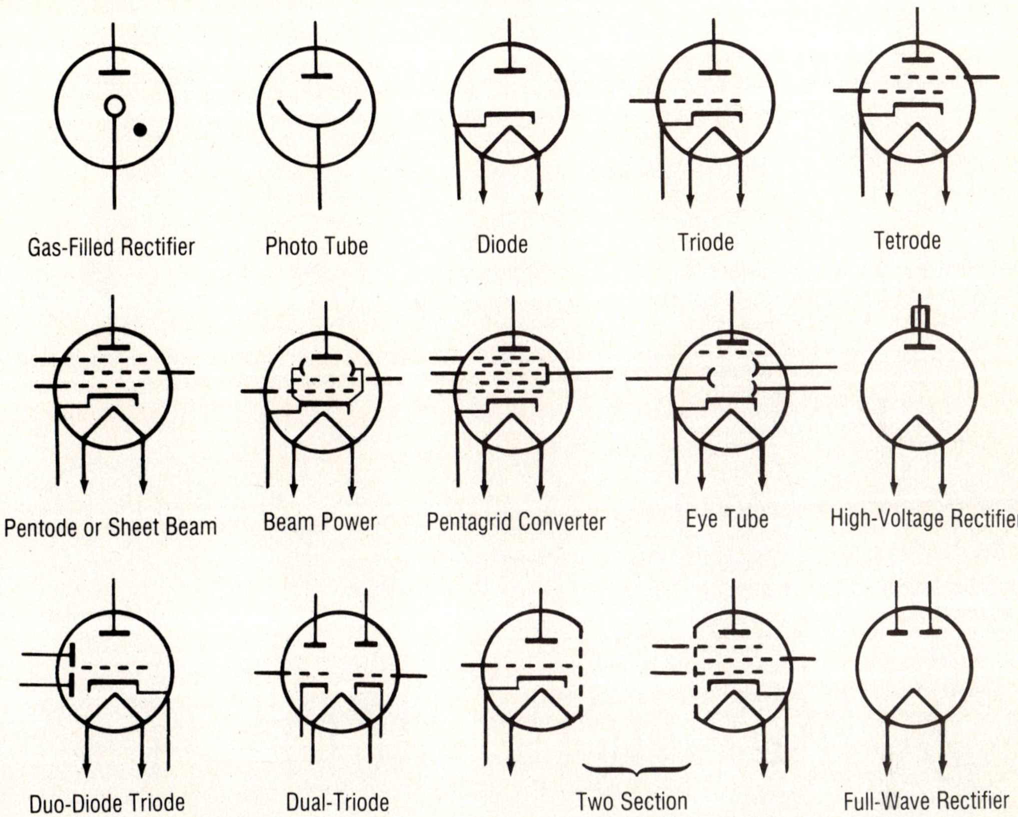

COMPUTER SYMBOLS

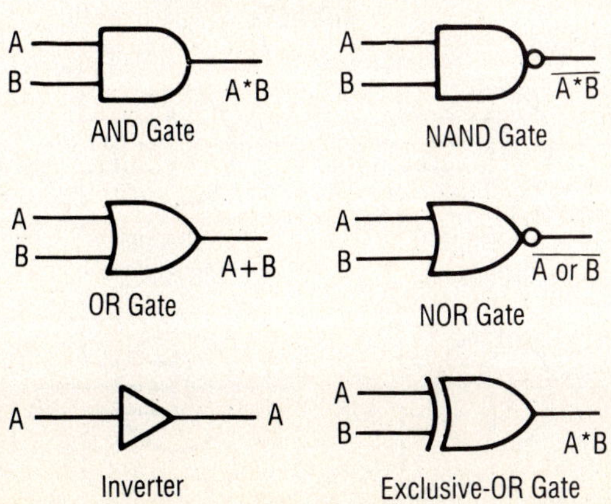

CATHODE-RAY TUBE

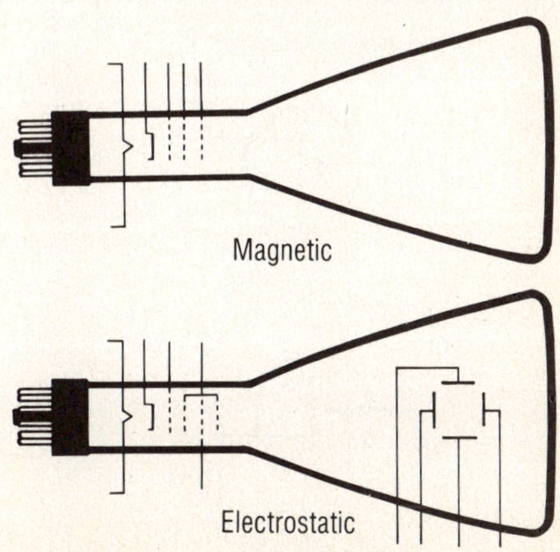

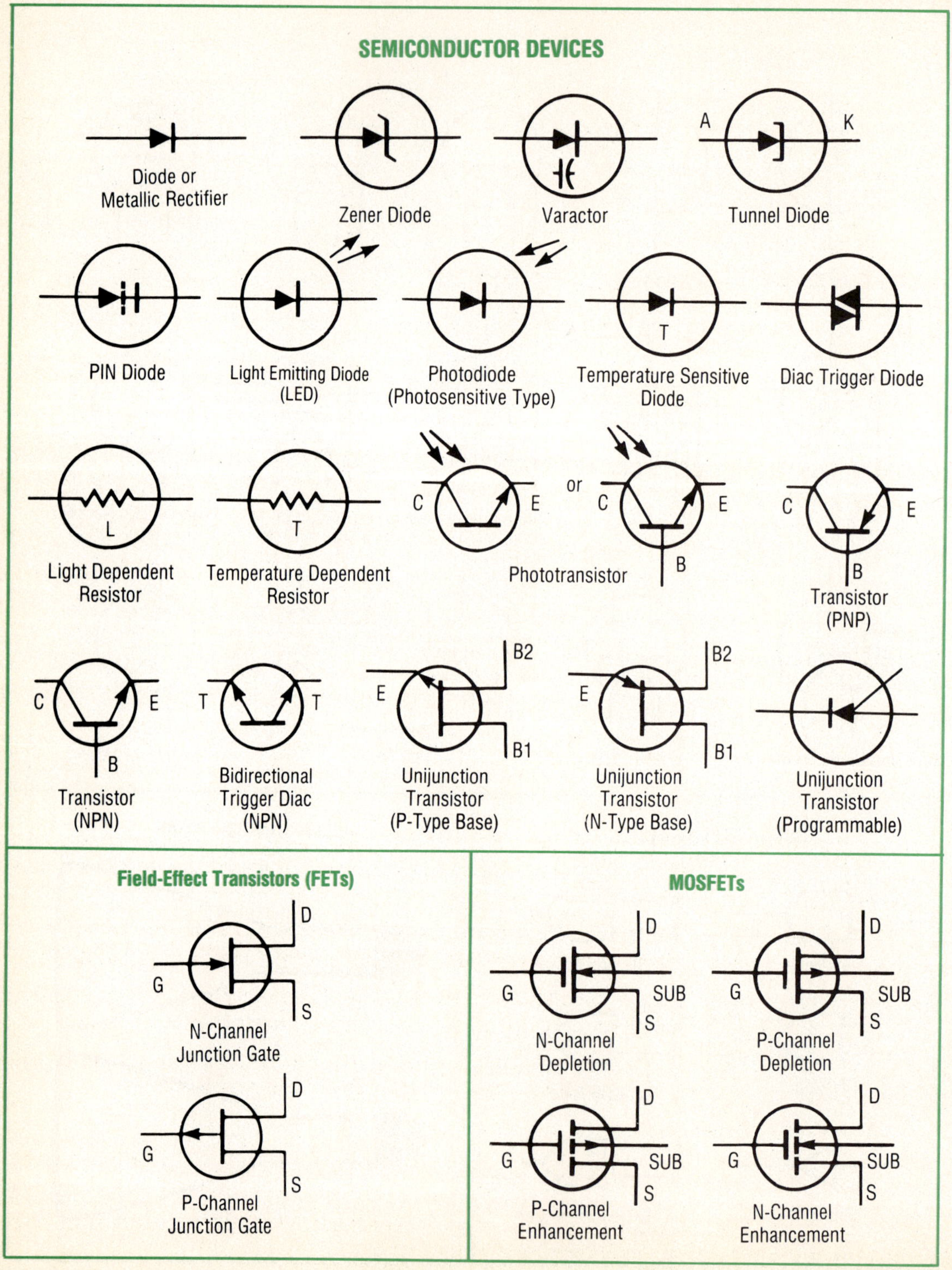
SEMICONDUCTOR DEVICES
Diode or Metallic Rectifier
Zener Diode
Varactor
A
K
Tunnel Diode
PIN Diode
Light Emitting Diode (LED)
Photodiode (Photosensitive Type)
T
Temperature Sensitive Diode
Diac Trigger Diode
L
Light Dependent Resistor
T
Temperature Dependent Resistor
C
E
or
C
E
B
Phototransistor
C
E
B
Transistor (PNP)
C
E
B
Transistor (NPN)
T
T
Bidirectional Trigger Diac (NPN)
E
B2
B1
Unijunction Transistor (P-Type Base)
E
B2
B1
Unijunction Transistor (N-Type Base)
Unijunction Transistor (Programmable)
Field-Effect Transistors (FETs)
D
G
S
N-Channel Junction Gate
D
G
S
P-Channel Junction Gate
MOSFETs
D
G
SUB
S
N-Channel Depletion
D
G
SUB
S
P-Channel Depletion
D
G
SUB
S
P-Channel Enhancement
D
G
SUB
S
N-Channel Enhancement

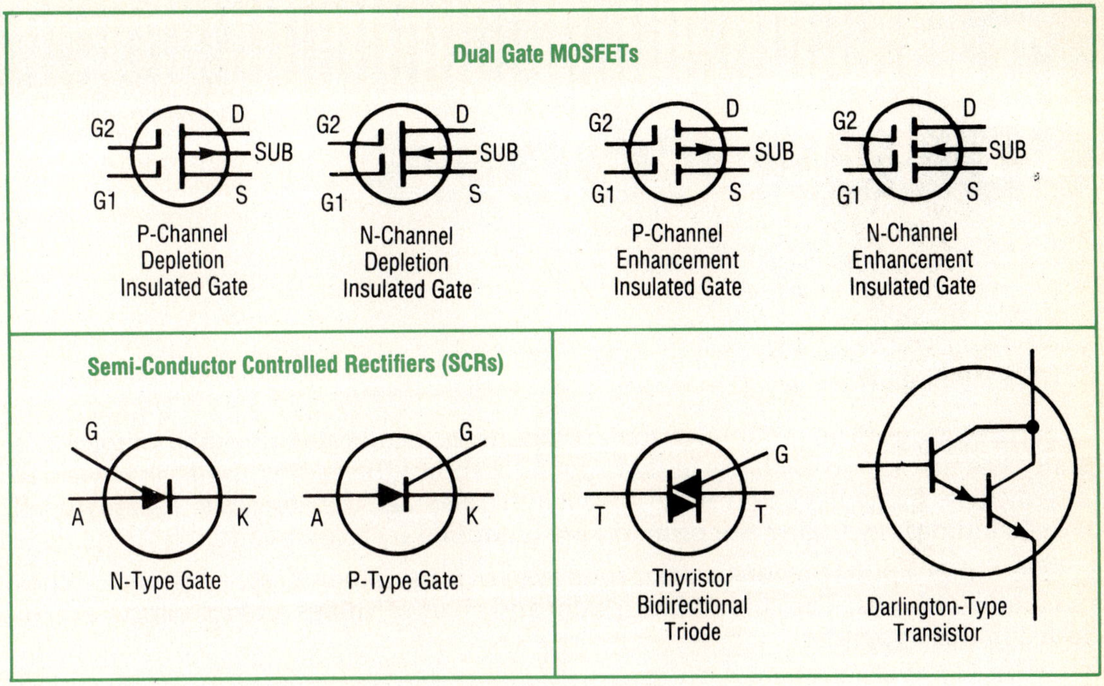

Recommended Practices for Drawing Schematic Symbols

1. A connecting line to a symbol should be either vertical or horizontal wherever possible. A slanted connecting line may be used if necessary and, unless otherwise noted, has no particular significance.
2. No more than two different sizes of symbols should be used on any one schematic. Two sizes may be used to create a contrast or differentiation among symbols.
3. Symbols should be drawn in proportion to the rest of the diagram. However, they may be in any size that is suitable for use in any particular drawing.
4. Although the standard symbol for a terminal (0) may be added to any component symbol, it should not be considered an actual part of the component symbol.
5. The meaning of a symbol is not affected by its position on a diagram.
6. The weight of the line used to draw a symbol does not affect the meaning of the symbol.
7. When a symbol is modified to suit a particular industrial practice or policy, it should be identified by a note accompanying the drawing.

COLOR CODES

Color codes are used on certain electronic components to indicate values and/or tolerance limits. The first group of charts in this appendix shows the coding used for power, IF (intermediate frequency), and audio output and interstage transformers. As shown, all of the windings, including the primary, are identified by a color code.

For composition resistors, a color code system is used to indicate resistance values and tolerance limits. The color band system and chart of values and tolerances are included in this appendix.

A color code is used to indicate capacitance and tolerance ratings for capacitors. The last chart shows the coding for tantalum capacitors.

POWER TRANSFORMERS

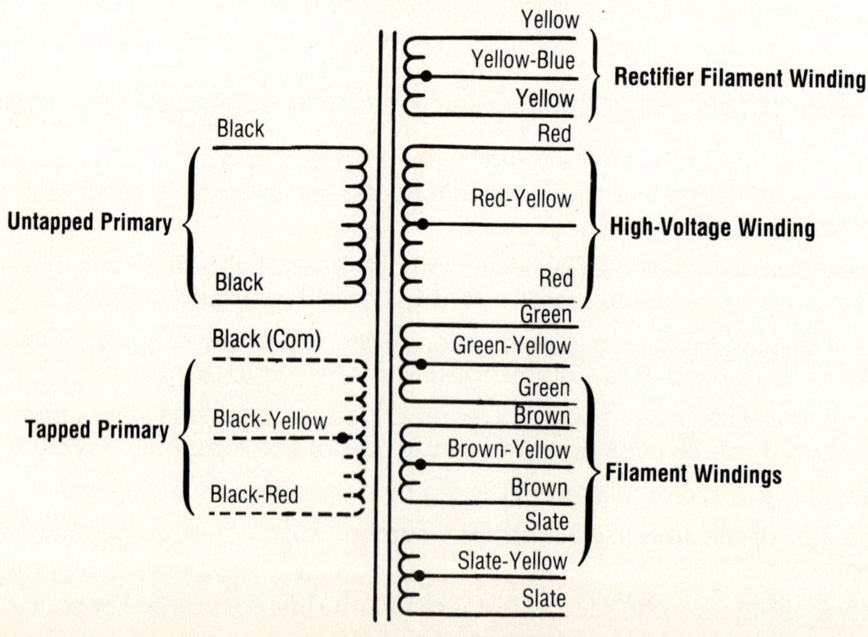

IF TRANSFORMERS

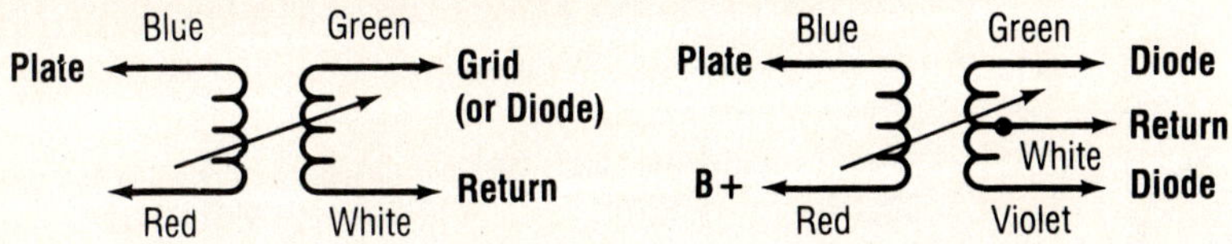

AUDIO OUTPUT AND INTERSTAGE TRANSFORMERS

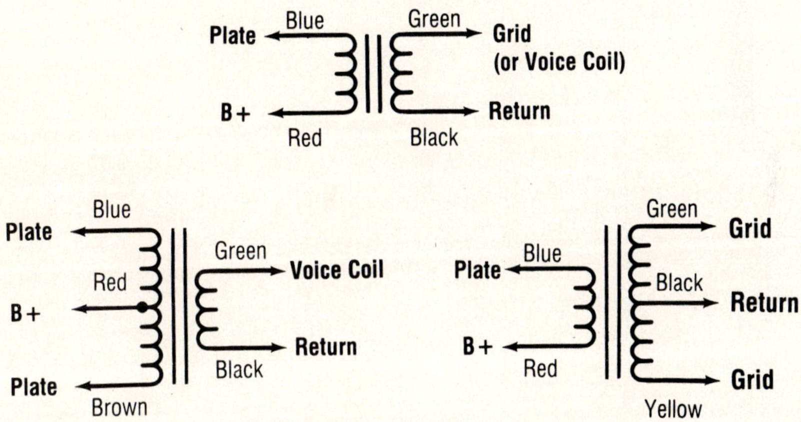

COLOR BAND SYSTEM

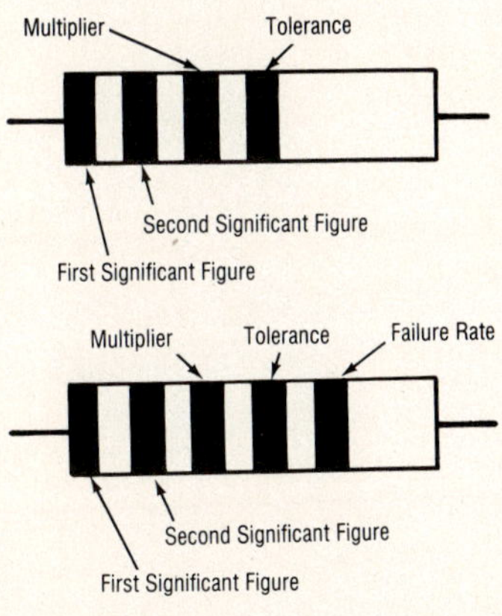

Resistors with black bodies are composition, non-insulated.
Resistors with colored bodies are composition, insulated.
Wire-wound resistors have double-width first color bands.

RESISTOR COLOR CODE

COLOR	FIRST AND SECOND SIGNIFICANT FIGURES	MULTIPLIER	TOLERANCE	FAILURE RATE*
Black	0	1		
Brown	1	10	±1%	1.0
Red	2	100	±2%	0.1
Orange	3	1000	±3%	0.01
Yellow	4	10000	±4%	0.001
Green	5	100000		
Blue	6	1000000		
Violet	7	10000000		
Gray	8	100000000		
White	9			Solderable**
Gold		0.1	±5%	
Silver		0.01	±10%	
No Color			±20%	

*When used on composition resistors, indicates percent failure per 1,000 hours.
**On film resistors, a white band indicates solderable terminal.

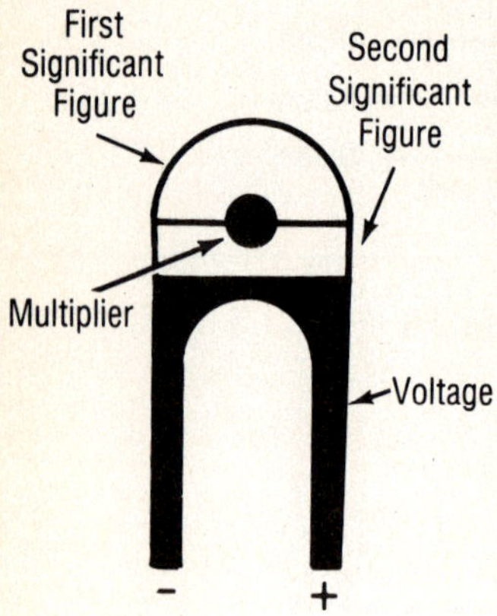

COLOR CODES FOR TANTALUM CAPACITORS
(CAPACITANCE GIVEN IN μF)

COLOR	FIRST SIGNIFICANT FIGURE	SECOND SIGNIFICANT FIGURE	MULTIPLIER	RATED DC VOLTAGE
Black		0	1	10
Brown	1	1	10	
Red	2	2	100	
Orange	3	3		
Yellow	4	4		6.3
Green	5	5		16
Blue	6	6		20
Violet	7	7		
Gray	8	8	0.01	25
White	9	9	0.1	3
Pink				35

PERIODIC TABLE OF ELEMENTS

All substances are made up of one or more elements. Of these, 92 occur naturally in the universe. At least 11 more have been created artificially since the discovery of nuclear fission. The table below shows the relationship of the atomic number of an element to its atomic mass. The atomic mass is the relative weight of an atom of an element as compared to the weight of an atom of oxygen. The atomic number refers to the number of protons in the nucleus of the atom.

ATOMIC NUMBER	ELEMENT	SYMBOL	ATOMIC MASS
1	Hydrogen	H	1.008
2	Helium	He	4.003
3	Lithium	Li	6.939
4	Beryllium	Be	9.012
5	Boron	B	10.81
6	Carbon	C	12.01
7	Nitrogen	N	14.01
8	Oxygen	O	16.00
9	Fluorine	F	19.00
10	Neon	Ne	20.18
11	Sodium	Na	22.99
12	Magnesium	Mg	24.31
13	Aluminum	Al	26.98
14	Silicon	Si	28.09
15	Phosphorus	P	30.97
16	Sulfur	S	32.06
17	Chlorine	Cl	35.45
18	Argon	Ar	39.95
19	Potassium	K	39.10
20	Calcium	Ca	40.08

ATOMIC NUMBER	ELEMENT	SYMBOL	ATOMIC MASS
21	Scandium	Sc	44.96
22	Titanium	Ti	47.90
23	Vanadium	V	50.94
24	Chromium	Cr	52.00
25	Manganese	Mn	54.94
26	Iron	Fe	55.85
27	Cobalt	Co	58.93
28	Nickel	Ni	58.71
29	Copper	Cu	63.54
30	Zinc	Zn	65.37
31	Gallium	Ga	69.72
32	Germanium	Ge	72.59
33	Arsenic	As	74.92
34	Selenium	Se	78.96
35	Bromine	Br	79.91
36	Krypton	Kr	83.80
37	Rubidium	Rb	85.47
38	Strontium	Sr	87.62
39	Yttrium	Y	88.91
40	Zirconium	Zr	91.22
41	Niobium	Nb	92.91
42	Molybdenum	Mo	95.94
43	Technetium	Tc	(99)
44	Ruthenium	Ru	101.0
45	Rhodium	Rh	102.9
46	Palladium	Pd	106.4
47	Silver	Ag	107.9
48	Cadmium	Cd	112.4
49	Indium	In	114.8
50	Tin	Sn	118.7
51	Antimony	Sb	121.8
52	Tellurium	Te	127.6
53	Iodine	I	126.9
54	Xenon	Xe	131.3
55	Cesium	Cs	132.9
56	Barium	Ba	137.4
57	Lanthanum	La	138.9
58	Cerium	Ce	140.1
59	Praseodymium	Pr	140.9
60	Neodymium	Nd	144.2
61	Promethium	Pm	(147)
62	Samarium	Sm	150.4
63	Europium	Eu	152.0
64	Gadolinium	Gd	157.3
65	Terbium	Tb	158.9
66	Dysprosium	Dy	162.5
67	Holmium	Ho	164.9
68	Erbium	Er	167.3
69	Thulium	Tm	168.9
70	Ytterbium	Yb	173.0

ATOMIC NUMBER	ELEMENT	SYMBOL	ATOMIC MASS
71	Lutetium	Lu	175.0
72	Hafnium	Hf	178.5
73	Tantalum	Ta	181.0
74	Tungsten	W	183.9
75	Rhenium	Re	186.2
76	Osmium	Os	190.2
77	Iridium	Ir	192.2
78	Platinum	Pt	195.1
79	Gold	Au	197.0
80	Mercury	Hg	200.6
81	Thallium	Tl	204.4
82	Lead	Pb	207.2
83	Bismuth	Bi	209.0
84	Polonium	Po	(209)
85	Astatine	At	(210)
86	Radon	Rn	(222)
87	Francium	Fr	(223)
88	Radium	Ra	(226)
89	Actinium	Ac	(227)
90	Thorium	Th	232.0
91	Protactinium	Pa	(231)
92	Uranium	U	238.0
93	Neptunium	Np	(237)
94	Plutonium	Pu	(244)
95	Americium	Am	(243)
96	Curium	Cm	(247)
97	Berkelium	Bk	(247)
98	Californium	Cf	(251)
99	Einsteinium	Es	(254)
100	Fermium	Fm	(257)
101	Mendelevium	Md	(256)
102	Nobelium	No	(254)
103	Lawrencium	Lr	(257)

TRIGONOMETRIC FUNCTIONS

Trigonometry deals with the relationship between the angles and the sides of a right triangle. Electricity and electronics make extensive use of these relationships, especially in the analysis of AC circuits. The relationships between the sides of a right triangle are called *functions*. These functions are stated as ratios between the sides. The angle between two sides of a right triangle is expressed in terms of one of these functions. When the mathematical equivalent of a function is known, the angle can be found by referring to a table of natural trigonometric functions. Or, when an angle and one side are known, the remaining side can be determined: $a = c \times \sin$ of angle A; $c = a/\sin$ of angle A; and $b = c \times \cos$ of angle A.

ANGLE	SIN	COS	TAN	ANGLE	SIN	COS	TAN
0°	0.0000	1.000	0.0000				
1	.0175	.9998	.0175	16°	.2756	.9613	.2867
2	.0349	.9994	.0349	17	.2924	.9563	.3057
3	.0523	.9986	.0524	18	.3090	.9511	.3249
4	.0698	.9976	.0699	19	.3256	.9455	.3443
5	.0872	.9962	.0875	20	.3420	.9397	.3640
6	.1045	.9945	.1051	21	.3584	.9336	.3839
7	.1219	.9925	.1228	22	.3746	.9272	.4040
8	.1392	.9903	.1405	23	.3907	.9205	.4245
9	.1564	.9877	.1584	24	.4067	.9135	.4452
10	.1736	.9848	.1763	25	.4226	.9063	.4663
11	.1908	.9816	.1944	26	.4384	.8988	.4877
12	.2079	.9781	.2126	27	.4540	.8910	.5095
13	.2250	.9744	.2309	28	.4695	.8829	.5317
14	.2419	.9703	.2493	29	.4848	.8746	.5543
15	.2588	.9659	.2679	30	.5000	.8660	.5774

ANGLE	SIN	COS	TAN	ANGLE	SIN	COS	TAN
31°	.5150	.8572	.6009	61°	.8746	.4848	1.8040
32	.5299	.8480	.6249	62	.8829	.4695	1.8807
33	.5446	.8387	.6494	63	.8910	.4540	1.9626
34	.5592	.8290	.6745	64	.8988	.4384	2.0503
35	.5736	.8192	.7002	65	.9063	.4226	2.1445
36	.5878	.8090	.7265	66	.9135	.4067	2.2460
37	.6018	.7986	.7536	67	.9205	.3907	2.3559
38	.6157	.7880	.7813	68	.9272	.3746	2.4751
39	.6293	.7771	.8098	69	.9336	.3584	2.6051
40	.6428	.7660	.8391	70	.9397	.3420	2.7475
41	.6561	.7547	.8693	71	.9455	.3256	2.9042
42	.6691	.7431	.9004	72	.9511	.3090	3.0777
43	.6820	.7314	.9325	73	.9563	.2924	3.2709
44	.6947	.7193	.9657	74	.9613	.2756	3.4874
45	.7071	.7071	1.0000	75	.9659	.2588	3.7321
46	.7193	.6947	1.0355	76	.9703	.2419	4.0108
47	.7314	.6820	1.0724	77	.9744	.2250	4.3315
48	.7431	.6691	1.1106	78	.9781	.2079	4.7046
49	.7547	.6561	1.1504	79	.9816	.1908	5.1446
50	.7660	.6428	1.1918	80	.9848	.1736	5.6713
51	.7771	.6293	1.2349	81	.9877	.1564	6.3138
52	.7880	.6157	1.2799	82	.9903	.1392	7.1154
53	.7986	.6018	1.3270	83	.9925	.1219	8.1443
54	.8090	.5878	1.3764	84	.9945	.1045	9.5144
55	.8192	.5736	1.4281	85	.9962	.0872	11.43
56	.8290	.5592	1.4826	86	.9976	.0698	14.30
57	.8387	.5446	1.5399	87	.9986	.0523	19.08
58	.8480	.5299	1.6003	88	.9994	.0349	28.64
59	.8572	.5150	1.6643	89	.9998	.0175	57.29
60	.8660	.5000	1.7321	90	1.0000	.0000	∞

POWERS OF TEN

Raising a number to a power means multiplying a base number by itself as indicated by an exponent. Thus, 3^4 means $3 \times 3 \times 3 \times 3$ or 81. The table below gives the powers of 10, with both negative and positive exponents. Since the positive powers are greater than 1, the positive exponent indicates how many places the decimal point is to be moved to the right from 1.000 and so on. With negative powers, the values are less than 1, and the negative exponent indicates how many places the decimal point is moved to the left from 1.

10^{-10} = 0.000,000,000,1	10^{0} = 1
10^{-9} = 0.000,000,001	10^{1} = 10
10^{-8} = 0.000,000,01	10^{2} = 100
10^{-7} = 0.000,000,1	10^{3} = 1000
10^{-6} = 0.000,001	10^{4} = 10,000
10^{-5} = 0.000,01	10^{5} = 100,000
10^{-4} = 0.000,1	10^{6} = 1,000,000
10^{-3} = 0.001	10^{7} = 10,000,000
10^{-2} = 0.01	10^{8} = 100,000,000
10^{-1} = 0.1	10^{9} = 1,000,000,000
10^{0} = 1	10^{10} = 10,000,000,000

SQUARES, CUBES, ROOTS, AND RECIPROCALS

The square of a number is the number multiplied by itself. For the cube of a number, the number is muliplied by itself twice. Thus, 3^2 means 3×3 which equals 9 and 3^3 means $3 \times 3 \times 3$ which equals 27. To find the square root of a number, find the number, or root, which when raised to the second power will produce the square root number. Thus, $\sqrt{9} = 3$ because $3^2 = 9$. To find the cube root, find the root which, when raised to the third power, will produce the cube root number. Thus, $^3\sqrt{8} = 2$ because $2 \times 2 \times 2 = 8$. The reciprocal of a number is 1 or 1,000 divided by the number. The table below gives the square, cube, square root, cube root, and reciprocal of numbers 1 through 100.

n	n^2	n^3	$\sqrt{n}$	$\sqrt[3]{n}$	1 000/n
1	1	1	1.00	1.000	1 000.000
2	4	8	1.414	1.260	500.000
3	9	27	1.732	1.442	333.333
4	16	64	2.000	1.587	250.000
5	25	125	2.236	1.710	200.000
6	36	216	2.449	1.817	166.667
7	49	343	2.646	1.913	142.857
8	64	512	2.828	2.000	125.000
9	81	729	3.000	2.080	111.111
10	100	1 000	3.162	2.154	100.000
11	121	1 331	3.317	2.224	90.909
12	144	1 728	3.464	2.289	83.333
13	169	2 197	3.606	2.351	76.923
14	196	2 744	3.742	2.410	71.429

n	n^2	n^3	$\sqrt{n}$	$\sqrt[3]{n}$	$1\,000/n$
15	225	3 375	3.873	2.466	66.667
16	256	4 096	4.000	2.520	62.500
17	289	4 913	4.123	2.571	58.824
18	324	5 832	4.243	2.621	55.556
19	361	6 859	4.359	2.668	52.632
20	400	8 000	4.472	2.714	50.000
21	441	9 261	4.583	2.759	47.619
22	484	10 648	4.690	2.802	45.455
23	529	12 167	4.796	2.843	43.478
24	576	13 824	4.899	2.885	41.667
25	625	15 625	5.000	2.924	40.000
26	676	17 576	5.099	2.962	38.462
27	729	19 683	5.196	3.000	37.037
28	784	21 952	5.292	3.037	35.714
29	841	24 389	5.385	3.072	34.483
30	900	27 000	5.477	3.107	33.333
31	961	29 791	5.568	3.141	32.258
32	1 024	32 768	5.657	3.175	31.250
33	1 089	35 937	5.745	3.208	30.303
34	1 156	39 304	5.831	3.240	29.412
35	1 225	42 875	5.916	3.271	28.571
36	1 296	46 656	6.000	3.302	27.778
37	1 369	50 653	6.083	3.332	27.027
38	1 444	54 872	6.164	3.362	26.316
39	1 521	59 319	6.245	3.391	25.641
40	1 600	64 000	6.325	3.420	25.000
41	1 681	68 921	6.403	3.448	24.390
42	1 764	74 088	6.481	3.476	23.810
43	1 849	79 507	6.557	3.503	23.256
44	1 936	85 184	6.633	3.530	22.727
45	2 025	21 125	6.708	3.557	22.222
46	2 116	97 336	6.782	3.583	21.739
47	2 209	193 823	6.856	3.609	21.277
48	2 304	110 592	6.928	3.634	20.833
49	2 401	117 649	7.000	3.659	20.408
50	2 500	125 000	7.071	3.684	20.000
51	2 601	132 651	7.141	3.708	19.608
52	2 704	140 608	7.211	3.733	19.231
53	2 809	148 877	7.280	3.756	18.868
54	2 916	157 464	7.347	3.780	18.519

n	n^2	n^3	$\sqrt{n}$	$\sqrt[3]{n}$	$1\,000/n$
55	3 025	166 375	7.416	3.803	18.182
56	3 136	175 616	7.483	3.826	17.857
57	3 249	185 193	7.550	3.849	17.544
58	3 364	195 112	7.616	3.871	17.241
59	3 481	205 379	7.681	3.893	16.949
60	3 600	216 000	7.746	3.915	16.667
61	3 721	226 981	7.810	3.936	16.393
62	3 844	238 328	7.874	3.958	16.129
63	3 969	250 047	7.937	3.979	15.873
64	4 096	262 144	8.000	4.000	15.625
65	4 225	274 625	8.062	4.021	15.385
66	4 356	387 496	8.124	4.041	15.152
67	4 489	300 763	8.185	4.062	14.925
68	4 624	314 432	8.246	4.082	14.706
69	4 761	328 509	8.307	4.102	14.493
70	4 900	343 000	8.367	4.121	14.286
71	5 041	357 911	8.426	4.141	14.085
72	5 184	383 248	8.485	4.160	13.889
73	5 329	389 017	8.544	4.179	13.169
74	5 476	405 224	8.602	4.198	13.514
75	5 625	421 875	8.660	4.217	13.333
76	5 776	438 976	8.718	4.236	13.158
77	5 929	456 533	8.775	4.254	12.987
78	6 084	474 552	8.832	4.273	12.821
79	6 241	493 039	8.888	4.291	12.658
80	6 400	512 000	8.944	4.309	12.500
81	6 561	531 441	9.000	4.327	12.346
82	6 724	551 368	9.055	4.344	12.195
83	6 889	571 787	9.110	4.362	12.048
84	7 056	592 704	9.165	4.380	11.905
85	7 225	614 125	9.220	4.397	11.765
86	7 396	636 056	9.274	4.414	11.628
87	7 569	658 503	9.327	4.431	11.494
88	7 744	681 472	9.381	4.448	11.364
89	7 921	704 969	9.434	4.465	11.236
90	8 100	729 000	9.487	4.481	11.111
91	8 281	753 571	9.539	4.498	10.989
92	8 464	778 688	9.592	4.514	10.870
93	8 649	804 357	9.644	4.531	10.753
94	8 836	830 584	9.695	4.547	10.638

n	n^2	n^3	$\sqrt{n}$	$\sqrt[3]{n}$	$1\,000/n$
95	9 025	857 375	9.747	4.563	10.526
96	9 216	884 736	9.798	4.579	10.417
97	9 409	912 673	9.849	4.595	10.309
98	9 604	941 192	9.899	4.610	10.204
99	9 801	970 299	9.950	4.626	10.101
100	10 000	1 000 000	10.000	4.642	10.000

ALPHABETIC SYMBOLS AND ABBREVIATIONS

Letters of the Greek alphabet are commonly used as symbols to designate various electronic properties or quantities. The first chart below contains the entire Greek alphabet, showing both capital and lowercase letters, and indicates those that are used as electronic symbols.

The second table lists the letters and abbreviations used for electronic characteristics, or properties. The basic unit of measure and its abbreviation is given for each characteristic.

GREEK ALPHABET AND ELECTRONIC SYMBOLS

NAME	GREEK LETTER CAPITAL	GREEK LETTER LOWERCASE	ELECTRONIC SYMBOL USE
Alpha	A	α	α for the current gain, or forward-current transfer ratio, of a transistor connected in common-base mode; also for angles and for amplifier circuits
Beta	B	β	β for the current gain, or amplification of a common-emitter transistor—the ratio of the collector current to the base current; also for angles and for amplifier circuits
Gamma	Γ	γ	
Delta	Δ	δ	Δ for small change in value
Epsilon	E	ϵ	ϵ for permittivity; also as base of natural logarithms
Zeta	Z	ζ	
Eta	H	η	η for the intrinsic standoff ratio
Theta	Θ	θ or ϑ	θ for phase angle

GREEK ALPHABET AND ELECTRONIC SYMBOLS

NAME	GREEK LETTER CAPITAL	LOWERCASE	ELECTRONIC SYMBOL USE
Iota	I	ι	
Kappa	K	$\varkappa$	$\varkappa$ for coupling coefficient K for dielectric constant (K_ϵ)
Lambda	Λ	λ	λ for wavelength
Mu	M	μ	μ for the prefix *micro*; for permeability; for the amplification factor
Nu	N	ν	
Xi	Ξ	ξ	
Omicron	O	o	
Pi	Π	π	π for the ratio of the circumference of a circle to its diameter
Rho	P	ϱ	
Sigma	Σ	s or σ	
Tau	T	τ	τ for a period of time T for the time constant
Upsilon	Υ	υ	υ for instantaneous voltage; for velocity
Phi	Φ	ϕ or φ	ϕ for magnetic flux; for angles
Chi	X	χ	
Psi	Ψ	ψ	
Omega	Ω	ω	Ω for ohms ω for work

ELECTRONIC CHARACTERISTIC	LETTER OR ABBREVIATION	BASIC UNIT OF MEASURE	UNIT SYMBOL
Alternating Current	AC		
Admittance	*Y*	siemens	S
Capacitance	*C*	farad	F
Capacitive Reactance	X_c	ohms	Ω
Charge	*Q*	coulomb	C
Conductance	*G*	siemens	S
Current	*I* or *i*	ampere	A
Direct Current	DC		
Electromotive Force	*E* or		
	emf	volt	V
Electrostatic Flux Density	*D*		
Frequency	*f*	hertz	Hz
Impedance	*Z*	ohm	Ω
Inductance	*L*	henry	H
Magnetic Flux Density	*B*	gauss	
Magnetomotive Force	mmf		
Period	*T*	second	s
Permeance	*P*		
Power	*P*	watt	W
Quality factor	*Q*		
Radius of a circle	*r*		
Reactance	*X*	ohm	Ω
Reluctance	$\mathcal{R}$		
Resistance	*R*	ohm	Ω
Susceptance	*B*	siemens	S
Voltage	*E* or *V*	volt	V

FREQUENCY SPECTRUM

The table below summarizes the main categories of electronic frequencies. Note that frequency and wavelength are inversely proportional: the higher the frequency, the shorter the wavelength; the shorter the frequency, the higher the wavelength.

FREQUENCY	BAND NO.	CLASSIFICATION	ABBREVIATION
3—30 kHz	4	Very low frequencies	vlf
30—300 kHz	5	Low frequencies	lf
300—3000 kHz	6	Medium frequencies	mf
3—30 MHz	7	High frequencies	hf
30—300 MHz	8	Very high frequencies	vhf
300—3000 MHz	9	Ultrahigh frequencies	uhf
3—30 GHz	10	Super-high frequencies	shf
30—300 GHz	11	Extremely high frequencies	ehf
300 GHz—3 THz	12	—	—

MKSA (SI) UNITS

The table below lists the multiples and submultiples of the Mksa (SI) units. The chart on the next page lists electronic qualities or properties, their symbols, and the associated SI units with their symbols and their common multiples and submultiples.

SI PREFIXES

PREFIX	SYMBOL	FACTOR BY WHICH THE UNIT IS MULTIPLIED	POWER
exa	E	1 000 000 000 000 000 000	$= 10^{18}$
peta	P	1 000 000 000 000 000	$= 10^{15}$
tera	T	1 000 000 000 000	$= 10^{12}$
giga	G	1 000 000 000	$= 10^{9}$
mega	M	1 000 000	$= 10^{6}$
kilo	k	1 000	$= 10^{3}$
hecto	h	100	$= 10^{2}$
deca	da	10	$= 10^{1}$
		1	$= 10^{0}$
deci	d	0.1	$= 10^{-1}$
centi	c	0.01	$= 10^{-2}$
milli	m	0.001	$= 10^{-3}$
micro	μ	0.000 001	$= 10^{-6}$
nano	n	0.000 000 001	$= 10^{-9}$
pico	p	0.000 000 000 001	$= 10^{-12}$
femto	f	0.000 000 000 000 001	$= 10^{-15}$
atto	a	0.000 000 000 000 000 001	$= 10^{-18}$

QUANTITY NAME AND SYMBOL	SI UNIT AND ITS SYMBOL	SOME COMMON MULTIPLES AND SUBMULTIPLES
Active Power *P*	watt (W)	kW mW μW nW
Admittance *Y*	siemens (S)	kS mS μS nS
Apparent Power P_a	volt ampere (VA)	kVA μVA nVA pVA
Capacitance *C*	farad (F)	μF nF pF
Common Temperature, Celsius Temperature *t*, *θ*	degree Celsius (°C)	
Conductance *G*	siemens (S)	MS kS mS μS
Electric Charge *Q*	coulomb (C)	kC mC μC nC pC
Electric Current *I*	ampere (A)	kA mA μA nA pA
Electric Potential *V* Electromotive Force emf, *E*	volt (V)	kV mV μV nV
Electrical Energy	joule (J)	kJ mJ

QUANTITY NAME AND SYMBOL	SI UNIT AND ITS SYMBOL	SOME COMMON MULTIPLES AND SUBMULTIPLES
Frequency f	hertz (Hz)	THz GHz MHz kHz
Impedance Z	ohm (Ω)	MΩ kΩ mΩ
Magnetic Field Strength H	ampere per metre (A/m)	kA/m
Magnetic Flux ϕ	weber (Wb)	mWb
Magnetic Flux Density B	tesla (T)	mT μT nT
Magnetomotive Force mmf	ampere (A)	kA mA
Mutual Inductance M, L_m	henry (H)	mH μH nH pH
Permeability μ	henries per metre (H/m)	mH/m μH/m nH/m
Permeance P	henry (H)	—
Permittivity ϵ	farads per metre (F/m)	μF/m nF/m pF/m
Reactance X	ohm (Ω)	MΩ kΩ mΩ
Reluctance $\mathcal{R}$	reciprocal henry (H^{-1})	—
Resistance R	ohm (Ω)	MΩ kΩ mΩ μΩ

QUANTITY NAME AND SYMBOL	SI UNIT AND ITS SYMBOL	SOME COMMON MULTIPLES AND SUBMULTIPLES
Self-Inductance L	henry (H)	mH μH nH pH
Susceptance B	siemens (S)	kS mS μS
Time τ Time of One Cycle, Periodic time T	second (s)	ks ms μs ns ps
Velocity v	metre per second (m/s)	cm/s mm/s
Wavelength λ	metre (m)	mm μm nm pm

LOGARITHM TABLE

Logarithms are used to simplify complicated problems involving powers and square roots. Certain formulas, such as those used to determine the power output of microphones, use logarithms.

Below is a table of common logarithms. The logarithm (log) of a given number is the power (exponent) to which another number (the base) has to be raised in order to equal the given number. The number 10 is used as the base for common logarithms. The logarithm of a number to the base 10 may be written as $\log_{10}$. Unless some other base is specified, the logarithm of any number is assumed to be to the base 10.

Logarithms of numbers are divided into parts. The whole number part is called the *characteristic*. The decimal part is called the *mantissa*. The characteristic of the logarithm of any number greater than 1 is always positive and 1 fewer than the number of digits to the left of the decimal point in the number. The mantissa of a logarithm is found by referring to a table of common logarithms. To find the logarithm of 15 (log 15) look down the *N* column to 15 and to the *0* column to find the mantissa. Log 15 = 1.1761. Similarly to find Log 15.5, look to the *5* column. Log 15.5 = 1.1903.

N	0	1	2	3	4	5	6	7	8	9	PROPORTIONAL PARTS								
											1	2	3	4	5	6	7	8	9
10	0000	0043	0086	0128	0170	0212	0253	0294	0334	0374	4	8	12	17	21	25	29	33	37
11	0414	0453	0492	0531	0569	0607	0645	0682	0719	0755	4	8	11	15	19	23	26	30	34
12	0792	0828	0864	0899	0934	0969	1004	1038	1072	1106	3	7	10	14	17	21	24	28	31
13	1139	1173	1206	1239	1271	1303	1335	1367	1399	1430	3	6	10	13	16	19	23	26	29
14	1461	1492	1523	1553	1584	1614	1644	1673	1703	1732	3	6	9	12	15	18	21	24	27
15	1761	1790	1818	1847	1875	1903	1931	1959	1987	2014	3	6	8	11	14	17	20	22	25
16	2041	2068	2095	2122	2148	2175	2201	2227	2253	2279	3	5	8	11	13	16	18	21	24
17	2304	2330	2355	2380	2405	2430	2455	2480	2504	2529	2	5	7	10	12	15	17	20	22
18	2553	2577	2601	2625	2648	2672	2695	2718	2742	2765	2	5	7	9	12	14	16	19	21
19	2788	2810	2833	2856	2878	2900	2923	2945	2967	2989	2	4	7	9	11	13	16	18	20

N	0	1	2	3	4	5	6	7	8	9	PROPORTIONAL PARTS 1	2	3	4	5	6	7	8	9
20	3010	3032	3054	3075	3096	3118	3139	3160	3181	3201	2	4	6	8	11	13	15	17	19
21	3222	3243	3263	3284	3304	3324	3345	3365	3385	3404	2	4	6	8	10	12	14	16	18
22	3424	3444	3464	3483	3502	3522	3541	3560	3579	3598	2	4	6	8	10	12	14	15	17
23	3617	3636	3655	3674	3692	3711	3729	3747	3766	3784	2	4	6	7	9	11	13	15	17
24	3802	3820	3838	3856	3874	3892	3909	3927	3945	3962	2	4	5	7	9	11	12	14	16
25	3979	3997	4014	4031	4048	4065	4082	4099	4116	4133	2	3	5	7	9	10	12	14	15
26	4150	4166	4183	4200	4216	4232	4249	4265	4281	4298	2	3	5	7	8	10	11	13	15
27	4314	4330	4346	4362	4378	4393	4409	4425	4440	4456	2	3	5	6	8	9	11	13	14
28	4472	4487	4502	4518	4533	4548	4564	4579	4594	4609	2	3	5	6	8	9	11	12	14
29	4624	4639	4654	4669	4683	4698	4713	4728	4742	4757	1	3	4	6	7	9	10	12	13
30	4771	4786	4800	4814	4829	4843	4857	4871	4886	4900	1	3	4	6	7	9	10	11	13
31	4914	4928	4942	4955	4969	4983	4997	5011	5024	5038	1	3	4	6	7	8	10	11	12
32	5051	5065	5079	5092	5105	5119	5132	5145	5159	5172	1	3	4	5	7	8	9	11	12
33	5185	5198	5211	5224	5237	5250	5263	5276	5289	5302	1	3	4	5	6	8	9	10	12
34	5315	5328	5340	5353	5366	5378	5391	5403	5416	5428	1	3	4	5	6	8	9	10	11
35	5441	5453	5465	5478	5490	5502	5514	5527	5539	5551	1	2	4	5	6	7	9	10	11
36	5563	5575	5587	5599	5611	5623	5635	5647	5658	5670	1	2	4	5	6	7	8	10	11
37	5682	5694	5705	5717	5729	5740	5752	5763	5775	5786	1	2	3	5	6	7	8	9	10
38	5798	5809	5821	5832	5843	5855	5866	5877	5888	5899	1	2	3	5	6	7	8	9	10
39	5911	5922	5933	5944	5955	5966	5977	5988	5999	6010	1	2	3	4	5	7	8	9	10
40	6021	6031	6042	6053	6064	6075	6085	6096	6107	6117	1	2	3	4	5	6	8	9	10
41	6128	6138	6149	6160	6170	6180	6191	6201	6212	6222	1	2	3	4	5	6	7	8	9
42	6232	6243	6253	6263	6274	6284	6294	6304	6314	6325	1	2	3	4	5	6	7	8	9
43	6335	6345	6355	6365	6375	6385	6395	6405	6415	6425	1	2	3	4	5	6	7	8	9
44	6435	6444	6454	6464	6474	6484	6493	6503	6513	6522	1	2	3	4	5	6	7	8	9
45	6532	6542	6551	6561	6571	6580	6590	6599	6609	6618	1	2	3	4	5	6	7	8	9
46	6628	6637	6646	6656	6665	6675	6684	6693	6702	6712	1	2	3	4	5	6	7	7	8
47	6721	6730	6739	6749	6758	6767	6776	6785	6794	6803	1	2	3	4	5	5	6	7	8
48	6812	6821	6830	6839	6848	6857	6866	6875	6884	6893	1	2	3	4	4	5	6	7	8
49	6902	6911	6920	6928	6937	6946	6955	6964	6972	6981	1	2	3	4	4	5	6	7	8
50	6990	6998	7007	7016	7024	7033	7042	7050	7059	7067	1	2	3	3	4	5	6	7	8
51	7076	7084	7093	7101	7110	7118	7126	7135	7143	7152	1	2	3	3	4	5	6	7	8
52	7160	7168	7177	7185	7193	7202	7210	7218	7226	7235	1	2	2	3	4	5	6	7	7
53	7243	7251	7259	7267	7275	7284	7292	7300	7308	7316	1	2	2	3	4	5	6	6	7
54	7324	7332	7340	7348	7356	7364	7372	7380	7388	7396	1	2	2	3	4	5	6	6	7
55	7404	7412	7419	7427	7435	7443	7451	7459	7466	7474	1	2	2	3	4	5	5	6	7
56	7482	7490	7497	7505	7513	7520	7528	7536	7543	7551	1	2	2	3	4	5	5	6	7
57	7559	7566	7574	7582	7589	7597	7604	7612	7619	7627	1	2	2	3	4	5	5	6	7
58	7634	7642	7649	7657	7664	7672	7679	7686	7694	7701	1	1	2	3	4	4	5	6	7
59	7709	7716	7723	7731	7738	7745	7752	7760	7767	7774	1	1	2	3	4	4	5	6	7
60	7782	7789	7796	7803	7810	7818	7825	7832	7839	7846	1	1	2	3	4	4	5	6	6
61	7853	7860	7868	7875	7882	7889	7896	7903	7910	7917	1	1	2	3	4	4	5	6	6
62	7924	7931	7938	7945	7952	7959	7966	7973	7980	7987	1	1	2	3	3	4	5	6	6
63	7993	8000	8007	8014	8021	8028	8035	8041	8048	8055	1	1	2	3	3	4	5	5	6
64	8062	8069	8075	8082	8089	8096	8102	8109	8116	8122	1	1	2	3	3	4	5	5	6
65	8129	8136	8142	8149	8156	8162	8169	8176	8182	8189	1	1	2	3	3	4	5	5	6
66	8195	8202	8209	8215	8222	8228	8235	8241	8248	8254	1	1	2	3	3	4	5	5	6
67	8261	8267	8274	8280	8287	8293	8299	8306	8312	8319	1	1	2	3	3	4	5	5	6
68	8325	8331	8338	8344	8351	8357	8363	8370	8376	8382	1	1	2	3	3	4	4	5	6
69	8388	8395	8401	8407	8414	8420	8426	8432	8439	8445	1	1	2	2	3	4	4	5	6

N	0	1	2	3	4	5	6	7	8	9	PROPORTIONAL PARTS								
											1	2	3	4	5	6	7	8	9
70	8451	8457	8463	8470	8476	8482	8488	8494	8500	8506	1	1	2	2	3	4	4	5	6
71	8513	8519	8525	8531	8537	8543	8549	8555	8561	8567	1	1	2	2	3	4	4	5	5
72	8573	8579	8585	8591	8597	8603	8609	8615	8621	8627	1	1	2	2	3	4	4	5	5
73	8633	8639	8645	8651	8657	8663	8669	8675	8681	8686	1	1	2	2	3	4	4	5	5
74	8692	8698	8704	8710	8716	8722	8727	8733	8739	8745	1	1	2	2	3	4	4	5	5
75	8751	8756	8762	8768	8774	8779	8785	8791	8797	8802	1	1	2	2	3	3	4	5	5
76	8808	8814	8820	8825	8831	8837	8842	8848	8854	8859	1	1	2	2	3	3	4	5	5
77	8865	8871	8876	8882	8887	8893	8899	8904	8910	8915	1	1	2	2	3	3	4	4	5
78	8921	8927	8932	8938	8943	8949	8954	8960	8965	8971	1	1	2	2	3	3	4	4	5
79	8976	8982	8987	8993	8998	9004	9009	9015	9020	9025	1	1	2	2	3	3	4	4	5
80	9031	9036	9042	9047	9053	9058	9063	9069	9074	9079	1	1	2	2	3	3	4	4	5
81	9085	9090	9096	9101	9106	9112	9117	9122	9128	9133	1	1	2	2	3	3	4	4	5
82	9138	9143	9149	9154	9159	9165	9170	9175	9180	9186	1	1	2	2	3	3	4	4	5
83	9191	9196	9201	9206	9212	9217	9222	9227	9232	9238	1	1	2	2	3	3	4	4	5
84	9243	9248	9253	9258	9263	9269	9274	9279	9284	9289	1	1	2	2	3	3	4	4	5
85	9294	9299	9304	9309	9315	9320	9325	9330	9335	9340	1	1	2	2	3	3	4	4	5
86	9345	9350	9355	9360	9365	9370	9375	9380	9385	9390	1	1	2	2	3	3	4	4	5
87	9395	9400	9405	9410	9415	9420	9425	9430	9435	9440	0	1	1	2	2	3	3	4	4
88	9445	9450	9455	9460	9465	9469	9474	9479	9484	9489	0	1	1	2	2	3	3	4	4
89	9494	9499	9504	9509	9513	9518	9523	9528	9533	9538	0	1	1	2	2	3	3	4	4
90	9542	9547	9552	9557	9562	9566	9571	9576	9581	9586	0	1	1	2	2	3	3	4	4
91	9590	9595	9600	9605	9609	9614	9619	9624	9628	9633	0	1	1	2	2	3	3	4	4
92	9638	9643	9647	9652	9657	9661	9666	9671	9675	9680	0	1	1	2	2	3	3	4	4
93	9685	9689	9694	9699	9703	9708	9713	9717	9722	9727	0	1	1	2	2	3	3	4	4
94	9731	9736	9741	9745	9750	9754	9759	9763	9768	9773	0	1	1	2	2	3	3	4	4
95	9777	9782	9786	9791	9795	9800	9805	9809	9814	9818	0	1	1	2	2	3	3	4	4
96	9823	9827	9832	9836	9841	9845	9850	9854	9859	9863	0	1	1	2	2	3	3	4	4
97	9868	9872	9877	9881	9886	9890	9894	9899	9903	9908	0	1	1	2	2	3	3	4	4
98	9912	9917	9921	9926	9930	9934	9939	9943	9948	9952	0	1	1	2	2	3	3	4	4
99	9956	9961	9965	9969	9974	9978	9983	9987	9991	9996	0	1	1	2	2	3	3	3	4

CONSTANTS, CONVERSION FACTORS, AND FORMULAS

This appendix section is intended to serve as a convenient reference for electronic measurement constants, conversions, and formulas. Ohm's law formulas for DC circuits can be derived from the illustration below. On the pages following are included formulas for AC circuits, resistors, coils, inductive reactance, capacitors, capacitive reactance, conductance, susceptance, admittance, reactance, and resonant frequency.

CONVERSION FACTORS AND CONSTANTS

π	=	3.1416	2π	=	6.2832
π^2	=	9.8696	$2\pi^2$	=	39.478
e	=	2.7183	$\sqrt{2}$	=	1.4142
$\sqrt{3}$	=	1.7321	LOG π	=	0.4972

1 METRE = 39.37 INCHES = 3.28 FEET
1 KILOMETRE = 0.621 MILE (ABOUT 3/5 MILE)
1 INCH = 2.54 CENTIMETRES
1 KILOGRAM = 2.2 POUNDS
1 LITRE = 1.06 QUARTS
1 OUNCE = 28.35 GRAMS
1 HORSEPOWER = 746 WATTS

OHM'S LAW FORMULAS FOR DC CIRCUITS

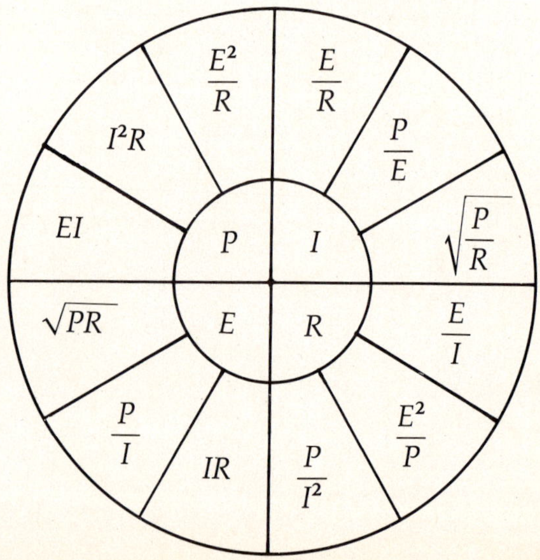

OHM'S LAW FORMULAS FOR AC CIRCUITS

SERIES

$$Z = \sqrt{R^2 + (X_L - X_C)^2}$$

$$E = IZ$$

$$I = \frac{E}{Z}$$

$$P_a = I^2Z$$

$$P = I_{R^2}R$$

$$PF = \frac{R}{Z}$$

$$E_R = IR$$

$$E_L = IX_L$$

$$E_C = IX_C$$

$$I_R = \frac{E_R}{R}$$

$$I_L = \frac{E_L}{X_L}$$

$$I_C = \frac{E_C}{X_C}$$

$$E = \sqrt{E_{R^2} + (E_L - E_C)^2}$$

PARALLEL

$$Z = \frac{E_A}{I_T}$$

$$E = IZ$$

$$I_T = \sqrt{I_{R^2} + (I_L - I_C)^2}$$

$$P_a = E_AI_T$$

$$P = I_{R^2}R$$

$$PF = \frac{P}{P_a}$$

$$E_R = I_RR$$

$$E_L = I_LX_L$$

$$E_C = I_CX_C$$

$$I_R = \frac{E}{R}$$

$$I_L = \frac{E}{X_L}$$

$$I_C = \frac{E}{X_C}$$

$$E = I_TZ_T$$

RESISTORS IN SERIES

$$R_T = R_1 + R_2 + R_3 + \ldots + R_N$$

TWO RESISTORS IN PARALLEL

$$R_T = \frac{R_1R_2}{R_1 + R_2} \qquad R_1 = \frac{R_TR_2}{R_2 - R_T}$$

EQUAL RESISTORS IN PARALLEL

$$R_T = \frac{R}{N}$$

WHERE N IS THE NUMBER OF RESISTORS

RESISTORS IN PARALLEL, GENERAL FORMULA

$$R_T = \frac{1}{\frac{1}{R_1} + \frac{1}{R_2} + \frac{1}{R_3} + \cdots + \frac{1}{R_N}}$$

COUPLED INDUCTANCE

K = COEFFICIENT OF COUPLING M = MUTUAL INDUCTANCE

$$L_T = L_1 + L_2 \pm 2M$$

$$K = \frac{M}{\sqrt{L_1 L_2}}$$

SINUSOIDAL VOLTAGES AND CURRENTS

EFFECTIVE VALUE = 0.707 × PEAK VALUE
AVERAGE VALUE = 0.637 × PEAK VALUE
PEAK VALUE = 1.414 × EFFECTIVE VALUE
EFFECTIVE VALUE = 1.11 × AVERAGE VALUE
PEAK VALUE = 1.57 × AVERAGE VALUE
AVERAGE VALUE = 0.9 × EFFECTIVE VALUE

CONDUCTANCE, SUSCEPTANCE, AND ADMITTANCE

$$G = \frac{1}{R} \quad \text{(FOR DC CIRCUIT)}$$

$$G = \frac{R}{R^2 + X^2} \quad \text{(FOR AC CIRCUIT)}$$

$$B = \frac{1}{X} \quad \text{(WHEN RESISTANCE IS 0)}$$

$$B = \frac{X}{R^2 + X^2}$$

$$Y = \frac{1}{Z} = \frac{1}{\sqrt{R^2 + X^2}}$$

REACTANCE FORMULAS

$$X_C = \frac{1}{2\pi f C} \qquad C = \frac{1}{2\pi f X_C}$$

$$X_L = 2\pi f L \qquad L = \frac{X_L}{2\pi f}$$

RESONANT FREQUENCY FORMULAS

$$f = \frac{1}{2\pi\sqrt{LC}}$$

$$L = \frac{1}{4\pi^2 f^2 C}$$

$$C = \frac{1}{4\pi^2 f^2 L}$$

IMPEDANCE FORMULAS

FOR SERIES CIRCUIT: $Z = \sqrt{R^2 + (X_L - X_C)^2}$

FOR *R* AND *X* IN PARALLEL: $Z = \frac{RX}{\sqrt{R^2 + X^2}}$

POWER FACTOR

$PF = \cos\theta$ WHERE θ IS THE ANGLE OF LEAD OR LAG

$$PF = \frac{\text{POWER}}{\text{APPARENT POWER}} = \frac{P}{EI}$$

$$PF = \frac{R}{Z}$$

Q FACTOR

$$Q = \frac{X_L}{R} \quad \text{OR} \quad \frac{X_C}{R}$$

FORMULAS FOR COILS

COILS CONNECTED IN SERIES:

$$L_T = L_1 + L_2 + L_3 + \ldots + L_N$$

(NO MUTUAL INDUCTANCE)

COILS CONNECTED IN PARALLEL:

$$L_T = \frac{1}{\frac{1}{L_1} + \frac{1}{L_2} + \frac{1}{L_3} + \cdots + \frac{1}{L_N}}$$

(NO MUTUAL INDUCTANCE)

INDUCTIVE REACTANCE

SERIES:

$$X_{L_T} = X_{L_1} + X_{L_2} + X_{L_3} + \ldots + X_{L_N}$$

PARALLEL:

$$X_{L_T} = \frac{1}{\frac{1}{X_{L_1}} + \frac{1}{X_{L_2}} + \frac{1}{X_{L_3}} + \cdots + \frac{1}{X_{L_N}}}$$

FORMULAS FOR CAPACITORS

CAPACITORS CONNECTED IN SERIES:

$$C_T = \frac{1}{\frac{1}{C_1} + \frac{1}{C_2} + \frac{1}{C_3} + \cdots + \frac{1}{C_N}}$$

CAPACITORS CONNECTED IN PARALLEL:

$$C_T = C_1 + C_2 + C_3 + \ldots + C_N$$

CAPACITIVE REACTANCE

SERIES:

$$X_{C_T} = X_{C_1} + X_{C_2} + X_{C_3} + \ldots + X_{C_N}$$

PARALLEL:

$$X_{C_T} = \frac{1}{\frac{1}{X_{C_1}} + \frac{1}{X_{C_2}} + \frac{1}{X_{C_3}} + \cdots + \frac{1}{X_{C_N}}}$$

TRANSFORMER RELATIONSHIPS

$$\frac{N_P}{N_S} = \frac{E_P}{E_S} = \frac{I_S}{I_P} = \sqrt{\frac{Z_P}{Z_S}}$$

DECIBEL FORMULAS

WHEN IMPEDANCES ARE EQUAL:

$$\text{dB} = 10\ \text{LOG}\ \frac{P_1}{P_2} = 20\ \text{LOG}\ \frac{E_1}{E_2} = 20\ \text{LOG}\ \frac{I_1}{I_2}$$

WHEN IMPEDANCES ARE UNEQUAL:

$$\text{dB} = 10\ \text{LOG}\ \frac{P_1}{P_2} = 20\ \text{LOG}\ \frac{E_1\sqrt{Z_2}}{E_2\sqrt{Z_1}} = 20\ \text{LOG}\ \frac{I_1\sqrt{Z_1}}{I_2\sqrt{Z_2}}$$

$$\text{dB} = 10\ \text{LOG}\ \frac{P_1}{P_2} = 20\ \text{LOG}\ \frac{E_1}{E_2} + 10\ \text{LOG}\ \frac{Z_2}{Z_1}$$

$$\text{dB} = 20\ \text{LOG}\ \frac{I_1}{I_2} + 10\ \text{LOG}\ \frac{Z_1}{Z_2}$$

DECIBEL TABLE

dB	POWER RATIO	VOLTAGE OR CURRENT RATIO	dB	POWER RATIO	VOLTAGE OR CURRENT RATIO
0	1.00	1.00	10	10.0	3.2
0.5	1.12	1.06	15	31.6	5.6
1.0	1.26	1.12	20	100	10
1.5	1.41	1.19	25	316	18
2.0	1.58	1.26	30	1,000	32
3.0	2.00	1.41	40	10,000	100
4.0	2.51	1.58	50	10^5	316
5.0	3.16	1.78	60	10^6	1,000
6.0	3.98	2.00	70	10^7	3,162
7.0	5.01	2.24	80	10^8	10,000
8.0	6.31	2.51	90	10^9	31,620
9.0	7.94	2.82	100	10^{10}	10^5

FREQUENCY AND WAVELENGTH

$$f\,\text{kHz} = \frac{3 \times 10^5}{\lambda\ \text{METRE}} \qquad \lambda\ \text{METRE} = \frac{3 \times 10^5}{f\,\text{kHz}}$$

$$f\,\text{MHz} = \frac{3 \times 10^4}{\lambda\ \text{CENTIMETRE}} \qquad \lambda\ \text{CENTIMETRE} = \frac{3 \times 10^4}{f\,\text{MHz}}$$

$$f\,\text{MHz} = \frac{984}{\lambda\ \text{FEET}} \qquad \lambda\ \text{FEET} = \frac{984}{f\,\text{MHz}}$$

TIME DURATIONS OF ONE CYCLE

100 kHz = 10 μ sec
200 kHz = 5 μ sec
250 kHz = 4 μ sec
1 MHz = 1 μ sec
4 MHz = 0.25 μ sec
10 MHz = 0.1 μ sec

TRIGONOMETRIC FORMULAS

$$\sin A = \frac{a}{c} = \frac{\text{OPPOSITE SIDE}}{\text{HYPOTENUSE}}$$

$$\cos A = \frac{b}{c} = \frac{\text{ADJACENT SIDE}}{\text{HYPOTENUSE}}$$

$$\tan A = \frac{a}{b} = \frac{\text{OPPOSITE SIDE}}{\text{ADJACENT SIDE}}$$

$$\text{cosec } A = \frac{c}{a} = \frac{1}{\sin A}$$

$$\sec A = \frac{c}{b} = \frac{1}{\cos A}$$

$$\cot A = \frac{b}{a} = \frac{1}{\tan A}$$

$$c^2 = a^2 + b^2$$

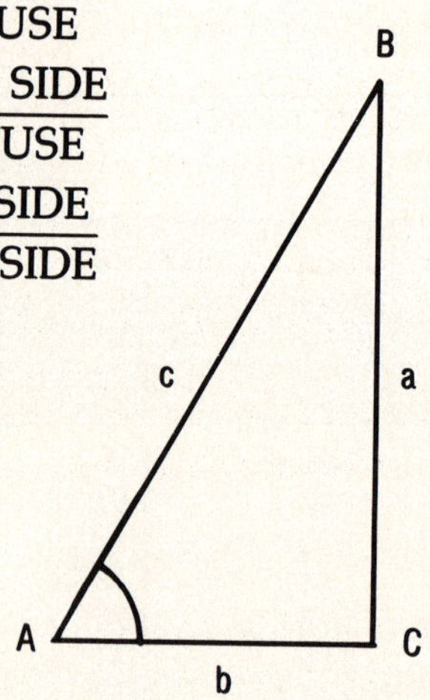

ELECTRICAL SAFETY

Any person working with electricity needs to understand and observe safe practices at all times. Electricity should be approached with respect. Improper handling and carelessness can cost lives. Any electrical current over 24 volts can be dangerous and potentially lethal. Because of this, industrial safety officers often have the authority to fire any persons guilty of unsafe behavior.

Always be careful to protect your own safety if you have to remove a victim from an electric circuit. Do it as quickly as possible as any delay could prove fatal. If necessary, start artificial resuscitation immediately and continue until the victim is revived or pronounced dead by a doctor.

Most accidents can be avoided through proper safety precautions and common sense. No list of rules can cover every possible unsafe situation; however, the list below can serve as a guide to help develop good safety habits. Most important, any time you handle electrical wires or circuits, always assume that they are dangerous and beware of a false sense of security.

1. Don't work on electrical equipment when you are completely alone and never work on "live" equipment when you are tired.
2. Unless it is impossible to avoid, don't work on "live" circuits. Always unplug devices before working on them. If you are working on wiring, turn off all power. Turn off the circuit breaker or fuse—or master switch, if necessary—and make sure power cannot be restored accidentally.

3. If you cannot turn off the power and must work on ''live'' or high-voltage equipment, always keep one hand in your pocket as you work. This will prevent you from establishing a grounding connection. You can avoid shocks and other accidents if you follow this rule. Follow this same rule when you work with capacitors or other devices that store electricity.
4. Always be careful when you work around transformers as they can step voltages up high enough to kill you. Since autotransformers do not isolate the ground from the primary in home or industrial circuits, you can get a shock if you touch a water pipe or ground at the same time that you are in contact with the ''hot'' side of the transformer. Keep one hand in your pocket when working around an energized transformer.
5. Never touch electrical equipment while standing on a damp or metal floor. Also, never handle electrical equipment when you or the equipment is wet or damp.
6. Ground all high-voltage points—remember, a capacitor can store a charge that can kill you. Don't handle electrical equipment that is not grounded and never remove equipment grounds.
7. Never touch two pieces of electrical equipment at the same time.
8. Never touch even one wire of an electrical line, as that may be the ''live'' one, and never touch even a ''dead'' circuit line with your open hand. Instead use the back of your fingers with your arm grounded.
9. Move slowly. Make sure your feet are placed for good balance. Never reach suddenly or lunge for falling tools.
10. Before working on electrical equipment, remove all watches or jewelry.
11. Use only the right tool for every job and use the necessary protective devices that are provided. Never tamper with or disconnect safety devices. Make sure the equipment you are using has no loose wires or frayed insulation on cords. Repair all equipment before using it.
12. Use the proper protective clothing or safety equipment for each job.
13. Never use water to put out an electrical fire and always have a fire extinguisher available. Be sure it is the correct kind for the work you are doing.

14. Always follow the directions for using chemicals and cleaning solvents, as improper use can cause explosions or fires. Never use a DC electrolytic on AC, as it will explode.
15. Use the proper meter movement for AC or DC. Do not use an ohmmeter on a circuit with power applied. Never handle the probes of a meter if the voltage is greater than 100 volts. Always attach the probes to the source and then turn on the circuit to be measured. Also never touch the metal tips of the probes when you are using electrical meters.
16. Always wear safety goggles when working with electrolytics, soldering, or operating any kind of power machinery.
17. Never pick up a charged capacitor. Always discharge a capacitor before handling it. Do this by shorting the terminals with a screwdriver or shorting wire.
18. Be careful not to spill the acid electrolyte from a lead-acid cell on your clothes or skin.
19. Always keep clear of an operating series motor, as it can explode if run without a load. Never use a belt to connect a DC series motor to a load because the motor could blow up if the belt breaks while the motor is running.
20. Make sure switches are rated at the right voltages and current for their circuits and don't use knife switches for voltages over 25 volts. Make sure the power is removed before touching any part of a circuit that has knife switches for contacts.
21. Always wear safety goggles when working close to a TV or cathode ray tube. These tubes can implode and, once the vacuum is released on cracking, glass is scattered widely. Remember, too, that vacuum tubes use high voltages, so always use standard safety precautions when working on such equipment.
22. Be careful not to touch the enclosure of a transistor or diode, as the heat generated can burn you.
23. Be careful with wirecutters. Always shield the wire when cutting so the cut piece won't fly into your eye.
24. Never put solder in your mouth when soldering. The soldering iron will probably have a voltage leakage.

Read Instructions, Obey Warnings, and Use Common Sense.

SOLDER AND SOLDERING

Solder is an alloy composed of tin and lead, both of which have high melting points. When these two metals are combined, the new alloy formed (solder) has a relatively low melting point. Solder melts at between 450 and 600 degrees F (232.2 and 315.5 degrees C).

There are many alloy mixtures for solder. These mixtures are given as percentages of tin to lead. The most common mixture used in electronics work is 60/40 solder—the alloy consisting of 60 percent tin and 40 percent lead.

Soldering is the method by which two or more electrical or electronic connections are joined together by melted solder which forms a coating over the connection point. In addition to joining the connections, solder also prevents the connections from oxidizing.

Most solder used in electrical or electronic work contains a chemical, called *rosin-flux.* This type of solder is called *rosin-core solder.* Rosin-flux melts faster than solder and, thus, moves ahead of the solder. It cleans the area of the connection and enables the solder to flow easily. In addition, flux prevents oxidation.

There is another type of flux available called *acid flux.* This is used for sheet metal soldering and should never be used for electrical or electronic work because it will damage copper and electronic parts.

CONTACT SOLDERING

The most common method of applying solder is called *contact soldering.* In this method, a soldering iron is used to melt the solder to the point where it is needed. A series of steps are followed to form a solder connection:

1. Make sure the connection to be soldered is clean.
2. Bring the soldering iron or pencil to full heat and clean it with a wet sponge. The iron must be hot to be cleaned properly.
3. Tin the iron. Do this by touching the tip of the iron with rosin-core solder. If you are using solder without flux, dip the tip of the solder wire into rosin-flux first. A small spot of solder will form on the tip of the iron. Wipe this spot with a rag, making the solder coat or ''tin'' the tip.
4. Touch the connection with the tip of the soldering iron lightly, letting the solder flow into the connection.

A good solder connection should be clean, shiny, and uncracked. If the solder is dull or cracked, the connection is not good. It means that the iron was not hot enough or the connection was not preheated properly. This ''cold connection'' should be reheated.

Another problem that occurs when soldering is not done properly is called *bridging*. When this happens, solder runs across copper strips of a printed circuit board. This can cause a short-circuit.

GLOSSARY

A. The abbreviation for ampere.

AC. The abbreviation for alternating current.

AC generator. *See* **alternator**.

active component. An electronic device or element that is able to control voltages or currents.

admittance. The reciprocal of impedance; represented by the symbol *Y*. Refers to the ease with which AC can flow in a circuit.

air-core choke. Coil of wire around a nonmetallic core.

air-core transformer. A device, made of wire, that is wound around a nonmetallic core and employs the principles of mutual induction to convert variations of current in one circuit into variations of voltage and current in another circuit.

alnico. An alloy made of aluminum, nickel, cobalt, and iron. Alnico has very high permeability and magnetic retention.

alpha. The current gain, or forward-current transfer ratio, of a transistor connected in the common-base mode; indicated by the Greek letter α.

alternating current. The flow of electricity that changes in magnitude and direction periodically; abbreviated AC.

alternator. A mechanical device that converts mechanical energy into AC.

ammeter. A device used to measure electric current.

ampere. The basic unit of measurement for electric current;

abbreviated A. One ampere equals 6.28×10^{18} electrons flowing past a given point in one second.

ampere-turn. A unit of measurement for magnetomotive force, represented by the symbol *NI*. One ampere-turn is the mmf produced by one ampere flowing through a coil of one turn.

ampere-turn/inch. The English system unit of measure for magnetizing force.

ampere-turn/metre. The Mksa system unit of measurement for magnetizing force.

amplification factor. The ratio of the change in plate voltage resulting from a change in control grid voltage, indicated by the Greek letter μ.

amplifier. An electronic device that increases the voltage, current, or power output of a circuit.

amplitude. (1) The magnitude of a parameter. (2) The height of a square wave (or sine wave, etc.) that can indicate the amount of voltage applied.

analog. (1) Refers to devices or circuits for which a variation in input results in a corresponding variation in output. (2) Refers to a computer that operates with a physical representation of directly measurable quantities, such as voltage.

anode. The positive terminal of an electronic device, such as a vacuum tube, battery, or semiconductor diode.

antenna. A device used to receive and/or transmit electromagnetic waves.

astable multivibrator. A type of circuit that produces a digital signal that can be used for some timing and synchronizing applications. It requires no input signal and, therefore, is also called a *free-running multivibrator*.

atom. The smallest unit of any element that still retains all the properties characteristic of that element.

attenuation. A decrease in amplitude.

audio frequency. Sound waves audible to the human ear, ranging in frequency from 20 Hz to 20 kHz; abbreviated AF.

autotransformer. A device having only a single winding—part of which acts as primary and part, as secondary—that employs the principles of mutual induction to convert variations of current in one circuit into variations of voltage and current in another circuit.

B

B. Symbol representing (1) the base of a transistor, (2) magnetic flux density, and (3) susceptance.

B+. The symbol used to indicate the positive terminal of a battery or power source.

back current. The reverse flow of an electrical charge when reverse bias is applied to the junction point of a semiconductor.

band. The range of frequencies between given upper and lower limits.

band pass filter. A device that allows only a given set of frequencies to pass, attenuating frequencies on either side of the set limit.

bandwidth. The width of a band of frequencies.

barrier voltage. The electromotive force that is necessary to overcome the resistance of a junction point in a semiconductor so that current may flow.

base. The region between the emitter and the collector of a transistor; represented by B.

battery. A DC voltage source, based on chemical energy, that consists of two or more cells.

beta. The current gain, or amplification of a common-emitter transistor—the ratio of the collector current to the base current; indicated by the Greek letter β.

B/H curve. The plotted curve which shows the relationship of applied magnetism to induced magnetism for a particular magnetic material. Also called a *hysteresis loop*.

bias. The electrical reference level voltage applied to establish the operation level of a device.

binary. Mathematical representation of circuit conditions *on* and *off*—representing the numbers 1 and 0, respectively. Binary arithmetic is the foundation of all digital computer operations.

Boolean algebra. A mathematical system that deals with logic relationships; invented by the English mathematician and logician George Boole and used to describe the behavior of on-off circuit elements in electronics.

bus lines. Strips, or rods, of metal used to distribute current within electronic or electrical circuits or devices.

C

C. Symbol representing (1) capacitance, (2) a capacitor, (3) the collector of a transistor, (4) Celsius, (5) a coulomb, (6) candle, and (7) the velocity of light.

capacitance. The property that allows electrons to be stored as an electric charge; represented by C and measured in farads (F).

capacitive reactance. The opposition that a capacitor presents to AC; represented by the symbol X_C and measured in ohms (Ω).

capacitor. A device, consisting of a dielectric sandwiched between two conducting plates, that is able to store electric charge. Represented by *C*.

cascade. An arrangement of two or more similar circuits that are connected so the output of one drives the input of the next one.

cathode. The negative electrode of tubes, transistors, diodes, SCRs, etc.; represented by the symbol *K*.

cathode ray tube. A special type of vacuum tube that displays an image on its front surface.

cell. A single unit that produces a voltage through chemical action.

Cgs system. The centimetre-gram-second system of units for the measurement of magnetism.

charge. An electrical force that binds atoms together. Electrical charges are identified as positive (+) and negative (−).

chip. Name given to semiconductors made on a single substrate.

choke. Inductance used to impede pulsating DC or AC.

circuit. A number of electronic devices connected together to form paths for electron flow.

circular mil. Term used to define the cross-sectional area of a wire.

coil. An insulated wire wound around a core; represented by the symbol *L*.

collector. The electrode carrying the primary electron flow in a transistor.

complex plane. A graphic representation of the mathematical values of the properties of a circuit, giving both the magnitude and direction of a vector.

conductance. The readiness with which a material transmits an electric current; represented by the symbol G. Conductance is the reciprocal of resistance.

conductor. A material through which electricity passes easily.

copper. One of the natural elements that is commonly used for wire because it has good electrical properties.

cosine. A trigonometric function. In a right triangle, the cosine of the acute angle is the ratio between the side adjacent to the angle and the hypotenuse.

coulomb. A unit of electric charge; defined as the charge that must flow for one second to produce one ampere of current; abbreviated C.

coupling coefficient. The relationship between inductive coils; represented by the Greek letter $\varkappa$.

CRT. A cathode ray tube.

current. Electrons moving along a conductor; represented by I. Current flow is from positive to negative. Electron flow is always electrons moving from a negative source (or terminal) towards a positive sink (or terminal).

cycle. One alternation, measured from any point on a waveform to the next corresponding point on the waveform.

D

D. (1) Symbol used to indicate electrostatic flux density. (2) Drain electrode in an FET.

d'Arsonval meter movement. The component of a measuring instrument consisting of a fixed permanent magnet and a moving electromagnetic coil and pointer that indicates the measured value. It is the most common type of instrument movement.

DC. The abbreviation for direct current.

decibel. The unit of transmission gain or loss. Abbreviated dB.

delta. Greek letter; capital (Δ) is used to indicate change in value.

denominator. The number below the line in a fraction,

showing how many parts into which the whole or unit is divided.

diac. A two-lead semiconductor device used as an AC switch.

dielectric. The insulating nonconductive material used between the two plates of a capacitor.

dielectric constant. A ratio that measures insulation properties in relation to a vacuum; also known as *relative permittivity*; represented by K_ϵ.

dielectric strength. Ability of a material to withstand voltage without breaking down.

differential amplifier. An amplifier with two similar input circuits, connected in such a way as to cause the output to respond to the difference between the two inputs.

diffusion. A process used in making semiconductors in which one material is made to intermingle with another as a result of thermal agitation.

digital. A signal or device that operates on the basis of on-off pulses.

diode. A device with two electrodes—a cathode and an anode—that converts AC to pulsating DC.

direct current. Constant-value current that flows in one direction only; abbreviated DC.

dopant. An additive material used to adjust the conductive properties of a semiconductor.

drift. The movement of carriers when voltage is applied to a semiconductor.

E

E. Symbol representing (1) voltage or electromotive force and (2) the emitter of a transistor.

eddy currents. Currents induced in a conducting material by variations in a magnetic flux.

effective value. The AC power that will produce the same amount of heat in a resistor as DC; also known as *rms* (root mean square).

electricity. A form of energy based on movement of electrons.

electrode. The conducting element of tubes, cells, semiconductors, and other such electronic devices.

electrolyte. A chemical substance that has the ability to conduct electricity as ion flow, from one electrode to another.

electrolytic capacitor. A capacitor having two conducting electrodes, one of which becomes coated with a metal oxide film during operation. The electrode which becomes coated with the metal oxide is the anode. The film formed works as the dielectric.

electromagnet. A core of metal surrounded by a coil of insulated wire through which an electric current is passed to cause the metal to act as a magnet and attract certain other metals. The core is usually made of metal to concentrate the magnetic lines of force. This type of magnet is called a *temporary magnet*.

electromagnetism. The magnetic field formed around a wire or a coil of wire when electric current passes through the wire.

electromotive force. The force that causes electricity to flow in a circuit; abbreviated as emf and represented by E.

electron. One of the basic parts of an atom. It carries a negative electric charge.

electronics. The technology involving systems or devices in which electrons flow through gas, vacuums, semiconductors, or conductors.

element. An atom of any substance. There are 93 elements that exist naturally on Earth.

emf. The abbreviation for electromotive force; also represented by E.

energy. The ability to do work; represented by W or ω.

English units system. A system of units for the measurement of magnetism using inches, pounds, and ampere turns.

epsilon. The Greek letter ϵ used as the symbol for permittivity.

eta. The Greek letter η used as the symbol for the intrinsic standoff ratio.

exponent. A numeral written above and to the right of a term, to indicate a power of the term.

F

F. Symbol representing: (1) farad, (2) Fahrenheit, (3) fuse, and (4) filament.

farad. A unit of measurement used in capacitance; abbreviated as F. It is the capacitance of the capacitor where a charge of one coulomb produces a potential difference of one volt.

feedback. System in which a portion of the output is returned to the input.

ferrous. A material composed of or containing iron. Ferrous metals have magnetic characteristics.

FET. Abbreviation for a field-effect transistor.

field-effect transistor. A semiconductor device having three terminals: a gate, a source, and a drain. Flow of current is from source to drain with the gate controlling current flow.

filament. A small coil of resistance wire in light bulbs and vacuum tubes.

filter. A network of resistors, inductors, or capacitors that blocks certain frequencies while passing others.

flip-flop circuit. A device used for a variety of digital applications, including counters.

flow. The passage of electrons or current from one point to another through a vacuum, semiconductor, or conductor.

four-layer diode. A semiconductor diode that has four alternating layers of P and N material with three junctions.

frequency. The number of times a current (or voltage) changes sign per unit of time; represented by *f* and expressed in the units hertz (Hz).

fuse. A protective or safety element designed to open if an excess current flows through a circuit.

G

G. Symbol representing (1) conductance, (2) the grid of a tube, (3) ground, (4) generator, (5) gate of a field-effect transistor, and (6) giga (10 to the power of 9).

gain. The ratio of output to input.

galvanometer. A measuring device used to indicate amount and direction of current flow through a circuit.

gauge. A unit of measure for the cross section of a wire; also a measurement device for the diameter of a wire.

gauss. The Cgs system unit of measurement for magnetic flux density.

generator. A device that changes mechanical energy into electrical energy.

germanium. A grayish-white, brittle metallic element; used as a basic semiconductor material.

gilbert. The unit of measure for magnetomotive force in the Cgs system.

ground. The common connection between circuits and the earth.

H

H. Symbol representing (1) henry, (2) magnetizing force or electromagnetic field intensity, and (3) heater.

henry. A unit of measurement for inductance, abbreviated H. A coil has an inductance of one henry when a current of one ampere, changing at the rate of one cycle per second, produces a counter emf of one volt across the coil terminals. A microhenry (μH) is 0.000001 H; a millihenry (mH) is 0.001 H.

hertz. The unit of measurement of frequency, abbreviated Hz.

hole circuit. The positive charges (called *holes*) flowing in the P region of a semiconductor.

hot carrier diode. A special type of semiconductor device used for high frequencies; also called a *Schottky diode*.

hydrogen atom. The atom with the most basic atomic structure, consisting of one proton, one neutron, and one electron.

hysteresis. Relationship between residual magnetism and energy loss in magnetic material.

hysteresis loop. The relationship of applied magnetism to residual magnetism for a particular magnetic material. Also called a *B/H curve*.

I

I. Symbol representing current.

IMPATT diode. A special semiconductor device formed around a PN junction and operated with a heavy back bias that produces an avalanche breakdown.

impedance. Total amount of circuit opposition to an alternating current of a given frequency; represented by the symbol Z.

induced. The effect produced by the an electric or magnetic field.

inductance. The property of a circuit which opposes any change in alternating current; represented by the symbol *L*.

inductive reactance. The opposition to AC by an inductor.

inductor. A device made of turns of insulated wire wound around a core. Also called a *coil*.

inert atom. A nonreactive atom such as the neon atom.

in phase. Two waves that are of the same frequency which pass through their minimum and maximum values at the same time.

input. The energy applied to a circuit, device, or system.

insulator. A nonconductive material which prevents electricity from flowing through it.

integrated circuit. A combination of circuit elements on one continuous substrate.

intrinsic standoff ratio. Ratio of the supply voltage to the voltage across a resistor, indicated by the Greek letter η.

J

J. The abbreviation for joule.

j-operator. The imaginary number or component of complex numbers in mathematical formulas.

joule. The unit of work needed to produce one volt.

junction. The area where semiconductor materials are joined.

K

K. Symbol representing (1) cathode, (2) kelvin, and (3) kilo.

kappa. The Greek letter $\varkappa$ representing coupling coefficient. The capital letter (K) is used to represent the dielectric constant, K_ϵ.

kilo. Prefix meaning one thousand (1,000); abbreviated K.

kilowatt hour. Energy supplied at the rate of 1,000 watts per hour.

Kirchhoff's laws. Statements describing voltage and current distribution in circuits.

L

L. Symbol representing (1) a coil and (2) inductance.

lag. Term used to indicate a current or voltage delay in a circuit with capacitance, inductance, or both.

lambda. Greek letter λ used to represent wavelength.

LCD. A liquid crystal display diode.

lead. (1) To move ahead of; the opposite of *lag*. (2) A wire used to or from circuit elements.

leakage current. A small but undesirable amount of current which flows through a semiconductor in a direction opposite to the main flow.

LED. A light-emitting diode.

Lenz's law. Proven statement, or theorem, on electromagnetic induction which says that a current caused by a change in a magnetic flux will be directed so as to oppose the act which caused the current.

light-emitting diode. A solid-state display device that glows when current is applied.

linear. An output that varies in direct proportion to input.

line/inch^2. A unit of measure for magnetic flux density in the English system.

line of force. A unit of measure for magnetic flux in the English system.

link. A strip of wire made of resistive foil; used as the working element in a fuse.

liquid crystal display diode. A solid-state device that goes on and off when current is applied.

load. An electrical device that absorbs power.

lodestone. A naturally occurring magnetic mineral.

logic probe. An instrument, resembling a pencil, used to read voltage levels and test signals in logic circuits.

M

magnet. Type of metal or metal ore that attracts certain other metals. These include iron and related metals, such as steel.

magnetomotive force. Force produced by current flowing through a coil of wire, abbreviated mmf.

magnetic field. The force lines that surround a magnet.

magnetic flux. *See* **magnetic field**.

magnetic flux density. Measure of how concentrated the magnetic flux is; represented by B.

magnetism. Physical phenomenom by which certain metals exert a force or attraction on other nearby metallic materials.

maxwell. A unit of measure for magnetic flux in the Cgs system, abbreviated Mx.

meter. An instrument used to measure. Electrical meters include ammeters, voltmeters, ohmmeters, and wattmeters.

mho. An old term used to describe a unit of conductance.

micro. Prefix for one-millionth (0.000001); represented by the Greek letter μ.

milli. Prefix for one-thousandth (0.001); abbreviated m.

Mksa system. A system of units for the measurement of magnetism using metres, kilograms, seconds, and amperes (the Systeme International d'Unites or SI units).

mmf. Abbreviation for magnetomotive force.

mu. The Greek letter μ used as a symbol for (1) the prefix *micro*, (2) permeability, and (3) amplification factor.

mutual induction. The phenomenon that occurs when two conductors are placed close to each other and the changing magnetic field of one affects the other, creating a voltage.

Mx. Abbreviation for maxwell.

N

N. Symbol representing (1) the north pole of a magnet and (2) the number of turns in a coil.

negative. Electrical condition in which there is an excess of electrons; represented by a minus (−) sign.

neon lamp. A lamp filled with neon (an inert gas) that has two or more insulated electrodes.

network. Various combinations of circuits or electrical elements.

neutron. An elementary particle found in the nucleus of an atom; has the same atomic weight as a proton, but no electrical charge.

NI. The symbol for ampere-turn.

See **square wave**.

N-type semiconductor. Semiconductor material having a high conduction-electron density.

nucleus. The central portion of an atom in which there are protons and neutrons.

numerator. In a fraction, the numeral above the line.

O

oersted. A unit of measure for magnetizing force in the Cgs system.

ohm. The unit of measure for electrical resistance; symbolized by the Greek letter omega (Ω).

Ohm's law. Basic electrical law stating the relationship between voltage, current, and resistance.

omega. Greek letter Ω used as a symbol for ohm. The lowercase omega (ω) may be used to represent work.

oscillation. A signal or quantity that changes continually and symmetrically.

oscillator. An electronic circuit that can generate an AC output at a determined frequency.

oscilloscope. An instrument that measures and displays instantaneous values and signals of one or more electrical quantities. These signals can be steady or varying.

output. The power or energy produced by a device or system.

P

P. Symbol representing (1) permeance, (2) power, (3) the plate of a tube, and (4) a pico.

parallel. Circuits in which resistors or other parts are connected in such a way as to allow current to divide and flow through each individual device.

pass band. The group or limit of frequencies that a certain filter will allow to pass.

passive. An inert component, such as a resistor, which does not create or amplify energy.

periodic table. An arrangement of the properties of chemical elements.

permeability. The measure of how a given material compares to air when used as a path for magnetic lines of force; represented by the Greek letter μ. Also, the ability to conduct magnetic flux.

permeance. The reciprocal of magnetic reluctance; represented by P.

phase. The time relationship between voltage and current in an AC circuit.

phase angle. The amount or degree to which current lags or leads voltage; represented by the Greek letter θ.

phasor. A rotating vector. The phasor varies in time.

phi. The Greek letter ϕ used as a symbol for (1) magnetic flux and (2) angles.

photo. Prefix meaning sensitive to light.

photoelectric. A material that reacts to sunlight by producing electricity.

pi. The Greek letter π; used as a symbol for the ratio of the circumference of a circle to its diameter; having a constant value (to eight decimal places) of 3.14159265.

piezoelectricity. A property that creates electrical current when a twisting or pushing force is applied to certain types of crystal materials.

PIN diode. A solid-state resistor in which the resistance is controlled by the amount of current applied.

polarity. Term used to indicate negative or positive conditions.

pole. (1) One of the electrodes of a battery and (2) one of the ends of a magnet.

positive. The electrical condition in which there is a deficiency of electrons; represented by a plus (+) sign.

potential difference. The voltage that exists between two points.

potentiometer. An electrical resistance device having three contact points.

power. Rate of doing work; abbreviated *P* and measured in watts (W).

printed circuit. Conductive paths on the surface of a copper-clad circuit board.

PROM. Programmable read only memory.

proton. A positively charged particle found in the nucleus of an atom.

P-type semiconductor. A semiconductive material which has a high hole density.

pulse width. The time duration of a pulsed wave.

Pythagorean theorem. A proven mathematical statement that, in a right triangle, the square of the length of the hypotenuse is equal to the sum of the squares of the lengths of the other two sides.

Q

Q. The symbol used for (1) the quantity of an electric charge and (2) the quality factor of a circuit.

R

R. Symbol representing (1) resistance, (2) a resistor, and (3) reluctance ($\mathcal{R}$).

r. Symbol representing the radius of a circle.

radian. An angle that encloses an arc equal to its radius.

radius. The distance from the exact center of a circle to the circumference.

RAM. Random access memory.

reactance. The opposition to the flow of AC; represented by the symbol *X*.

reciprocal. A mathematical term in which a given value is used as a denominator under a numerator of one.

rectifier. A device that changes AC to DC by allowing current to flow in only one direction.

reluctance. Resistance to magnetic lines of force; represented by the symbol $\mathcal{R}$.

resist. A material used in processing circuit boards; used to cover areas that are not to be etched away by chemical action.

resistance. Opposition to the flow of electrons; represented by R.

resistor. A device that opposes the flow of current; represented by R.

resonance. A condition of a circuit in which inductive reactance is equal to capacitive reactance.

retentivity. The ability to retain magnetism.

rheostat. A variable resistor.

ROM. Read only memory.

root-mean-square. The AC power voltage that would produce the same amount of heat in a resistor as DC.

S

S. Abbreviation for siemens.

Schottky diode. A hot carrier diode.

selectivity. Property of a tuned circuit which permits only selected signals to pass.

semiconductor. Material that is able to conduct electricity under certain conditions. Materials such as silicon and germanium are preferred semiconductor materials.

series. Method of connecting one device to another in sequence, so that all available current will pass through each device.

Shockley diode. A special semiconductor device constructed of four layers that passes current in a single direction.

siemens. A unit of measure for admittance, conductance, and susceptance; abbreviated S.

silicon controlled rectifier. A four-layer PNPN semiconductor device used as an electronic switch.

sine. A trigonometric function; an acute angle in a right triangle, the ratio between the side opposite the angle and the hypotenuse; also abbreviated as sin.

sine wave. A graph that illustrates the flow pattern for periodic oscillations in which the amplitude of displacement at each point is proportional to the sine of the phase angle of the displacement; the wave pattern of the output of a rotating AC generator. It is the most common waveform for AC. Also called a *sinusoidal wave*.

sink. A point to which electrons flow.

sinusoidal wave. *See* **sine wave**.

SI units. Abbreviation for the Systeme International d'Unites.

solenoid. A tightly wound coil of wire in the shape of a cylinder. When current is applied, a magnetic field is created within and around the coil.

source. A point from which electrons flow.

square wave. A wave form having steep rise and fall with a constant level between rise and fall; one type of wave other than a perfect sine wave; one type of non-sinusoidal wave.

static electricity. A form of electricity having no electron flow and an electrical potential; present when there are two charges of opposite polarity in close proximity.

susceptance. The reciprocal of reactance; represented by the symbol *B* and measured in siemens.

Systeme International d'Unites. A system of units sanctioned as an international standard.

T

T. Symbol representing a transformer.

tangent. A trigonometric function; for a right triangle, the ratio between the side opposite the acute angle and the side adjacent; abbreviated tan.

tau. Greek letter; the lowercase τ used for a period of time, and the capital T used for the time constant.

terminal. Connecting point for wires at circuit points.

tesla. The Mksa system unit of measurement for magnetic flux density.

thermocouple. A device that generates electricity when heated; made of two metals of widely different electrical properties which are joined together mechanically.

theta. The Greek letter θ; used as the symbol for the phase angle of an *RL* circuit.

time constant. A measure of the time required to charge (or discharge) a capacitor or to build (or collapse) a magnetic field around a coil; represented by T.

transformer. An electric device that can induce electrical energy from one coil to another by magnetic means.

transistor. A semiconductor device that can control and amplify current.

triode. A three-electrode vacuum tube.

truth table. A numeric chart, or table, that describes a number of inputs and shows the output corresponding to each input.

tunnel diode. A semiconductor device that has a negative resistance region and can operate at high frequencies. These devices are often included in signal generators as part of the oscillator circuit.

U

upsilon. The Greek letter υ; used as the symbol for (1) instantaneous voltage and (2) velocity.

V

V. Symbol representing (1) volt, (2) voltmeter, and (3) vacuum tube.

valence. The outermost electron shell of an atom.

varactor diode. A semiconductor device designed with special capacitance characteristics.

velocity. The speed of travel for a signal in a given medium; represented by the Greek letter υ.

volt. A unit of measure for voltage or electromotive force; abbreviated V.

voltage. Electromotive force that causes the flow of electricity, represented by E.

voltage drop. The amount of voltage across any resistor in a circuit.

voltage vector. A representation for voltage quantity and phase, which requires a magnitude and an angle to specify the value completely.

voltaic cell. A device that generates electricity by chemical means.

W

W. Symbol representing (1) watt, (2) work, and (3) power.

watt. A unit of electrical power, abbreviated W.

waveform. A graph that illustrates the flow pattern in time of a given type of current or voltage.

wavelength. The length of an AC cycle; represented by the Greek letter λ.

Wb. Abbreviation for weber.

weber. A unit of measure for magnetic flux in the Mksa system; abbreviated Wb.

wire gauge. An instrument for measuring the diameter of wires.

X

X. Symbol for reactance.

Y

Y. Symbol for admittance.

Z

Z. Symbol for impedance.

zener diode. A semiconductor diode that breaks down at a predetermined voltage.

ANSWERS TO ODD-NUMBERED REVIEW PROBLEMS

REVIEW OF CHAPTERS 1, 2, AND 3

Below are the answers to the odd-numbered problems in this review:

1. 16
3. 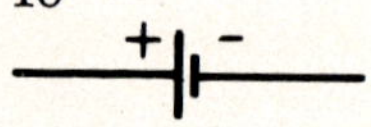

5. 0.02 S
7. 5 V
9. 20 V
11. 2 A
13. $R_T = 6\ \Omega$, $E = 12$ V, $I_{R_1} = 1.2$ A, $I_{R_2} = 0.6$ A, $I_{R_3} = 0.2$ A
15. $R_{Th} = 6.67\ \Omega$, $E_{Th} = 20$ V
17.

Truth Table 3 input NAND			
Input			Output
A	B	C	D
1	1	1	0
1	1	0	1
1	0	0	1
0	0	0	1
0	0	1	1
0	1	1	1
0	1	0	1
1	0	1	1

REVIEW OF CHAPTERS 4 AND 5

Below are the answers to the odd-numbered problems in this review:

1. 24 gauge
3. 2,030–5,080
5. 73 Ω ±5 percent
7. 
9. 0.8 Ω
11. 1.5 V

REVIEW OF CHAPTERS 6 AND 7

Below are the answers to the odd-numbered problems in this review:

1. Nickel
3. 100
5. 25,000 G
7. 400 ampere-turns
9. 502.8 Gb
11. 60 cps
13. 64.28 V
15. 333.33 cm
17. $E_{pp} = 282.8$ V
19. $E_{pp} = 80$ V
21. $E_{pp} = 707$ V

REVIEW OF CHAPTERS 8 AND 9

Below are the answers to the odd-numbered problems in this review:

1. 12 H
3. 0.5
5. 22
7. 7 μH
9. 0.75 Ω
11. 40 V
13. 500 V
15. 0.6 V
17. 5 μF
19. 16 μF
21. 1,327 Ω

REVIEW OF CHAPTERS 10 AND 11

Below are the answers to the odd-numbered problems in this review:

1. 90°
3. 62.8K Ω
5. 26.93 Ω
7. 0.015 A
9. 25 Ω
11. 0.0125
13. 12.5 W
15. $2j\sqrt{3}$
17. $11 - j$
19. $10 \angle 42°$
21. 31.6 V
23. 180 μA
25. 145.29 Hz
27. 10 Ω
29. 50 kHz
31. 2

REVIEW OF CHAPTERS 12, 13, AND 14

Below are the answers to the odd-numbered problems in this review:

1. 24 mA
3. 1.042 Ω
5. 20
7. 0.2 μmhos
9. 0.167 A
11. 400
13. 10 V
15. 1.12

REVIEW OF CHAPTERS 15 AND 16

Below are the answers to the odd-numbered problems in this review:

1. 50
3. 9
5. 100000
7.

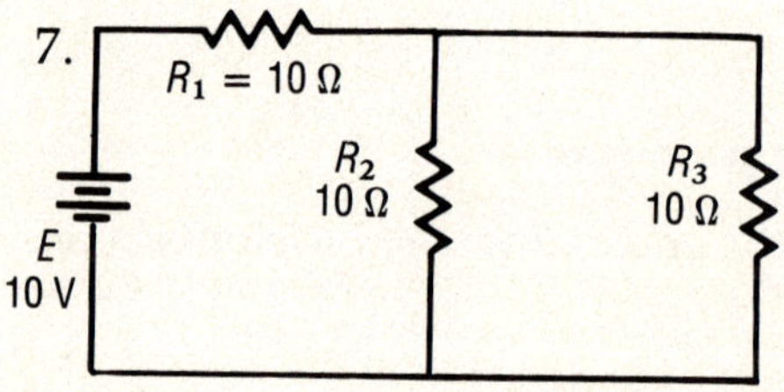

INDEX

Electrostatic flux density, 566